Control of Messenger RNA Stability

Control of Messenger RNA Stability

Edited by

Joel G. Belasco
Department of Microbiology and Molecular Genetics
Harvard Medical School
Boston, Massachusetts

George Brawerman
Department of Biochemistry
Tufts University
Boston, Massachusetts

ACADEMIC PRESS, INC.
Harcourt Brace Jovanovich, Publishers

San Diego New York Boston London Sydney Tokyo Toronto

This book is printed on acid-free paper. ∞

Academic Press, Inc.
1250 Sixth Avenue, San Diego, California 92101-4311

United Kingdom Edition published by
Academic Press Limited
24–28 Oval Road, London NW1 7DX

Library of Congress Cataloging-in-Publication Data

Control of messenger RNA stability / edited by Joel Belasco and George Brawerman.
p. cm.
Includes bibliographical references and index.
ISBN 0-12-084782-5
1. Messenger RNA--Metabolism. 2. Genetic regulation.
I. Belasco, Joel. II. Brawerman, George
QP623.5.M47C66 1993
574.87'3283--dc20 92-43555
CIP

PRINTED IN THE UNITED STATES OF AMERICA
93 94 95 96 97 98 QW 9 8 7 6 5 4 3 2 1

Contents

3

5′ mRNA Stabilizers

David Bechhofer

4

RNA Processing and Degradation by RNase K and RNase E

Öjar Melefors, Urban Lundberg, and Alexander von Gabain

5

RNA Processing and Degradation by RNase III

Donald Court

6

Translation and mRNA Stability in Bacteria: A Complex Relationship

Carsten Petersen

PART II

EUKARYOTES

7

mRNA Degradation in Eukaryotic Cells: An Overview

George Brawerman

8

Hormonal and Developmental Regulation of mRNA Turnover

David L. Williams, Martha Sensel, Monica McTigue, and Roberta Binder

9

Control of the Decay of Labile Protooncogene and Cytokine mRNAs

Michael E. Greenberg and Joel G. Belasco

10

Translationally Coupled Degradation of Tubulin mRNA

Nicholas G. Theodorakis and Don W. Cleveland

11

Iron Regulation of Transferrin Receptor mRNA Stability

Joe B. Harford

12

Degradation of a Nonpolyadenylated Messenger: Histone mRNA Decay

William F. Marzluff and Roberta J. Hanson

13

mRNA Turnover in *Saccharomyces cerevisiae*

Stuart W. Peltz and Allan Jacobson

14

Control of mRNA Degradation in Organelles

Wilhelm Gruissem and Gadi Schuster

15

Control of Poly(A) Length

Ellen J. Baker

16

mRNA Decay in Cell-Free Systems

Jeff Ross

17

Eukaryotic Nucleases and mRNA Turnover

Audrey Stevens

PART III

METHODS OF ANALYSIS

18

Experimental Approaches to the Study of mRNA Decay

Joel G. Belasco and George Brawerman

Contributors

Numbers in parentheses indicate the pages on which the authors' contributions begin.

Ellen J. Baker (367), Department of Biology, University of Nevada, Reno, Reno, Nevada 89517

David Bechhofer (31), Department of Biochemistry, Mount Sinai School of Medicine, New York, New York 10029

Joel G. Belasco (3, 199, 475), Department of Microbiology and Molecular Genetics, Harvard Medical School, Boston, Massachusetts 02115

Roberta Binder (161), Department of Pharmacological Sciences, State University of New York at Stony Brook, Stony Brook, New York 11794

George Brawerman (149, 475), Department of Biochemistry, Tufts University, Boston, Massachusetts 02111

Helen C. Causton (13), ICRF Laboratories, Institute of Molecular Medicine, University of Oxford, John Radcliffe Hospital, Oxford OX3 9DU, United Kingdom

Don W. Cleveland (219), Department of Biological Chemistry, Johns Hopkins University School of Medicine, Baltimore, Maryland 21205

Donald Court (71), Laboratory of Chromosome Biology, Molecular Control and Genetic Section, ABL-Basic Research Program, NCI/Frederick Cancer Research and Development Center, Frederick, Maryland 21702

Geoffrey S. C. Dance (13), ICRF Laboratories, Institute of Molecular Medicine, University of Oxford, John Radcliffe Hospital, Oxford OX3 9DU, United Kingdom

Michael E. Greenberg (199), Department of Microbiology and Molecular Genetics, Harvard Medical School, Boston, Massachusetts 02115

Wilhelm Gruissem (329), Department of Plant Biology, University of California, Berkeley, California 94720

Roberta J. Hanson (267), Program in Molecular Biology and Biotechnology, University of North Carolina, Chapel Hill, North Carolina 27599

Joe B. Harford (239), Cell Biology and Metabolism Branch, National Institute of Child Health and Development, National Institutes of Health, Bethesda, Maryland 20892

Christopher F. Higgins (13), ICRF Laboratories, Institute of Molecular Medicine, University of Oxford, John Radcliffe Hospital, Oxford OX3 9DU, United Kingdom

Allan Jacobson (291), Department of Molecular Genetics and Microbiology, University of Massachusetts Medical School, Worcester, Massachusetts 01655

Urban Lundberg[1] (53), Department of Bacteriology, Karolinska Institute, S-10401 Stockholm, Sweden

William F. Marzluff (267), Program in Molecular Biology and Biotechnology, University of North Carolina, Chapel Hill, North Carolina 27599

Monica McTigue (161), Department of Pharmacological Sciences, State University of New York at Stony Brook, Stony Brook, New York 11794

Öjar Melefors[2] (53), Department of Bacteriology, Karolinska Institute, S-10401 Stockholm, Sweden

Elisabeth A. Mudd (13), ICRF Laboratories, Institute of Molecular Medicine, University of Oxford, John Radcliffe Hospital, Oxford OX3 9DU, United Kingdom

Stuart W. Peltz (291), Department of Molecular Genetics and Microbiology, University of Massachusetts Medical School, Worcester, Massachusetts 01655

Carsten Petersen (117), University Institute of Microbiology, University of Copenhagen, DK-1353, Copenhagen K, Denmark

Jeff Ross (417), McArdle Laboratory for Cancer Research, University of Wisconsin, Madison, Madison, Wisconsin 53706

Gadi Schuster (329), Department of Plant Biology, University of California, Berkeley, Berkeley, California 94720

Martha Sensel (161), Department of Pharmacological Sciences, State University of New York at Stony Brook, Stony Brook, New York 11794

Audrey Stevens (449), Biology Division, Oak Ridge National Laboratory, Oak Ridge, Tennessee 37831

Nicholas G. Theodorakis (219), Department of Biological Chemistry, Johns Hopkins University School of Medicine, Baltimore, Maryland 21205

[1]*Present address:* Department of Biology, Yale University, New Haven, Connecticut 06511.
[2]*Present address:* European Molecular Biology Laboratory, 6900 Heidelberg, Germany.

Alexander von Gabain (53), Department of Bacteriology, Karolinska Institute, S-10401 Stockholm, Sweden

David L. Williams (161), Department of Pharmacological Sciences, State University of New York at Stony Brook, Stony Brook, New York 11794

Preface

The importance of post-transcriptional genetic control processes has become increasingly apparent in recent years. Among these processes, one that has begun to receive considerable attention is the control of mRNA stability. With the growing recognition that mRNA degradation has a profound impact on gene expression and that rates of mRNA decay can be modulated in response to environmental and developmental signals, a vigorous research effort aimed at understanding this process is now taking place. Significant progress has been made, but there has not been any comprehensive review of this field of research. By filling this void, this volume is intended to promote interest and progress in the field of mRNA degradation and to complement the numerous books and symposia that deal with other aspects of gene regulation.

To achieve these goals, we have compiled a book to serve both as a useful reference for specialists in the field of mRNA degradation and as a general introduction for the broader community of biological scientists. Each chapter is a general review of one important aspect of mRNA degradation. The chapters are intended to integrate a variety of findings on each topic into a unified conceptual whole. The authors have attempted to describe research in detail sufficient to indicate the experimental basis for the conclusions derived from it. At the same time, they have sought to ensure that the key take-home messages come through clearly without being obscured by technical minutiae.

A major virtue of this volume is that it brings together studies on mRNA degradation in both prokaryotes and eukaryotes. There are likely to be fundamental principles that govern mRNA decay in all organisms. For example, one important parallel between eukaryotes and prokaryotes is the apparent need for terminal structures that can protect mRNA from nonspecific degradation. Another is the relationship between translation and mRNA decay. It is our hope that integrating information on prokaryotic and eukaryotic mRNA decay in a single volume will encourage cross-fertilization of concepts and techniques. Thus, the greater understanding of enzymes and mechanisms controlling mRNA decay in bacteria provides useful paradigms that could aid in the formulation of models for mRNA degradation in eukaryotic cells. Conversely, the more advanced knowledge of how mRNA lifetimes are modulated in eukaryotic cells could serve as a guide for research on the regulation of prokaryotic mRNA decay rates.

This volume is subdivided into three parts. Parts I and II deal with various aspects of mRNA degradation in prokaryotic and eukaryotic cells. Part III provides a critical examination of experimental procedures for studying mRNA decay and is intended primarily as a guide for newcomers to the field. Parts I and II on prokaryotic and eukaryotic mRNA decay are each introduced by a succinct overview that integrates current knowledge within that subfield and provides a conceptual framework for understanding the chapters that follow. Chapters on prokaryotic mRNA degradation (Part I) are organized to emphasize RNA elements and ribonucleases thought to be of general importance for controlling rates of mRNA decay in bacteria. Chapter 6 addresses the complex relationship of translation to mRNA decay in bacterial cells. For the most part, the chapters in Part II on eukaryotic mRNA degradation address the decay of particular classes of mRNAs and experimental systems for studying mRNA decay. Part II includes chapters on eukaryotic mRNAs with unique and interesting decay characteristics (Chapters 8–12) as well as chapters on biochemical approaches to identifying ribonucleases and other protein factors that contribute to mRNA degradation in eukaryotic cells (Chapters 16 and 17). Chapter 13 addresses mRNA decay in yeast, an organism that offers the possibility of employing genetic selection in the study of eukaryotic mRNA decay. Because of the potential importance of poly(A) tails for promoting eukaryotic mRNA stability, we have included Chapter 15, which deals with the control of poly(A) length. Part II also includes a review of mRNA decay in mitochondria and chloroplasts (Chapter 14), which in many ways resembles prokaryotic mRNA decay.

While the focus of each chapter is well defined, some degree of overlap in subject matter is unavoidable between chapters on related topics. This allows each chapter to be a thorough review that can be read and understood as an independent unit. It also helps to unify the book and permits the expression of diverse viewpoints on controversial topics.

The knowledge collected in this volume makes clear both how far our understanding of mRNA degradation has come in the last few years and how much remains to be discovered about this important genetic regulatory process. It is our hope that this volume will stimulate interest in this burgeoning area of study and serve as a source of inspiration for scientists entering this field.

PART I

PROKARYOTES

1

mRNA Degradation in Prokaryotic Cells: An Overview

JOEL G. BELASCO

I. Introduction

The degradation of messenger RNA is one of the principal means by which genes are regulated in prokaryotic organisms. Remarkably, the instability of mRNA was anticipated by Jacques Monod and François Jacob before the actual discovery of mRNA. A key prediction of their operon model was that instructions for protein synthesis are conveyed from genes to ribosomes by a labile RNA transcript. As they pointed out (Jacob and Monod, 1961),

> the structural message must be carried by a very short-lived intermediate both rapidly formed and rapidly destroyed during the process of information transfer. This is required by the kinetics of induction. . . . [T]he addition of inducer . . . provokes the synthesis of enzyme at maximum rate within a matter of a few minutes, while the removal of inducer . . . interrupts the synthesis within an equally short time. Such kinetics are incompatible with the assumption that the repressor–operator interaction controls the rate of synthesis of *stable* enzyme-forming templates.

This prediction was soon confirmed by the discovery of a highly labile RNA class with characteristics expected for messenger RNA (Brenner *et al.*, 1961; Gros *et al.*, 1961). That mRNA and its lability *in vivo* were discovered simultaneously was no accident; indeed, the high turnover rate of mRNA was instrumental to the detection of this class of RNAs, which comprise only 4% of the RNA in *Escherichia coli*.

Jacob and Monod's crucial insight that a rapid genetic response to a changing cellular environment requires an unstable molecular messenger provides the biological imperative that explains the widespread instability

Control of Messenger RNA Stability

of mRNA in all organisms. In the absence of regulation at the translational or post-translational level, the half-time for maximal response to an environmental stimulus that induces or represses transcription can be no shorter than the mRNA half-life, as the longevity of mRNA limits the rate at which the cellular concentration of a transcript can increase or decline to a new steady-state level. Furthermore, even under conditions of continuous gene expression, the instability of mRNA has a major impact on protein synthesis through its effect on the steady-state concentration of mRNA, which is as sensitive to the rate of mRNA degradation as it is to the rate of mRNA synthesis.

Rates of mRNA degradation can vary widely within a single cell. In *E. coli*, for example, half-lives of individual messages can be as short as 20–30 sec or as long as 50 min, with lifetimes typically between 2 and 4 min (Pedersen *et al.*, 1978; Nilsson *et al.*, 1984; Donovan and Kushner, 1986; Emory and Belasco, 1990; Baumeister *et al.*, 1991). Marked differences in mRNA stability are commonly observed both for unrelated gene transcripts and for different translational units within a polycistronic transcript (Blundell *et al.*, 1972; Belasco *et al.*, 1985; Newbury *et al.*, 1987; Båga *et al.*, 1988).

Generally, mRNA half-lives do not exceed the cell doubling time. A longer mRNA lifetime would not significantly increase or prolong gene expression, as the effective mRNA "turnover" rate equals the rate of mRNA decay plus the rate of mRNA dilution by cell growth. Therefore, no matter how slowly a message is degraded, this effective turnover rate can be no slower than the cell growth rate. Along with the need of rapidly proliferating microorganisms to adapt swiftly to environmental changes in order to compete successfully for survival, this practical limit probably explains why mRNA half-lives typically are so much shorter in bacteria (0.5–50 min; about 3 min on average) and yeast (3–100 min; about 15 min on average) than in vertebrate cells (15 min to a few weeks; several hours on average). Consistent with this hypothesis, the longest half-life yet reported for a bacterial message (5–7 hr) was measured for *hoxS* mRNA in *Alcaligenes eutrophus* cells growing with a long doubling time of 20 hr (Oelmuller *et al.*, 1990). An interesting exception is the *hok* gene transcript of plasmid R1, which mediates killing of *E. coli* progeny cells that have failed to inherit the R1 plasmid. This long-lived mRNA must survive cell division in order to perform its function of plasmid maintenance, as it is selectively translated in plasmid-free segregants to produce a highly toxic protein (Gerdes *et al.*, 1990).

The longevity of a given mRNA need not be fixed, but can vary in response to a variety of growth conditions and environmental signals. For example, excessive accumulation of OmpA protein in the outer membrane of *E. coli* is prevented under conditions of slow cell growth by accelerated degradation of the *ompA* transcript (Nilsson *et al.*,1984). Some *Bacillus subtilis* mRNAs that are synthesized primarily in vegetative cells (e.g., *sdh* mRNA, which encodes succinate dehydrogenase) or in cells entering stationary phase (e.g., *aprE* mRNA, which encodes subtilisin) are suddenly destabilized when cells are shifted to another growth stage (Melin *et al.*, 1989; Resnekov *et al.*, 1990). Furthermore, mRNAs that encode proteins important for nitrogen fixation (*nif* in *Klebsiella pneumoniae*) and photosynthesis (*puf* in *Rhodobacter capsulatus*), processes induced in these bacteria only when oxygen is scarce, are degraded more slowly under anaerobic conditions than in the presence of oxygen (Collins *et al.*, 1986; Klug, 1991). Similarly, the *Staphylococcus aureus ermC* and *ermA* mRNAs, which encode resistance to erythromycin, are stabilized when expression of these genes is induced by the presence of low concentrations of erythromycin (Bechhofer and Dubnau, 1987; Sandler and Weisblum, 1988). Thus, modulation of gene expression can be achieved through changes in mRNA stability.

Despite its importance to gene expression, mRNA degradation has been the slowest of the principal gene regulatory processes in bacteria to be elucidated. Only now, 30 years after the discovery of the instability of mRNA, are we beginning to glimpse the structural determinants of mRNA stability and instability, the cellular factors that degrade mRNA, and the molecular mechanisms by which this process occurs. There are a number of explanations of why progress in understanding mRNA degradation has come later than major advances in the fields of prokaryotic transcription and translation regulation. These include the structural complexity of mRNA and the general failure of genetic selection to aid in the identification of cis- and trans-acting genetic loci that control mRNA lifetimes. Only with the advent of recombinant DNA technology has it become possible to address these questions effectively and to obtain detailed information about mRNA degradation *in vivo*.

II. Ribonucleases

Although many different ribonucleases have been identified in *E. coli*, only a small number of these have been shown to participate in mRNA degradation. These include a few endonucleases (RNase E, RNase K, RNase III) and two 3′ exonucleases [RNase II and polynucleotide phosphorylase (PNPase)]. To date, no ribonuclease that removes single nucleotides sequentially from the 5′ end of RNA has been identified in any prokaryotic organism.

The cleavage-site specificity of RNases E, K, and III is as yet rather poorly defined. RNases E and K, which appear to be interrelated, preferentially cleave certain short, single-stranded AU-rich sequences that are located in an appropriate secondary/tertiary structural context (see Chapter 4). Biochemical characterization of RNases E and K, which have not yet been purified to homogeneity, is rather sparse. Most conclusions as to the importance of RNase E for mRNA degradation have relied on genetic evidence obtained by thermally inactivating this ribonuclease in *E. coli* strains with a temperature-sensitive mutation in the *rne* gene (also known as the *ams* gene). It now appears that cleavage by RNase E is the rate-determining step in the degradation of many *E. coli* mRNAs. RNase III (the product of the *E. coli rnc* gene), which participates in the degradation of a much more limited number of mRNAs, cleaves RNA at certain double-helical sites (see Chapter 5). Its substrates include most very long RNA duplexes that are perfectly paired and some RNA stem–loop structures. The key features of RNA hairpins cleaved by RNase III that causes them to be substrates for this enzyme remain ill-defined. In addition to their role in mRNA degradation, RNase E and RNase III also participate in the processing of stable RNAs, such as ribosomal RNA.

Exonucleases RNase II and PNPase (the products of the *E. coli rnb* and *pnp* genes) are enzymes that readily degrade RNA that is not base-paired at the 3′ end (see Chapter 2). RNA digestion by these two 3′ exonucleases is impeded when a significant 3′ stem–loop structure is encountered.

It is likely that other ribonucleases, including some still undiscovered, also participate in mRNA degradation in bacteria. Candidates include *E. coli* RNase M, which cleaves at some Py-A dinucleotides, and RNase I*, an endonuclease specific for RNA oligonucleotides (Cannistraro and Kennell, 1989; Cannistraro and Kennell, 1991). RNase I* appears to be a cytoplasmic form of the periplasmic *E. coli* endonuclease RNase I. In addition, at least one phage T4-encoded ribonuclease contributes to mRNA degradation in T4-infected cells. This is the bacteriophage T4 RegB protein, which cleaves RNA endonucleolytically at the sequence GGAG (Uzan *et al.*, 1988; Ruckman *et al.*, 1989). As this tetranucleotide is often present within signals for translation initiation (i.e., within Shine–Dalgarno elements), cleavage there provides T4 with a direct means of functionally inactivating mRNA.

III. Structural Determinants of mRNA Stability and Instability

A number of cis-acting elements in mRNA have been identified that control the lifetime of mRNA. These can be categorized as mRNA stabilizers or destabilizers. Some of these cis-acting decay determinants are predictable from our present knowledge of the substrate specificity of the ribonucleases known to participate in mRNA degradation in *E. coli*, whereas others suggest hitherto unknown specificities of these or other enzymes.

There are two classes of prokaryotic mRNA stabilizers. One class comprises a group of 5′-terminal segments of naturally long-lived mRNAs that, under appropriate conditions, can enhance the stability of otherwise labile heterologous messages fused downstream (see Chapter 3). To date, five such 5′ mRNA stabilizers have been identified. The 5′ UTRs of the vast majority of long-lived mRNAs have not yet been tested for this property; most of these 5′ UTRs may also be capable of stabilizing labile messages to which they are joined. Message stabilization by at least two of the known 5′ stabilizers (the *ompA* and *ermC* leader regions) is not a result of highly efficient translation initiation and consequent steric occlusion of mRNA by ribosomes densely packed along the entire length of the protein-coding region (Emory and Belasco, 1990; Bechhofer and Dubnau, 1987). Instead, the ability of these relatively small 5′-terminal RNA elements to protect an entire transcript from degradation suggests the existence in bacteria of an important mRNA degrading enzyme that can be impeded by RNA structural features or ribosome stalling near the mRNA 5′ end. Recent evidence suggests that in *E. coli* this key ribonuclease may be RNase E, as internal cleavage by this endonuclease can be influenced by base-pairing at the extreme 5′ terminus (Bouvet and Belasco, 1992).

The other class of protective elements comprises 3′-terminal and intercistronic stem–loop structures, which can impede the initiation or propagation of 3′ exonuclease digestion (see Chapter 2). As a general rule, a 3′-terminal stem–loop structure is essential for a completed RNA transcript to survive in the cytoplasm for more than a few seconds. Therefore, it is not surprising that almost all detectable prokaryotic mRNAs end in such a structure, which often doubles as a signal for transcription termination. For the same reason, polycistronic mRNAs whose 5′-proximal coding units are markedly more stable than the rest of the transcript invariably have an intercistronic stem–loop structure at the segmental boundary. After the 3′-terminal hairpin of these polycistronic transcripts is severed by endonucleolytic cleavage within the more labile 3′ mRNA segment, this intercistronic stem–loop prevents rapid decay of the 3′ segment from being propagated exonucleolytically into the long-lived 5′ segment.

RNA elements that facilitate degradation include sites of endonucleolytic initiation of mRNA decay by RNase E and RNase III (see Chapters 4 and 5). For an endonuclease cleavage site to be classified as a destabilizing element, its mutation should prolong the lifetime of the RNA that contains it. This criterion may often be difficult to satisfy, as elimination of one such site might not significantly affect the half-life of an RNA containing functionally redundant cleavage sites. Nevertheless, destabilizing elements that are endonuclease targets have been identified unambiguously in several RNAs, including the λ P_L transcript, which is cleaved at the *sib* locus by RNase III, and pBR322 RNA I, whose decay begins with RNase E cleavage at a site five nucleotides from the 5′ end (Schmeissner *et al.*, 1984; Lin-Chao and Cohen, 1991). The lack of an inverse correlation between mRNA length and half-life indicates that mRNA degradation is not controlled by a nonspecific endonuclease that cleaves randomly (Chen and Belasco, 1990).

A second class of RNA destabilizing elements comprises poor ribosome binding sites that initiate translation inefficiently and very early nonsense mutations that cause premature translation termination (see Chapter 6). Messages containing such an element frequently decay rapidly with a half-life of ≤1 min (Nilsson *et al.*, 1987; Baumeister *et al.*, 1991). The mechanism linking the efficiency and extent of translation to the rate of mRNA decay is not yet clear.

IV. Mechanisms of mRNA Degradation

In principle, a number of different mRNA decay mechanisms can be imagined in prokaryotic cells. These can be classified according to the type of enzyme (endonuclease or exonuclease) that initiates degradation and the scavenger enzymes that degrade the RNA products generated by the initial degradative event. Purely exonucleolytic degradation models invoke the exclusive action of either 3′ exonucleases or a hypothetical 5′ exonuclease in the overall decay of mRNA (Figs. 1A and 1B). It is unlikely that the disparate stabilities of most mRNAs that end in a stem–loop results from differential susceptibility of these terminal stem–loops to penetration by 3′ exonucleases, as labile messages generally are not stabilized by replacing their 3′-terminal hairpin with that of a stable transcript (Belasco *et al.*, 1986; Wong and Chang, 1986; Chen *et al.*, 1988) and as the average half-life of bulk *E. coli* mRNA is not altered by overproduction of RNase II or by the absence of either RNase II or PNPase (Donovan and Kushner, 1983, 1986) (see Chapter 2). Therefore, 3′-exonucleolytic initiation of RNA decay probably is rare except in the case of labile RNAs lacking a substantial 3′ hairpin and long-lived RNAs resistant to attack by all other types of ribonucleases.

As a bacterial 5′ exoribonuclease that removes nucleotides sequentially from the 5′ end has never been detected, a degradation mechanism that relies on such an enzyme activity must be considered speculative. A priori, a purely 5′-exonucleolytic decay mechanism cannot account for the degradation of polycistronic transcripts whose 3′ segment is first to decay, just as 3′-exonuclease digestion is insufficient to explain RNAs whose 5′ segment is most labile. Purely endonucleolytic models for mRNA decay also seem implausible, as the efficient degradation of RNA to mononucleotides requires the action of a completely nonspecific ribonuclease in the final stages of decay, and no such endoribonuclease is known to exist in the cytoplasm of bacteria.

Consequently, the degradation of most mRNAs is likely to involve the combined action of endonucleases and exonucleases, with endonuclease cleavage at one or more sites initiating the decay process (Fig. 1C) (Apirion, 1973; Kennell, 1986; Belasco and Higgins, 1988). Unprotected by a 3′-terminal stem–loop structure, the resulting 5′ cleavage products could be rapidly degraded by 3′ exonuclease digestion, which would proceed virtu-

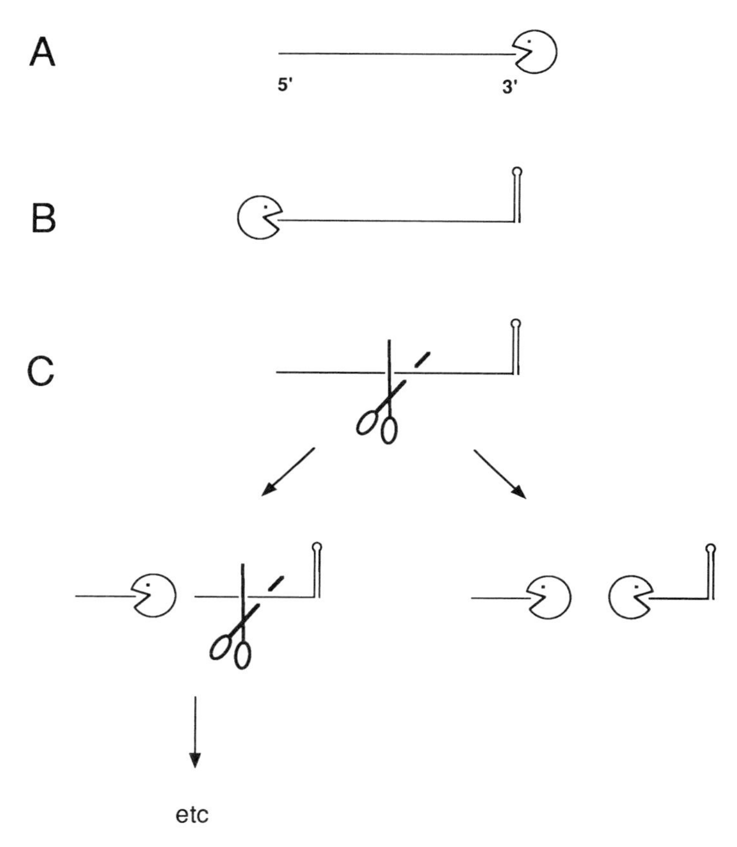

FIGURE 1 Theoretically possible mechanisms of mRNA decay in prokaryotic cells. (A) Degradation exclusively by a 3′ exonuclease. (B) Degradation exclusively by a 5′ exonuclease. (C) Degradation initiated by endonuclease cleavage and propagated by the combined action of either an endonuclease and a 3′ exonuclease (left) or a 5′ exonuclease and a 3′ exonuclease (right). Scissors, endonuclease. Leftward-facing fresser, 3′ exonuclease. Rightward-facing fresser, 5′ exonuclease. RNAs ending with a 3′-terminal stem–loop are resistant to 3′-exonuclease digestion.

ally to the 5′ end unless an untranslated stem–loop structure of significant thermodynamic stability was encountered (see Chapter 2). As processive digestion by RNase II and PNPase stops when the RNA length has been reduced to three to eight nucleotides (Singer and Tolbert, 1965; Thang *et al.*, 1967; Nossal and Singer, 1968; Klee and Singer, 1968), complete degradation to mononucletides may also require the participation of a ribonuclease that can digest oligonucleotides, such as RNase I* (Cannistraro and Kennell, 1991).

Rapid degradation of the downstream products of endonucleolytic cleavage, which often are more labile than the full-length transcript, is more difficult to account for. In principle, the initial cleavage could accelerate subsequent endonucleolytic attack on the 3′ fragment either by cutting away its ribosome binding site and thereby preventing further translation initiation or by inducing it to isomerize to a more vulnerable conformation. Alternatively, swift decay of the 3′ cleavage product could be explained if the enzyme responsible for the initial cleavage event can act processively in a 5′-to-3′ manner to chop the transcript into multiple fragments (Belasco and Higgins, 1988; Emory and Belasco, 1990) or if there exists an as yet undiscovered 5′ exoribonuclease that specifically digests the 3′ products of endonuclease cleavage (Emory and Belasco, 1990; Lin-Chao and Cohen, 1991).

As rapid progress is now being made in elucidating the mechanisms and structural determinants of mRNA degradation in prokaryotic organisms, there is good reason to hope for answers soon to many of the riddles that have long perplexed investigators studying this important biological process. For example, the recent discovery that endonucleolytic cleavage by RNase E can be influenced by 5′-terminal RNA structure may explain how rates of mRNA degradation can be controlled by elements near the RNA 5′ end despite the apparent absence of bacterial 5′ exoribonucleases. But many fundamental questions remain unanswered. How does the RNA 5′ terminus affect the rate of RNase E cleavage at internal sites located well downstream? How are widely spaced translating ribosomes able to contribute to mRNA stability, and why are stalled ribosomes especially effective at protecting mRNA from degradation? What is the structural basis for the cleavage-site selectivity of RNase E, RNase K, and RNase III? Which other ribonucleases initiate mRNA degradation in bacteria, and what determines their specificity? Are most prokaryotic mRNAs degraded via one or another of a few basic pathways, or do a large number of different ribonucleases with disparate specificities and mechanisms of action contribute to mRNA decay in bacteria? Why are some 3′ products of endonucleolytic cleavage by RNase E extremely labile, while others are long-lived? What are the mechanisms by which mRNA stability is regulated by environmental signals such as growth rate, growth stage, and oxygen availability? The answers to these important questions will enhance our understanding of a key aspect of gene regulation and lead to a greater appreciation of the variety of molecular mechanisms by which prokaryotic organisms thrive.

Acknowledgment

I gratefully acknowledge the support of a research grant from the National Institutes of Health (GM35769).

References

Apirion, D. (1973). Degradation of RNA in *Escherichia coli*: A hypothesis. *Mol. Gen. Genet.* **122,** 313–322.

Båga, M., Goransson, M., Normark, S., and Uhlin, B. E. (1988). Processed mRNA with differential stability in the regulation of *E. coli* pilin gene expression. *Cell* **52,** 197–206.

Baumeister, R., Flache, P., Melefors, Ö, von Gabain, A., and Hillen, W. (1991). Lack of a 5′ non-coding region in Tn*1721* encoded *tetR* mRNA is associated with a low efficiency of translation and a short half-life in *Escherichia coli*. *Nucleic Acids Res.* **19,** 4595–4600.

Bechhofer, D. H., and Dubnau, D. (1987). Induced mRNA stability in *Bacillus subtilis*. *Proc. Natl. Acad. Sci. USA* **84,** 498–502.

Belasco, J. G., Beatty, J. T., Adams, C. W., von Gabain, A., and Cohen, S. N. (1985). Differential expression of photosynthesis genes in *R. capsulata* results from segmental differences in stability within the polycistronic *rxcA* transcript. *Cell* **40,** 171–181.

Belasco, J. G., and Higgins, C. F. (1988). Mechanisms of mRNA decay in bacteria: A perspective. *Gene* **72,** 15–23.

Belasco, J. G., Nilsson, G., von Gabain, A., and Cohen, S. N. (1986). The stability of *E. coli* gene transcripts is dependent on determinants localized to specific mRNA segments. *Cell* **46,** 245–251.

Blundell, M., Craig, E., and Kennell, D. (1972). Decay rates of different mRNA in *E. coli* and models of decay. *Nature New Biol.* **238,** 46–49.

Bouvet, P., and Belasco, J. G. (1992). Control of RNase E-mediated RNA degradation by 5′-terminal base pairing in *E. coli*. *Nature* (*London*) **360,** 488–491.

Brenner, S., Jacob, F., and Meselson, M. (1961). An unstable intermediate carrying information from genes to ribosomes for protein synthesis. *Nature* (*London*) **190,** 576–581.

Cannistraro, V. J., and Kennell, D. (1989). Purification and characterization of ribonuclease M and mRNA degradation in *Escherichia coli*. *Eur. J. Biochem.* **181,** 363–370.

Cannistraro, V. J., and Kennell, D. (1991). RNase I*, a form of RNase I, and mRNA degradation in *Escherichia coli*. *J. Bacteriol.* **173,** 4653–4659.

Chen, C.-Y. A., Beatty, J. T., Cohen, S. N., and Belasco, J. G. (1988). An intercistronic stem-loop structure functions as an mRNA decay terminator necessary but insufficient for *puf* mRNA stability. *Cell* **52,** 609–619.

Chen, C.-Y. A., and Belasco, J. G. (1990). Degradation of *pufLMX* mRNA in *Rhodobacter capsulatus* is initiated by nonrandom endonucleolytic cleavage. *J. Bacteriol.* **172,** 4578–4586.

Collins, J. J., Roberts, G. P., and Brill, W. J. (1986). Post-transcriptional control of *Klebsiella pneumoniae nif* mRNA stability by the *nifL* product. *J. Bacteriol.* **168,** 173–178.

Donovan, W. P., and Kushner, S. R. (1983). Amplification of ribonuclease II (*rnb*) activity in *Escherichia coli* K-12. *Nucleic Acids Res.* **11,** 265–275.

Donovan, W. P., and Kushner, S. R. (1986). Polynucleotide phosphorylase and ribonuclease II are required for cell viability and mRNA turnover in *Escherichia coli* K-12. *Proc. Natl. Acad. Sci. USA* **83,** 120–124.

Emory, S. A., and Belasco, J. G. (1990). The *ompA* 5′ untranslated RNA segment functions in *Escherichia coli* as a growth-rate-regulated mRNA stabilizer whose activity is unrelated to translational efficiency. *J. Bacteriol.* **172,** 4472–4481.

Gerdes, K., Thisted, T., and Martinussen, J. (1990). Mechanism of post-segregational killing by the *hok*/*sok* system of plasmid R1: the *sok* antisense RNA regulates the formation of a *hok* mRNA species correlated with killing of plasmid free cells. *Mol. Microbiol.* **4,** 1807–1818.

Gros, F., Hiatt, H., Gilbert, W., Kurland, C. G., Risebrough, R. W., and Watson, J. D. (1961). Unstable ribonucleic acid revealed by pulse labelling of *Escherichia coli*. *Nature* (*London*) **190,** 581–585.

Jacob, F., and Monod, J. (1961). Genetic regulatory mechanisms in the synthesis of proteins. *J. Mol. Biol.* **3,** 318–356.

Kennell, D. E. (1986). The instability of messenger RNA in bacteria. *In* Maximizing Gene Expression (W. S. Reznikoff and L. Gold, Eds.), pp. 101–142. Butterworths, Stoneham, Massachusetts.

Klee, C. B., and Singer, M. F. (1968). The processive degradation of individual polyribonucleotide chains. *J. Biol. Chem.* **243,** 923–927.

Klug, G. (1991). Endonucleolytic degradation of *puf* mRNA in *Rhodobacter capsulatus* is influenced by oxygen. *Proc. Natl. Acad. Sci. USA* **88,** 1765–1769.

Lin-Chao, S., and Cohen, S. N. (1991). The rate of processing and degradation of antisense RNA I regulates the replication of Col E1-type plasmids in vivo. *Cell* **65,** 1233–1242.

Melin, L., Rutberg, L., and von Gabain, A. (1989). Transcriptional and posttranscriptional control of the *Bacillus subtilis* succinate dehydrogenase operon. *J. Bacteriol.* **171,** 2110–2115.

Newbury, S. F., Smith, N. H., Robinson, E. C., Hiles, I. D., and Higgins, C. F. (1987). Stabilization of translationally active mRNA by prokaryotic REP sequences. *Cell* **48,** 297–310.

Nilsson, G., Belasco, J. G., Cohen, S. N., and von Gabain, A. (1984). Growth-rate dependent regulation of mRNA stability in *Escherichia coli. Nature* (*London*) **312,** 75–77.

Nilsson, G., Belasco, J. G., Cohen, S. N., and von Gabain, A. (1987). Effect of premature termination of translation on mRNA stability depends on the site of ribosome release. *Proc. Natl. Acad. Sci. USA* **84,** 4890–4894.

Nossal, N. G., and Singer, M. F. (1968). The processive degradation of individual polyribonucleotide chains. *J. Biol. Chem.* **243,** 913–922.

Oelmüller, U., Schlegel, H. G., and Friedrich, C. G. (1990). Differential stability of mRNA species of *Alcaligenes eutrophus* soluble and particulate hydrogenases. *J. Bacteriol.* **172,** 7057–7064.

Pedersen, S., Reeh, S., and Friesen, J. D. (1978). Functional mRNA half-lives in *E. coli. Mol. Gen. Genet.* **166,** 329–336.

Resnekov, O., Rutberg, L., and von Gabain, A. (1990). Changes in the stability of specific mRNA species in response to growth stage in *Bacillus subtilis. Proc. Natl. Acad. Sci. USA* **87,** 8355–8359.

Ruckman, J., Parma, D., Tuerk, C., Hall, D. H., and Gold, L. (1989). Identification of a T4 gene required for bacteriophage mRNA processing. *New Biol.* **1,** 54–65.

Sandler, P., and Weisblum, B. (1988). Erythromycin-induced stabilization of *ermA* messenger RNA in *Staphylococcus aureus* and *Bacillus subtilis. J. Mol. Biol.* **203,** 905–915.

Schmeissner, U., McKenney, K., Rosenberg, M., and Court, D. (1984). Removal of a terminator structure by RNA processing regulates *int* gene expression. *J. Mol. Biol.* **176,** 39–53.

Singer, M. F., and Tolbert, G. (1965). Purification and properties of a potassium-activated phosphodiesterase (RNAase II) from *Escherichia coli. Biochemistry* **4,** 1319–1330.

Thang, M. N., Guschlbauer, W., Zachau, H. G., and Grunberg-Manago, M. (1967). Degradation of transfer ribonucleic acid by polynucleotide phosphorylase. *J. Mol. Biol.* **26,** 403–421.

Uzan, M., Favre, R., and Brody, E. (1988). A nuclease that cuts specifically in the ribosome binding site of some T4 mRNAs. *Proc. Natl. Acad. Sci. USA* **85,** 8895–8899.

Wong, H. C., and Chang, S. (1986). Identification of a positive retroregulator that stabilizes mRNAs in bacteria. *Proc. Natl. Acad. Sci. USA* **83,** 3233–3237.

2

The Role of the 3′ End in mRNA Stability and Decay

CHRISTOPHER F. HIGGINS, HELEN C. CAUSTON,
GEOFFREY S. C. DANCE, AND ELISABETH A. MUDD

I. Introduction

Studies over the past 20 or so years have elucidated a number of general features of mRNA degradation in bacteria (reviewed by Belasco and Higgins, 1988). Nevertheless, a detailed understanding of the degradation pathways, and particularly their regulation, still presents a considerable challenge. It should also not be forgotten that our concepts of mRNA turnover in prokaryotes are derived almost entirely from studies of *Escherichia coli* and its phage. Although it is not unreasonable to assume that similar principles will apply to other prokaryotic species, some important details are bound to differ, and there is little doubt that exotic regulatory mechanisms will be uncovered.

mRNA turnover is mediated by a combination of endoribonucleolytic and exoribonucleolytic activities (Figs. 1 and 2) (reviewed by Belasco and Higgins, 1988). Two 3′-to-5′ (3′–5′) exonucleases, polynucleotide phosphorylase (PNPase) and ribonuclease II (RNase II), are especially important for degrading mRNA to ribonucleotides. These two enzymes degrade RNA from the 3′ end, removing a nucleotide at a time as they proceed toward the 5′ end of the molecule (Fig. 1a). The end products of such degradation are the released mononucleotides and a small 5′-terminal oligonucleotide, 10 to 20 ribonucleotides long, which is relatively resistant to digestion by either enzyme. An additional enzyme, RNase I*, may complete the process by degrading these short oligonucleotides (Cannistraro and Kennell, 1991). It should be noted that no 5′–3′ processive exonucleases that degrade mRNA have been identified in bacteria.

Control of Messenger RNA Stability

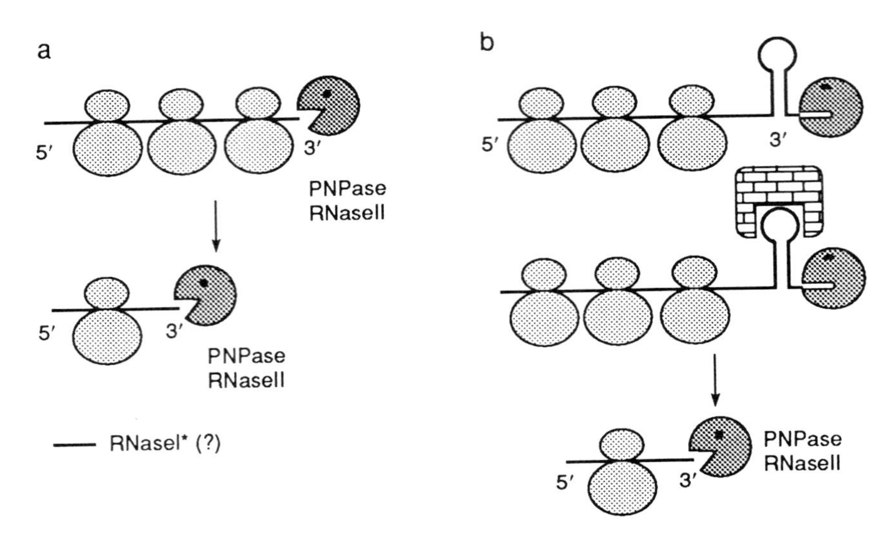

FIGURE 1 (a) In the simplest circumstances, unstructured mRNA molecules are degraded from the 3′ end by RNaseII and PNPase. The small oligoribonucleotides remaining after such degradation may be degraded to monoribonucleotides by RNaseI*. (b) In an RNA with a structured 3′ end, the processive activity of PNPase and RNaseII is inhibited by the structure alone or by the structure together with a stem–loop binding protein. Sometimes the exonucleases may be able to "break through" this barrier. More often than not, however, it appears that this barrier is effectively permanent and further degradation requires the action of endonucleases (Figure 2).

For some messages (probably a small minority), the pathway outlined above (Fig. 1a) may provide a complete description of their degradation; certainly, this simple mode of degradation can be mimicked *in vitro* (McLaren *et al.*, 1991). Nevertheless, for the majority of messages the 3′ end is protected from exonucleolytic attack by RNA secondary structure (stem–loops) (Fig. 1b). For such messages, decay is more complex and proceeds by one of several alternative pathways. In some cases, the 3′–5′ exonucleases may overcome the blockage presented by the 3′ terminal stem–loop and continue processive degradation (Fig. 1b). For most mRNAs, however, it seems that decay is initiated by endonucleolytic cleavage (Fig. 2). Either the terminal structure is removed, exposing a free 3′ end that is accessible to exonucleolytic decay (Fig. 2a), or endonucleolytic cleavage (or some other event) at the 5′ end renders the message susceptible to further endonucleolytic cleavage throughout its length (a process that is still poorly understood), generating fragments with free 3′ ends that are then degraded by the 3′–5′ processive exonucleases (Fig. 2b).

This chapter addresses the role of the 3′ end of mRNA molecules in determining and regulating mRNA half-lives. When considering the

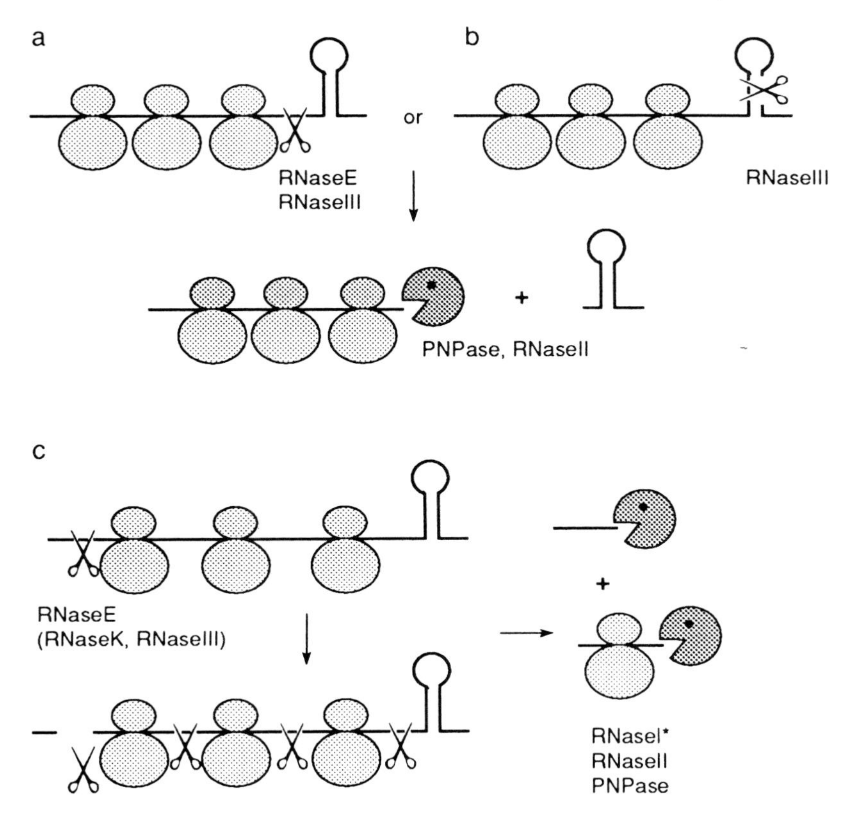

FIGURE 2 The "Pacman" model for RNA degradation in *E. coli*, adapted from Belasco and Higgins (1988). The block to processive exonucleases can be overcome by removal of the stem–loop, either by an enzyme that recognizes and cleaves the structure itself (b) or by cleavage immediately upstream (5′) of the structure (a). Alternatively, degradation can be initiated by cleavage near the 5′ end of the message, for example, by RNaseE. In some manner yet to be determined, this initial cleavage leads to degradation of the remainder of the message by a combination of endo- and exonucleolytic activities. See text for further details.

regulation of mRNA degradation, it is pertinent to distinguish between "rate-limiting steps" and "rate-determining pathways." A metabolic pathway consists of a number of steps in series. If a single step is limiting, increasing the rate of this rate-limiting step will increase the rate of the entire pathway; increasing the rate of any other step will have little or no effect. In contrast, mRNA degradation can, potentially, be mediated by any one of several parallel pathways (for example, 5′ endonucleolytic cleavage, Fig. 2b; degradation by 3′–5′ exonucleases, Fig. 1b). Increasing the rate of any one of these pathways could increase the rate of degradation; all are potentially rate-determining. What is crucial as far as mRNA degradation is concerned is the step that normally initiates the decay of a given mRNA molecule.

II. The 3′–5′ Exoribonucleases: PNPase and RNase II

The role of the 3′ end in mRNA degradation is, of course, intimately associated with the activity of any enzyme for which it is a substrate. Several 3′–5′ exoribonuclease activities have been identified in bacteria (Deutscher, 1985). Only two of these, PNPase and RNase II, play a general role in mRNA turnover; other known activities are restricted to the processing of tRNA and other stable RNA molecules.

A. Polynucleotide Phosphorylase

PNPase was first identified in 1955 and has been extensively characterized (reviewed by Godefroy-Colburn and Grunberg-Manago, 1972; Littauer and Soreq, 1982). Although this enzyme can catalyze untemplated RNA synthesis from nucleoside diphosphates, conditions in the cell favor the reverse, degradative reaction. The catalytic (α) subunit of PNPase has a molecular weight of 86 kDa and functions as a trimer. The gene encoding this subunit (*pnp*) has been cloned and sequenced (Regnier *et al.*, 1987). Some preparations of PNPase also include a second (β) subunit of 48 kDa that is not required for catalytic activity (Portier, 1975); the function of the β-subunit is unknown. PNPase degrades RNA phosphorolytically from the 3′ end, removing one nucleotide at a time to yield nucleoside diphosphates as products. *In vitro*, the activity is processive in that the enzyme degrades one RNA molecule to completion (i.e., to a short oligonucleotide end product) before it is free to initiate decay of another. PNPase degrades essentially any polynucleotide, irrespective of its sequence or base composition, although the progress of the enzyme is impeded when it encounters secondary structures in the RNA (Godefroy-Colburn and Grunberg-Manago, 1972; Littauer and Soreq, 1982; McLaren *et al.*, 1991; Chen *et al.*, 1991).

Together with ribosomal protein S15 (*rpsO*), the gene encoding PNPase forms an operon located at 69 min on the *E. coli* chromosomal map. These two proteins are differentially expressed following processing of the *rpsO–pnp* transcript by both RNase E and RNase III (Portier *et al.*, 1987; Regnier and Hajnsdorf, 1991).

B. Ribonuclease II

RNase II is similar to PNPase in many respects (Nossal and Singer, 1968; Singer and Tolbert, 1965; Gupta *et al.*, 1977). It too is a processive, 3′–5′ exoribonuclease whose activity is sequence-independent although sensitive to RNA secondary structure. If anything, the degradative activity of RNaseII is impeded by RNA secondary structure to an extent greater than that of PNPase (Nossal and Singer, 1968; Gupta *et al.*, 1977; McLaren *et al.*, 1991). The *rnb* gene, which encodes the 80-kDa RNase II polypeptide, is located at 28 min on the *E. coli* chromosome (Donovan and Kushner, 1983) and has recently been sequenced (Zilhao *et al.*, 1993). RNaseII differs from PNPase in that it is hydrolytic rather than phosphorolytic and requires potassium ions for full activity.

C. PNPase, RNase II, and mRNA Degradation

PNPase and RNase II have been known to play a role in mRNA degradation for many years (e.g., Kaplan and Apirion, 1974; Kinscherf and Apirion, 1975). Mutations in the *pnp* and *rnb* genes made it possible to demonstrate that these are the two major 3′–5′ exoribonucleases that degrade mRNA in *E. coli* (Donovan and Kushner, 1986; Deutscher and Reuven, 1991). About 90% of the exoribonucleolytic activity of *E. coli* is RNase II, while PNPase makes up the remainder. In contrast, PNPase appears to be the predominant and perhaps the only exoribonuclease that degrades mRNA in *Bacillus subtilis* (Deutscher and Rueven, 1991). Strains of *E. coli* deficient in either one of these two enzymes are viable and degrade mRNA at essentially normal rates. However, *pnp rnb* double mutants are nonviable (Donovan and Kushner, 1986), and intermediates in mRNA decay accumulate *in vivo* when both of these enzyme activities are absent (Arraiano *et al.*, 1988). Thus, in *E. coli* the enzymes appear to be in excess and to be functionally redundant.

Why, then, do cells contain two 3′–5′ exonucleases? There are two alternative explanations. First, the enzymes may exhibit greater specificity than they are normally assigned: the degradation of some RNA molecules differs in *pnp* and *rnb* mutants, implying some specificity of function. For example, a specific degradation intermediate of the ribosomal S20 mRNA accumulates in a *pnp* strain but not in an *rnb* strain (Mackie, 1989). Degradation of phage λ *int* mRNA following RNase III cleavage is defective in *pnp* strains but not in *rnb* strains (Guarneros *et al.*, 1988; Guarneros and Portier, 1990). These differences may be a consequence of a greater sensitivity of RNase II, versus PNPase, to RNA secondary structure (see, for example, Guarneros and Portier, 1990; McLaren *et al.*, 1991). Another possible explanation is based on the fact that PNPase is a phosphorolytic enzyme while RNase II is hydrolytic. As phosphorolysis releases nucleoside diphosphates, it is an energy conserving process and might be advantageous under conditions of energy limitation. Differential use of the two enzymes may reflect an adaptation to changing energy status (Deutscher and Reuven, 1991). This model is supported by the finding that *B. subtilis,* which normally inhabits energy-poor environments, uses PNPase exclusively, while RNase II is the predominant 3′ exoribonuclease in *E. coli* (Deutscher and Reuven, 1991).

III. 3′ Stem–Loop Structures Stabilize Upstream mRNA

Stem–loop structures that can potentially form in mRNA with an appropriate "inverted repeat" sequence have two distinct influences on mRNA degradation: (1) a "passive" function as impediments to the processive action of 3′–5′ exonucleases and (2) an "active" function as sites or determinants of cleavage by specific endonucleases, such as RNase III or RNaseE (Mudd *et al.*, 1988; Ehretsmann *et al.*, 1992; Krinke and Wulff, 1990).

The degradative activities of both RNase II and PNPase have long been known to be sensitive to RNA secondary structure. In particular, their processive activities are impeded by 3′ stem–loop structures (illustrated in Fig. 1b). Direct evidence that stem–loop structures play a role in determining the rate of mRNA decay *in vivo* came initially from studies on the *trp* operon of *E. coli* (Mott *et al.*, 1985). The 3′ end of *trp* mRNA *in vivo* was mapped to the 3′ base of a stem–loop structure (*trpt*) and was assumed to be the product of rho-independent transcription termination there. However, genetic data and *in vitro* studies showed that most transcription of the *trp* operon actually terminates further downstream at a rho-dependent terminator (*trpt′*); the *in vivo* 3′ end is derived largely from 3′–5′ exonucleolytic processing of the primary transcript back to the *trpt* stem–loop structure.

Many studies have now shown that stem–loop structures can stabilize upstream mRNA *in vivo* (Hayashi and Hayashi, 1985; Panyatatos and Truong, 1985; Wong and Chang, 1986; Newbury *et al.*, 1987a,b; Mackie, 1987; Klug *et al.*, 1987; Chen *et al.*, 1988; Plamann and Stauffer, 1990). The majority of these studies were on gram-negative species. However, since stem–loop structures can also stabilize upstream mRNA in *B. subtilis* (Wong and Chang, 1986), mRNA degradative pathways in gram-positive bacteria may not be too different.

Of particular interest are the highly conserved REP (repetitive extragenic palindromic) sequences, about 500 copies of which are present on the *E. coli* genome (Higgins *et al.*, 1982; Stern *et al.*, 1984; Gilson *et al.*, 1984). As RNA, REP sequences form stem–loop structures which can stabilize upstream mRNA: a typical REP stem–loop is shown in Fig. 3. In at least some operons, this REP stem–loop can serve an important regulatory function (Newbury *et al.*, 1987a,b; Higgins *et al.*, 1988). In this regard, REP sequences are not unique, and they can be replaced functionally by other stem–loop structures. Thus, the sequence conservation of REP elements is not simply a consequence of their ability to stabilize mRNA (Higgins *et al.*, 1988). This illustrates a general principle: the ability of a stem–loop to

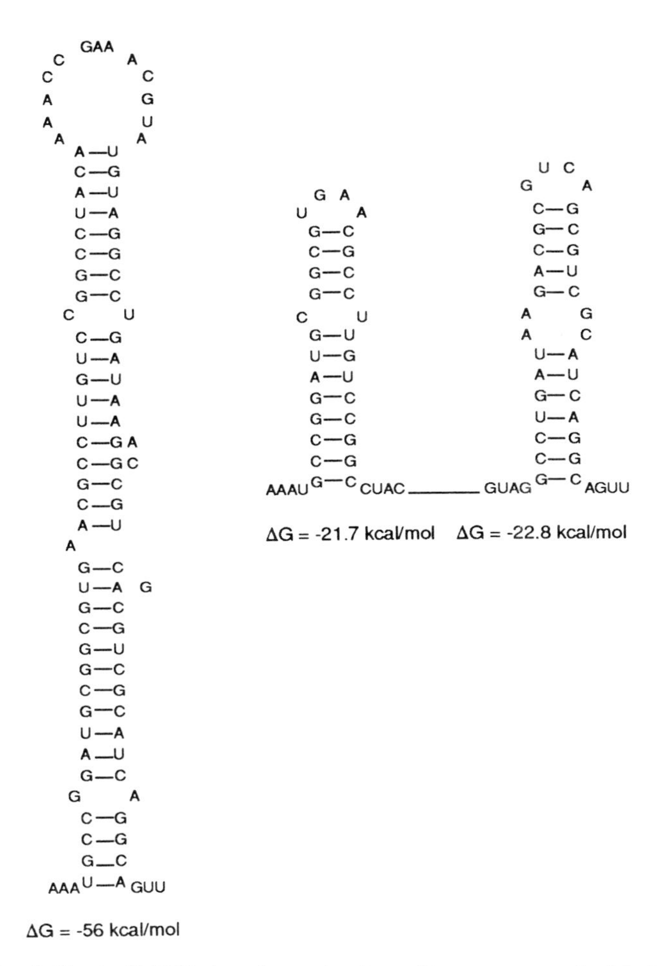

FIGURE 3 A "typical" REP stem–loop structure. The sequences depicted are from the *malE–malF* intergenic region. In this region there are two copies of the REP sequence, inverted in orientation with respect to each other. Thus, they can potentially fold to form a single large hairpin structure or, alternatively, the two REP sequences can fold independently to form two smaller structures. [Adapted from Higgins *et al.* (1988).]

stabilize mRNA is relatively independent of its sequence, and substitution of one stable stem–loop for another at the 3′ end of a gene has little or no effect on mRNA half-life (Belasco *et al.*, 1986; Newbury *et al.*, 1987b; Chen *et al.*, 1988; McLaren *et al.*, 1991).

There are three circumstances in which a stem–loop structure is not expected to stabilize upstream mRNA. First, a stem–loop structure will not stabilize mRNA if sequences upstream of the structure are particularly susceptible to endonucleolytic cleavage. Therefore, if a stem–loop is inserted downstream of the principal site(s) of endonucleolytic initiation of mRNA decay, it will not prolong the lifetime of the upstream RNA segment (Chen and Belasco, 1990). This may provide an explanation for the failure of stem–loop structures to stabilize *cat* mRNA (DeFranco and Schottel, 1989). Second, a stem–loop structure will not stabilize upstream mRNA if the stem–loop itself serves as a site for endonucleolytic cleavage that disrupts the structure. Indeed, endonucleolytic cleavage of stem–loop structures at the 3′ end of certain transcripts can serve a regulatory role (see below). RNase III recognizes many large stem–loop structures (Krinke and Wulff, 1990), and secondary structure plays a role in RNaseE recognition (Mudd *et al.*, 1988; Ehretsmann *et al.*, 1992). Stem–loop structures that stabilize mRNA *in vivo,* such as REP sequences, have presumably been selected for during evolution because they are not recognized by such enzymes (Stern *et al.*, 1984). Third, if a stem–loop is positioned so that translating ribosomes can disrupt the structure, it is much less efficient at stabilizing upstream mRNA (Chen and Belasco 1990; Klug and Cohen, 1990). This is important in cases where stem–loop structures are located within protein-coding sequences, as these stem–loops might otherwise stabilize mRNA fragments that direct the synthesis of truncated, nonfunctional polypeptides.

IV. How Do Stem–Loop Structures Stabilize Upstream mRNA?

It is generally assumed that stem–loop structures stabilize upstream mRNA by impeding 3′–5′ processive exonucleolytic degradation (Fig. 4). Formal proof for this is lacking, although the large body of available data is entirely consistent with such a model, and there is little doubt it is correct. (1) Stem–loop structures that stabilize mRNA *in vivo* can impede the processive activity of both PNPase and RNase II *in vitro* (Mott *et al.*, 1985; McLaren *et al.*, 1991). (2) The 3′ ends of many mRNAs *in vivo* are at the 3′ base of such structures, as would be expected if they are generated by impeding 3′–5′ exonucleolytic activity (e.g., Stern *et al.*, 1984; Belasco *et al.*, 1985; Mott *et al.*, 1985; Newbury *et al.*, 1987a; McLaren *et al.*, 1991). For *malE* mRNA, the end points identified *in vivo* have been shown to be similar to those generated on the same transcript *in vitro* by PNPase and RNase II (McLaren *et al.*, 1991). (3) Deletion of the REP stem–loop at the 3′ end of the *glyA* gene destabilizes upstream mRNA only in strains in which PNPase or RNase II are functional (Plamann and Stauffer, 1990).

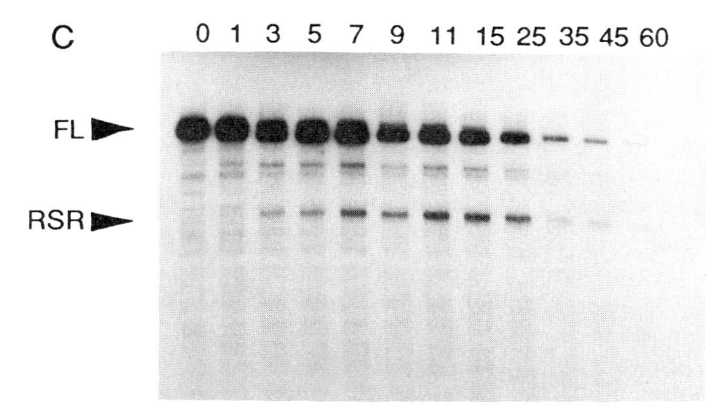

FIGURE 4 *In vitro* degradation of mRNA containing a REP (stem–loop) sequence. Full-length mRNA (FL), synthesized *in vitro*, was incubated for the indicated periods of time with purified PNPase. The products of these reactions were separated on a denaturing polyacrylamide gel. As the full-length mRNA is degraded, an intermediate species appears transiently (RSR). The end point of this intermediate species is at the base of the stem–loop structure formed by the REP sequence and is due to temporary blockage of the processive activity of PNPase by this structure. [For further details of such experiments see McLaren *et al.* (1991).]

V. 3′ Stem–Loop Structures Influence Gene Expression

Stem–loop structures that stabilize upstream mRNA can, potentially, increase gene expression. The first clear demonstration of this phenomenon was for the *trpt* terminator, which protects upstream mRNA against exonucleolytic degradation (see previous section). Deletion of this stem–loop structure reduced expression of upstream genes *in vivo* (Mott *et al.*, 1985). Many additional studies have now confirmed that deletion or disruption of 3′ stem–loop structures can reduce expression of upstream genes; similarly, addition of a stem–loop can increase expression (e.g., Guarneros *et al.*, 1982; Plamann and Stauffer, 1985; Becerril *et al.*, 1985; Panyatatos and Truong, 1985; Newbury *et al.*, 1987a,b; Klug *et al.*, 1987). In certain cases it has been demonstrated that this phenomenon is indeed a result of altered mRNA stability (Newbury *et al.*, 1987a,b; Klug *et al.*, 1987), and it seems reasonable to assume that this explanation is generally applicable. Indeed, it would now seem that an important function of the stem–loop structures of rho-independent terminators, irrespective of any role in termination itself, is to protect the completed transcipt from exonucleolytic degradation.

Not only are 3′ stem–loops important in determining the overall stability of a transcript, but they also can make a crucial contribution to differential stability of segments within a polycistronic transcript and, hence, to differential expression of genes within an operon. For example, for both the *pufBALMX* transcript of *Rhodobacter capsulatus* and the *malEFG* transcript of *E. coli*, the 5′ mRNA segment is more stable than the 3′ segment. An intercistronic stem–loop is found precisely at the 3′ end of the stable segment, and, in each instance, deletion of the intercistronic stem–loop decreases the stability of the upstream segment (Newbury *et al.*, 1987a,b; Chen *et al.*, 1988; Belasco and Chen, 1988). In the case of the *malEFG* maltose transport operon, the periplasmic MalE protein encoded by the 5′ segment of the transcript is expressed at 30 times the level of the membrane-associated MalF and MalG proteins. Deletion of the intercistronic stem–loop (REP sequence) between the *malE* and *malF* segments destabilizes the upstream *malE* message and reduces *malE* expression without affecting expression of the downstream *malF* and *malG* genes. This reduction in MalE protein synthesis results in reduced maltose transport and chemotaxis (Newbury *et al.*, 1987b). Similarly, deletion of an intercistronic stem–loop from *pufBALMX* mRNA reduces synthesis of the light-harvesting polypeptides encoded by the upstream *pufBA* RNA segment and lowers the rate of photosynthetic cell growth (Klug *et al.*, 1987). Thus, differential mRNA stability in a polycistronic transcript, mediated by an intercistronic stem–loop structure, can play a physiologically important role in determining relative levels of gene expression.

It should be noted that even when the physical stability of a message is increased or decreased by adding or deleting a 3′ stem–loop, protein synthesis is not necessarily affected. Translation may be limited by factors other than the amount of mRNA [e.g., by secretion of cotranslationally exported proteins (Newbury *et al.*, 1987a,b)], and a fraction of some mRNAs is translationally inactive (Hayashi and Hayashi, 1985; Mackie, 1987; Newbury *et al.*, 1987a). Furthermore, the translational half-life of a message (i.e., the length of time during which it can direct protein synthesis after cessation of transcription) is not necessarily the same as its physical half-life (Schwartz *et al.*, 1970; Gupta and Schlessinger, 1976; Yamamoto and Imamoto, 1975; Shen *et al.*, 1981) (see Chapter 6). A recent example of this phenomenon is the demonstration that the physical half-life of *E. coli* mRNA is substantially increased in an RNaseE-deficient strain, while the translational half-life of these messages (and therefore protein synthesis) is only slightly increased (Mudd *et al.*, 1990). The mechanisms of translational inactivation are unclear. Thus, although the use of a stem–loop structure to stabilize upstream mRNA and increase gene expression is an attractive concept, many factors need to be considered when attempting to employ such a strategy.

A. Removal of a Stem–Loop Can Serve a Regulatory Function

As stem–loop structures can stabilize upstream mRNA and influence gene expression, specific mechanisms for removing stem–loop structures could, potentially, have a regulatory function. Several examples of such mechanisms are now known. One of the best examples is the expression of the phage λ *int* (intergrase) gene. When phage λ grows lytically, the *int* gene is transcribed as part of the long leftward transcript of λ. RNase III cleavage of a stem–loop structure immediately downstream of *int* allows rapid exonucleolytic degradation of *int* mRNA. The resulting reduction in integrase synthesis, relative to other proteins encoded by this large polycistronic transcript, helps to suppress integration of the λ genome into the *E. coli* chromosome (Guarneros *et al.*, 1982; Schmeissner *et al.*, 1984).

Regulating the endonucleolytic removal of 3′ stem–loops could, potentially, regulate gene expression in response to environmental signals. As both RNase III and RNase E process ribosomal RNA, variations in their activity are unlikely to be significant in this context. Nevertheless, there may be specific, regulated endonucleases that have not yet been identified and that serve such a role; such activities have been implicated in mRNA decay in chloroplasts (Adams and Stern, 1990) (see Chapter 14). An alternative to a regulated endonuclease is regulation of the accessibility of a 3′ stem–loop to an endonuclease. This could be achieved by alternative folding of mRNA or by interaction of mRNA with translating ribosomes, antisense RNA (transposon Tn*10* transposase mRNA base pairs with an antisense transcript to create a duplex that is a substrate for RNase III; Case *et al.*, 1990), or stem–loop binding proteins (see later). Thus, there is considerable potential for regulating mRNA stability by modulation of events at the 3′ end; whether this provides a general regulatory mechanism remains to be established.

VI. Degradation of mRNA Stabilized by 3′ Stem–Loops

The 3′ stem–loop structures stabilize upstream mRNA by impeding the processive activities of RNaseII and PNPase. However, all messages are eventually degraded, and it is important to consider how the barrier presented by these stem–loops is overcome or bypassed. There are two possibilities. Certain messages may be sufficiently resistant to endonucleolytic cleavage that decay is initiated by exonucleases that breach the stem–loop barrier and degrade the mRNA in a 3′–5′ direction (Fig. 1b).

However, it is increasingly apparent for the majority of messages that the 3′ stem–loop structure provides a sufficiently long-term block to exonuclease digestion that degradation is initiated by upstream endonucleolytic cleavage (Fig. 2) (see Chapter 3). As some messages have half-lives of around 20 min, stem–loops must be able to impede 3′–5′ exonucleases for at least this long. Thus, for most messages, which have shorter half-lives, the block provided by the 3′ stem–loop must be bypassed, presumably by endonucleolytic cleavage. This conclusion is supported by experiments showing that replacement of the 3′ stem–loop of an unstable mRNA with that from a stable transcript does not increase the mRNA half-life (Belasco *et al.*, 1986). In contrast, exchanging the 5′ ends of transcripts without altering the 3′ stem–loop can dramatically affect their half-lives (Yamamoto and Imamoto, 1975; Gorski *et al.*, 1985; Belasco *et al.*, 1986; Chen *et al.*, 1988). For the *puf* and *ompA* messages, for example, there is direct evidence that decay of messages stabilized by 3′ stem–loops is initiated by upstream endonucleolytic cleavage rather than by exonucleases penetrating the stem–loop (Melefors and vonGabain, 1988; Chen and Belasco, 1990; Klug and Cohen, 1990). Finally, the decay intermediates that have been detected for many messages are not consistent with simple 3′–5′ degradation, but support the view that decay involves multiple endonucleolytic cleavages (e.g., Arraiano *et al.*, 1988; G. S. C. Dance and C. F. Higgins, unpublished data). The nature and control of these endonucleolytic cleavages, and their role in determining mRNA half-life, are discussed elsewhere in this volume.

VII. Stem–Loop-Binding Proteins

Messages with stem–loop structures at the 3′ end are protected from the processive actions of PNPase and RNase II both *in vivo* and *in vitro*. *In vitro*, both enzymes can eventually overcome stem–loop barriers (Mott *et al.*, 1985; McLaren *et al.*, 1991). This breaching appears to be an all-or-nothing event: a ladder of intermediates as the enzyme "unzips" the stem–loop structure is not observed (McLaren *et al.*, 1991). The finding that PNPase and RNaseII have different sensitivities to secondary structure implies that the enzymes themselves play an active role in overcoming the barrier; if their progress depended solely on spontaneous melting of the RNA structure, the two enzymes would be impeded to the same extent. Also consistent with this view is the observation that stem–loop structures with very different calculated thermodynamic stabilities, which would be expected to exhibit different melting characteristics, do not necessarily differ in their ability to impede the 3′–5′ exonucleases.

Importantly, the length of time for which a stem–loop structure can impede PNPase and RNase II *in vitro* is relatively short in comparison with the *in vivo* half-lives of mRNAs that end with the same structure (McLaren *et al.*, 1991). For example, REP sequences impede PNPase and RNase II for 1 to 2 min *in vitro* (McLaren *et al.*, 1991), yet can resist degradation at least four to eight times longer *in vivo* (Newbury *et al.*, 1987a,b; Merino *et al.*, 1987). The 3′ stem–loops from the T4 gene 32 and *E. coli ompA* messages also block exonucleases for a relatively short period of time *in vitro*, even though these messages have half-lives of over 15 min *in vivo* (Belasco *et al.*, 1986; Belin *et al.*, 1987). Evidence that the degradation of most messages (including those with long half-lives, such as *ompA*) is initiated by endonucleolytic cleavage also implies that stem–loop structures can provide an effective long-term block to 3′–5′ exonuclease digestion *in vivo*.

The differences between data obtained *in vivo* and those obtained *in vitro* imply that an additional factor may be involved in impeding the passage of endonucleases through stem–loop structures *in vivo*. Theoretically, stem–loops bound by such a factor could impede 3′–5′ exonuclease degradation for sufficiently long that decay would be initiated endonucleolytically. As the ability of a stem–loop to stabilize upstream mRNA *in vivo* is essentially independent of the stem–loop sequence, this putative factor would also be expected to bind in a sequence-independent manner. Using gel retardation of labeled RNA as an assay, a protein with the appropriate characteristics has now been identified and partially characterized (H. C. Causton and C. F. Higgins, unpublished results). This protein binds to every stem–loop structure tested, irrespective of its sequence. *In vitro*, the partially purified protein impedes the progress of PNPase through stem–loop structures. Interestingly, this stem–loop binding protein copurifies with the catalytic subunit of PNPase. The possibility that it is the β-subunit of PNPase (which has not yet been assigned a function; see previous text) is attractive but remains to be tested.

The identification of stem–loop binding proteins provides scope for regulating mRNA stability and gene expression. The activity or specificity of such a protein could be influenced by external conditions, and there may even be a family of such proteins with different binding characteristics. Whether such mechanisms have been adopted by bacterial cells remains to be established.

VIII. Chloroplasts

Since chloroplasts are of prokaryotic origin, it is relevant to consider the mechanisms of RNA degradation in these organelles (see also Chapter 14). In many respects, chloroplast mRNAs appear to be degraded by mechanisms similar to those of prokaryotes. Stem–loop structures at the 3′ end of chloroplast transcripts are involved in the stabilization of upstream RNA both *in vitro* and *in vivo* (Stern and Gruissem, 1987; Stern *et al.*, 1989; Adams and Stern, 1990), probably by protecting against 3′–5′ exonucleolytic attack (Stern and Gruissem, 1987). PNPase is found in chloroplasts (Godefroy-Colburn and Grunberg-Manago, 1972). Furthermore, a 28-kDa protein that interacts with the 3′ stem–loop of various chloroplast mRNAs has been identified (Schuster and Gruissem, 1991). As the abundance of this protein correlates with mRNA stability under different light regimes, it may be involved in the regulation of stability and, hence, gene expression. Finally it appears, at least for some chloroplast messages, that degradation occurs by endonucleolytic removal of the 3′ stem–loop (Adams and Stern, 1990;

Schuster and Gruissem, 1991). Thus, the role of the 3′ end of messages in chloroplasts appears to be very similar to that in prokaryotic cells. The finding that specific regulation of mRNA stability in chloroplasts in response to environmental signals is mediated by events at the 3′ end supports the idea that such regulatory mechanisms may also operate in bacteria.

IX. Concluding Remarks: The Role of the 3′ End in Regulation

The degradation of mRNA to ribonucleotides is mediated by enzymes that attack the 3′ end. Thus, the nature of the 3′ end of a message is important in determining its absolute rate of decay. This does not, however, mean that the 3′ end necessarily plays a regulatory role, influencing the stability of specific mRNA species, and hence gene expression, in response to external stimuli.

The primary function of RNase II and PNPase, the two 3′–5′ exonucleases that attack the 3′ end of messages, is to remove degradation intermediates (and prematurely terminated messages), thereby recycling ribonucleotides and preventing the wasteful synthesis of incomplete polypeptides. These ribonucleases are, therefore, present in excess, and any subtle control of their activity seems unlikely. Additionally, because these enzymes are in excess, the 3′ ends of intact messages must be protected if these messages are to survive for any significant length of time. Thus, for the majority of messages the 3′ end is blocked by stem–loop structures and, possibly, associated proteins that make it inaccessible to these exonucleases. Other means of protecting mRNA 3′ ends have also been suggested, such as pseudoknots or poly(A) tails (Taljanidisz *et al.*, 1987), although no such role has yet been demonstrated for these structures in bacteria. Thus, the primary role of stem–loops at the 3′ end of mRNA appears to be to ensure stability, rather than to regulate mRNA longevity.

The only scope for modulation of mRNA stability by the 3′ end is to alter its accessibility to exonuclease digestion. This could be achieved by endonuclease cleavage of the 3′ stem–loop, by alternate folding of the mRNA, or potentially by altering interactions of the stem–loop with binding proteins. While such events occur, none are known to be regulated in response to specific signals. Instead, the few regulatory events that have been studied involve rate-determining endonucleolytic cleavages controlled by the 5′ end of the message (Melefors and vonGabain, 1988; Emory and Belasco, 1990), and this seems likely to be the main site at which regulation of mRNA stability occurs. Nevertheless, cells normally make use of a wide variety of regulatory mechanisms, and it is not unlikely that examples of regulation involving the mRNA 3′ end will be discovered. The identification of proteins that bind to 3′ stem–loops and influence their resistance to exonuclease attack may shed light on such regulation. Further developments are awaited with interest.

Acknowledgments

We are grateful to many colleagues for stimulating discussions. The authors' research is supported by the Imperial Cancer Research Fund.

References

Adams, C. C., and Stern, D. B. (1999). Control of mRNA stability in chloroplasts by 3′ inverted repeats: effects of stem and loop mutations on degradation of *psbA* mRNA *in vitro*. *Nucleic Acids Res.* **18,** 6003–6010.

Arraiano, C. M., Yancey, S. D., and Kushner, S. R. (1988). Stabilization of discrete mRNA breakdown products in *ams pnp rnb* multiple mutants of *Escherichia coli* K-12. *J. Bacteriol.* **170,** 4625–4633.

Becerril, B., Valle, F., Merino, E., Riba, L., and Bolivar, F. (1985). Repetitive extragenic palindromic (REP) sequences in the *Escherichia coli gdhA* gene. *Gene* **37,** 53–62.

Belasco, J. G., Beatty, J. T., Adams, C. W., von Gabain, A., and Cohen, S. N. (1985). Differential expression of photosynthesis genes in *R. capsulata* results from segmental differences in stability within the polycistronic *rxcA* transcript. *Cell* **40,** 171–181.

Belasco, J. G., and Chen, C-Y. A. (1988). Mechanism of *puf* mRNA degradation: The role of an intercistronic stem–loop structure. *Gene* **72,** 109–117.

Belasco, J. G., Nilsson, G., von Gabain, A., and Cohen, S. N. (1986). The stability of *E. coli* gene transcripts is dependent on determinants localized to specific mRNA segments. *Cell* **46,** 245–251.

Belasco, J. G., and Higgins, C. F. (1988). Mechanisms of mRNA decay bacteria: A perspective. *Gene* **72,** 15–23.

Belin, D., Mudd, E. A., Prentki, P., Yu, Y-Y., and Krisch, H. M. (1987). Sense and antisense transcription of bacteriophage T4 gene 32. *J. Mol. Biol.* **194,** 231–243.

Cannistraro, V. J., and Kennell, D. (1991). RNase I*, a form of RNase I, and mRNA degradation in *Escherichia coli*. *J. Bacteriol.* **173,** 4653–4659.

Case, C. C., Simons, E. L., and Simons, R. W. (1990). The IS*10* transposase mRNA is destabilized during antisense control. *EMBO J.* **9,** 1259–1266.

Chen, C-Y. A., Beatty, J. T., Cohen, S. N., and Belasco, J. G. (1988). An intercistronic stem–loop structure functions as an mRNA decay terminator necessary but insufficient for *puf* mRNA stability. *Cell* **52,** 609–619.

Chen, C-Y. A., and Belasco, J. G. (1990). Degradation of *pufLMX* mRNA in *Rhodobacter capsulatus* is initiated by nonrandom endonucleolytic cleavage. *J. Bacteriol.* **172,** 4578–4586.

Chen, L-H., Emory, S. A., Bricker, A. L., Bouvet, P., and Belasco, J. G. (1991). Structure and function of a bacterial mRNA stabilizer: Analysis of the 5′ untranslated region of *ompA* mRNA. *J. Bacteriol.* **173,** 4578–4586.

De Franco, C., and Schottel, J. L. (1989). Terminal sequences do not contain the rate-limiting decay determinants of *E. coli cat* mRNA. *Nucleic Acids Res.* **17,** 1139–1157.

Deutscher, M. P. (1985). *E. coli* RNases: Making sense of alphabet soup. *Cell* **40,** 731–732.

Deutscher, M. P., and Reuven, N. B. (1991). Enzymatic basis of hydrolytic versus phosphorolytic mRNA degradation in *Escherichia coli* and *Bacillus subtilis*. *Proc. Natl. Acad. Sci. USA* **88,** 3277–3280.

Donovan, W. P., and Kushner, S. R. (1983). Amplification of ribonuclease II (*rnb*) activity in *Escherichia coli* K-12. *Nucleic Acids Res.* **11,** 265–275.

Donovan, W. P., and Kushner, S. R. (1986). Polynucleotide phosphorylase and ribonuclease II are required for cell viability and mRNA turnover in *Escherichia coli* K-12. *Proc. Natl. Acad. Sci. USA* **83,** 120–124.

Ehretsmann, C. P., Carpousis, A. J., and Krisch, H. M. (1992). Specificity of *E. coli* endoribonuclease RNase E: *In vivo* and *in vitro* analysis of mutants in a bacteriophage T4 mRNA processing site. *Genes Dev.* **6,** 149–159.

Emory, S. A., and Belasco, J. G. (1990). The *ompA* 5′ untranslated RNA segment functions in *Escherichia coli* as a growth-rate-regulated mRNA stabilizer whose activity is unrelated to translational efficiency. *J. Bacteriol.* **172,** 4472–4481.

Gilson, E., Clement, J-M., Brutlag, D., and Hofnung, M. (1984). A family of dispersed repetitive extragenic palindrome DNA sequences in *E. coli*. *EMBO J.* **3,** 1417–1422.

Godefroy-Colburn, T., and Grunberg-Manago, M. (1972). Polynucleotide phosphorylase. *In* The Enzymes (P. D. Boyer, Ed.), Vol. 7, pp. 533–574. Academic Press, New York.

Gorski, K., Roch, J-M., Prentki, P., and Krisch, H. M. (1985). The stability of bacteriophage T4 gene 32 mRNA: A 5′ leader sequence that can stabilize mRNA transcripts. *Cell* **43,** 461–469.

Guarneros, G., Kameyama, L., Orozco, L., and Velazquez, F. (1988). Retroregulation of a *int–lacZ* gene fusion in a plasmid system. *Gene* **72,** 129–130.

Guarneros, C., and Porter, C. (1990). Different specificities of ribonuclease II and polynucleotide phosphorylase in 3′ mRNA decay. *Biochimie* **72,** 771–777.

Gupta, R. S., and Schlessinger, D. (1976). Coupling of rates of transcription, translation and messenger ribonucleic acid degradation–Streptomycin-dependent mutants of *Escherichia coli*. *J. Bacteriol.* **125,** 84–93.

Gupta, R. S., Kasai, T., and Schlessinger, D. (1977). Purification and some novel properties of RNase II. *J. Biol. Chem.* **252,** 8945–8951.

Hayashi, M. N., and Hayashi, M. (1985). Cloned DNA sequences that determine mRNA stability of bacteriophage ϕX174 *in vivo* are functional. *Nucleic Acids Res.* **13,** 5937–5948.

Higgins, C. F., Ames, G. F-L., Barnes, W. M., Clement, J. M., and Hofnung, M. (1982). A novel intercistronic regulatory element of prokaryotic operons. *Nature* (*London*) **298,** 760–762.

Higgins, C. F., McLaren, R. S., and Newbury, S. F. (1988). Repetitive extragenic palindromic sequences, mRNA stability and gene expression: Evolution by gene conversion?—A review. *Gene* **72,** 3–14.

Kaplan, R., and Apirion, D. (1974). The involvement of ribonuclease I, ribonuclease II, and polynucleotide phosphorylase in the degradation of stable ribonucleic acid during carbon starvation in *E. coli*. *J. Biol. Chem.* **249,** 149–151.

Kinscherf, T. G., and Apirion, D. (1975). Polynucleotide phosphorylase can participate in decay of mRNA in *Escherichia coli* in the absence of ribonuclease II. *Mol. Gen. Genet.* **139,** 357–362.

Klug, G., Adams, C. W., Belasco, J., Doerge, B., and Cohen, S. N. (1987). Biological consequences of segmental alterations in mRNA stability: Effects of deletion of the intercistronic hairpin loop region of the *R. capsulatus puf* operon. *EMBO J.* **6,** 3515–3520.

Klug, G., and Cohen, S. N. (1990). Combined actions of multiple hairpin loop structures and sites of rate-limiting endonucleolytic cleavage determine differential degradation rates of individual segments within polycistronic *puf* operon mRNA. *J. Bacteriol.* **172,** 5140–5146.

Krinke, L., and Wulff, D. L. (1990). The cleavage specificity of RNase III. *Nucleic Acids Res.* **18,** 4809–4815.

Littauer, U. Z., and Soreq, H. (1982). Polynucleotide phosphorylase. In The Enzymes (P. D. Boyer, Ed.), Vol. 17, pp. 517–553. Academic Press, New York.

Mackie, G. A. (1987). Posttranscriptional regulation of ribosomal protein S20 and stability of the S20 mRNA species. *J. Bacteriol.* **169,** 2697–2701.

Mackie, G. A. (1989). Stabilization of the 3′ one-third of *Escherichia coli* ribosomal protein S20 mRNA in mutants lacking polynucleotide phosphorylase. *J. Bacteriol.* **171,** 4112–4120.

McLaren, R. S., Newbury, S. F., Dance, G. S. C., Causton, H. C., and Higgins, C. F. (1991). mRNA degradation by processive 3′–5′ exoribonucleases *in vitro* and the implications for prokaryotic mRNA decay *in vivo*. *J. Mol. Biol.* **221,** 81–95.

Melefors, O., and von Gabain, A. (1988). Site-specific endonucleolytic cleavages and the regulation of stability of *E. coli ompA* mRNA. *Cell* **52,** 893–901.

Merino, E., Becerril, B., Valle, F., and Bolivar, F. (1987). Deletion of a repetitive extragenic palindromic (REP) sequence downstream from the structural gene of *Escherichia coli* glutamate dehydrogenase affects the stability of its mRNA. *Gene* **58,** 305–309.

Mott, J. E., Galloway, J. C., and Platt, T. (1985). Maturation of *Escherichia coli* tryptophon operon mRNA: Evidence for 3′ exonucleolytic processing after rho-dependent termination. *EMBO J.* **4,** 1887–1891.

Mudd, E. A., Prentki, P., Belin, D., and Krisch, H. M. (1988). Processing of unstable bacteriophage T4 gene 32 mRNA into a stable species requires *Escherichia coli* ribonuclease E. *EMBO J.* **7,** 3601–3607.

Mudd, E. A., Krisch, H. M., and Higgins, C. F. (1990). RNase E, an endoribonuclease, has a general role in the chemical decay of *Escherichia coli* mRNA: Evidence that *rne* and *ams* are the same genetic locus. *Mol. Microbiol.* **4,** 2127–2135.

Newbury, S. F., Smith, N. H., Robinson, E. C., Hiles, I. D., and Higgins, C. F. (1987a). Stabilization of translationally active mRNA by prokaryotic REP sequences. *Cell* **48,** 297–310.

Newbury, S. F., Smith, N. H., and Higgins, C. F. (1987b). Differential mRNA stability controls relative gene expression within a polycistronic operon. *Cell* **51,** 1131–1143.

Nilsson, G., Lundberg, U., and von Gabain, A. (1988). *In vivo* and *in vitro* identity of site-specific cleavages in the 5′ non-coding region of *ompA* and *bla* mRNA in *Escherichia coli*. *EMBO J.* **7,** 2269–2275.

Nossal, N. G., and Singer, M. F. (1968). The processive degradation of individual polynucleotide chains. *J. Biol. Chem.* **243,** 913–922.

Panyatatos, N., and Truong, K. (1985). Cleavage within an RNase III site can control mRNA stability and protein synthesis *in vivo*. *Nucleic Acids Res.* **10,** 2227–2240.

Plamann, M. D., and Stauffer, G. V. (1985). Characterization of a cis-acting regulatory mutation that maps at the distal end of the *Escherichia coli glyA* gene. *J. Bacteriol.* **161,** 650–654.

Plamann, M. D., and Stauffer, G. V. (1990). *Escherichia coli glyA* mRNA decay: The role of secondary structure and the effects of the *pnp* and *rnb* mutations. *Mol. Gen. Genet.* **220,** 301–306.

Portier, C. (1975). Quaternary structure of polynucleotide phosphorylase from *Escherichia coli:* Evidence of a complex between two types of polypeptide chains. *Eur. J. Biochem.* **55,** 573–582.

Portier, C., Dondon, L., Grunberg-Manago, M., and Regnier, P. (1987). The first step in the functional inactivation of the *Escherichia coli* polynucleotide phosphorylase messenger is a ribonuclease III processing at the 5′ end. *EMBO J.* **6,** 2165–2170.

Regnier, P., Grunberg-Manago, M., and Portier, C. (1987). Nucleotide sequence of the *pnp* gene of *Escherichia coli* encoding polynucleotide phosphorylase. *J. Biol. Chem.* **262,** 63–68.

Regnier, P., and Hajnsdorf, E. (1991). Decay of mRNA encoding ribosomal protein S15 of *Escherichia coli* is initiated by an RNase E-dependent endonucleolytic cleavage that removes the 3′ stabilizing stem and loop structure. *J. Mol. Biol.* **217,** 283–292.

Schmeissner, U., McKenney, K., Rosenberg, M., and Court, D. (1984). Removal of a terminator structure by RNA processing regulates *int* gene expression. *J. Mol. Biol.* **176,** 39–53.

Schuster, G., and Gruissem, W. (1991). Chloroplast mRNA 3′ end processing requires a nuclear-encoded RNA-binding protein. *EMBO J.* **10,** 1493–1502.

Schwartz, T., Craig, E., and Kennell, D. (1970). Inactivation and degradation of messenger ribonucleic acid from the lactose operon of *Escherichia coli*. *J. Mol. Biol.* **54,** 299–311.

Shen, V., Cynamon, M., Daugherty, B., Kung, H-F., Schlessinger, D. (1981). Functional inactivation of *lac* α-peptide mRNA by a factor that purifies with *Escherichia coli* RNase III. *J. Biol. Chem.* **256,** 1896–1902.

Singer, M. F., and Tolbert, G. (1965). Purification and properties of a potassium-activated phosphodiesterase (RNase II) from *Escherichia coli. Biochemistry* **4,** 1319–1330.

Stern, D. B., and Gruissem, W. (1987). Control of plastid gene expression: 3 inverted repeats act as mRNA processing and stabilizing elements, but do not terminate transcription. *Cell* **51,** 1145–1157.

Stern, D. B., Jones, H., and Gruisserm, W. (1989). *J. Biol Chem.* **264,** 18742–18750.

Stern, M. J., Ames, G. F-L., Smith, N. H., Robinson, E. C., and Higgins, C. F. (1984). Repetitive extragenic palindromic sequences: a major component of the bacterial genome. *Cell* **37,** 1015–1026.

Taljanidisz, J., Karnik, P., and Sarkar, N. (1987). Messenger ribonucleic acid for the lipoprotein of the *Escherichia coli* outer membrane is polyadenylated. *J. Mol. Biol.* **193,** 507–575.

Wong, H. C., and Chang, S. (1986). Identification of a positive retroregulator that stabilizes mRNA's in bacteria. *Proc. Natl. Acad. Sci USA* **83,** 3233–3237.

Yamamoto, T., and Imamoto, F. (1975). Differential stability of *trp* messenger RNA synthesized originating at the *trp* promoter and P_L promoter of lambda *trp* phage. *J. Mol. Biol.* **92,** 289–309.

Zilhao, R., Camelo, L. and Arraiano, C. M. (1993). DNA sequencing and expression of the gene *rnb* encoding *Escherichia coli* ribonuclease II. *Mol. Microbiol.,* in press.

3

5′ mRNA Stabilizers

DAVID BECHHOFER

I. Introduction

The prevalence of 5′-to-3′ directionality in nucleic acid biology is a unifying feature in molecular genetics. The fact that RNA synthesis, DNA synthesis, and translation of messenger RNA all proceed in the 5′-to-3′ direction imparts to the field of molecular biology a pleasing orientation, as it seems that a few simple line diagrams emphasizing upstream to downstream, 5′ to 3′, or left to right are sufficient to give the basic outline of the discipline.

One would think that mRNA decay should also follow this pattern. Since translation occurs in the 5′-to-3′ direction, initiation of mRNA decay from the 5′ end would result in rapid functional inactivation of a message (or of the promoter-proximal cistron of polycistronic messages), whereas decay from the 3′ end or internal sites would leave partially degraded mRNA molecules that could serve as templates for abortive translation (Kennell, 1986). However, it has so far not been possible to describe the process of mRNA decay in simple terms, even for prokaryotes. Studies on the degradation of many individual mRNAs has not yielded a universal directionality to the decay process. Furthermore, it appears that different mRNAs may be subject to alternate modes of decay, with several ribonucleases participating in the decay process. A diversity of decay mechanisms would be an indication of the critical role that regulation of mRNA stability plays in the cell. Nevertheless, it is expected that current efforts to elucidate the mechanism of decay of specific bacterial messenger RNAs will

Control of Messenger RNA Stability

furnish a small set of unifying concepts that will clarify this aspect of gene expression.

In this chapter, several examples of bacterial "5′ stabilizers" are discussed. A 5′ stabilizer is defined as a 5′ region of a messenger RNA that can confer stability to heterologous sequences that are fused downstream of it. Since mRNAs are synthesized in an environment that includes endoribonucleases and 3′-to-5′ exoribonucleases, the presence of such 5′ stabilizers might not have been predicted, and the way in which 5′ stabilizers perform their function has, until recently, been difficult to explain. The very existence of these elements has important implications for our understanding of the nature of bacterial mRNA decay.

II. Function of a 5′ Stabilizer

A. Definition

Our current picture of the initation of mRNA decay is based on earlier studies that showed that mRNA decay proceeds with first-order kinetics (Kepes, 1969; Mosteller *et al.*, 1970) and that initiation of decay is nonrandom (i.e., not every phosphodiester bond is susceptible to a cleavage event that initiates decay) (Achord and Kennell, 1974; Blundell and Kennell, 1974). Thus, mRNA decay commences with a nucleolytic cleavage at one or several specific sites, and the half-life of a message is determined by the number and the susceptibility of sites at which mRNA decay can initiate. If initiation of decay could occur only at the 3′ end of a message or at some internal cleavage site, one would not expect that a 5′ mRNA sequence could have much effect on the rate of decay. The observation that RNA sequences located in the 5′ region of a message can stabilize diverse downstream RNA sequences suggests that initiation of decay can occur in the 5′ region. (In this discussion, the term "5′ region" is used to denote the segment of RNA that is proximal to the +1 nucleotide; the term "5′ end" is used to denote the +1 nucleotide itself.)

Measuring the half-lives of RNAs that contain a 5′ stabilizer may yield information about the mechanism of decay initiation. Consider a message that has a 20-min half-life that is determined by the 5′ region of the message. Suppose that a number of transcriptional fusions are constructed such that this 5′ region is transcribed as a leader region for many different RNAs that have different intrinsic decay rates. If it is found that the presence of this 5′ region confers a 20-min half-life on **any** downstream RNA sequence, one would conclude that the rate of initiation of decay of all mRNAs was determined solely by the nature of the 5′ region. In fact, when such experiments are done, what is usually found is that, while the presence of a particular 5′ stabilizer does confer increased stability to heterologous RNA sequences, the actual half-life of these fusion RNAs varies. Two possible conclusions can be drawn from these kinds of observations. First, the recognition of some site in the 5′ region may be important for the initiation of decay, but may not be the only such site. Other decay initiation sites may exist on an mRNA that are recognized and used despite the presence of the 5′ stabilizer. The effect of replacing a 5′ region that is a susceptible target for initiation of decay with a 5′ region (the 5′ stabilizer) that is less susceptible may be a reduction in the total number of decay initiation sites, resulting in a prolonged half-life. A second possibility is that decay initiation begins solely at or near the 5′ end of a message and that the presence of a 5′ stabilizer serves to slow the initiation of decay. The variable half-life that is observed when the 5′ stabilizer is placed upstream of different RNAs may be due not to the presence of additional decay initiation sites but to the influence of downstream sequences on the efficacy of the 5′ stabilizer. For example, one could surmise that sequences immediately downstream of the 5′ stabilizer region could alter its secondary structure such that the functioning of the 5′ stabilizer would vary depending on the particular RNA to which it was fused. Thus, a 5′ stabilizer is defined as an RNA segment that confers stability on heterologous RNAs, but not necessarily to the same extent for every RNA.

B. Initiation of Decay in the 5′ Region

Despite clear demonstrations of the importance of 5′ regions in determining the decay rate of many messages, the mechanism by which overall mRNA decay can initiate with an event in a 5′ region remains to be determined. Because a 5′-to-3′ exoribonuclease has not been found in *Escherichia coli* and, by extension, is thought not to exist in other prokaryotes, it is unlikely that the decay initiation event is binding of an exonuclease to the 5′ end. Rather, it has been assumed that in these cases a specific endonuclease cleavage near the 5′ end is the initiating step in overall decay. The target site for endonuclease cleavage could be a primary sequence, a secondary structure, or perhaps a site bound by a cellular protein that either marks the site for cleavage or facilitates ribonuclease activity at the site (cf. Brewer, 1991). Messenger RNAs that have an accessible endonuclease cleavage site in their 5′ regions would decay faster than those mRNAs that either lack such a recognition site or contain a recognition site that is inaccessible due to occlusion by RNA folding or binding of some factor (e.g., a host protein or a ribosome).

However, in the absence of any known 5′-to-3′ exoribonuclease activity, the notion of decay commencing with an endonucleolytic cleavage near the 5′ end needs further explanation. Viewed simply, such endonucleolytic cleavage would merely generate an mRNA molecule with a new 5′ end whose rate of decay would be governed by the same considerations that

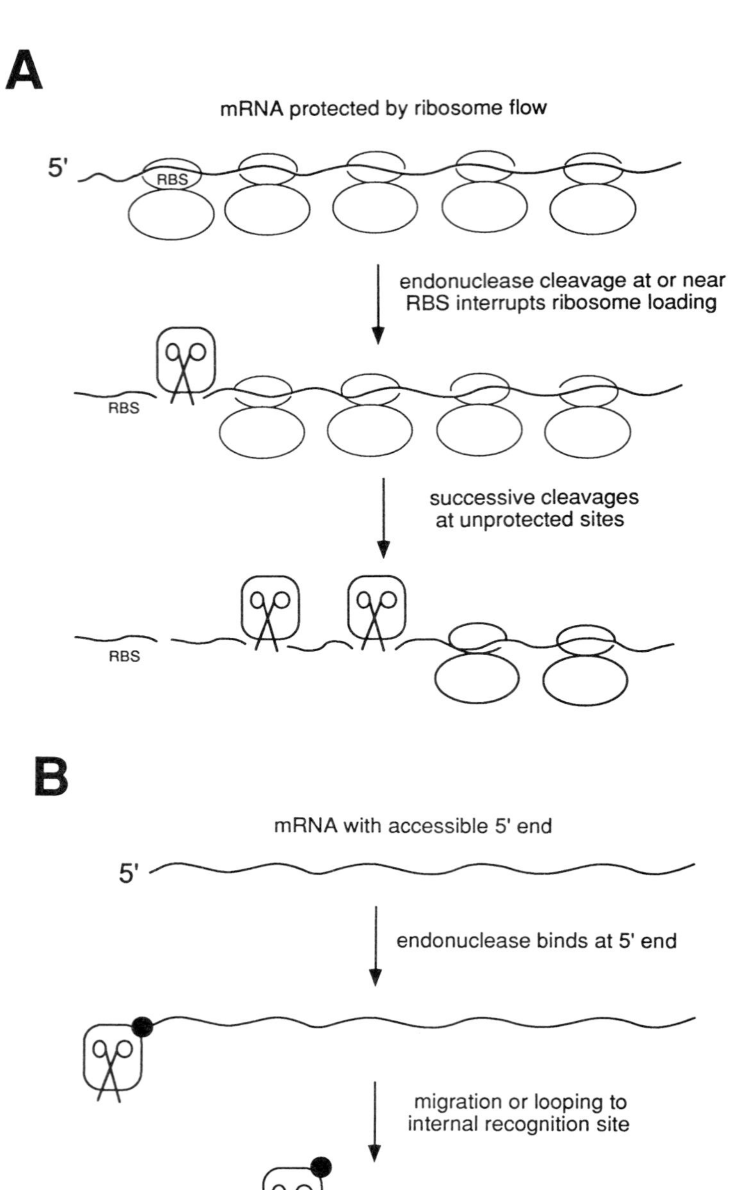

FIGURE 1 see legend on following page.

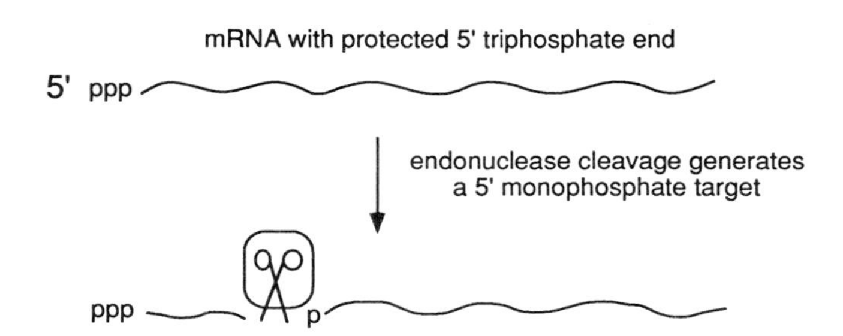

FIGURE 1 Models for initiation of decay from the 5′ region. (A) Model for decay of *lac* mRNA proposed by Cannistraro *et al.* (1986). Numerous endonucleolytic cleavage sites are protected by translating ribosomes. When ribosome binding is interrupted by a random cleavage event near the ribosome binding site (RBS), a wave of cleavages in the 5′-to-3′ direction can occur following the last translating ribosome. (B) The endonuclease contains a binding site (closed circle) for the 5′ end of an mRNA. After 5′-end binding, the endonuclease can migrate or loop to a downstream cleavage site (Emory *et al.*, 1992). (C) Internal cleavage exposes a monophosphate 5′ end, which is a target for rapid decay (Lin-Chao and Cohen, 1991).

applied before the initial cleavage. How then would initiation of overall decay begin by cleavage in a 5′ region? Several models have been put forth to solve this problem. Early studies suggested a relationship between mRNA decay rates and occupancy on mRNA by translating ribosomes (Cremer *et al.*, 1974; Walker *et al.*, 1976; Schneider *et al.*, 1978). This led to a model for *lac* mRNA decay (Cannistraro *et al.*, 1986) in which the initial event of decay is a cleavage in or near the ribosome binding site, which is located close to the 5′ end (Fig. 1A). Such a cleavage could prevent further ribosome binding and disrupt the wave of translating ribosomes that normally protect the mRNA from endonucleolytic attack. As the 5′-most translating ribosome moves down the message, the exposed mRNA segments are subject to successive endonucleolytic cleavages. Although this model may explain the observed pattern of *lac* mRNA decay, it is clearly not universally applicable, since more recent studies on decay of translated vs. untranslated mRNA regions have demonstrated that untranslated segments of mRNA are not necessarily more labile than translated segments (Bechhofer and Dubnau, 1987; Lundberg *et al.*, 1988; Emory and Belasco, 1990).

Several other models to explain initiation of overall decay by endonucleolytic cleavage in the 5′ region have been proposed (Belasco and Higgins, 1988; Régnier and Grunberg-Manago, 1990). In one variation, mRNA decay proceeds via successive cleavages by an endonuclease migrating down the message in the 5′-to-3′ direction (Fig. 1B). The fragments generated by these successive cleavages are rapidly degraded by 3′-to-5′ exonucleases. According to this model, the endonuclease responsible for initiation of decay in the 5′ region can be thought of as a 5′-binding endonuclease, which must bind at or near the 5′ end and can subsequently "search" in the 5′-to-3′ direction for its specific cleavage sites. The rate-determining step in decay would be the association of the endonuclease either at a site in the 5′ region or at the 5′ end, and the half-life of a message would then be a function of the accessibility of the 5′ binding site, which would be determined by local secondary structure or the presence of trans-acting factors. An alternative model is that a single cleavage by an endonuclease in the 5′ region creates a unique 5′ end (either a 5′ end with a specific terminal sequence or perhaps a monophosphate end that differs from the triphosphate end present in the primary transcript) that is a substrate for an as yet unidentified 5′ exonuclease (Fig. 1C). Of course, once one proposes the existence of such a 5′ exonuclease, one need not think in terms of prior endonuclease cleavage as rate-limiting; the accessibility of the 5′ end of the primary transcript to exonuclease attack could be rate-determining.

C. Protection at the 3′ End

To conclude this general discussion on 5′ stabilizers, I raise the issue of how 5′ stabilizers function in a cellular environment such as that of *E. coli*, which is known to contain at least two 3′-to-5′ exoribonucleases (RNase II and polynucleotide phosphorylase) that can act on mRNA. One would not expect a 5′ stabilizing region to have any influence on possible decay initiation at a remote 3′ end, unless a novel interaction between the 5′ and 3′ ends of a message was proposed. The observation that certain 5′ regions can act to stabilize messages suggests that 3′-to-5′ exonucleases usually do not initiate mRNA decay, possibly due to the presence of the 3′ stem–loop structure that is characteristic of Rho-independent transcription termination sites. The function of 3′ exonucleases in mRNA decay would instead be to act as scavengers of RNA fragments that arise by endonucleolytic cleavage or by Rho-mediated, premature transcription termination. Recent results from Higgins and colleagues have suggested that the ability of 3′ stem–loop structures to act as a barrier to exonucleolytic decay *in vivo* may be augmented by protein factor(s) that bind to these structures and prevent the transit of 3′ exonucleases (McLaren *et al.*, 1991) (see Chapter 2). Thus, a message that does not have internal endonucleolytic cleavage sites and that has a 5′ stabilizer region and a 3′ stem–loop would enjoy substantial protection from cellular decay enzymes.

In the following sections, I present a few examples of 5′ stabilizers. The 5′ regions of the *ompA* and *erm* genes are 5′ stabilizers that have been well characterized at the molecular level and that confer resistance to decay in different ways. These examples are preceded by a brief review of studies on *trp* mRNA decay, which serves as an appropriate introduction to the subjects of 5′ stabilizers and initiation of decay from the 5′ region.

III. *Escherichia coli trp* mRNA

A. 5′-to-3′ Directionality

Several papers from the late 1960s and early 1970s were concerned with the decay of *E. coli trp* mRNA (e.g., Morikawa and Imamoto, 1969; Morse *et al.*, 1969; Forchhammer *et al.*, 1972). Decay of polycistronic *trp* mRNA was followed by pulse-labeling nascent RNA and hybridizing RNA isolated at various times after the pulse to *trp* DNA fragments bound on nitrocellulose filters. By this method, the half-lives of portions of the message could be assessed, and it was found that the 5′-proximal portions had a shorter half-life than the distal portions. A model in which decay proceeds exclusively in a 5′-to-3′ wave of decay was favored at the time, perhaps because this fitted well with the 5′-to-3′ directionality of other processes in gene expression. In a somewhat later study (Schlessinger *et al.*, 1977), *trp* mRNA was labeled specifically at its 3′ end by introducing labeled nucleotide after inhibition of transcription initiation. The analysis of *trp* mRNA decay under these conditions led to the conclusion that decay was initiated at the 5′ end and proceeded in the 3′ direction, with pauses in the decay process occurring at intercistronic boundaries.

Biochemical characterization of *E. coli* ribonucleases that could be involved in mRNA decay was also taking place at this time. After initial studies suggesting the existence of a 5′-to-3′ exoribonuclease in *E. coli* (called "RNase V"; Kuwano *et al.*, 1969, 1970), further experimentation showed that the presumed RNase V activity was probably due to residual RNase II activity (Bothwell and Apirion, 1971; Holmes and Singer, 1971). Since then, the assumption has been that a 5′-to-3′ exoribonuclease does not exist in *E. coli*. Therefore, the 5′-to-3′ directionality of *trp* mRNA decay was explained by an initial endonucleolytic cleavage, perhaps at the 5′-proximal ribosome binding site, which would deprive the mRNA of ribosomes and thereby leave it exposed to sequential endonucleolytic cleavages in the 5′-to-3′ direction.

B. 5′ Stabilizer of the *N* Gene

In addition to focusing attention on the 5′ region as the primary initiation site for decay, *trp* mRNA was also involved in the earliest demonstration of a 5′ stabilizer. While the study of 5′ stabilizers is currently carried out by making gene fusions between a DNA segment that encodes a specific 5′ region and one that encodes heterologous sequences, this was not possible until the advent of gene cloning. A precursor to this way of studying 5′ stabilizers was provided fortuitously by the isolation of a λ*trp* transducing phage in which *trp* sequences were fused to the λp_L promoter such that *trp* mRNA could be transcribed either from its own promoter (p_T *trp* mRNA) or from the phage p_L promoter (p_L *trp* mRNA). In p_L *trp* mRNA, transcription started with the *N* gene (Fig. 2A). It was found that during the course of phage infection, p_L *trp* mRNA demonstrated a 10-fold greater chemical stability than p_T *trp* mRNA, which decayed at the same rate as *trp* mRNA in uninfected cells (Fig. 2B) (Yamamoto and Imamoto, 1975). This finding suggested that the rate-limiting step in mRNA decay initiation is determined by a sequence located at or near the 5′ end of the messenger RNA.

This report should be considered in light of a paper published several years later (Kano and Imamoto, 1979) in which it was shown that an endonucleolytic cleavage site is present in *trp* leader mRNA and that the downstream RNA fragment resulting from this cleavage (i.e., the bulk of *trp* mRNA) decays two to three times faster than the upstream RNA fragment (i.e., the 5′-proximal two-thirds of *trp* leader mRNA). The suggestion was made that this endonucleolytic cleavage may be the initial event in *trp* mRNA decay and that exposure of internal *trp* sequences by endonucleolytic cleavage allows the sequential 5′-to-3′ wave of decay to commence. Importantly, it was noted that this endonucleolytic cleavage does not occur in p_L *trp* mRNA. Thus, although the same *trp* leader endonuclease cleavage site is present in p_T *trp* mRNA and p_L *trp* mRNA, stabilization of p_L *trp* mRNA does occur. Either the location of the endonuclease cleavage site is important in determining its function as an initiation site for decay or the fusion to upstream p_L promoter sequences prevents this site from being used as an initiation site for decay. The fact that stabilization of the p_L *trp* mRNA increased with time after infection was evidence that some phage-encoded factor may be required for stabilization. In retrospect, the obvious candidate for such a factor would be the N protein itself, which has been shown to bind at the *nut* site of the p_L-promoted transcript (Lazinski *et al.*, 1989).

Unfortunately, it appears that this set of tools to study initiation of decay at the 5′ end has not been taken advantage of in recent times, in which the ability to clone specific segments of the *trp* 5′ region, as well as to regulate expression of a plasmid-encoded *N* gene product without the complication of phage infection, would allow for a clarification of the mechanism of this interesting mRNA stabilization in *E. coli*. As far as can be ascertained, the only more recent report on *trp* mRNA decay is one in which the effect of point mutations upstream of the *trpE* ribosome binding site on mRNA stability was studied (Cho and Yanofsky, 1988). However, in that report the *trpE* gene was fused directly to the *trp* promoter, and the leader sequences that may play a role in the initiation of decay were not present.

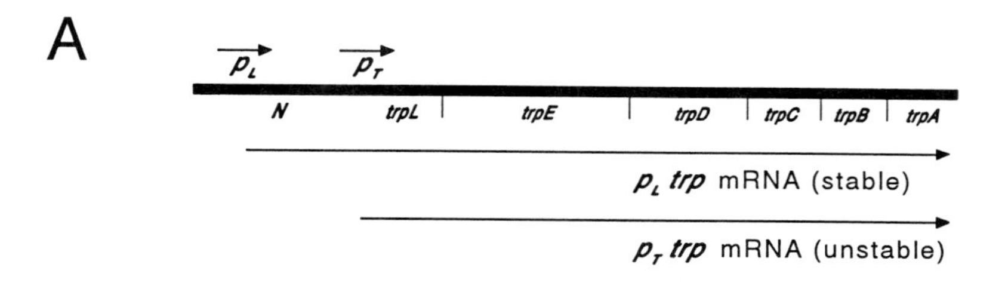

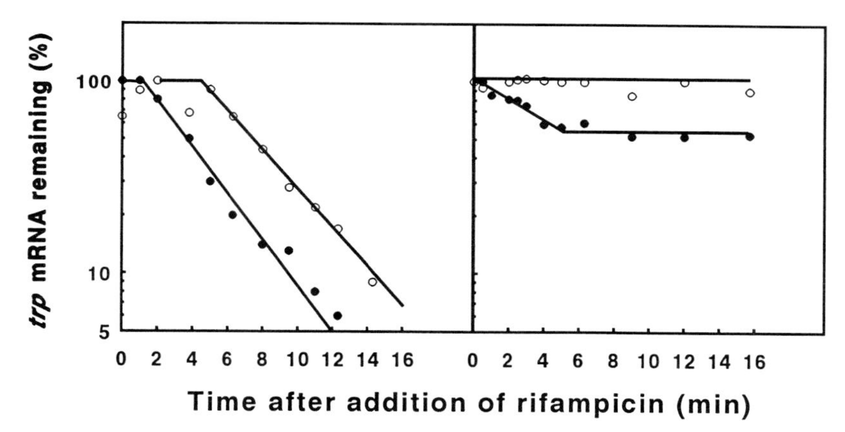

FIGURE 2 (A) Schematic diagram of p_L *trp* mRNA and p_T *trp* mRNA. The thick line represents the portion of the λ*trp* phage that encodes these two mRNAs. For consistency with other figures, transcription is shown from left to right. The *N* gene of λ, the *trp* leader (*trp*L), and other *trp* genes are shown, but are not drawn to scale. (B) Decay of p_T *trp* mRNA (left) and p_L *trp* mRNA (right) encoded by infecting λtrp phage. Data are from Yamamoto and Imamoto (1975). To measure p_L *trp* mRNA specifically, a strain containing a deletion of the *trp* operon but retaining the *trpR* (*trp* repressor) gene was infected in the presence of tryptophan. To measure p_T *trp* mRNA specifically, a similar strain that was lysogenic for λ, and therefore contained λ repressor, was infected in the absence of tryptophan. RNA was pulse-labeled for 1 min at 15 min after infection, followed by addition of rifampicin. Remaining *trp* mRNA was measured by hybridization to filter-bound *trpED* (●) DNA or *trpCBA* (○) DNA. The slight difference in slope between the decay of *trpED* mRNA and that of *trpCBA* mRNA translates into half-lives of 2.6 and 3.0 min, respectively, demonstrating the 5′-to-3′ directionality of decay.

IV. *Escherichia coli ompA* Message

A. Segmental Decay

The most extensively characterized 5′ stabilizer in *E. coli* is that of the *ompA* gene, which encodes a major outer membrane protein and whose mRNA has an unusually long half-life of about 17 min. The first detailed analysis of *ompA* mRNA decay (von Gabain *et al.*, 1983) used a method that differed significantly from the older methods described above for *trp* mRNA decay. Previously, segmental decay was analyzed by isolating labeled RNA at increasing times after inhibition of transcription and by hybridizing the RNA to filter-bound DNA fragments that represented discrete regions of the gene. With this method, mRNA sequences from a specific region of a transcript would hybridize and give a positive signal, but no information could be gained as to the size of the hybridizing mRNA fragments and whether endonucleolytic cleavages had occurred in the particular region. The new method described by von Gabain *et al.* used uniformly labeled single-stranded DNA probes that were hybridized to RNA in solution and then subjected to S1 nuclease digestion. The sizes of the complementary DNAs were different for upstream, middle, and downstream regions of the transcript. The amount of complementary DNA remaining at times after rifampicin addition, determined by the intensity of the protected DNA fragment running at a certain position on a gel, was an indication of stability. In principle, single endonucleolytic cleavages would be scored as a decay event since smaller protected probe fragments would not migrate in the gel at the same position as totally protected probe. The analysis of *ompA* mRNA decay using this method showed that the 3′-proximal segments decayed faster than the 5′-proximal segment, suggesting that the directionality of decay was in the 3′-to-5′ direction. This could mean either that decay proceeded from the 3′ end, presumably due to the action of a 3′-to-5′ exoribonuclease, or that endonucleolytic cleavages were occurring preferentially in the 3′ portion of the message. In the same report, decay of the short-lived β-lactamase (*bla*) message, which has a 3-min half-life, was examined and all segments of the message appeared to decay at the same rate.

B. *ompA–bla* Transcriptional Fusions

Subsequently, various fusions between portions of the *ompA* and *bla* genes were constructed in an effort to identify mRNA segments that confer stability to *ompA* mRNA or instability to *bla* mRNA (Belasco *et al.*, 1986). The critical finding was that when the 5′ 147-nucleotide (nt) segment of the *ompA* gene transcriptional unit was placed upstream of the entire *bla* transcriptional unit, the resultant message displayed a prolonged half-life relative to wild-type *bla* mRNA. The 147-nt *ompA* segment contained a 133-nt untranslated leader region (Fig. 3) followed by the first 14 nucleotides of the *ompA* coding sequence. This 147-nt segment was positioned upstream of the *bla* transcriptional unit such that ribosomes initiating at the *ompA* ribosome binding site could translate the *bla* coding sequence in frame. Another observation was that when a translational stop signal was inserted

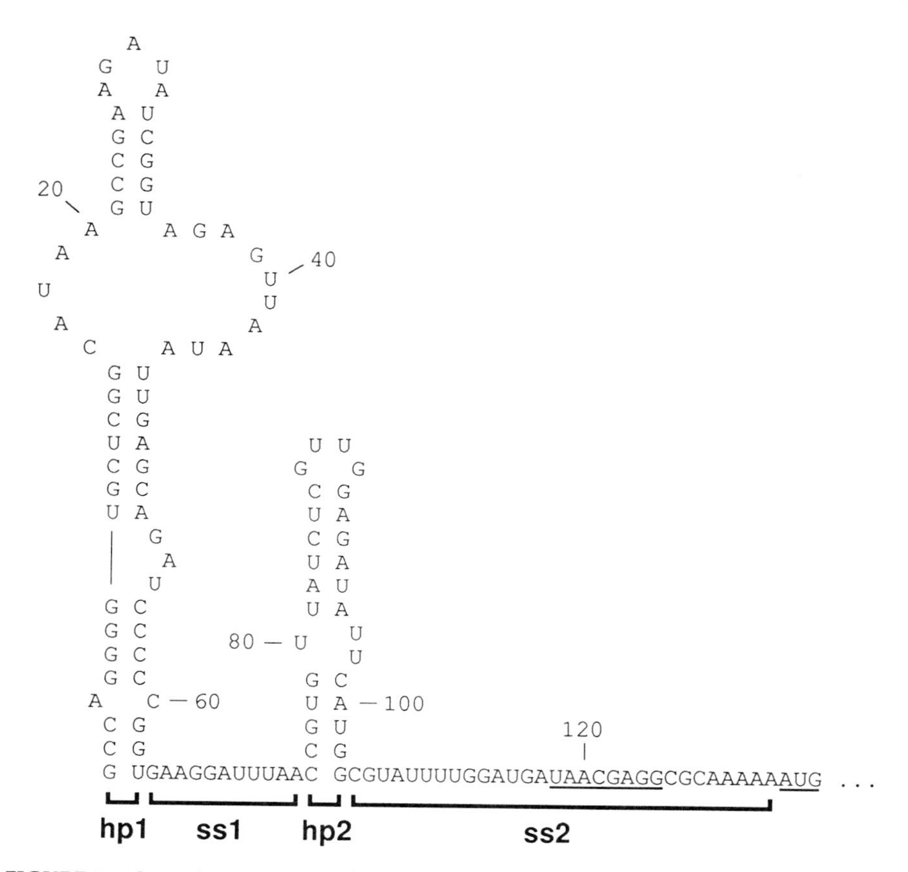

FIGURE 3 Secondary structure of the *ompA* 5′ untranslated region. Brackets delineate the boundaries of the four secondary structure domains within this RNA segment (Chen *et al.*, 1991). hp, hairpin structure; ss, single-stranded region. The Shine–Dalgarno element and translation initiation codon are underlined. [Reproduced from Emory *et al.* (1992).]

in the short *ompA* coding sequence of this construct, the resultant message displayed *bla*-like instability. This was not necessarily due to some property of the *bla* sequence that required translation for stability because a similar result was obtained with *ompA* message itself: introduction of a translational stop signal at the 29th codon of otherwise wild-type *ompA* mRNA resulted in a 10-fold decrease in *ompA* mRNA stability (Nilsson *et al.*, 1987). The significance of the requirement for ribosome transit in the N-terminal coding region remains to be clarified, since the stability of *ompA* mRNA stability is not merely a function of translation rate (Emory and Belasco, 1990). This finding may reflect a property of 5′ stabilizers that was discussed above, namely, that sequences immediately downstream of the 5′ region may influence the efficacy of a 5′ stabilizer. Perhaps translation of the adjacent sequences is required to keep the 5′ stabilizer in the conformation that is resistant to initiation of decay.

The *ompA–bla* fusion results indicated that the 5′ end of the *ompA* message could act as a cis-dominant mRNA stabilizer. In the construct described above, the 147-nt *ompA* 5′ region was appended upstream of the entire *bla* transcriptional unit and did not replace *bla* 5′ sequences. This shows again (as with p_L *trp* mRNA; see earlier) that 5′ stabilizers act in a dominant fashion when placed upstream of intact mRNA sequences that are intrinsically unstable. These types of results point to the 5′ terminal sequences as the crucial elements in determining mRNA half-life.

C. 5′ Secondary Structure

A recent report from Belasco and colleagues focused on the *ompA* 5′ sequences in an effort to determine which domains of this region actually function as the 5′ stabilizer (Emory *et al.*, 1992). In a study that may mark a milestone in the field of bacterial mRNA decay, it was shown that the 5′-most stem–loop structure (hp1 in Fig. 3) and a 30-nt sequence surrounding the *ompA* ribosome binding site (ss2) were sufficient to confer wild-type stability to *ompA* mRNA. The stabilizing effect of the 5′-proximal stem–loop structure was independent of its sequence; either hp2 (Fig. 3) or a synthetic stem–loop structure could, in conjunction with the ss2 sequence, confer wild-type stability on *ompA* mRNA. The same synthetic stem–loop structure was placed upstream of *bla* mRNA and caused a twofold increase in *bla* mRNA half-life. Furthermore, the 5′ stabilizer function of hp1 was dependent on its proximity to the 5′ end. When as few as five unpaired nucleotides were added at the 5′ end of an *ompA* transcript beginning with hp2, the half-life of *ompA* mRNA was reduced about threefold. An analysis of the decay of several mutant *ompA* messages showed that internal *ompA* mRNA segments decayed faster than the 5′ segment.

The straightforward, and far-reaching, conclusion from these results is that initiation of mRNA decay in *E. coli* is dependent on access to the message by a nuclease that requires a single-stranded 5′ terminal segment for binding. Binding of this nuclease is followed by initiation of decay at a cleavage site downstream of the 5′ region, suggesting migration of the nuclease in the 5′-to-3′ direction until a suitable recognition site is reached. In this model, the 5′-proximal stem–loop structure serves to restrict access of the decay-initiating nuclease and predicts that the sequence of the stem–loop structure is not important in conferring stability. This is supported by the results of a previous report (Chen *et al.*, 1991) in which the 5′ regions of *ompA* mRNAs from *Serratia marcescens* and *Enterobacter aerogenes* were able to confer mRNA stability in *E. coli*, despite considerable sequence differences in their 5′ stem–loop structures. In addition, it was noted that

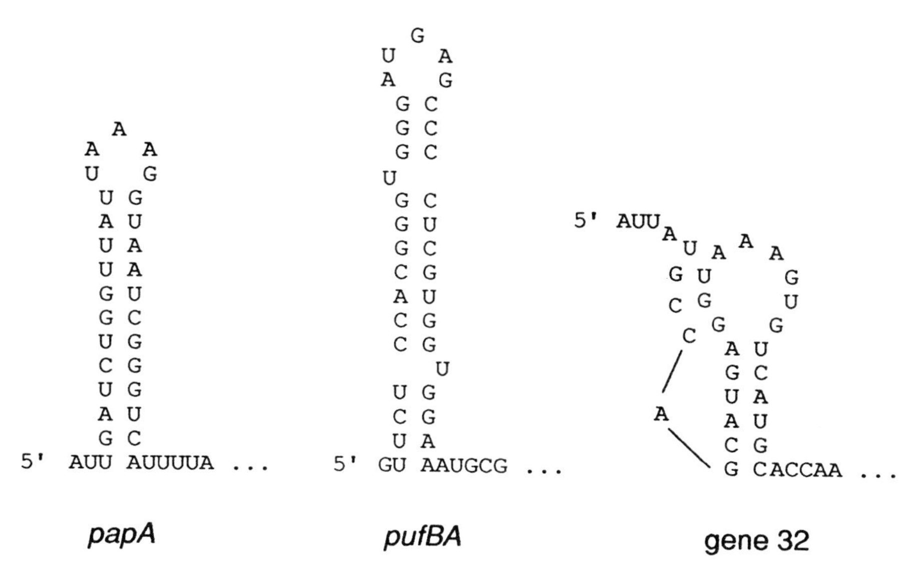

FIGURE 4 Stem–loop structures near the 5′ end of stable prokaryotic mRNAs. The sequence and likely secondary structure is shown for a 5′ terminal untranslated segment of *E. coli papA* mRNA, *Rhodobacter capsulatus pufBA* mRNA, and T4 gene *32* monocistronic mRNA. [Reproduced from Emory *et al.* (1992).]

several other mRNAs that have long half-lives in *E. coli* contain a 5′-proximal stem–loop or pseudoknot structure located no more than a few nucleotides from the 5′ end (Fig. 4). In fact, the 5′ region of the gene *32* monocistronic mRNA has been shown to function as a 5′ stabilizer by Krisch and colleagues (Gorski *et al.*, 1985).

The model for initiation of decay proposed on the basis of these data is the one depicted in Fig. 1B, in which access to an internal cleavage site is dependent on binding to an unstructured 5′ terminus. Identification of the 5′-binding endonuclease that possesses these characteristics will be critical to our understanding of initiation of decay. Based on recent reports of the importance of RNase E in the decay of many *E. coli* mRNAs (see Chapter 4), it has been proposed that this endonuclease plays a major role in the initiation of decay. Most recently, Bouvet and Belasco have been studying the small RNA I encoded by plasmid pBR322, whose decay is initiated by RNase E cleavage at a site located five nucleotides from the 5′ end (Lin-Chao and Cohen, 1991). They have shown that decay of RNA I is slowed by addition of a 5′-terminal stem–loop structure (Bouvet and Belasco, 1992). Rapid, RNase E-initiated decay is restored by the addition of several unpaired nucleotides at the 5′ end. These results suggest strongly that RNase E is the 5′-binding endonuclease responsible for the mode of mRNA decay initation that is controlled by the structural nature of the 5′ region.

In principle, one could envision that the various 5′ stem–loop structures function as stabilizers not solely by obviating 5′ terminal single-stranded RNA but by acting as binding sites for proteins that recognize these structures (analogous to the binding of proteins to 3′ stem structures, as has been shown by Higgins and colleagues) (see Chapter 2). The bound proteins could impede progression or association of a nuclease. However, the results of dimethylsulfate (DMS) methylation protection studies by Belasco and colleagues (Chen *et al.*, 1991) indicate that this is not so, at least for *ompA* mRNA. When the *E. coli ompA* 5′ region was probed with DMS *in vivo* and *in vitro*, similar patterns of protection were obtained. This suggests that the 5′ RNA segment is not stably bound *in vivo* by a ribosome or a cellular protein, since one would expect such binding to alter the methylation pattern substantially.

D. Growth-Rate Dependence

Another fascinating aspect of *ompA* mRNA, which could be important in elucidating the mechanism of *E. coli* mRNA decay, is that its stability is growth rate dependent (Nilsson *et al.*, 1984). The 17-min half-life of *ompA* mRNA is observed only in fast-growing cells. In slow-growing cells the half-life is reduced to about 4 min. The relationship between stability and growth rate was shown to be determined by the 5′ segment of *ompA* mRNA and was not a function of the rate of *ompA* translation (Lundberg *et al.*, 1988; Emory and Belasco, 1990).

The finding of differential stability of a message under different growth conditions raises interesting possibilities about the regulation of mRNA decay and the function of 5′ stabilizers. If the stability of *ompA* mRNA is due to a 5′-proximal structure that disallows nuclease binding, one would not expect this inherent property of the RNA to change under different growth conditions. Growth-rate-dependent stability might therefore be an indication of a change in the host mechanism of mRNA decay, e.g., expression or activation of a different ribonuclease activity that is capable of cleaving the *ompA* 5′ sequence or expression of a factor that binds to and disrupts the *ompA* 5′-proximal RNA structure. In this respect, it would be of interest to know the results of *in vivo* structure probing studies (DMS methylation) performed on *ompA* mRNA in slow-growing cells, in which the *ompA* mRNA half-life is reduced by at least threefold. If the methylation pattern that was observed in fast-growing cells was also observed in slow-growing cells, this would suggest that the change in stability was not due to a change in the presentation of the *ompA* 5′ region but that, in fact, a change in the decay mechanism itself was being induced under conditions of slow growth. The prospect of discovering regulation at the level of nuclease expression or factors that affect nuclease activity is an exciting one.

Recent data from von Gabain and colleagues showing that the growth-rate-dependent change from a long to a short half-life occurs within 1 min indicate that macromolecular synthesis is not required for this change (Georgellis *et al.*, 1992).

E. Directionality of Decay

One additional point to consider briefly is the previously cited finding that internal segments of *ompA* mRNA variants are degraded faster than the 5′ segment (Emory *et al.*, 1992). These recent experiments are consistent with earlier experiments in which wild-type *ompA* mRNA appeared to decay in the 3′-to-5′ direction (von Gabain *et al.*, 1983). If initiation of *ompA* mRNA decay is via a nuclease that binds at or near the 5′ end and migrates downstream "in search" of cleavage sites (Fig. 1B), the first such site in the case of *ompA* could be quite some distance from the 5′ region. Internal cleavage would generate an upstream fragment that is a substrate for 3′-to-5′ exonucleases and a downstream fragment containing an unprotected 5′ end that could be bound by the same 5′-binding nuclease.

V. *erm* Genes of Gram-Positive Bacteria

A. mRNA Decay in Bacteria Other Than *E. coli*

The discussion to this point has centered around the process of mRNA decay as it occurs in *E. coli*. What has been learned about mRNA decay in other organisms, including gram-positive bacteria such as *Bacillus subtilis*, can be integrated with the broad base of knowledge that has come from studying *E. coli*. The basic elements of gene expression (replication, transcription, and translation) have all been found to be conserved among diverse bacterial systems, and the same can be expected for mRNA turnover. Comparisons of mRNA decay mechanisms in diverse organisms could be instructive at two levels. First, the nuclease activities of other organisms may differ from those of *E. coli*. For example, the major 3′-to-5′ exoribonuclease activity in *E. coli*, the hydrolytic RNase II, has been shown not to be a significant activity in *B. subtilis* extracts (Deutscher and Reuven, 1991). This sort of observation would suggest that RNase II is a redundant activity in *E. coli*, which, in fact, appears to be the case (Donovan and Kushner, 1983). Second, examining decay of specific mRNAs that are encoded by homologous genes in different organisms or that are introduced into different hosts could reveal important events in decay pathways. For example, conservation of an endonuclease cleavage site on a specific mRNA would point to this endonucleolytic event as possibly an important step in the initiation of decay.

B. Induced Stability of *erm* mRNA

The induced stability of messenger RNAs encoded by erythromycin resistance genes (*erm* genes), originally isolated from *Staphylococcus aureus* and subsequently studied in *S. aureus* and *B. subtilis*, has been the subject

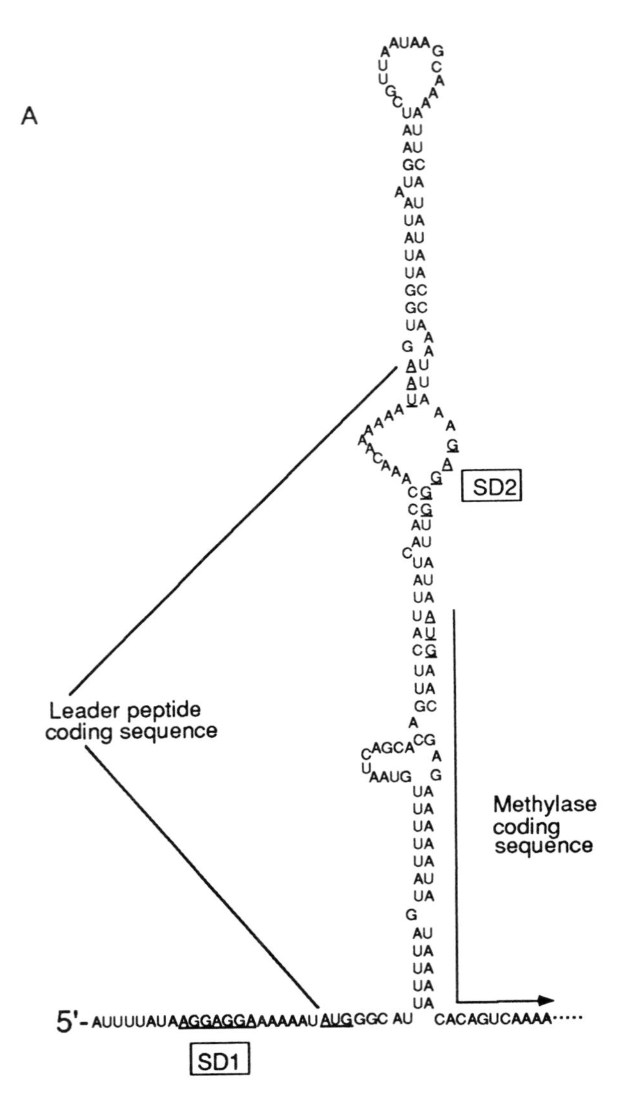

FIGURE 5 (A) *ermC* mRNA 5′ secondary structure. Proposed secondary structure of the *ermC* leader region in the absence of induction, showing the extent of the leader peptide coding sequence and the beginning of the methylase coding sequence. Translational signals are underlined. SD, Shine–Dalgarno element. (B) Schematic diagram of the unstable and stable (with erythromycin (Em)-bound stalled ribosome) *ermC* mRNAs (continues).

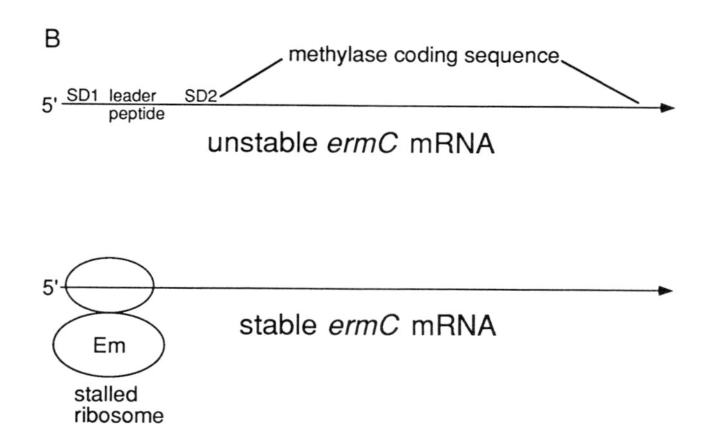

FIGURE 5 (continued).

of much investigation. The 5′ regions of the *ermC* and *ermA* genes function as 5′ stabilizers, but, unlike the case of *ompA* mRNA, this is not a property of the RNA sequence itself but of binding by a ribosome, as is explained later.

Both *ermA* and *ermC* encode ribosomal RNA methylases, and expression of these genes is induced post-transcriptionally by the addition of subinhibitory concentrations of erythromycin (Bechhofer, 1990). A ribosome to which erythromycin is bound stalls while translating a leader peptide sequence, which results in melting of the mRNA leader region secondary structure to allow high-level translation of the methylase coding sequence (Fig. 5). The functional stability of *ermC* mRNA under conditions of erythromycin induction was first reported when it was shown that synthesis of *ermC* methylase in minicells continued well after the addition of rifampicin (Shivakumar *et al.*, 1980). In a later study, the physical half-life of induced *ermC* mRNA was estimated to be about 30–40 min and was shown to require ribosome stalling in the leader peptide coding sequence (Bechhofer and Dubnau, 1987). Importantly, translation of the *ermC* coding sequence was not required for induced stability of *ermC* mRNA.

The mechanism by which ribosome stalling can cause mRNA stability has been investigated for *ermA* and *ermC* (Sandler and Weisblum, 1989; Bechhofer and Zen, 1989). Relocation of the ribosome stalling site to a position distal to the 5′ end of the message resulted in RNAs in which the segment upstream of the stall site was unstable while the segment downstream of the stall site was relatively stable. These and other experiments suggested that the effect of the stalled ribosome was to act as a barrier to degradation in the 5′-to-3′ direction, either by a 5′-to-3′ exoribonuclease or by an endonuclease that binds at the 5′ end and moves downstream. Results with deletion mutations in the *ermC* leader region have shown that the stalled ribosome does not act by protecting the 5′ region from endonucleolytic cleavage at a site formed by the leader region secondary structure (Hue and Bechhofer, 1991). Instead, a stalled ribosome restricts binding to the 5′ end by a nuclease(s), which is the principal factor in determining the rate of decay.

C. 5′ Stabilizers of *erm* Genes

Induced chemical stability of RNAs encoded by an *ermC–lacZ* transcriptional fusion, which contained the 140-nt *ermC* leader region and 220 nucleotides of the *ermC* coding sequence fused to the coding sequence of *lacZ*, showed that the 5′ region of *ermC* functions as a 5′ stabilizer. This stabilization occurs also in the complete absence of translation of the *ermC–lacZ* fusion protein. Fusions of the *ermC* leader region to coding sequences for several *B. subtilis* genes have demonstrated the capacity of the *ermC* 5′ region to confer induced stability to diverse messages (J.F. DiMari and D.H. Bechhofer, unpublished). The *ermA* 5′ region also stabilizes downstream *lacZ* sequences (Sandler and Weisblum, 1988), as well as RNA derived from the coding sequence of the *cat-86* chloramphenicol resistance gene of *Bacillus pumilus* (Sandler and Weisblum, 1989). Recently, it has been shown that ribosome stalling in the 5′ leader region of mRNA encoded by the *cat* gene of pUB112 (a plasmid isolated from *S. aureus* and studied in *B. subtilis*) causes a 15-fold increase in *cat* mRNA stability (Dreher and Matzura, 1991). However, it has not yet been demonstrated that the pUB112 *cat* 5′ region can also confer stability to heterologous downstream sequences.

The fact that the *ermC* and *ermA* 5′ regions confer stability to downstream sequences with diverse 3′ ends indicates again that the 3′ ends of mRNAs usually do not constitute decay initiation targets. The proposal that protein binding at mRNA 3′ stem–loop structures enhances protection against 3′-to-5′ exonuclease activity in *E. coli* (see Chapter 2) may be true of *B. subtilis* as well. In light of the above-mentioned study of Deutscher and Reuven (1991), in which RNase II-like activity was not observed to be significant in *B. subtilis* extracts, it is possible that the details of how this protection of the 3′ end works may differ in these two organisms. In any event, the observation of 5′ stabilizers in both *E. coli* and *B. subtilis* is a strong indication that initiation of mRNA decay in the 5′ region is a common theme. It will be of interest to determine whether an RNase E-like activity is present also in *B. subtilis*.

D. Initiation of Decay at an Internal Site

The lack of substantial information about *B. subtilis* ribonucleases (e.g., might a 5′-to-3′ exonuclease be found in *B. subtilis*?) makes it easier to propose that initiation of decay in this organism may occur exclusively from the 5′ region. This hypothesis would predict that the presence of the *ermC* 5′ stabilizer could confer increased stability on any downstream sequence

under inducing conditions (i.e., during ribosome stalling in the leader peptide), since initiation of decay might begin with an obligatory "inspection" of the 5′ end of the message. Recent results from our laboratory, however, provide evidence that the dominant stabilizing effect of a 5′ stabilizer can be overcome. Insertion into the *ermC* coding sequence of an rRNA sequence, which was shown previously to be a site for endonucleolytic cleavage by Bs-RNase III (a *B. subtilis* enzyme that resembles *E. coli* RNase III in its site specificity; Panganiban and Whiteley, 1983), resulted in an *ermC* mRNA that could not be stabilized upon addition of erythromycin. Preliminary experiments indicate that internal cleavage in the inserted rRNA region occurs and leaves a downstream RNA fragment that has an unprotected 5′ end and an upstream RNA fragment that has a 3′ end unlike those that result from transcription termination (J. F. DiMari and D. H. Bechhofer, unpublished). These findings suggest that the ability of a 5′ region to function as a 5′ stabilizer is dependent on the susceptibility of the downstream sequences to alternative decay initiation events and that inspection of the 5′ region is not an obligatory step in the initiation of decay. While for some messages an initial event in the 5′ region (or at the 5′ end) may be the rate-determining step in decay and this event is blocked or at least slowed by RNA secondary structure (*ompA*, gene *32*), protein binding (λp_L perhaps), or ribosome stalling (*ermA*, *ermC*), the initiation of decay at alternative sites in other messages could negate the effect of a 5′ stabilizer.

Initiation of decay by *E. coli* RNase III cleavage in a leader region is thought to be the mechanism for the control of mRNA stability of three operons (Régnier and Grunberg-Manago, 1990). However, the specificity of endoribonucleases such as *E. coli* RNase III and Bs-RNase III and the fact that bulk mRNA decay appears to be unaffected in an RNase III$^-$ mutant (Apirion and Watson, 1975; Apirion *et al.*, 1976) make it unlikely that this type of endonucleolytic cleavage is a major factor in initiation of decay. Taken together with the proposal of Belasco and colleagues that cleavage by enzymes such as *E. coli* RNase E is preceded by a 5′ binding event (Emory *et al.*, 1992), this would suggest that decay initiation of most messenger RNAs will not occur by an internal cleavage event. Rather, 5′ binding events will be the rate-determining step in the rapid decay of most mRNAs, and those messages that have 5′ stabilizers will be protected from this rapid decay.

VI. Conclusions

The properties of a few bacterial 5′ stabilizers have been described in this chapter. Undoubtedly many more 5′ stabilizers exist and will come to light as research efforts on the control of mRNA stability continue. Studying these elements has proven valuable in understanding the mechanism of mRNA decay, and more will be learned about initiation of mRNA decay by examining mutant versions of these 5′ stabilizing regions.

A practical consequence of the function of 5′ stabilizers could be their utility in discovering sites for RNA processing. The lack of stability conferred by the *ermC* 5′ stabilizer when an endonucleolytic cleavage site was present in the transcript (see earlier) suggests that the construction of transcriptional units containing 5′ stabilizers could be used to identify internal RNA processing sites. RNAs that contain such internal sites would not be stabilized by the presence of the 5′ stabilizer. If a genetic selection technique could be devised that could differentiate between cells containing an unstable and those containing a stable version of a particular message, this would constitute a powerful method for isolating randomly cloned DNA fragments that encode RNA processing sites. However, such a selection system has not yet been reported.

In the opening lines of this chapter, the established directionality of replication, transcription, and translation was contrasted with the undetermined directionality of the mRNA decay process. Through the study of 5′ stabilizers, the notion that initiation of mRNA decay often occurs at or near the 5′ end has gained credence. The recent findings cited here, and more detailed molecular analysis of other 5′ stabilizers, may prove the decades-old concept of 5′-to-3′ directionality in bacterial mRNA decay to be entirely correct.

References

Achord, D., and Kennell, D. (1974). Metabolism of messenger RNA from the *gal* operon of *Escherichia coli*. *J. Mol. Biol.* **90,** 581–599.

Apirion, D., Neil, J., and Watson, N. (1976). Consequences of losing ribonuclease III on the *Escherichia coli* cell. *Mol. Gen. Genet.* **144,** 185–190.

Apirion, D., and Watson, N. (1975). Unaltered stability of newly synthesized RNA in strains of *Escherichia coli* missing a ribonuclease specific for double-stranded ribonucleic acid. *Mol. Gen. Genet.* **136,** 317–326.

Bechhofer, D. (1990). Triple post-transcriptional control. *Mol. Microbiol.* **4,** 1419–1423.

Bechhofer, D. H., and Dubnau, D. (1987). Induced mRNA stability in *Bacillus subtilis*. *Proc. Natl. Acad. Sci. USA* **84,** 498–502.

Bechhofer, D. H., and Zen, K. (1989). Mechanism of erythromycin-induced *ermC* mRNA stability in *Bacillus subtilis*. *J. Bacteriol.* **171,** 5803–5811.

Belasco, J. G., and Higgins, C. F. (1988). Mechanisms of mRNA decay in bacteria: A perspective. *Gene* **72,** 15–23.

Belasco, J. G., Nilsson, G., von Gabain, A., and Cohen, S. N. (1986). The stability of *E. coli* gene transcripts is dependent on determinants localized to specific mRNA segments. *Cell* **46,** 245–251.

Blundell, M., and Kennell, D. (1974). Evidence for endonucleolytic attack in decay of *lac* messenger RNA in *Escherichia coli*. *J. Mol. Biol.* **83,** 143–161.

Bothwell, A. L. M., and Apirion, D. (1971). Is RNase V a manifestation of RNase II? *Biochem. Biophys. Res. Commun.* **44,** 844–851.

Bouvet, P., and Belasco, J. G. (1992). Control of RNase E-mediated RNA degradation by 5′-terminal base pairing in *E. coli*. *Nature* (*London*) **360,** 488–491.

Brewer, G. (1991). An A + U-rich element RNA-binding factor regulates c-*myc* mRNA stability in vitro. *Mol. Cell. Biol.* **11,** 2460–2466.

Cannistraro, V. J., Subbarao, M. N., and Kennell, D. (1986). Specific endonucleolytic cleavage sites for decay of *Escherichia coli* mRNA. *J. Mol. Biol.* **192,** 257–274.

Chen, L.-H., Emory, S. A., Bricker, A. L., Bouvet, P., and Belasco, J. G. (1991). Structure and function of a bacterial mRNA stabilizer: Analysis of the 5′ untranslated region of *ompA* mRNA. *J. Bacteriol.* **173,** 4578–4586.

Cho, K. O., and Yanofsky, C. (1988). Sequence changes preceding a Shine–Dalgarno region influence *trpE* mRNA translation and decay. *J. Mol. Biol.* **204,** 51–60.

Cremer, K. J., Silengo, L., and Schlessinger, D. (1974). Polypeptide formation and polyribosomes in *Escherichia coli* treated with chloramphenicol. *J. Bacteriol.* **118,** 582–589.

Deutscher, M. P., and Reuven, N. B. (1991). Enzymatic basis for hydrolytic versus phosphorolytic mRNA degradation in *Escherichia coli* and *Bacillus subtilis*. *Proc. Natl. Acad. Sci. USA* **88,** 3277–3280.

Donovan, W. P., and Kushner, S. R. (1983). Amplification of ribonuclease II (*rnb*) activity in *Escherichia coli* K12. *Nucleic Acids Res.* **11,** 265–275.

Dreher, J., and Matzura, H. (1991). Chloramphenicol-induced stabilization of *cat* messenger RNA in *Bacillus subtilis*. *Mol. Microbiol.* **5,** 3025–3034.

Emory, S. A., and Belasco, J. G. (1990). The *ompA* 5′ untranslated RNA segment functions in *Escherichia coli* as a growth-rate-regulated mRNA stabilizer whose activity is unrelated to translational efficiency. *J. Bacteriol.* **172,** 4472–4481.

Emory, S. A., Bouvet, P., and Belasco, J. G. (1992) A 5′-terminal stem-loop structure can stabilize mRNA in *E. coli*. *Genes Dev.* **6,** 135–148.

Forchhammer, J., Jackson, E. N., and Yanofsky, C. (1972). Different half-lives of messenger RNA corresponding to different segments of the tryptophan operon of *Escherichia coli*. *J. Mol. Biol.* **71,** 687–699.

Georgellis, D., Arvidson, S., and von Gabain, A. (1992). Decay of *ompA* mRNA and processing of 9*S* RNA are immediately affected by shifts in growth rate, but in opposite manners. *J. Bacteriol.* **174,** 5382–5390.

Gorski, K., Roch, J.-M., Prentki, P., and Krisch, H. M. (1985). The stability of bacteriophage T4 gene *32* mRNA: A 5′ leader sequence that can stabilize mRNA transcripts. *Cell* **43,** 461–469.

Holmes, R. K., and Singer, M. F. (1971). Inability to detect RNase V in *Escherichia coli* and comparison of other ribonucleases before and after infection with coliphage T7. *Biochem. Biophys. Res. Commun.* **44,** 837–843.

Hue, K. H., and Bechhofer, D. H. (1991). Effect of *ermC* leader region mutations on induced mRNA stability. *J. Bacteriol.* **173,** 3732–3740.

Kano, Y., and Imamoto, F. (1979). Evidence for endonucleolytic cleavage at the 5′-proximal segment of the *trp* messenger RNA in *Escherichia coli*. *Mol. Gen. Genet.* **172,** 25–30.

Kennell, D. E. (1986). The instability of messenger RNA in bacteria. *In* Maximizing Gene Expression (W. Reznikoff and L. Gold, Eds.), pp. 101–142. Butterworth, Stoneham, Massachusetts.

Kepes, A. (1969). Transcription and translation in the lactose operon of *Escherichia coli* studied by *in vivo* kinetics. *Prog. Biophys. Mol. Biol.* **19,** 199–236.

Kuwano, M., Kwan, C. N., Apirion, D., and Schlessinger, D. (1969). Ribonuclease V of *Escherichia coli*. I. Dependence on ribosomes and translation. *Proc. Natl. Acad. Sci. USA* **64,** 693–700.

Kuwano, M., Schlessinger, D., and Apirion, D. (1970). Ribonuclease V of *Escherichia coli*. IV. Exonucleolytic cleavage in the 5′ to 3′ direction with production of 5′-nucleotide monophosphates. *J. Mol. Biol.* **51,** 75–82.

Lazinski, D., Grzadzielska, E., and Das, A. (1989). Sequence-specific recognition of RNA hairpins by bacteriophage antiterminators requires a conserved arginine-rich motif. *Cell* **59,** 207–218.

Lin-Chao, S. and Cohen, S. N. (1991). The rate of processing and degradation of antisense RNAI regulates the replication of ColE1-type plasmids in vivo. *Cell* **65,** 1233–1242.

Lundberg, U., Nilsson, G., and von Gabain, A. (1988). The differential stability of the *Escherichia coli ompA* and *bla* mRNA at various growth rates is not correlated to the efficiency of translation. *Gene* **72,** 141–149.

McLaren, R. S., Newbury, S. F., Dance, G. S. C., Causton, H. C., and Higgins, C. F. (1991). mRNA degradation by processive 3′-to-5′ exoribonucleases *in vitro* and the implications for prokaryotic mRNA decay *in vivo*. *J. Mol. Biol.* **221,** 81–95.

Morikawa, N., and Imamoto, F. (1969). Degradation of tryptophan messenger. *Nature (London)* **223,** 37–40.

Morse, D. E., Mosteller, R., Baker, R. F., and Yanofsky, C. (1969). Direction of *in vivo* degradation of tryptophan messenger RNA—A correction. *Nature (London)* **223,** 40–43.

Mosteller, R. D., Rose, J. K., and Yanofsky, C. (1970). Transcription initiation and degradation of *trp* mRNA. *Cold Spring Harbor Symp. Quant. Biol.* **35,** 461–466.

Nilsson, G., Belasco, J. G., Cohen, S. N., and von Gabain, A. (1984). Growth-rate dependent regulation of mRNA stability in *Escherichia coli. Nature (London)* **312,** 75–77.

Nilsson, G., Belasco, J. G., Cohen, S. N., and von Gabain, A. (1987). Effect of premature termination of translation on mRNA stability depends on the site of ribosome release. *Proc. Natl. Acad. Sci. USA* **84,** 4890–4894.

Panganiban, A. T., and Whiteley, H. R. (1983). Purification and properties of a new *Bacillus subtilis* RNA processing enzyme. *J. Biol. Chem.* **258,** 12487–12493.

Régnier, P., and Grunberg-Manago, M. (1990). RNase III cleavage in non-coding leaders of *Escherichia coli* transcripts control mRNA stability and genetic expression. *Biochimie* **72,** 825–834.

Sandler, P., and Weisblum, B. (1988). Erythromycin-induced stabilization of *ermA* messenger RNA in *Staphylococcus aureus* and *Bacillus subtilis. J. Mol. Biol.* **203,** 905–915.

Sandler, P., and Weisblum, B. (1989). Erythromycin-induced ribosome stall in the *ermA* leader: A barricade to 5′-to-3′ nucleolytic cleavage of the *ermA* transcript. *J. Bacteriol.* **171,** 6680–6688.

Schlessinger, D., Jacobs, K. A., Gupta, R. S., Kano, Y., and Imamoto, F. (1977). Decay of individual *Escherichia coli trp* messenger RNA molecules is sequentially ordered. *J. Mol. Biol.* **110,** 421–439.

Schneider, E., Blundell, M., and Kennell, D. (1978). Translation and mRNA decay. *Mol. Gen. Genet.* **160,** 121–129.

Shivakumar, A. G., Hahn, J., Grandi, G., Kozlov, Y., and Dubnau, D. (1980). Posttranscriptional regulation of an erythromycin resistance protein specified by plasmid pE194. *Proc. Natl. Acad. Sci. USA* **77,** 3903–3907.

von Gabain, A., Belasco, J. G., Schottel, J. L., Chang, A. C. Y., and Cohen, S. N. (1983). Decay of mRNA in *Escherichia coli*: Investigation of the fate of specific segments of transcripts. *Proc. Natl. Acad. Sci. USA* **80,** 653–657.

Walker, A. C., Walsh, M. L., Pennica, D., Cohen, P. S., and Ennis, H. L. (1976). Transcription–translation and translation–messenger RNA decay coupling: Separate mechanisms for different messengers. *Proc. Natl. Acad. Sci. USA* **73,** 1126–1130.

Yamamoto, T., and Imamoto, F. (1975). Differential stability of *trp* messenger RNA synthesized originating at the *trp* promoter and p_L promoter of lambda *trp* phage. *J. Mol. Biol.* **92,** 289–309.

4

RNA Processing and Degradation by RNase K and RNase E

ÖJAR MELEFORS, URBAN LUNDBERG, AND ALEXANDER VON GABAIN

I. Introduction

There is considerable evidence that degradation of mRNA and other labile RNAs in bacteria is generally initiated by endonucleolytic cleavage rather than by exonucleolytic attack at the 5′ or 3′ end. Among the small number of endoribonucleases that have been implicated in initiating RNA decay in *Escherichia coli,* two—RNase E and RNase K—have attracted particular interest of late. This interest has been prompted by the relationship of these two endonuclease activities to the product of the *ams/rne* gene, which appears to control the decay of a major fraction of *E. coli* mRNAs.

In this chapter we will discuss the properties of these two nuclease activities, their relationship to each other, their connection with the *ams/rne* gene product, and their role in cleavages that initiate RNA degradation (catabolic cleavages) or that generate relatively stable RNA processing products (anabolic cleavages). RNase E was originally described as a ribonuclease activity that processes the 9S ribosomal RNA precursor to yield 5S ribosomal RNA. RNase K was first identified as an enzyme activity that catalyzes cleavages in the 5′ untranslated region (UTR) of *ompA* mRNA, which encodes a major outer membrane protein. These RNase K cleavages appear to promote accelerated degradation of this message in slowly growing cells. Cleavage of 9S RNA and *ompA* mRNA at these sites is impaired in *E. coli*

This chapter is dedicated to the memory of Dr. David Apirion, whose untimely death is a great loss for the whole field. His curiosity and vision initiated and developed research on RNase E, which plays a central role in RNA metabolism.

Control of Messenger RNA Stability

strains with mutations in the *ams/rne* gene. Recent data suggest that there may be a common control mechanism regulating these different types of endonucleolytic cleavage.

II. Original Phenotypes of the *Escherichia coli* ams^{ts} and rne^{ts} Mutations

In order to find mutations that impair mRNA degradation, Kuwano *et al.* (1977) mutagenized *E. coli*, scored for conditional mutants temperature-sensitive for growth, and subsequently identified mutant strains with retarded decay of bulk mRNA at 42°C. In this way, they isolated a temperature-sensitive mutant, HAK117, in which the bulk mRNA half-life was significantly increased at the nonpermissive temperature (Ono and Kuwano, 1979). The mutation was attributed to a locus called *ams* (altered mRNA stability). The phenotype was not an indirect manifestation of impaired translation, as protein synthesis was maintained even at the nonpermissive temperature (Ono and Kuwano, 1979). The first attempt to clone the wild-type *ams* gene by complementation led to the isolation of a "false" clone that turned out to be a part of the *groEL* gene, which encodes a chaperonin protein (Chanda *et al*, 1985). Some years later, a clone comprising the *ams* gene was reported by Claverie-Martin *et al.* (1989). It is noteworthy that cloning the *ams* gene seemed to succeed only when low-copy-number plasmids were used, suggesting that overexpression of the *ams* gene product may not be tolerated by *E. coli* cells (Claverie-Martin *et al.*, 1989; Melefors and von Gabain, 1991).

Analysis of the decay of the *E. coli ompA* transcript led to the discovery of *ams*-dependent site-specific cleavages in its 5′ region (Melefors and von Gabain, 1988; Lundberg *et al.*, 1990). For the most part, the cleavage sites map upstream of the *ompA* ribosome binding site. Relative to the full-length *ompA* transcript, truncated *ompA* messages resulting from these cleavages are more abundant in slowly growing cells; thus, the relative steady-state concentration of these cleavage products correlates with the rate of degradation of *ompA* mRNA, which accelerates at slow bacterial growth rates (Nilsson *et al.*, 1984, Melefors and von Gabain, 1988). Three of the endonucleolytic cleavages identified in the 5′ region of *ompA* mRNA are inhibited in the ams^{ts} strain, HAK117, at the nonpermissive temperature (Lundberg *et al.*, 1990). After shifting the ams^{ts} strain to the nonpermissive temperature, the cleavage products extending from these sites to the 3′ end were found to decay far more rapidly than the intact transcript, whose half-life was concomitantly prolonged. These latter results, together with the correlation between the cleavage activity and growth-controlled degradation of *ompA* mRNA, strongly suggest that these cleavages can initiate the decay of the *ompA* transcript (Melefors and von Gabain, 1988; Lundberg *et al*, 1990). In the case of *trx* mRNA, an accumulation of mRNA breakdown products was observed at the nonpermissive temperature in an *ams*^{ts} strain bearing additional mutations in the *pnp* and *rnb* genes, which both encode 3′-to-5′exonucleases. It was therefore suggested that the *ams* gene product affects later stages of *trx* mRNA degradation but not the initial cleavage of the full-length transcript (Arraiano *et al.* 1988).

The ability to reproduce *in vitro* some of the *in vivo* cleavages of *ompA* mRNA opened the way for the isolation of an endoribonuclease, RNase K, which cleaves *ompA* mRNA at the *ams*-dependent sites (Nilsson *et al.*, 1988; Lundberg *et al.*, 1990). The seemingly straightforward interpretation that the *ams* gene encodes RNase K was complicated by the fact that cleavage at these *ompA* sites *in vivo* was also inhibited in a different *E. coli* strain (N3431) that carries a temperature-sensitive mutation in the *rne* gene (Melefors and von Gabain, 1991).

The temperature-sensitive *rne*^{ts} mutation was originally defined by the loss of capacity to process 9S RNA into 5S RNA at 42°C (Ghora and Apirion, 1978). It was suggested that the mutation resides in a gene encoding a ribonuclease, designated RNase E. RNase E cleavage has subsequently been demonstrated *in vitro* for various RNAs, including 9S RNA and RNA I, the antisense RNA that regulates colE1 plasmid replication (Table 1). The cleavage activity of RNase E has been found to be thermosensitive in extracts prepared from *rne*^{ts} mutants (Misra and Apirion, 1980; Tomcsanyi and Apirion, 1985). A DNA fragment from *E. coli* was cloned on the basis of its ability to complement the *rne*^{ts} mutation (Dallmann *et al.*, 1987). *In vivo* studies with an *rne*^{ts} strain subsequently revealed that the decay and the processing of certain mRNA species are impeded at the nonpermissive temperature. This finding that RNase E can control mRNA degradation suggested a more general role for this enzyme (Mudd *et al.*, 1988; Carpousis *et al.*, 1989; Mudd *et al.*, 1990a; Regnier and Hajnsdorf, 1991).

III. The *ams*^{ts} and *rne*^{ts} Mutations Map to the Same Gene

As both the *ams*^{ts} and *rne*^{ts} mutations inhibit mRNA degradation, it was of interest to see how they are related. Indeed comparative studies showed that they have almost identical phenotypes, i.e., longer mRNA half-lives and inhibited 9S RNA processing (Mudd *et al.*, 1990b; Melefors and von Gabain, 1991). In each of four studies, a cloned *E. coli* chromosomal fragment was identified which complemented both the *ams*^{ts} and the *rne*^{ts} mutations (Mudd *et al.*, 1990b; Babitzke and Kushner, 1991; Taraseviciene *et al.*, 1991, Melefors and von Gabain, 1991). The cloned fragments all contained the *ams* gene previously isolated by Claverie-Martin *et al.* (1989). Complementation of either of the two mutations with these clones completely restored the *ompA* mRNA cleavages, the processing of 9S RNA, and the degradation of bulk mRNA. Thus, it was concluded that *ams*^{ts} and *rne*^{ts} are mutations in the same gene, referred to hereafter as the *ams/rne* gene. It should be noted that the *rne* gene cloned by complementation of the *rne*^{ts} mutation (Dallmann *et al.*, 1987) is identical to the *ams* gene cloned by Claverie-Martin *et al.* (1989).

A comparison of the reports on the *ams*^{ts} and *rne*^{ts} phenotypes shows that the *ams/rne* gene product is involved in controlling the processing and degradation of a large number of RNAs (Table 1). In most cases, the effects of the *ams* and *rne* mutations on RNA cleavage and RNA turnover have been found to be the same at both permissive temperatures and nonpermissive temperatures. There are, however, examples of RNAs whose degradation is differentially affected by the two mutations, e.g., RNA I and *tetR* mRNA (Lin Chao and Cohen, 1991; Baumeister *et al.*, 1991). Therefore *ams*^{ts} and *rne*^{ts} appear to be different mutations in a single gene.

As the *ams/rne* mutations seem to inhibit the activity of RNase E both *in vivo* and *in vitro*, RNase E would appear to be the key enzyme controlling mRNA decay in *E. coli*. However, it has not yet been possible to obtain a homogeneously pure preparation of RNase E with which to verify that this enzyme in fact cuts at the *ams/rne*-sensitive sites identified *in vivo*. This caveat is important in light of the ability of *ams/rne* mutations to inhibit RNA cleavage *in vivo* at some sites thought to be targets of other nucleases, such as RNase K, RNase F, RNase P, and RNase III (Pragai and Apirion, 1982; Gurewitz *et al.*, 1983; Srivastava *et al.*, 1990; Lundberg *et al.*, 1990).

IV. The Endoribonucleases RNase K and RNase E

Apart from their association with the *ams/rne* gene, the relationship between RNase K and RNase E is not clear. Comparison of the published reports on these enzymes is difficult because each deals with the purification and characterization of only one of the two ribonuclease activities (Misra and Apirion, 1979; Roy and Apirion, 1983; Chauhan *et al.*, 1991; Nilsson *et al.*, 1988; Lundberg *et al.*, 1990). In these *in vitro* studies, RNase E and RNase K were partially purified by assaying for cleavage of either 9S RNA or *ompA* RNA, respectively. Reports of the molecular mass and subunit composition of RNase E vary; in one of the reports a size of 66 kDa was suggested (Misra and Apirion, 1979; Roy and Apirion, 1983; Chauhan *et al.*, 1991). Sufficiently pure preparations for determination of an RNase E amino acid sequence have not yet been obtained. The purest preparation of RNase K consisted of five major polypeptides of sizes ranging from 21 to 62 kDa, when analyzed on a silver-stained denaturing gel (Lundberg *et al.*, 1990). It is not yet clear which of these polypeptides are needed for RNase K activity, which elutes as a 55- to 60-kDa enzyme upon gel filtration under nondenaturing conditions (Lundberg *et al.*, 1990). The activity of both

TABLE 1 Nucleotide Sequence of *ams/rne*-Dependent Cleavage Sites

Substrate	Cleavage site	Cleavage product	
		Upstream	Downstream
ompA A[1] (RNase K)	UCAGA CUUUA	Unstable	Unstable
ompA C[2] (RNase K)	GAAGG AUUUA	Unstable	Unstable
ompA D[2] (RNase K)	GGCGU AUUUU	Unstable	Unstable
9Sa[3] (RNase E)	ACAGA AUUUG	Unstable	Stable
9Sb[3] (RNase E)	AUCAA AUAAA	Stable	Unstable
RNAI (RNase E)	ACAGU AUUUG[4]	Unstable[18]	Unstable[18]
T4GENE32[5] (RNase E)	UGCGA AUUAU	Unstable	Stable
T4D24[6] (RNase F)	nd	nd	nd
10Sa[7] (RNase III)	CAGCU CCACC	Stable	Unstable
10Sb (M1)[8]	CACCU GAUUU	Stable	Unstable
T4 GENE59[9]	CUAUG AUUAA	Unstable	Stable
gltX[10]	CCAGG AUUUG	Unstable	Stable
gltX[10]	CUUAA UUUUU	Unstable	Stable
dicF[11]	UCAAU UUUCU	Unstable	Stable
Phage f1 C[12]	AAAAC UUCUU	Stable	Stable
Phage f1 D[12]	UUAUG UAUCU	Stable	Stable
Phage f1 E[12]	UGAAU CUUUC	Stable	Stable
Phage f1 F[12]	UAGAU UUUUC	Stable	Stable
S15 M[13]	UUCAA GCUGA	Unstable	Unstable
S15 M2[13]	GCGAG UUUCA	Unstable	Unstable
S20 (191)[14]	UAAGC ACAAC	Unstable	Unstable
S20 (301)[14]	AACCG AUCGU	Unstable	Unstable
papB-A[15]	UUUGU AUUGA	Unstable	Stable
atpE 1[16]	UUAAU UUACC	Unstable	Unstable
atpE 2[16]	UACGU UUUAA	Unstable	Unstable
tetR[17]	nd	nd	nd

Note Left column depicts the designation of the *ams/rne*-dependent cleavage sites. Superscripts indicate reference articles. In cases where the ribonuclease that cleaves at the site *in vitro* has been identified, the identity of the ribonuclease is stated in parentheses. The sequence gap shows the reported site of cleavage. Right columns indicate whether the resulting upstream and downstream cleavage products are stable or unstable *in vivo* relative to the RNA precursor. nd, not determined. 1, Lundberg, 1991; 2, Lundberg *et al.*, 1990; 3, Roy and Apirion, 1983; 4, Tomcsanyi and Apirion, 1985; 5, Ehretsmann *et al.*, 1992; 6, Pragai and Apirion, 1983; 7, Srivastava *et al.*, 1990; 8, Gurewitz *et al.*, 1983; 9, Carpousis *et al.*, 1989; 10, Brun *et al.*, 1990; 11, Faubladier *et al.*, 1990; 12, Kokoska *et al.*, 1990; 13, Regnier and Hajnsdorf, 1991; 14, Mackie, 1991; 15, Nilsson and Uhlin, 1991; 16, Gross, 1991; 17, Baumeister *et al.*, 1991; 18, Lin-Chao and Cohen, 1991.

RNase E and RNase K is dependent on Mg^{2+}, but RNase E seems to depend on K^{+} as well (Misra and Apirion, 1979; Nilsson *et al.*, 1988; Lundberg *et al.*, 1990). Another difference is that RNase E precipitates at a lower ammonium sulfate concentration than RNase K (Misra and Apirion, 1979; Lundberg *et al.*, 1990).

More recent, mostly not yet published, studies have involved purification protocols in which the cleavage of both 9S and *ompA* RNA was assayed. From these preliminary reports, it is apparent that RNase E and RNase K copurify to some extent (Sohlberg and von Gabain 1991; Carpousis, 1991; Sohlberg *et al.*, in press). Moreover, both RNases cleave at AU-rich motifs in the RNA substrates tested (Table 1). Nevertheless, it is possible to partially separate RNase K activity (cleavage of an *ompA* mRNA substrate) from RNase E activity (cleavage of a 9S RNA substrate) by ion-exchange chromatography (Sohlberg and von Gabain, 1991; Sohlberg *et al.*, in press) (Fig. 1). Very recently, RNase E was found to copurify with the *E. coli* chaperonin protein GroEL (Sohlberg *et al.*, in press). Subsequent analysis of the nuclease activity associated with GroEL supported the conclusion that GroEL functionally interacts *in vivo* and *in vitro* with RNase E activity but not with RNase K activity (Sohlberg *et al.*, in press). The conclusion is remarkable in light of the previous finding that the *ams*^{ts} mutation could be complemented by a truncated *groEL* gene (Chanda *et al.*, 1985). The binding of RNase E to GroEL, which itself forms a high molecular weight complex, could also explain the observation that this nuclease activity is found to elute in the void volume of Sephadex G150- and G100-size fractionation columns (Sohlberg and von Gabain 1991; Carpousis, 1991; Sohlberg *et al.*, in press), indicating that the enzyme is large or part of large aggregates. As a possible rationalization to reconcile the common and distinct features of RNase E and RNase K, it has been suggested that RNase K may be a proteolytic fragment of RNase E (C. Higgins, personal communication). Another possibility is that the two enzymes may share a common core but differ with respect to accessory factors. Only the purification of RNase E and RNase K to homogeneity will clarify their relationship.

V. The *ams/rne* Gene Product and Its Relation to RNase K and RNase E

The *ams/rne* gene specifies a transcript of about 3.5 kb, as determined by Northern blot analysis (Melefors, 1991). In contrast to *ompA* mRNA, this transcript can barely be detected in wild-type cells or in *ams/rne*^{ts} mutant strains at the permissive temperature, but it accumulates to relatively high levels in the mutant strains at the nonpermissive temperature (Melefors, 1991; von Gabain *et al.*, unpublished data). The promoter of the *ams/rne* gene has been identified; the site of transcription initiation maps 0.36 kb upstream of the *ams/rne* translation initiation codon (Claverie-Martin *et al.*, 1991). It has been suggested that the low level of expression of this gene may, at least in part, be a consequence of post-transcriptional feedback regulation involving cleavage by RNase E (Claverie-Martin *et al.*, 1991). In

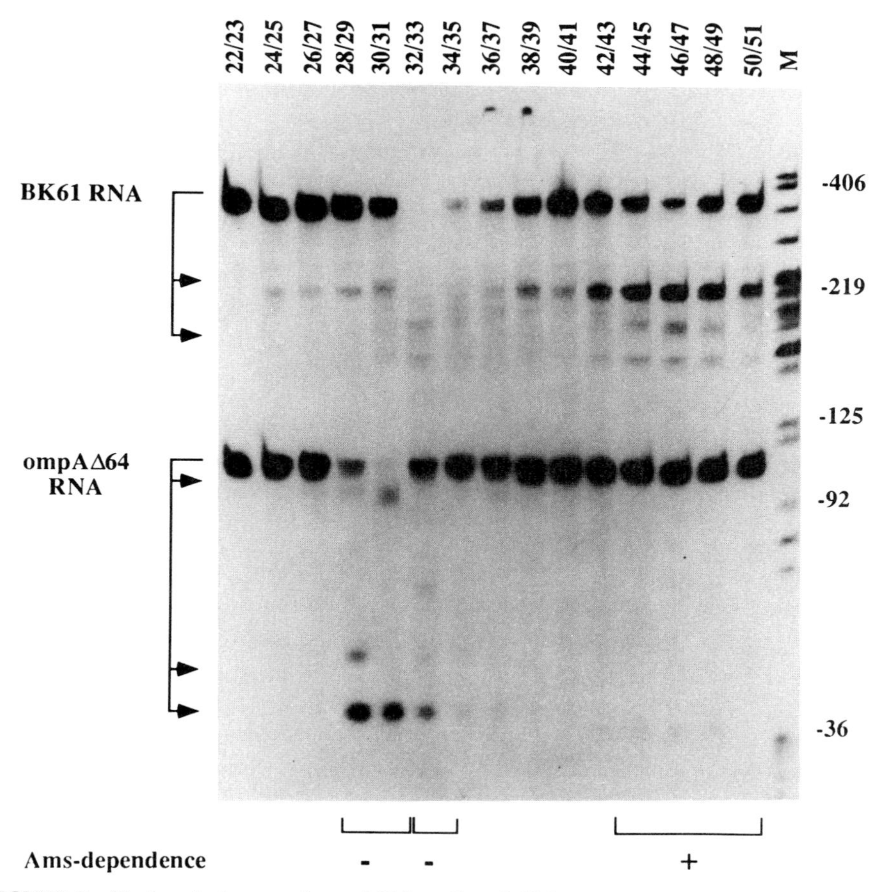

FIGURE 1 Biochemical separation of RNase E and RNase K activities. An extract from *E. coli* polysomes was fractionated by anion-exchange chromatography on a Mono-Q column. Fractions were assayed simultaneously for RNase E and RNase K activity by incubation with a mixture of a 5S rRNA precursor (BK61 RNA) and an *ompA* 5′ UTR derivative (ompAΔ64 RNA), which had been both synthesized and uniformly radiolabeled *in vitro*. RNA substrates and cleavage products were then analyzed by denaturing gel electrophoresis and autoradiography. RNase K activity eluted primarily in fractions 28–31, and RNase E activity eluted in fractions 42–51. A third ribonuclease activity eluted just after RNase K and before RNase E. Arrows on the left point from the full-length RNA substrates to their respective cleavage products. On the right is indicated the length of selected size markers (lane M) in nucleotides. The *ams* dependence of each ribonuclease activity was determined in a separate experiment by examining its temperature sensitivity when isolated from isogenic ams^{ts} versus ams^{+} *E. coli* strains. (This figure shows a previously unpublished experiment by L.-H. Chen, U. Lundberg, A. von Gabain, and J. G. Belasco.)

addition, the *ams/rne* transcript contains a poor Shine–Dalgarno sequence that may be occluded by secondary structure (Claverie-Martin *et al.*, 1991; Chauhan *et al.*, 1991).

Three research groups have independently sequenced the *ams/rne* open reading frame (ORF) (Claverie-Martin *et al.*, 1991; Chauhan *et al.*, 1991; Casaregola *et al.*, 1992). The sequence data in one study suggested an *ams/rne* ORF that can code for a 91-kDa polypeptide (Claverie-Martin *et al.*, 1991). However, another study identified an ORF that has the same 5′ terminus but has an earlier termination codon, yielding a polypeptide with a predicted size of 62 kDa (Chauhan *et al.*, 1991). The product of the *ams/rne* gene, identified by overexpressing the cloned gene in *E. coli*, has been reported to migrate as a 110-kDa band on denaturing SDS gels (Claverie-Martin *et al.*, 1989; Chauhan *et al.*, 1991), which is larger than the 91-kDa size predicted by Claverie-Martin *et al.* (1991) and much larger than the 62-kDa protein proposed by Chauhan *et al.* (1991). To explain the latter discrepancy, Chauhan *et al.* (1991) have suggested that the protein product of the *ams/rne* gene may associate with an RNA subunit.

The characterization of the *ams/rne* gene and its gene product took a new turn when a study was initiated with the aim of identifying a myosinlike contractile protein in *E.coli* (Casaregola *et al.*, 1990). By using a monoclonal antibody against the heavy chain myosin from *Saccharomyces cerevisiae*, a cross-reacting *E. coli* protein, Hmp1 (high-molecular-weight protein), was identified that migrates as a 180-kDa band on denaturing SDS gels (Casaregola *et al.*, 1990). Cloning and sequencing of the *hmp1* gene, which encodes a 114-kDa protein, revealed that it is identical to the previously isolated *ams/rne* gene (Casaregola *et al.*, 1992). However, compared with the previously published *ams/rne* sequences (Claverie-Martin *et al.*, 1991; Chauhan *et al.*, 1991), *hmp1* specifies a polypeptide which is longer at its carboxy terminus (Casaregola *et al.*, 1992). The same study also demonstrated that the "180-kDa" polypeptide is the primary gene product and that the slow electrophoretic migration of this protein is not due to aggregation with other molecules. The data indicate that the highly charged carboxy-terminal region is responsible for the anomalous migration of this protein when it is analyzed by SDS–PAGE (Casaregola *et al.*, 1992). A simple explanation for discrepancies with the previous sequences of the *ams/rne* gene is to assume a one-nucleotide change in the study by Chauhan *et al.* (1991) and an omission of two bases in the study by Claverie-Martin *et al.* (1991). The higher electrophoretic mobility of the polypeptide analyzed by Claverie-Martin *et al.* (1989) and Chauhan *et al.* (1991) apparently resulted from overexpression of a truncated *ams/rne* gene fragment that, while incomplete, can nevertheless complement the ams^{ts} and rne^{ts} mutations.

Analysis of the sequence of the *ams/rne* (*hmp1*) ORF disclosed a number of interesting features that suggest multiple functions for the Ams/Hmp1 protein (Claverie-Martin *et al.*, 1991, Casaregola *et al.*, 1992). A putative nucleotide binding site and a transmembrane domain were identified in the amino-terminal half of the Ams/Hmp1 molecule (Casaregola *et al.*, 1992). In the carboxy-terminal region, a homology of 18% was found between Ams/Hmp1 and a ribosomal protein from *Neurospora crassa* (Claverie-Martin *et al.*, 1991). Furthermore, in the carboxy-terminal half, which appears to constitute a separate domain, a short region of 86 amino acid residues is 29% homologous with the highly conserved 70-kDa protein of the human U1 small nuclear ribonucleoprotein particle (snRNP) (Claverie-Martin *et al.*, 1991, Casaregola *et al.*, 1992). Thus, the *ams/rne* gene product shows some sequence characteristics of an RNA binding protein, although this bacterial protein lacks the most conserved block of 70-kDa snRNP residues, which are specifically concerned with binding RNA (Casaregola *et al.*, 1992).

The prevailing view is that the *ams/rne* gene product is identical to RNase E (Mudd *et al.*, 1990b; Babitzke and Kushner, 1991; Taraseviciene *et al.*, 1991). The strongest argument in favor of this interpretation of the present data is that the 9S RNA cleavage activity is absent at 44°C when extracts containing the activity are prepared from rne^{ts} cells (Misra and Apirion, 1980; Tomcsanyi and Apirion, 1985) and that membrane material from rne^{ts} cells bearing the wild-type *ams/rne* gene is capable of cleaving 9S RNA at the elevated temperature (Chauhan and Apirion, 1991). However, these data indicate only that the *ams/rne* gene product is necessary for RNase E activity, not that it alone is sufficient.

Some experiments suggest that RNase E may be structurally complex. When the cleavage of 9S RNA is examined either *in vivo* in rne^{ts} cells growing exponentially or *in vitro* with RNase E partially purified from log-phase rne^{ts} cells, RNase E activity is suppressed at the nonpermissive temperature relative to rne^{+} cells (Misra and Apirion, 1979; Tomcsanyi and Apirion, 1985; A. von Gabain *et al.*, unpublished data). However, when these experiments are repeated with rne^{ts} cells turning into stationary phase, no reduction in RNase E activity is observed at the nonpermissive temperature, both *in vivo* and *in vitro* (A. von Gabain *et al.*, unpublished data). Thus, the effect of this mutation on RNase E activity seems to be limited to certain growth stages, an observation that weakens the assertion that the *ams/rne* gene product is the nuclease itself. Furthermore, an inverse relationship between RNase E and RNase K activity is observed *in vivo* when fast-growing and slow-growing cells are compared. After a shift from fast to slow growth, RNase E activity (as assayed by 9S RNA processing) decreases, whereas RNase K activity (as assayed by *ompA* mRNA decay) increases (Georgellis *et al.*, 1992). Although such changes in endonuclease cleavage might be mediated by hypothetical cellular factors that bind in a growth-rate-regulated manner to 9S RNA and *ompA* mRNA, it is attractive to attribute these changes instead to factors that bind to, or modify, the nuclease(s) themselves. One may well imagine that RNase E and RNase K share a core nuclease and that the *ams/rne* gene product is an auxiliary factor that differentially regulates the activity of this core nuclease toward the two different substrates (9S RNA and *ompA* mRNA).

The original conclusion that RNase K is distinct from RNase E was based on the fact that, unlike RNase E, RNase K activity (as assayed by *ompA* RNA cleavage) is not temperature-sensitive *in vitro* when crude extracts from the *rne*^{ts} strain N3431 and its isogenic wild-type counterpart (N3433) are compared (Nilsson *et al.*, 1988). RNase K also seems to be *ams*/*rne*-independent *in vitro* when the enzyme is assayed in extracts from rapidly growing log-phase cultures of the *ams*^{ts} strain HAK117 (Fig. 1). On the other hand, cleavage at these *ompA* mRNA sites is temperature-sensitive *in vivo* when examined in rapidly growing HAK117 (*ams*^{ts}) and N3431 (*rne*^{ts}) cells (Lundberg *et al.*, 1990; Melefors and von Gabain, 1991). The basis for this discrepancy in *ams*/*rne* dependence between the *in vivo* and the *in vitro* results is not yet clear, but it might be explained in a number of ways. For instance, RNase E may be able to cleave *ompA* mRNA at the same sites as RNase K, only much less efficiently. If so, cleavage of *ompA* mRNA in rapidly growing cells might be due primarily to attack by RNase E, since under these growth conditions RNase E activity may predominate over RNase K activity (Georgellis *et al.*, 1992). Conversely one may argue that, in the absence of RNase E, RNase K may to some degree be capable of processing 9S RNA at the RNase E cleavage sites and that cleavage by RNase K might thereby account for the comparatively slow processing of 9S RNA that occurs in slowly growing cells (Georgellis *et al.*, 1992). This would explain why cleavage of *ompA* mRNA is *ams*/*rne*-dependent under conditions of rapid cell growth and why RNase E activity does not seem to be thermolabile in *rne*^{ts} cells turning into stationary phase (von Gabain *et al.*, unpublished data). An increase in RNase K activity and a decrease in RNase E activity in slowly growing cells could then explain the observed increase in the *ompA* mRNA decay rate and decrease in 9S RNA processing, respectively (Georgellis *et al.*, 1992). This hypothesis makes the testable prediction that cleavage of *ompA* mRNA should be *ams*/*rne*-independent in slowly growing cells. Whether these two enzyme activities can indeed substitute for one another at different growth rates or growth stages will require further experimental study. In this connection, it is noteworthy that decay of *ompA* mRNA, degradation of bulk mRNA, and processing of 9S RNA are all retarded when cells cultured at a constant slow growth rate are switched from aerobic to anaerobic growth (Georgellis *et al.*, submitted for publication). Thus, the activity of both RNase E and RNase K seems to be reduced when slow growing cells are cultured in the absence of oxygen. The precise relationship among RNase E, RNase K, and the *ams*/*rne* gene product will be decided only when the enzymes have been purified to homogeneity and their protein sequences compared with that of the *ams*/*rne* gene. For the time being, it seems wise to retain the working definitions that RNase E catalyzes efficient 9S RNA cleavage and RNase K catalyzes efficient *ompA* mRNA cleavage.

VI. *ams/rne*-Dependent Cleavage Sites

To date, all RNA sites classified as targets for cleavage by RNase E or RNase K have been identified on the basis of *in vivo* analysis of *ams/rne* mutants or *in vitro* assays with partially purified enzymes (Table 1). A comparison of *ams/rne*-dependent cleavage sites led Ehretsmann *et al.* (1992) to suggest *R*AUU*W* (*R*=G or A; *W*=U or A) as a consensus RNase E cleavage site. *In vivo* and *in vitro* studies of the effect of point mutations on RNase E cleavage of phage T4 gene 32 mRNA are consistent with this consensus sequence and also suggest that cleavage by RNase E may be enhanced by a downstream stem–loop flanking the cleavage site (Fig. 2) (Ehretsmann *et al.*, 1992). However, an investigation of RNase E cleavage of RNA I variants *in vivo* does not support the concept of a canonical RNase E cleavage sequence (Lin-Chao and Cohen, 1991b).

Recently, Bouvet and Belasco have obtained *in vivo* evidence that efficient endonucleolytic cleavage of RNA I by RNase E can be facilitated by the presence of several unpaired nucleotides at the RNA 5′ end (Bouvet and Belasco, 1992; see Chapter 3). This apparent 5′-end dependence of RNase E could explain how 5′-terminal secondary structure can impede degradation of mRNA in *E. coli* (Emory *et al.*, 1992).

In vitro analysis of the sites of cleavage by partially purified RNase K in the 5′ untranslated region of *ompA* mRNA has identified G*YX*UUU (*Y*=G, A, or U; *X*=A or C) as a possible RNase K consensus sequence (Lundberg, 1991). Secondary structure analysis of the *ompA* 5′ UTR suggests that these RNase K cleavage sites reside in regions that are not base-paired (Chen *et al.*, 1991; Rosenbaum *et al.*, in press).

It is important to bear in mind that because no RNase E or RNase K cleavage site has yet been defined using homogeneously pure enzyme, definitive assignment of these cleavages to a particular enzyme may be difficult. Sites of *ams/rne*-dependent cleavage that have been detected only *in vivo* may be particularly prone to erroneous assignment. Indeed, some of these sites are believed actually to be targets not of RNase E or RNase K, but of RNase III, RNase P, or RNase F (Pragai and Apirion, 1982; Gurewitz *et al.*, 1983; Srivastava *et al.*, 1990). Such ambiguity might result in cases where prior cleavage by RNase E or RNase K elsewhere in the RNA molecule must occur before another ribonuclease can recognize and cut the RNA at the site in question.

Like RNase III (see Chapter 5), RNase E (and maybe RNase K) is both an anabolic and a catabolic ribonuclease, and cleavage by this enzyme can

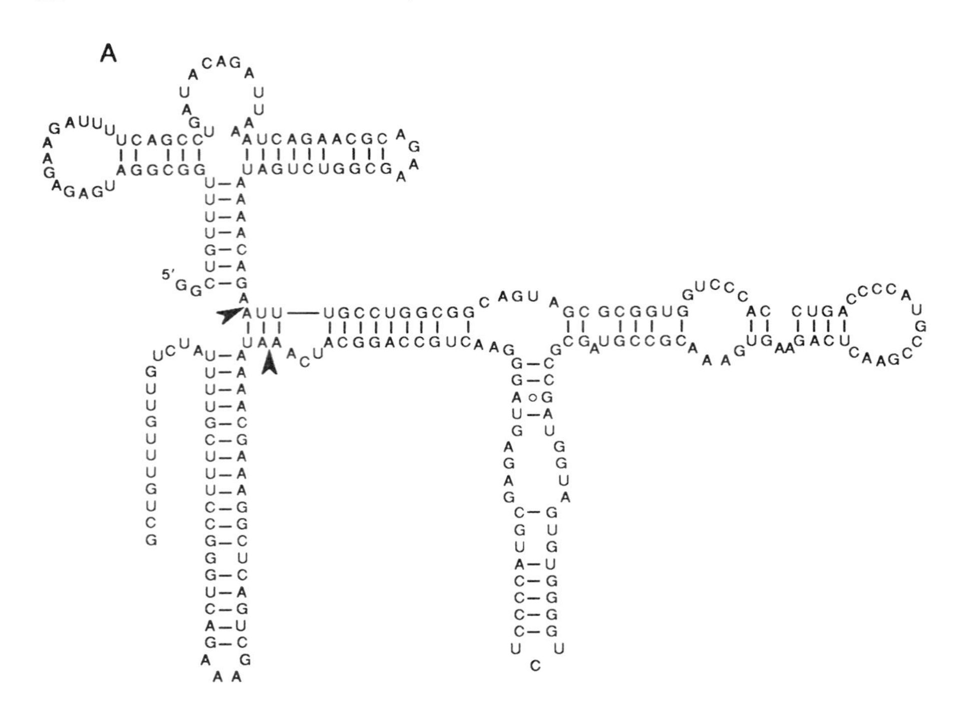

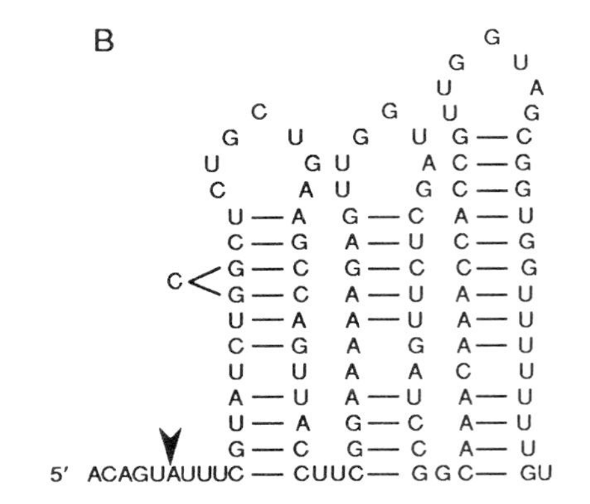

FIGURE 2 see legend on following page.

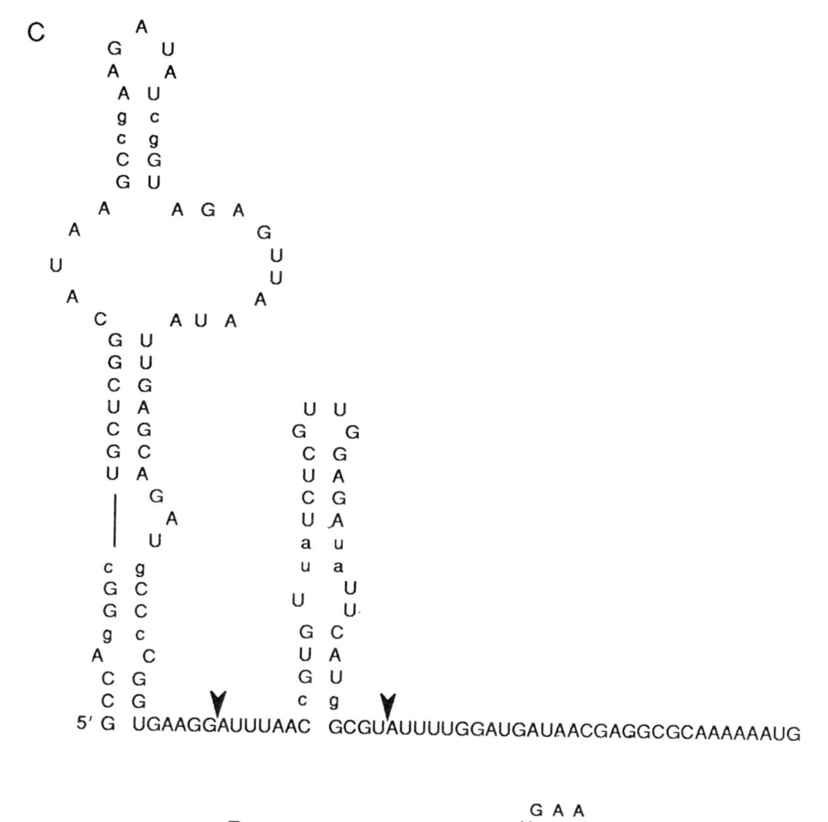

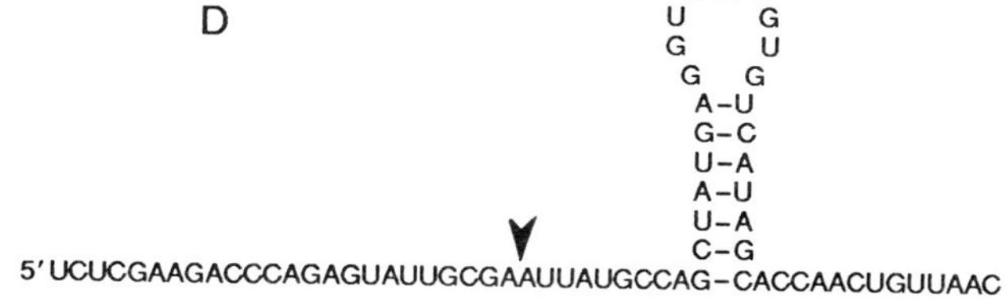

FIGURE 2 RNase E and RNase K cleavage sites and RNA secondary structures. Figures 2A, 2B, and 2D indicate the RNase E cleavage sites (arrows) relative to the published secondary structures of T4 gene32 RNA (Ehretsman *et al.*, 1992), RNA I (Lin-Chao and Cohen, 1991), and 9S RNA (Roy and Apirion, 1983; Christiansen, 1988), respectively. Figure 2C indicates the RNase K cleavage sites relative to the published secondary structure of the 5′ UTR of *ompA* mRNA (Chen *et al.*, 1991). The depicted structure of the *ompA* 5′ UTR is identical to one of the three coexisting structures that were found in a study by Rosenbaum *et al.* (in press).

lead to relatively stable upstream and downstream processing products or to rapid degradation of both cleavage products. There are even examples of *ams*/*rne*-dependent cleavages that initiate decay of the upstream but not the downstream cleavage product, and vice versa (Table 1). The structural basis for selective degradation of upstream cleavage products seems straightforward. These presumably are readily digested by 3′-to-5′ exonucleases (RNase II and polynucleotide phosphorylase) unless protected by a hairpin structure upstream of the cleavage site (Mott *et al.*, 1985; Donovan and Kushner, 1986; Plamann and Stauffer, 1990; McLaren *et al.*, 1991) (see Chapter 2). It is more difficult to explain why downstream cleavage products are often susceptible to further rapid degradation, as no 5′-to-3′ exoribonucleases have been found so far in bacteria (Deutscher and Zhang, 1990). For example, it has been shown that a downstream cleavage product that differs from a primary transcript by only eight 5′-terminal nucleotides [i.e., the downstream product of A-site cleavage of the *tacompA* transcript versus wild-type *ompA* mRNA (Lundberg *et al.*, 1990)] or by as little as a 5′ triphosphate versus a 5′ monophosphate [i.e., the pRNA I_{-5} cleavage product of RNA I versus its triphosphate homolog pppRNA I_{-5} (Lin-Chao and Cohen, 1991)] can be much less stable despite the seemingly minor structural difference. The mere absence of a 5′-terminal triphosphate is not adequate to explain the relatively rapid degradation of these downstream cleavage products, as other cleavage events result in 5′-monophosphorylated downstream products that are relatively stable (Table 1). Instead, cleavages that initiate decay may be coupled mechanistically to the degradation of the 3′ cleavage product. For example, certain structural features downstream of a cleavage site might facilitate or impede subsequent digestion by an as yet undiscovered 5′-to-3′ exonuclease or by an endonuclease that "scans" RNA in a 5′-to-3′ direction. Coupling might be achieved if this hypothetical scavenger enzyme is RNase E or RNase K, which might act processively, or if it forms a complex with one or both of these ribonucleases.

VII. Concluding Summary

The relationship between two *E. coli* mutations that affect RNA processing and decay, *ams*^{ts} (controlling mRNA decay) and *rne*^{ts} (controlling RNase E), has only recently been recognized. These mutations were found to reside in the same gene, referred to here as the *ams*/*rne* gene, which was independently identified as the *hmp1* gene. The *ams*/*rne* (*hmp1*) gene product resembles a transmembrane protein with multiple functional domains. The product of this gene apparently interferes with two types of endonucleolytic activities: catabolic cleavages that initiate RNA decay and anabolic cleavages that process precursor RNA molecules into functional units. Most of these cleavages have been attributed to endoribonucleases RNase E and RNase K, which have been partially purified. The relationship between these two endoribonuclease activities is not yet clear. Although they copurify to some degree, they can be separated by ion-exchange chromatography, and they display distinct metal ion requirements. Furthermore, the chaperonin GroEL interacts functionally with RNase E activity but not with RNase K activity, and RNase K activity seems not to be thermosensitive *in vitro* when extracts of *ams*^{ts} mutant strains are tested. Finally, the activities of RNase E and RNase K *in vivo* are differentially regulated in response to changes in growth rate. There is no agreement yet as to the precise function of the *ams/rne* (*hmp1*) gene product. The *rne*^{ts} mutation seems to render the activity of RNase E heat-sensitive, supporting the prevailing view that this gene product is identical to RNase E, which appears to be the key enzyme that is primarily responsible for controlling mRNA decay rates in *E. coli*. Only sequence analysis of RNase E and RNase K purified to homogeneity will reveal the true connection between these two enzymes and the *ams/rne* gene product.

Acknowledgments

We thank S. N. Cohen and T. Barlow for critically reading the manuscript and B. Sohlberg for helping to prepare the manuscript. We are grateful to the editors, J. Belasco and G. Brawerman, for their never-ending patience to improve the quality of this chapter. This review has been supported by grants to A.v.G. from the Swedish Cancer Society (RmC), the Swedish Board of Technical Development (STU), and the Swedish Foundation of Biotechnology (SBF). Ö.M. and U.L. have been supported by stipends from the Swedish Cancer Society (RmC) and the Swedish Foundation of Biotechnology (SBF).

References

Arraiano, C. M., Yancey, S. D., and Kushner, S. R. (1988). Stabilization of discrete mRNA breakdown products in *ams, pnp rnb* multiple mutants of *Escherichia coli*. *J. Bacteriol.* **170,** 4625–4633.

Babitzke, P., and Kushner, S. R. (1991). The *ams* (altered mRNA stability) protein and ribonuclease E are encoded by the same structural gene of *Escherichia coli*. *Proc. Natl. Acad. Sci. U.S.A.* **88,** 1–5.

Baumeister, R., Flache, P., Melefors, Ö, von Gabain, A., and Hillen, W. (1991). Lack of a 5′ non-coding region in Tn1721 encoded *tetR* mRNA is associated with a low efficiency of translation and a short half-life in *Escherichia coli*. *Nucleic Acids Res.* **19,** 4595–4600.

Bouvet, P., and Belasco, J. G. (1992). Control of RNase E-mediated RNA degradation by 5′-terminal base pairing in *E. coli*. *Nature (London)* **360,** 488–491.

Brun, Y. V., Sanfacon, H., Breton, R., and Lapointe, J. (1990). Closely spaced and divergent promoters for an aminoacyl-tRNA synthetase gene and a tRNA operon in *Escherichia coli*. *J. Mol. Biol.* **214,** 845–864.

Carpousis, A. J., Mudd, E. A., and Krisch, H. M. (1989). Transcription and mRNA processing upstream of bacteriophage T4 gene 32. *Mol. Gen. Genet.* **219,** 39–48.

Carpousis, A. J. (1991). Presented at the 1991 ASM Meeting on mRNA Decay and RNA Processing.

Casaregola, S., Norris, V., Goldberg, M., and Holland, I. B. (1990). Identification of a 180 kDa protein in *Escherichia coli* related to the yeast heavy chain myosin. *Mol. Microbiol.* **4,** 505–511.

Casaregola, S., Jacq, A., Laoudj, D., McGurk, G., Margarson, S., Tempete, M., Norris, V., and Holland, I. B. (1992). Cloning and anlysis of the entire *Escherichia coli ams* gene: *ams* is identical to *hmp1* and encodes a 114 kDa protein that migrates as a 180 kDA protein. *Mol. Biol.* **228,** 30–40.

Chanda, P. K., Ono, M., Kuwano, M., and Kung, H.-F. (1985). Cloning, sequence analysis, and expression of alteration of the mRNA stability gene (*ams*) of *Escherichia coli. J. Bacteriol.* **161,** 446–449.

Chauhan, A. K., and Apirion, D. (1991). The *rne* gene is the structural gene for the processing endoribonuclease RNase E of *Escherichia coli. Mol. Gen. Genet.* **228,** 49–54.

Chauhan, A. K., Miczak, A., Taraseviciene, L., and Apirion, D. (1991). Sequencing and expression of the *rne* gene of *Escherichia coli. Nucleic Acids Res.* **19,** 125–129.

Chen, L.-H., Emory., S. A., Bricker, A. L., Bouvet, P., and Belasco, J. (1991). Structure and function of a bacterial mRNA stabilizer: Analysis of the 5′ untranslated region of *ompA* mRNA. *J. Bacteriol.* **173,** 4578–4586.

Claverie-Martin, F., Diaz-Torres, M. R., Yancey, S. D., and Kushner, S. R. (1989). Cloning the altered mRNA stability (*ams*) gene of *Escherichia coli* K-12. *J. Bacteriol.* **171,** 5479–5486.

Claverie-Martin, F., Diaz-Torres, M. R., Yancey, S. D., and Kushner, S. R. (1991). Analysis of the altered mRNA stability (*ams*) gene of *Escherichia coli. J. Biol. Chem.* **266,** 2843–2851.

Christiansen, J. (1988). The 9S precursor of *Escherichia coli* 5S RNA has three structural domains: Implications for processing. *Nucleic Acids Res.* **16,** 7457–7476.

Dallmann, G., Dallmann, K., Sonin, A., Miczak, A., and Apirion, D. (1987). Expression of the gene for the RNA processing ribonuclease E in plasmids. *Mol. Gen. (Life Sci. Adv.)* **6,** 99–107.

Deutscher, M. P., and Zhang, J. (1990). *In* Post-Transcriptional Control of Gene Expression (J.E.G. McCarthy and M.F. Tuite, Eds.) NATO ASI Series Vol. 49, pp.1–12. Springer-Verlag, Berlin/Heidelberg.

Donovan, W. P., and Kushner, S. R. (1986). Polynucleotide phosporylase and ribonuclease II are required for cell viability and mRNA turnover in *Escherichia coli* K-12. *Proc. Natl. Acad. Sci. U.S.A.* **83,** 120–124.

Ehretsmann, C. P., Carpousis, A. J., and Krisch, H. M. (1992). *Escherichia coli* RNase E has a role in the decay of bacteriophage T4 mRNA. *Genes Dev.* **6,** 149–159.

Emory, S. A., Bouvet, P., and Belasco, J. G. (1992). A 5′-terminal stem–loop structure can stabilize mRNA in *Escherichia coli. Genes Dev.* **6,** 135–148.

Faubladier, M., Cam, K., and Bouche, J. P. (1990). *Escherichia coli* cell division inhibitor DicF-RNA of the *dicB* operon: Evidence for its generation *in vivo* by transcription termination and by RNase III and RNase E-dependent processing. *J. Mol. Biol.* **212,** 461–471.

Georgellis, D., Arvidsson, S., and von Gabain, A. (1992). Decay of *ompA* mRNA and processing of 9S RNA are immediately affected by shifts in growth rate, but in opposite manners. *J. Bacteriol.* **174,** 5382–5390.

Georgellis, D., Barlow. T., Arvidsson, S., and von Gabain, A. Retarded mRNA turnover in *Escherichia coli:* A means to maintain gene expression during anaerobiosis. Submitted for publication.

Ghora, B. K., and Apirion, D. (1978). Structural analysis and *in vitro* processing of 9S RNA to p5S rRNA of a RNA molecule isolatated from an *rne* mutant of *Escherichia coli. Cell* **15,** 1055–1066.

Gross, G. (1991). RNase E cleavage in the *atpE* leader region of *atp*/interferon beta hybrid transcripts in *Escherichia coli* causes enhanced rates of mRNA decay. *J. Biol. Chem.* **10,** 17885–17889.

Gurewitz, M., Jain, S. K., and Apirion, D. (1983). Identification of a precursor molecule for the RNA moiety of the processing enzyme RNase P. *Proc. Natl. Acad. Sci. U.S.A.* **80,** 4450–4454.

Kokoska, R. J., Blumer, K. J., and Steege, D. A. (1990). Phage f1 mRNA processing in *Escherichia coli* : Search for the upstream products of endonuclease cleavage, requirement for the product of altered mRNA stability (*ams*) locus. *Biochimie* **72,** 803–811.

Kuwano, M., Ono, M., Endo, H., Hori, K., Nakamura, K., Hirota, Y., and Ohnishi, Y. (1977). A gene affecting longevity of mRNA: A mutant of *Escherichia coli* with altered mRNA stability. *Mol. Gen. Genet.* **154,** 279–285.

Lin-Chao, S., and Cohen, S. N. (1991a). The rate of processing and degradation of antisense RNAI regulates the replication of colE1-type plasmids *in vivo*. *Cell* **65,** 1233–1242.

Lin-Chao, S., and Cohen, S. N. (1991b). Presented at the 1991 ASM Meeting on mRNA Decay and RNA Processing.

Lundberg, U., von Gabain, A., and Melefors, Ö. (1990). Cleavages in the 5′region of the *ompA* and *bla* mRNA control stability: Studies with an *Escherichia coli* mutant altering mRNA stability and a novel endoribonuclease. *EMBO. J.* **9,** 2731–2741.

Lundberg, U. (1991). *In* Purification and Characterization of an mRNA Specific Endoribonuclease in *Escherichia coli*. Thesis, Karolinska Institute, Stockholm.

McLaren, R. F., Newbury, S. F., Dance, G. S. C., Gauston, H. C., and Higgins, C. (1991). Degradation by processive 3′–5′ exoribonucleases *in vitro* and the implications for prokaryotic mRNA deacy *in vivo*. *J. Mol. Biol.* **221,** 81–95.

Mackie, G. A. (1991). Specific endonucleolytic cleavage of the mRNA for ribosomal protein S20 of *Escherichia coli* requires the product of the *ams* gene *in vivo* and *in vitro*. *J. Bacteriol.* **173,** 2488–2497.

Melefors, Ö. (1991). *In* The Mechanism of *ompA* mRNA Degradation in *Escherichia coli*. Thesis, Karolinska Institute, Stockholm.

Melefors, Ö., and von Gabain, A. (1988). Site-specific endoribonucleolytic cleavages and the regulation of stability of *Escherichia coli ompA* mRNA. *Cell* **52,** 893–901.

Melefors, Ö., and von Gabain, A. (1991). Genetic studies of cleavage-initated mRNA decay and processing of ribosomal 9S RNA show that *Escherichia coli ams* and *rne* loci are the same. *Mol. Micriobiol.* **5,** 857–864.

Misra, T. K., and Apirion D. (1979). RNase E , an RNA processing enzyme from *Escherichia coli*. *J. Biol. Chem.* **254,** 11154–11159.

Misra, T. K., and Apirion D. (1980). Gene *rne* affects the structure of the ribonucleic acid-processing enzyme ribonuclease RNase E of *Escherichia coli*. *J. Bacteriol.* **142,** 359–361.

Mott, J. E., Galloway, J. L., and Platt, T. (1985). Maturation of *Escherichia coli* tryptophan operon mRNA: Evidence for 3′ exonucleolytic processing after rho-dependent termination. *EMBO J.* **4,** 1887–1891

Mudd, E. A., Prentki, P., Belin, D., and Krisch, H. M. (1988). Processing of unstable bacteriophage T4 gene 32 mRNAs into stable species requires *Escherichia coli* ribonuclease E. *EMBO J.* **7,** 3601–3607.

Mudd, E. A., Carpousis, A. J., and Krisch, H. M. (1990a). *Escherichia coli* RNase E has a role in the decay bacteriophage T4 mRNA. *Genes Dev.* **4,** 873–881.

Mudd, E. A., Krisch, H. M., and Higgins, C. F. (1990b). RNase E, an endoribonuclease has a general role in the chemical decay of mRNA. *Mol. Micriobiol.* **4,** 2127–2135.

Nilsson, G., Belasco, J., Cohen, S. N., and von Gabain, A. (1984). Growth-rate dependent regulation of mRNA stability in *Escherichia coli*. *Nature (London)* **312,** 75–77.

Nilsson, G., Lundberg, U., and von Gabain, A. (1988). *In vivo* and *in vitro* identity of site-specific cleavages in the 5′noncoding region of *ompA* and *bla* mRNA in *Escherichia coli*. *EMBO J.* **7,** 2269–2275.

Nilsson, P., and Uhlin, B. E. (1991). Differential decay of a polycistronic *Escherichia coli* is initiated by RNase E- dependent endonucleolytic processing. *Mol. Microbiol.* **5,** 1791–1799.

Ono, M., and Kuwano, M. (1979). A conditional lethal mutation in an *Escherichia coli* strain with a longer chemical lifetime of messenger RNA. *J. Mol. Biol.* **129,** 343–357.

Plamann, M. D., and Stauffer, G. V. (1990). *Escherichia coli glyA* mRNA decay: The role of 3′ secondary structure and the effects of the *pnp* and *rnb* mutations. *Mol. Gen. Genet.* **220,** 301–306.

Pragai, B., and Apirion, D. (1982). Processing of bacteriophage T4 transfer RNAs, structural analysis of an *in vitro* processing of precursors that accumulate in RNase E strains. *J. Mol. Biol.* **154,** 465–484.

Regnier, P., and Hajnsdorf, E. (1991). Decay of mRNA encoding ribosomal protein S15 of *Escherichia coli* is initiated by an RNase E-dependent endonucleolytic cleavage that removes the 3′ stabilizing stem and loop structure. *J. Mol. Biol.* **217,** 283–292.

Rosenbaum, V., Klahn, T., Lundberg, U., Homgran, E., von Gabain,A. and Riesner, D. (1993). Co-existing structures of an mRNA stability determinant: The 5' region of the *Escherichia coli* and *Serratia marcescens ompA* mRNA. *J. Mol. Biol.* in press.

Roy, M. K., and Apirion D. (1983). Purification and properties of ribonuclease E, an RNA processing enzyme from *Escherichia coli*. *Biochim. Biophys. Acta* **747,** 200–208.

Sohlberg, B., and von Gabain, A. (1991). Presented at the 1991 ASM Meeting on mRNA Decay and RNA Processing.

Sohlberg, B., Lundberg, U., Hartl, U., and von Gabain, A. (1993). Functional interaction of heat shock protein GroEL with RNase E in *Escherichia coli*. *Proc. Nat. Acad. Sci USA,* in press.

Srivastava, R. K., Miczak, A., and Apirion, D. (1990). Maturation of precursor 10Sa RNA in *Escherichia coli* is a two-step process: The first reaction is catalyzed by RNase III in presence of Mn^{2+}. *Biochim.* **72,** 791–802.

Taraseviciene, L., Miczak, A., and Apirion, D (1991). The gene specifying RNase E (*rne*) and a gene affecting mRNA stability (*ams*) are the same gene. *Mol. Microbiol.* **5,** 851–855.

Tomcsanyi, T., and Apirion, D. (1985). Processing enzyme ribonuclease E specifically cleaves RNAI an inhibitor of primer formation in plasmid DNA synthesis. *J. Mol. Biol.* **185,** 713–720.

5

RNA Processing and Degradation by RNase III

DONALD COURT

I. Introduction

RNase III was first detected in *Escherichia coli* as an endoribonuclease that cleaves double-strand RNA molecules (Robertson *et al.*, 1968). The enzyme degrades both natural and synthetic RNA duplexes to small acid-soluble fragments, and it was suggested that RNase III might provide a possible cellular defense against infection by viruses. Although no specific antiviral role has yet been found for RNase III in *E. coli*, other roles for the protein have been discovered, namely, the processing of stable RNA species and certain mRNA molecules (Dunn and Studier, 1973a, 1973b; Nikolaev *et al.*, 1973). However, only a subset of the bacterial mRNA population is affected by RNase III processing; the stability of the majority of cellular mRNAs is essentially unchanged in mutants defective in RNase III (Talkad *et al.*, 1978; Apirion and Gitelman, 1980). In keeping with the theme of this book, the control of mRNA decay, I will discuss genes whose expression is known to be affected directly by RNase III, paying particular attention to the effects of RNase III on mRNA stability. I will also review the function of RNase III in stable RNA processing and outline the general properties of the protein, its RNA substrates, and its gene organization.

II. RNase III Enzyme and Its Substrates

A double-strand RNA structure is common to all *in vivo* RNase III processing sites for which a sequence has been determined. This was the original feature that was used to detect and purify the endonuclease (Robertson *et al.*, 1968); the enzyme degrades long stretches of perfect duplex RNA into acid-soluble segments of about 10 to 18 base pairs (Schweitz and Ebel, 1971; Robertson and Dunn, 1975). In *E. coli*, natural duplex RNA molecules that arise by annealing of antisense and sense RNA molecules are cleaved by RNase III (Krinke and Wulff, 1987). In addition to cleaving duplex RNAs, RNase III also cleaves specific regions of RNA that form base-paired stem–loop structures. Although such stem–loop structures are found in almost all RNA molecules, most are not cleaved by RNase III, and the specific structural and sequence elements required for RNase III-dependent cleavage remain largely unpredictable (Dunn, 1976; Krinke and Wulff, 1990a; Chelladurai *et al.*, 1991). The first structures in RNA transcripts that were shown to be substrates for RNase III were intercistronic regions of the 5000-nucleotide T7 early precursor mRNA and the 30S ribosomal precursor RNA. Normally, neither of these precursors is observed in wild-type cells, but they do accumulate in RNase III-defective mutants. This accumulation of precursors is a direct result of RNase III deficiency, as shown by their processing to normal mature forms when RNase III is provided *in vitro* (Dunn and Studier, 1973a, 1973b; Nikolaev *et al.*, 1973). The RNase III cleavage sites in these T7 and ribosomal RNAs are both site specific and occur in regions containing double-strand RNA (Rosenberg and Kramer, 1977; Robertson *et al.*, 1977; Young and Steitz, 1978; Bram *et al.*, 1980).

A. Purification and Cellular Location of RNase III

The purification of RNase III from *E. coli* revealed that the active form of RNase III is a dimer of two identical 25-kDa polypeptides (Robertson *et al.*, 1968; Dunn, 1976). Upon extraction from cells, RNase III sediments with a particulate fraction, and subsequent treatment with high salt elutes RNase III into the soluble fraction. Robertson and colleagues (1968, 1990) had suggested that RNase III is associated with ribosomes. However, Miczak *et al.* (1991) could separate the enzyme from the ribosome fraction in low salt and suggested from sedimentation studies that RNase III is bound to the inner membrane of the cell. Chen *et al.* (1990) purified RNase III to homogeneity from cells in which RNase III levels reached 40% of the total protein. This RNase III preparation was also in a particulate fraction following cell lysis and could be solubilized in 0.5 *M* NaCl. Interestingly, after purification, the protein aggregated upon removal of salt and precipitated from the solution; high salt again solubilized the pure protein (S. M. Chen and D. Court, unpublished). Thus, the particulate nature of RNase III is a real concern when trying to distinguish aggregation from membrane partitioning of the protein. In more dilute solutions, self-aggregation might be difficult even at low salt concentration, but the protein might exhibit nonspecific association with other factors such as ribosomes or membranes.

B. Types of RNA Cleavages Made by RNase III

Typically, RNase III makes staggered, double-strand cleavages *in vitro* and *in vivo* in duplex regions of rRNA precursors and in certain mRNAs. The staggered cut yields a two-base 3′ overhang (Schweitz and Ebel, 1971; Paddock *et al.*, 1976; Lozeron *et al.*, 1976; Young and Steitz, 1978; Bram *et al.*, 1980) and leaves 5′ phosphoryl and 3′ hydroxyl ends on the RNA products (Robertson *et al.*, 1968; Crouch, 1974). Unlike most cellular RNAs, the T7 precursor mRNA is not cut in both paired strands by RNase III. Five different sites in the precursor are processed by RNase III to yield five mature mRNA species encoding the early T7 functions (Fig. 1). Four (R1, R2, R3, and R4) of the five sites are cut only on the distal side of the stem in an internal loop region. One site (R5) is cut on both strands of the wild-type transcript (Dunn, 1976) but can be mutated such that only the distal strand is cut (Saito and Richardson, 1981). This mutational change, HS9, decreases base pairing in the region of the cut site at R5a (Fig. 2). Subsequent changes to the mutant site that enhance base pairing restore cutting to both strands (see WT and Rev2 in Fig. 2) (Saito and Richardson, 1981). Other Rev mutants that reduce base pairing of HS9 eliminate cleavage at R5b as well (Fig. 2) (Saito and Richardson, 1981). Thus, whether the product of RNase III action is a single-strand nick or double-strand break depends in part upon the degree of base pairing in the region of the cut sites.

Buffer and salt conditions that optimize *in vitro* processing of long duplex RNA (synthetic or natural) are similar to the conditions required for efficient processing of natural mRNA substrates of RNase III (Dunn, 1976; Rosenberg *et al.*, 1977; Robertson *et al.*, 1977). These optimal conditions include Na^+, K^+, or NH_4^+ concentrations between 100 and 300 m*M* and Mg^{2+} concentrations of 1 to 10 m*M* (Mn^{2+} at 0.1 m*M* can substitute for Mg^{2+}). The optimal pH is 7.4 to 9.6 (Robertson *et al.*, 1968; Dunn, 1976); the isoelectric point of RNase III that is predicted from the deduced protein sequence is near 10 (Nashimoto and Uchida, 1985). It has recently been shown that *in vitro* cleavage at one of the natural target sites on the *E. coli*

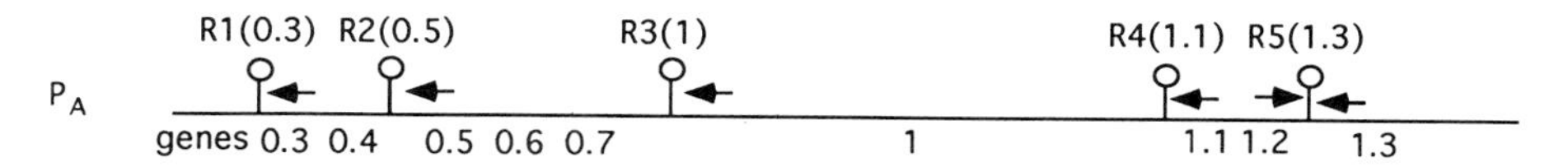

FIGURE 1 The T7 early gene transcripts. *Escherichia coli* RNA polymerase initiates transcription on the T7 DNA at three promoters (A_1, A_2, A_3) located to the left of gene *0.3* near the left end of the phage genome and indicated in the figure as P_A. The early genes *0.3, 0.4, 0.5, 0.6, 0.7, 1, 1.1, 1.2,* and *1.3* are indicated below the 5-kilobase transcript. Locations of the secondary structures processed by RNase III are indicated by the stem–loop regions on the transcript. The number of each site is indicated above the loop (R1–R5). In parentheses are alternative names for these RNase III sites (Dunn and Studier, 1983). Arrows indicate where the stems are cleaved by RNase III. Note that R5 is cleaved twice.

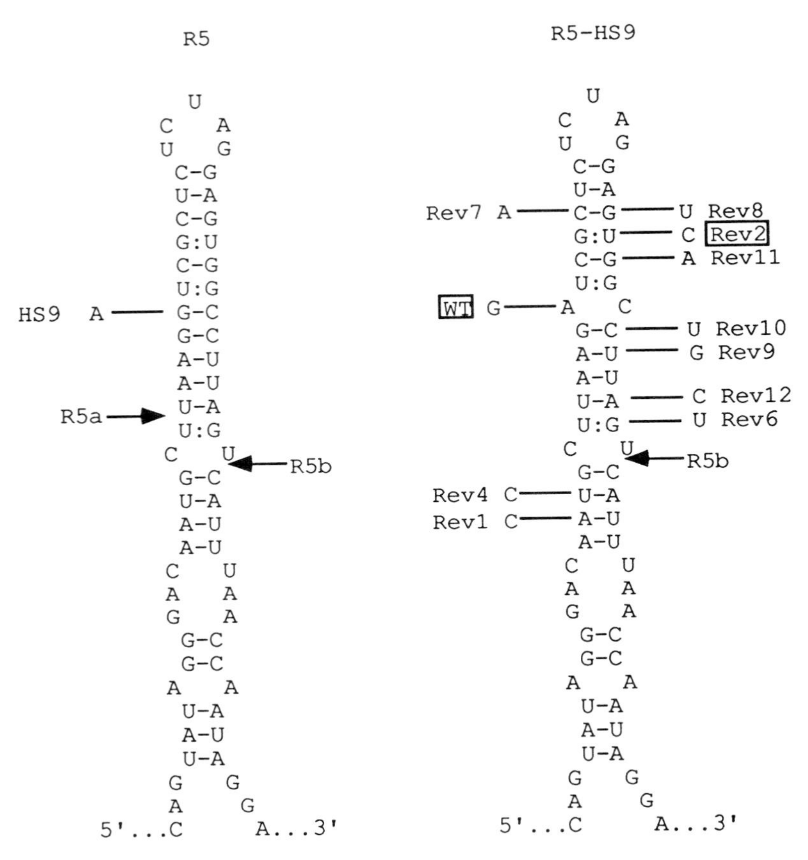

FIGURE 2 Mutations in the phage T7 wild-type and HS9 mutant R5 RNase III structures. The wild-type R5 structure is shown on the left with the base change that creates the HS9 mutant structure. The HS9 mutant structure is shown on the right. The HS9 mutation prevents expression of T7 functions 1.1 and 1.2 and thereby prevents phage replication. Mutants able to replicate were isolated from HS9 (Saito and Richardson, 1981), and the mutations are indicated beside the HS9 structure. R5a and R5b are cleavage sites for RNase III. Note that HS9 prevents the R5a cleavage. The boxed changes, WT and Rev2, in HS9 restore R5a cleavage. All other mutations shown in HS9 prevent RNase III cleavage at R5b as well as at R5a.

10Sa RNA requires Mn^{2+} instead of Mg^{2+} (Srivastava *et al.*, 1990), suggesting that different localized conditions within the cell might activate different processing events by RNase III.

C. Secondary Cleavages or "Star Activities" of RNase III

Reduced salt concentration (100 m*M*) or high enzyme concentration can cause cleavage events at secondary sites in RNA not normally cut or cut poorly *in vivo* (Dunn, 1976; Birenbaum *et al.*, 1978; Gross and Dunn, 1987). Some primary sites normally cleaved *in vivo* are also affected by these changed conditions. The R5 site in T7 early precursor mRNA is always cleaved *in vivo* on the distal side (R5b) of the stem but is cleaved only 40% of the time on the 5′ side (R5a) of the stem (Fig. 2). These same proportions of R5a and R5b cleavage are seen *in vitro* under optimized conditions; however, at low salt or high enzyme concentrations, RNase III cleaves both sides of the R5 stem nearly 100% of the time (Dunn, 1976; Robertson *et al.*, 1977). Thus, these conditions enhance cutting at weaker primary sites and allow new cleavage events at secondary sites. It is conceivable that secondary sites may be used *in vivo* under some conditions, for example, when the RNase III concentration in the cell might be abnormally high.

In competition studies in which various substrates of RNase III were tested against each other, it was found that duplex RNA and primary mRNA sites could compete with each other for RNase III processing at high salt concentrations, whereas secondary mRNA sites were unable to compete with either duplex RNA or primary mRNA sites (Dunn, 1976). Thus, in conditions simulating physiological salt concentrations, secondary sites do not appear to bind efficiently to RNase III.

The 3′ overhangs generated by RNase III cleavage are similar to the overhangs generated by some DNA restriction enzymes. By analogy to DNA restriction enzymes, the low salt effect on RNase III that causes secondary site cleavage may result from a change in the structure (March and Gonzalez, 1990) and activity of RNase III to make it less site specific, as occurs with "star activities" of certain restriction enzymes (Nasri and Thomas, 1986). High restriction enzyme concentrations also generate star-type cutting at secondary sites even at the optimal salt concentration. Other analogies between RNase III and DNA restriction enzymes can be made. Some restriction enzymes recognize interrupted palindromic sequences and cleave between them in regions lacking sequence specificity for the enzyme. A proposed consensus sequence for RNase III has a similar interrupted symmetry with cleavage in a nonconserved region (Krinke and Wulff, 1990a) (see Section II.E). Like RNase III, some restriction enzymes are able to cleave single-strand nucleic acids in regions of secondary structure (New England Biolabs, 1990–1991 Catalog). Although restriction enzymes and RNase III differ in terms of specificity for DNA and RNA and the requirement for sequence specificity when cleaving long duplex nucleic acids, the information gained about restriction enzymes may improve our understanding of RNase III.

D. Substrate Recognition Specificity of RNase III

Specific RNase III cleavage requires that the enzyme first recognize and bind to an appropriate RNA substrate. RNase III recognizes, among other features, double-strand RNA structures. With long duplex RNA, RNase III is essentially nonspecific in its ability to bind and cut. It can cleave synthetic poly(I)–poly(C) and poly(A)–poly(U) RNA duplexes as well as natural RNAs such as the reovirus RNA genome. These synthetic and natural duplexes are reduced to small fragments (Robertson *et al.*, 1968; Crouch, 1974; Robertson and Dunn, 1975). Krinke and Wulff (1990a) used a natural double-strand RNA from phage λ of *E. coli*, the oop/cII (antisense/sense) RNA duplex, to examine duplex RNA cleavage in finer detail. RNase III appears to measure and cleave 10 to 14 base pairs from the edge of a double-strand region (Robertson, 1982; Krinke and Wulff, 1990a). The variability in cleavage site indicated by the 10- to 14-base-pair length of the cleavage products was suggested by Krinke and Wulff (1990a) to be caused by a requirement for interaction of the enzyme with specific nucleotides. Since both poly(I)–poly(C) and poly(A)–poly(U) synthetic RNA duplexes are processed by RNase III, perfect matches to a consensus are obviously not required; however, some nucleotide juxtapositions may be favored over others.

RNase III can also process mRNA and rRNA, i.e., RNAs that do not contain extensive stretches of perfect duplex RNA, but do contain regions of secondary structure that are base paired to form stem–loops or hairpin structures. Although some of these RNA secondary structures are sites for RNase III processing, most are not. Structures that are processed generally contain stem regions of less than 20 consecutive base pairs that usually contain unpaired bases as bulges or internal loops (for example, see Fig. 2).

The RNase III substrate is contained within the RNA stem–loop structure. It is unlikely that tertiary interactions with sequences external to the stem–loop are required to render them substrates for RNase III. For example, pseudoknots between unpaired bases within the stem–loop structure and bases outside it are not important. This statement is supported by two observations. A minimal RNA molecule made *in vitro* that contains only the nucleotide sequence forming the RNA stem–loop is fully sensitive to processing by RNase III (Nicholson *et al.*, 1988; Chelladurai *et al.*, 1991). Furthermore, deletion analysis *in vivo* suggested that the sequence within the stem–loop is required for RNase III processing (Court *et al.*, 1983a). Thus, the stem–loop structures are both necessary and sufficient for recognition and cleavage by RNase III.

The double-strand stems containing RNase III-sensitive sites often contain unpaired bases (see Fig. 2). Model building predicts that these bulged bases may be aligned to form tertiary interactions with the major groove of the upper and lower base-paired stems, thereby forming hydrogen-bonded structures similar to double-strand RNA (Robertson and Barany, 1979; Robertson, 1982). The importance of such double-strand mimicry for RNase III recognition is questionable, since Chelladurai *et al.* (1991) mutated the looped-out bases that were postulated to make these tertiary interactions with the stem but found no specific effect on RNase III processing.

E. Sequence Specificity in RNase III Cleavage Sites

It has long been suggested that site-specific cleavage by RNase III of mRNA and rRNA is dependent in part upon sequence specificity (Dunn and Studier, 1973a, 1973b; Gegenheimer and Apirion, 1981; Daniels *et al.*, 1988; Bardwell *et al.*, 1989). Krinke and Wulff (1990a) compared the sequence of more than 20 RNase III processing sites to identify conserved nucleotides at positions surrounding the cut site. The derived consensus sequence is shown in Fig. 3 as a 26-base-pair double-strand RNA sequence with the staggered cutting sites in the middle. The symmetry of this sequence is compatible with the idea that each of the double-strand cuts is made by one of the identical subunits of the RNase III dimer. The length of the consensus

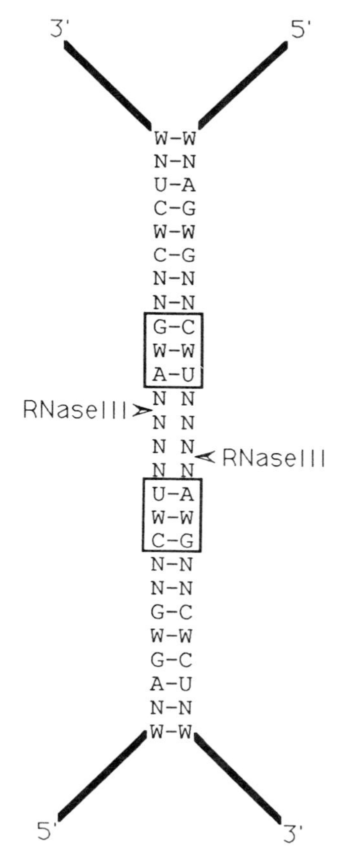

FIGURE 3 A consensus sequence proposed for RNase III recognition. The consensus sequence of Krinke and Wulff (1990a) is shown as double-strand RNA. N indicates A, U, G, or C, and W indicates A or U. Arrows indicate positions of RNase III cleavage. The boxed areas represent highly conserved sequences present in many RNase III sites (Gegenheimer and Apirion, 1981).

sequence is also compatible with the minimum length of the poly(A)–poly(U) synthetic RNA duplex that could be cut in half by RNase III (Crouch, 1974).

Within the consensus sequence, 16 positions are often conserved in individual sites. However, it is clear from examining individual RNase III sites that a functional site may contain as few as half of the conserved base pairs, and they are never base paired over the entire stem segment (Krinke and Wulff, 1990a). Thus, there remains a great deal of uncertainty as to the sequence that is required for cleavage. The consensus does indicate that the nucleotides at the cleavage site itself are not conserved; this was also suggested by analysis of terminal nucleotides at cleaved sites (Robertson and Dunn, 1975; Robertson, 1977) and by the lack of effect of mutations there (Chelladurai *et al.*, 1991).

Mutational changes in the substrates for RNase III that prevent or alter processing have not provided the critical information required to define the specificity determinants of these structures. Mutations that change pairs from G:C → G–A cause a greater defect in processing than those that change G:C → G–U (Guarneros *et al.*, 1982; Saito and Richardson, 1981), supporting the premise that the type of base pairing is important. In order to identify bases that are important for recognition by RNase III, compensatory mutations need to be constructed that maintain base pairing in the RNA stems, i.e., G:C → C:G; U:A → G:C; etc. Chelladurai *et al.* (1991) have initiated experiments of this type. They have made multiple changes in the T7 mRNA processing site R4 and have examined processing *in vitro*. The sequence at the R4 site matches the Krinke and Wulff (1990a) consensus sequence at 12 of 16 conserved positions and includes the highly conserved CWU sequence (boxed in Fig. 3) located just 5′ to the cut site (Gegenheimer and Apirion, 1981; Daniels *et al.*, 1988). However, when Chelladurai *et al.* (1991) converted the CUU in the upper stem of R4 to an AGG with compensatory changes on the complementary RNA strand to maintain base pairing, the resulting mutant RNA structure was processed as efficiently as the wild type *in vitro*. Among several changes of this type made in the upper stem of R4, only one had a negative effect on processing. In that case, a conserved G residue (Krinke and Wulff, 1990a), nine bases 5′ to the cut site, was changed to C, resulting in some reduction in processing. Despite the multiple changes in different constructs that were made by Chelladurai *et al.* (1991), a minimum of nine matches with the Krinke and Wulff consensus remained in the R4 modified sites. As Krinke and Wulff point out, several active processing sites contain nine and even fewer matches with their consensus; thus, to identify the specificity determinants, it may be necessary to examine an even larger array of multiple changes. It will also be important to examine the effects of the changes *in vivo*.

The Krinke and Wulff consensus indicates that symmetry is important. It is clear from other mutational analyses that base pairs on both sides of the cleavage site are important for cleavage (Studier *et al.*, 1979; Saito and Richardson, 1981; Guarneros *et al.*, 1982; Stark *et al.*, 1985). A question relating to this symmetry may be posed. Chellandurai *et al.* (1991) modified the sequence in contiguous three-base-pair segments to one side of the processing site. Since either peptide of the RNase III homodimer may be able to recognize a sequence flanking the cleavage site, genetic analysis of the recognition site may require that symmetric changes on both sides of the cleavage site be made. For example, the consensus CUU sequence that Chellandurai changed is also represented on the other side of the R4 cleavage site. Removal of part or all of this second CUU sequence may be critical to prevent RNase III binding and processing.

III. *rnc*: The Gene for RNase III

The first mutant defective in RNase III expression was isolated by Kindler *et al.* (1973) following extensive mutagenesis with nitrosoguanidine and screening for a strain defective in the enzyme. The mutation in this strain (*rnc105*) was mapped to the 55-min region of the *E. coli* chromosome (Apirion and Watson, 1975; Studier, 1975). Later, the *rnc* gene was cloned on a multicopy plasmid (Watson and Apirion, 1985), and the sequence of the wild-type *rnc* gene was determined (Nashimoto and Uchida, 1985; March *et al.*, 1985). Clones carrying only the region encoding the *rnc* gene complement the *rnc105* mutant strain for RNase III and encode a ~25-kDa polypeptide (Watson and Apirion, 1985; Nashimoto and Uchida, 1985; March *et al.*, 1985) corresponding in size to the RNase III monomer (Dunn, 1976) (Fig. 4).

A. The *rnc* Operon

The *rnc* gene is the first of three genes in the multifunctional *rnc* operon (Fig. 5). The *era* and *recO* genes are located downstream of *rnc* in the operon (Ahnn *et al.*, 1986; Takiff *et al.*, 1989). The product of the *era* gene is a GTP-binding protein with GTPase activity (March *et al.*, 1988; Chen *et al.*, 1990). The *era* gene is essential for the growth of *E. coli* (Takiff *et al.*, 1989; Inada *et al.*, 1989), although its function in *E. coli* remains unknown. The product of the *recO* gene takes part in the RecF recombination and repair pathway (Morrison *et al.*, 1989). Although related functions often cluster in the same operon in *E. coli*, the relationship between these three functions remains obscure.

The operon has been defined by polar mutations in the *rnc* gene that prevent expression of the distal *era* and *recO* genes and by polar mutations in *era* that prevent expression of *recO* (Takiff *et al.*, 1989; Inada *et al.*, 1989). S1 analysis indicates that all three genes are cotranscribed (Bardwell *et al.*,

```
rnc
                                       rnc14::Tn10
1   ATGAACCCCATCGTAATTAATCGGCTTCAACGGAAGCTGGGCTACACTTTTAATCATCAGGAACTGTTGCAGCAGGCATTAACTCATCGT
1 ▶ MetAsnProIleValIleAsnArgLeuGlnArgLysLeuGlyTyrThrPheAsnHisGlnGluLeuLeuGlnGlnAlaLeuThrHisArg
RNaseIII
                                            A - rnc105
91  AGTGCCAGCAGTAAACATAACGAGCGTTTAGAATTTTTAGGCGACTCTATTCTGAGCTACGTTATCGCCAATGCGCTTTATCACCGTTTC
31 ▶ SerAlaSerSerLysHisAsnGluArgLeuGluPheLeuGlyAspSerIleLeuSerTyrValIleAlaAsnAlaLeuTyrHisArgPhe
                                            Asp

181 CCTCGTGTGGATGAAGGCGATATGAGCCGGATGCGCGCCACGCTGGTCCGTGGCAATACGCTGGCGGAACTGGCGCGCGAATTTGAGTTA
61 ▶ ProArgValAspGluGlyAspMetSerArgMetArgAlaThrLeuValArgGlyAsnThrLeuAlaGluLeuAlaArgGluPheGluLeu

                                                                                   A - rnc70
271 GGCGAGTGCTTACGTTTAGGGCCAGGTGAACTTAAAAGCGGTGGATTTCGTCGTGAGTCAATTCTCGCCGACACCGTCGAAGCATTAATT
91 ▶ GlyGluCysLeuArgLeuGlyProGlyGluLeuLysSerGlyGlyPheArgArgGluSerIleLeuAlaAspThrValGluAlaLeuIle
                                                                                  Lys

361 GGTGGCGTATTCCTCGACAGTGATATTCAAACCGTCGAGAAATTAATCCTCAACTGGTATCAAACTCGTTTGGACGAAATTAGCCCAGGC
121 ▶ GlyGlyValPheLeuAspSerAspIleGlnThrValGluLysLeuIleLeuAsnTrpTyrGlnThrArgLeuAspGluIleSerProGly

451 GATAAACAAAAAGATCCGAAAACGCGCTTGCAAGAATATTTGCAGGGTCGCCATCTGCCGCTGCCGACTTATCTGGTAGTCCAGGTACGT
151 ▶ AspLysGlnLysAspProLysThrArgLeuGlnGluTyrLeuGlnGlyArgHisLeuProLeuProThrTyrLeuValValGlnValArg

541 GGCGAAGCGCACGATCAGGAATTTACTATCCACTGCCAGGTCAGCGGCCTGAGTGAACCGGTGGTTGGCACAGGTTCAAGCCGTCGTAAG
181 ▶ GlyGluAlaHisAspGlnGluPheThrIleHisCysGlnValSerGlyLeuSerGluProValValGlyThrGlySerSerArgArgLys

    T - rev3
631 GCTGAGCAGGCTGCCGCCGAACAGGCGTTGAAAAAACTGGAGCTGGAA
211 ▶ AlaGluGlnAlaAlaAlaGluGlnAlaLeuLysLysLeuGluLeuGlu
    Val
```

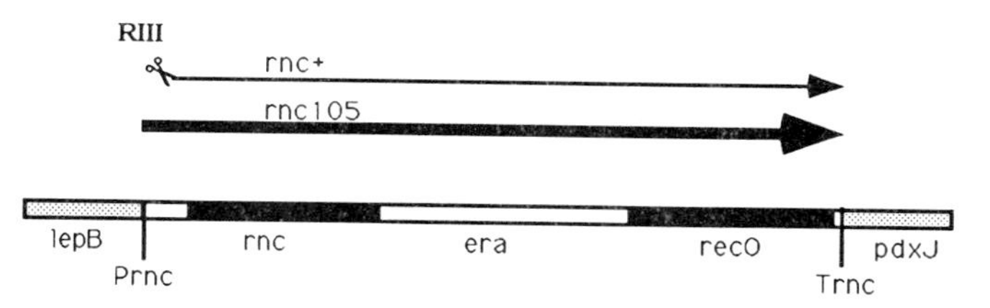

FIGURE 5 The *rnc* operon is shown. The open and closed rectangles represent the *rnc* operon genes named below them. Outside of the operon, the *lepB* (leader peptidase) gene of *E. coli* is indicated in the stippled area to the left; the *pdxJ* (pyridoxine biosynthesis) gene is indicated in the stippled area to the right. The *rnc* promoter (P*rnc*) and transcription terminator (T*rnc*) are shown (Takiff *et al.*, 1992). The arrows represent transcripts and their extents. Transcripts found for the wild-type operon and the *rnc105* mutant operon are indicated. The *rnc*+ transcript, which is shortened at the 5′ end by RNase III (RIII) processing, is shown as a thinner line to indicate reduced RNA levels resulting from more rapid decay (Bardwell *et al.*, 1989).

1989). Cotranscription of *rnc, era,* and *recO* genes terminates beyond *recO* and does not include the downstream *pdxJ* operon (Bardwell *et al.*, 1989; Takiff *et al.*, 1992). The polarity observed in the *rnc* operon may be caused, in part, by translational coupling between *rnc* and *era* (Bardwell *et al.*, 1989; Takiff *et al.* 1989).

B. Mutations in the *rnc* Gene

The *rnc* gene is not essential for bacterial survival. It can be disrupted without major harm to cell growth, despite the fact that RNase III takes part in the maturation of rRNA and processing of several cellular mRNAs (Takiff *et al.*, 1989).

One *rnc* null mutation, *rnc14::ΔTn10*, is caused by a mini-Tn10 insertion within the *rnc* structural gene. Although the mini-Tn10 blocks transcription

---

FIGURE 4 The *rnc* gene and RNase III protein sequences. Mutations indicated above the nucleotide sequence correspond to the underlined base(s). The *rnc14* mutation is a mini-Tn10 insertion at the nine-base target sequence indicated. The target sequence is duplicated on each side of the insertion. The amino acids created by point mutations are indicated below the protein sequence. The underlined protein sequences indicate homologies to other proteins discussed in Sections VII and VIII. A discrepancy exists in the coding sequence determined by March *et al.* (1985) and Nashimoto and Uchida (1985). Three extra thymidine residues in the DNA sequence reported by March *et al.* (1985) cause a frameshift difference (codons 168 to 196) between their coding region sequence and the sequence reported by Nashimoto *et al.* (1985), which is shown here. [T. Inada, Y. Nakamura, and D. Court (unpublished) have confirmed the sequence determined by Nashimoto and Uchida (1985).]

from the *rnc* promoter, the novel joint where the insertion occurred creates a new promoter to allow essential *era* expression (Takiff *et al.*, 1989, 1992).

A second mini-Tn10 insertion mutation, *rnc40::ΔTn10*, is located within the *rnc* leader region. This mutation also displays unusual features. The mini-Tn10 blocks transcription of the *rnc* operon from its normal promoter as described above for *rnc14*. However, in this case, the *tet* promoter within the mini-Tn10 element is sufficient to transcribe the operon. Since this internal promoter is induced only in the presence of tetracycline, the strain is tetracycline-dependent for expression of *era* and cell growth (Takiff *et al.*, 1992).

Three *rnc* point mutants with lesions in the *rnc* structural gene have been isolated (Fig. 4). The classical *rnc105* mutation described earlier in this section changes a glycine to an aspartic acid codon at position 44 (Nashimoto and Uchida, 1985). A second mutation, *rnc70*, was isolated after localized mutagenesis of the *rnc* region DNA, and it changes codon 117 of the *rnc* gene from a glutamic acid to an alanine codon (Fig. 4) (Inada *et al.*, 1989). The *rnc70* mutation confers the same phenotypes as the *rnc105* mutation; i.e., 30S rRNA precursors accumulate, phage λ N protein is underexpressed, and polynucleotide phosphorylase is overexpressed (T. Inada, Y. Nakamura, L. Fernandez, and D. Court, unpublished) (see Table 1 for a phenotypic comparison of wild-type *rnc* and different *rnc* mutants). Unlike

TABLE 1 Phenotypes Caused by *rnc* Mutations

Allele	rRNA[a]	Pnpase[b]	λN[c]	*rps118*[d]
rnc$^+$	16S, 23S, 5S	Low	High	t^S
rnc105	30S	High	Low	NT
rnc70	30S	High	Low	NT
rnc14::Tn10	30S	High	Low	NT
rev3	16S, 23S, 5S	NT[e]	NT[e]	t^R

[a] In *rnc*$^+$ and *rev3* only 16S, 23S, and 5S ribosomal RNA species are observed; in *rnc105*, *rnc70*, and *rnc14* the 30S precursor exists as well (Nashimoto *et al.*, 1985; T. Inada, Y. Nakamura, and D. Court, unpublished).

[b] Polynucleotide phosphorylase (70-kDa polypeptide) is visible on polyacrylamide gels and is overexpressed in isogeneic strains with *rnc* mutants (T. Inada, Y. Nakamura, and D. Court, unpublished).

[c] λ*N* gene expression is reduced in *rnc* mutants tested (Kameyama *et al.*, 1991; L. Fernandez and D. Court, unpublished).

[d] *rps118* causes a growth defect at elevated temperatures in *rnc*$^+$ that is suppressed by *rev3* (Nashimoto *et al.*, 1985; Nashimoto and Uchida, 1985).

[e] NT, not tested.

the Rnc105 mutant protein, the purified Rnc70 protein retains specific binding to RNase III substrates (L. Fernandez and D. Court, unpublished).

A third point mutation, *rev3*, does not cause a detectable RNase III processing defect (Table 1) (Nashimoto and Uchida, 1985). The *rev3* mutation changes position 211 of *rnc* from an alanine to a valine codon. This mutation is discussed in Section V.A in the context of its role in rRNA processing.

IV. Regulation of *rnc* Expression and RNase III Activity

The absence of RNase III in a cell has pleiotropic effects on the expression of many proteins. Approximately 10% of all proteins in the cell that are detectable by gel analysis are either under- or overproduced in an *rnc105* mutant strain (Gitelman and Apirion, 1980; Takata *et al.*, 1987). This large difference in gene expression between rnc^+ and *rnc105* strains suggests that RNase III should be considered to have a potential global regulatory function. If RNase III plays such a role, then the intracellular level or activity of RNase III might be expected to fluctuate with cellular growth conditions. These fluctuations could, in turn, regulate the expression of other genes either directly or indirectly.

A. Post-Transcriptional Control of *rnc* and *era* Expression: Autoregulation by RNase III

March *et al.* (1985) noticed a hairpin structure in the leader region of the *rnc* gene and proposed that it might be a site for autoprocessing by RNase III. Further studies by Bardwell *et al.* (1989) demonstrated *in vivo* and *in vitro* that RNase III processed this site in its own leader RNA and, furthermore, that this processing was responsible for a three- to sixfold reduction in protein and mRNA levels for both RNase III and Era. Thus, RNase III has the capability of controlling its own expression; i.e., the *rnc*-leader transcript would be in competition with other RNAs for processing by RNase III. Since the synthesis of RNase III represents 0.01% of total protein synthesis, the number of molecules per cell is of the same order as other regulatory proteins such as the σ factor (Bardwell *et al.*, 1989). This value corresponds well with the amount of RNase III purified from cells by Robertson *et al.* (1968) and Dunn (1976). If RNase III levels are limiting, as suggested by these numbers, then an increase in the intracellular concentration of competitor RNAs would lead to uncut *rnc*-leader mRNA and a subsequent increase in RNase III and Era expression. The major substrate for RNase III is ribosomal RNA; in rich medium, ribosomal RNA levels can increase to 85% of total RNA synthesized (Bremer and Dennis, 1987).

Shifting cultures from poor to rich media would increase rRNA synthesis. This rapid increase in precursor rRNA would be expected to titrate RNase III, resulting in an increased concentration of uncut *rnc* mRNA and an increase in RNase III production. Thus, RNase III levels would be expected to be regulated indirectly by growth-rate control systems to ensure that rRNA precursors are efficiently processed.

Rapidly growing cultures are expected to have faster rates of RNase III synthesis and higher levels of the protein than poorly growing cultures. This does not necessarily mean that, at these rapid growth rates and higher RNase III concentrations, RNase III will be equally available for processing all RNAs. 'In fact, RNase III sites are different from one another and are processed at different rates dependent upon RNase III concentration (Dunn, 1976). Thus, the relative concentration of RNase III and the number and type of its substrate sites present are expected to have significant effects on the processing at individual sites. This type of regulatory control is similar to that of RNA polymerase and its interaction with promoters of different strengths. Each promoter competes for a limited but autoregulated supply of polymerase (Bremer and Dennis, 1987). It will be interesting to understand the effects of cell growth and cellular conditions on RNase III expression; however, such studies have not been reported as yet.

B. Post-Translational Control of RNase III Activity

In addition to the autoregulatory control of RNase III synthesis, there is post-translational control of RNase III activity. Mayer and Schweiger (1983) demonstrated that following T7 infection, RNase III activity was stimulated fourfold by protein phosphorylation. This effect was seen within 10 min after infection and was dependent upon the serine/threonine-specific protein kinase encoded by the T7 gene *0.7* (Fig.1), an early T7 function (Rahmsdorf *et al.*, 1974; Mayer and Schweiger, 1983). Modified RNase III contained phosphorylated serine residues.

The T7 serine/threonine-specific kinase also phosphorylates six other proteins in *E. coli*, including the β subunit of RNA polymerase (Zillig *et al.*, 1975; Robertson and Nicholson, 1990). The newly phosphorylated proteins have the same chemical half-life as their unphosphorylated derivatives (Robertson and Nicholson, 1990). The proteins phosphorylated by the T7 kinase are also phosphorylated in uninfected cells but to a much lesser extent, indicating that *E. coli* may contain a kinase activity analogous to that of T7 (Robertson and Nicholson, 1990). The T7 gene *0.7* has recently been cloned and expressed in *E. coli* without toxic effects. When expressed from this plasmid, the kinase phosphorylates the same six proteins as those phosphorylated during phage infection, and the kinase protein is also visible on gels as a phosphorylated protein (Micalewicz and Nicholson, 1992).

V. Ribosomal RNA Processing

Each of the seven ribosomal RNA operons in *E. coli* contains one copy of the 16S, 23S, and 5S rRNA coding sequences along with spacer tRNA genes. In each rRNA operon of *E. coli* and other prokaryotes, transcription of the entire rRNA region occurs as one unit to generate a single large precursor (30S) molecule of about 7000 nucleotides from which the smaller 16S, 23S, and 5S RNA and tRNA molecules are processed (Nikolaev *et al.*, 1973; Dunn and Studier, 1973b). The 30S precursor molecule is present in mutants defective for RNase III, whereas this precursor is not detectable in wild-type cells (Nikolaev *et al.*, 1973; Dunn and Studier, 1973b). For a more complete description of rRNA processing and the roles of RNase III and other nucleases, see reviews by Gegenheimer and Apirion (1981) and King *et al.* (1986).

A. The 16S Ribosomal RNA

RNase III is the first processing enzyme known to act on the 30S rRNA precursor. Processing by RNase III is rapid and occurs during transcription of the rRNA operon (Hoffman and Miller, 1977). RNA polymerase traverses an entire ribosomal RNA operon in ~130 sec, moving at 42 nucleotides per second (Morgan *et al.*, 1978; Gotta *et al.*, 1991). Using these kinetic data, one can estimate that RNase III is able to recognize and cleave the 16S rRNA segment of the precursor transcript within 4 sec from the time that its RNA substrate becomes available during transcription. The speed with which RNase III is able to find and cleave its rRNA substrate is consistent with the short time required for RNase III to process a specific pulse-labeled mRNA substrate (Schmeissner *et al.*, 1984a). The time is also approximately the same (2 sec) as that between consecutive RNA polymerase initiations at an rRNA operon promoter in cells that are doubling every 25 min (Gotta *et al.*, 1991). Thus, rRNA precursor synthesis and RNase III processing of this transcript occur with essentially the same kinetics.

Processing of the 30S ribosomal RNA precursor occurs in stages by the action of several enzymes acting endo- and exonucleolytically (Gegenheimer and Apirion, 1981; King *et al.*, 1986). As the transcript encoding the 16S rRNA segment is completed, it contains at its 3′ end a sequence that has nearly perfect complementarity to the leader segment of the transcript (Young and Steitz, 1978). These two regions pair in a long stem structure, looping out the entire 16S sequence. RNase III cleaves across both strands of the stem, releasing a pre-16S rRNA from the still-to-be-transcribed tRNA, 23S, and 5S segments (Bram *et al.*, 1980). The pre-16S rRNA is further processed by other nucleases to yield the mature 16S species (King and Schlessinger, 1983).

In the absence of RNase III, the pre-16S stem structure is not rapidly processed and remains attached to the distal tRNA. In time, however, processing by other nucleases occurs, generating the normal 5′ and 3′ mature ends of 16S RNA, independent of RNase III processing (King and Schlessinger, 1983; Srivastava and Schlessinger, 1989a). Furthermore, the distal side of the stem structure necessary for RNase III processing may be completely deleted without interfering with 16S maturation (King *et al.*, 1986; Srivastava and Schlessinger, 1989b). Thus, neither RNase III processing nor the RNase III substrate is required to generate a mature 16S rRNA molecule.

If RNase III processing is not essential for 16S rRNA formation, then why is the processing site evolutionarily conserved (Woese, 1987)? It is thought that, although not essential for final maturation, RNase III processing reduces the time for maturation of 16S rRNA (Gegenheimer and Apirion, 1981). It is also possible that the interaction of RNase III with pre-16S rRNA may have an additional and more direct role in assembly of the 30S ribosomal subunit (Nashimoto and Uchida, 1985; King *et al.*, 1986). *In vitro* ribosomal assembly requires a higher activation energy to reconstitute 30S particles from mature 16S RNA (Traub and Nomura, 1969) than from precursor-RNA (Mangiarotti *et al.*, 1975). Assembly of the 30S subunit begins during transcription of the precursor (Cowgill de Narvaez and Schaup, 1979), a time during which RNase III processing is taking place. The *rpsL118* mutation causes a temperature-sensitive defect in an early step of assembly of the small ribosomal subunit (Nashimoto *et al.*, 1985). Second-site suppressors of this temperature-sensitive defect in ribosomal protein S12 were found either in the S4 ribosomal protein gene *rpsD* or in the *rnc* gene (Nashimoto *et al.* 1985; Nashimoto and Uchida, 1985). This RNase III suppressor mutant (Rev3), which retains processing activity, may suppress the S12 defect through a physical interaction with the 16S rRNA precursor. The *rev3* suppressor mutation of *rnc* is recessive to *rnc*$^{+}$. The *rev3* change occurs in a segment of RNase III that is found to be conserved in several other proteins that bind double-strand RNA (D. Court, unpublished; see Section VIII). Thus, a genetic link suggests that although RNase III may not be essential for 30S subunit assembly, since *rnc* null mutants are viable, RNase III may modulate the rate or affect the outcome of the assembly process. It is also clear that RNase III is not absolutely required for assembly *in vitro;* however, assembly may be stimulated by RNase III, and this could be tested directly during *in vitro* assembly.

B. The 23S Ribosomal RNA

The transcript segment that contains the 23S ribosomal RNA also is looped out by a stem structure that is formed by complementary sequences 5′ and 3′ to the 23S region. As in the case of 16S, this stem is also cleaved by RNase III, releasing the pre-23S segment (Bram *et al.*, 1980). The pre-23S

RNA is then processed by other nucleases to generate its mature 3′ ends (Sirdeshmukh and Schlessinger, 1985a, 1985b). This final maturation step occurs only after the 70S ribosome has become part of a polysome following translational initiation on mRNA. Thus, there are always two types of 70S ribosomes involved in translation: mature ribosomes and new, precursor ribosomes that are still undergoing processing by ribonucleases. It would be interesting to determine whether the same ribonucleases that complete maturation of the 23S RNA also take part in the degradation of the mRNA.

In a mutant lacking RNase III, mature 23S rRNA is never made. Instead, an abnormal RNA species is formed that retains the double-strand stem structure of the RNase III site. In fact, these pre-23S RNAs are packaged into 50S ribosomal particles that can be isolated and later processed with RNase III *in vitro* (King *et al.*, 1984; Clark and Lake, 1984; Srivastava and Schlessinger, 1988). 50S subunits of the ribosome made in RNase III mutant cells are capable of translation (King *et al.*, 1984). However, the slightly reduced growth rate of these cells may indicate that the ribosomes are not fully active. In fact, the pleiotropic effect on 10% of all *E. coli* proteins caused by the *rnc* mutations (Gitelman and Aprion, 1980; Takata *et al.*, 1987) could, in part, be attributed to these abnormal ribosomes and not to a direct RNase III defect in mRNA processing.

C. RNase III Removes Intervening Sequences

Some bacterial species, including *Salmonella typhimurium*, contain evolutionarily conserved 23S rRNA genes with intervening sequences. In *S. typhimurium*, several intervening RNA sequences are excised by RNase III during maturation of the pre-23S RNA species (Burgin *et al.*, 1990). The resulting "23S molecule" is not rejoined as one molecule, but is assembled as fragments into the 50S ribosomal subunit.

VI. Control of Gene Expression by RNase III

The level of gene expression is primarily determined by three factors: transcriptional rates, message stability, and efficiency of translation. The factors that influence transcription have been the object of intense study for decades and much is known about them. In contrast, the factors that regulate translation and affect mRNA stability are still poorly understood, even though they profoundly influence the synthesis rate of many proteins (Gold, 1988; Belasco and Higgins, 1988). Rates of message decay differ widely; in *E. coli*, message half-lives vary over at least a 50-fold range from seconds to 25 min (Blundell *et al.*, 1972; Pedersen *et al.*, 1978; Nilsson *et al.*, 1984) and are affected in a variety of host mutants (Ono and Kuwano, 1979; Donovan and Kushner, 1986). It appears that the decay of a transcript is often controlled by a rate-limiting initial cleavage step, which is followed by rapid degradation (Schmeissner *et al.*, 1984a; Belasco *et al.*, 1985; Cannistraro *et al.*, 1986; Portier *et al.*, 1987; Melefors and von Gabain, 1988; Srivastava *et al.*, 1992). In several cases, RNase III is implicated in the rate-limiting cleavage step that leads to mRNA decay. In addition to triggering decay of specific mRNAs, RNase III affects translation in more direct ways by changing the structure of mRNA and affecting its ability to initiate translation. Below, several bacterial and phage genes whose expression is affected by RNase III are discussed with respect to the way in which control of expression is exerted.

A. Inhibition of Gene Expression by RNase III Processing within the 5′ Noncoding Region

In three operons in *E. coli* [P1–*rpsO*–P2*pnp*; P–*rnc*–*era*–*recO*; and the P1–*metY*–P2*orf15A*–*nusA*–*infB* (Figs. 5 and 6)], primary transcripts are processed by RNase III, leading to a decrease in the stability of mRNA downstream of the processing site. This decreased stability is reflected in a lower mRNA level for the distal genes and reduced synthesis of the proteins encoded by the distal *rnc*, *era*, and *pnp* genes. Protein synthesis directed by the *orf15A* gene is also expected to be reduced, but this has not yet been measured. The effects of RNase III processing in these operon transcripts have been compared in wild-type (*rnc*$^+$) and mutant (*rnc105*) cells. A complete description of the effects of RNase III processing on the transcripts of these three operons has been compiled by Régnier and Grunberg-Manago (1990), and, for that reason, only a brief review will be presented here. In each of the three primary transcripts of the *rnc*, *pnp*, and *metY* operons (Fig. 7A), cleavage by RNase III removes RNA secondary structures that can form in a noncoding segment located 5′ of one or more genes. This processing by RNase III dramatically reduces the half-life of translational units beyond the RNase III cleavage sites in these three transcripts (Portier and Régnier, 1984; Takata *et al.*, 1985, 1987, 1989; Régnier and Portier, 1986; Portier *et al.*, 1987; Bardwell *et al.*, 1989; Régnier and Grunberg-Manago, 1989, 1990) (Table 2). Base-paired structures at the 5′ end of transcripts have been proposed to block 5′-to-3′ processive attack by nucleases, thereby protecting the adjoining downstream mRNA (Emory and Belasco, 1990). Removal of these structures by RNase III may facilitate the entry of such nucleases, which may respond to a new 5′ end and attack the unprotected downstream RNA segment, thereby reducing its half-life (see Table 2, Fig. 8). Models to explain the decay of mRNA from a 5′-to-3′ direction have been discussed (Kennell, 1986; Belasco and Higgins, 1988). Since 5′-to-3′ exonucleases have never been detected in *E. coli*, a special endonuclease, whose identity is still unknown (see Chapter 3), could initiate attack at or near the 5′ end and proceed toward the 3′ end, producing RNA fragments that could be rapidly

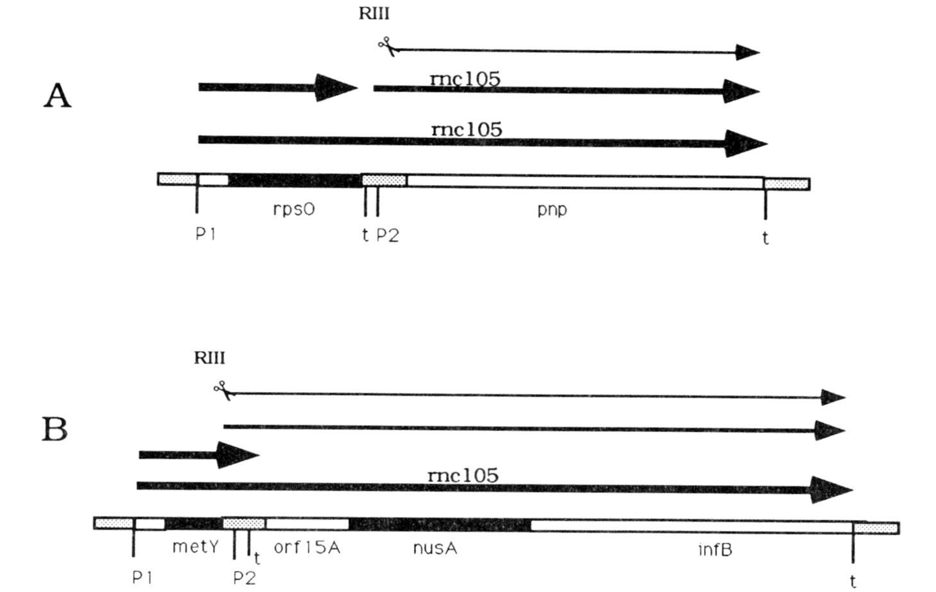

FIGURE 6 The *rpsO–pnp* and *metY–orf15A–nusA–infB* operons. The *rpsO–pnp* (A) and *metY–orf15A–nusA–infB* (B) operons are shown. The open and closed rectangles represent the genes in each operon. *rpsO* encodes the S15 protein of the 30S ribosomal subunit, *pnp* encodes the polynucleotide phosphorylase that degrades mRNA from the 3′-to-5′ direction, *metY* encodes fmet-tRNA, *orf15A* is an uncharacterized gene, *nusA* encodes a protein that complexes with RNA polymerase during transcription elongation, and *infB* encodes translation initiation factor 2. Promoters (P1 and P2) and terminators (t) are indicated. Transcripts are shown as arrows whose width indicates the level of RNA accumulation. RIII with scissors indicates the transcripts that have been cut by RNase III and the position of the cutting site relative to the primary transcripts shown below the processed transcript. Primary transcripts indicated by *rnc105* are found only in RNase III-defective strains. Note that in the *rpsO–pnp* operon, most transcription from promoter P1 terminated before *pnp*; however, some transcription extends through the terminator into *pnp* and, like the P1 transcript, is processed by RNase III. The P1 read-through transcript is also processed by RNaseE as shown in Fig. 7A. In the *metY* operon, transcription from P1 that extends through the terminator is processed by RNase III; P2 transcripts are not processed.

reduced to mononucleotides by the existing 3′-to-5′ processive exonucleases, polynucleotide phosphorylase and ribonuclease II (Kennell, 1986; Donovan and Kushner, 1986). The presence of RNA species containing 5′ ends within the coding sequence has been observed for all three operons. The fact that the production of these apparent degradation intermediates is dependent upon RNase III activity (Bardwell *et al.*, 1989; Régnier and Grunberg-Manago, 1989; Takata *et al.*, 1992) supports the idea that this enzyme initiates the processive decay.

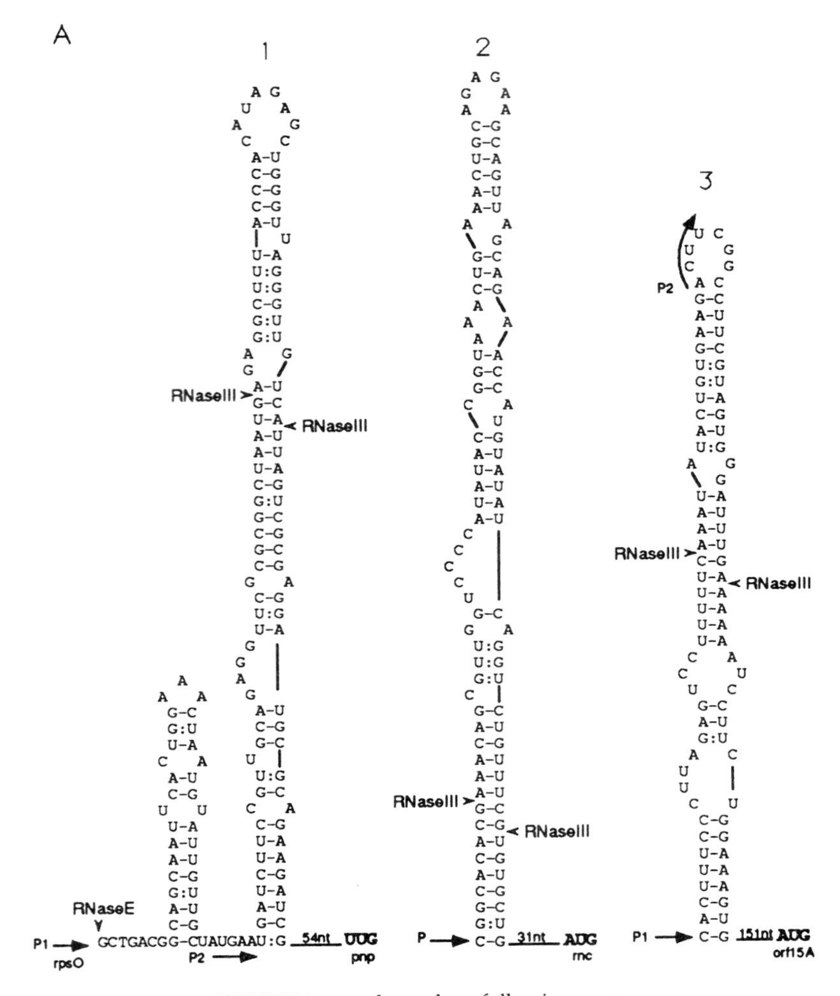

FIGURE 7 see legend on following page.

As proposed here and by other laboratories, the idea that RNase III processing directly initiates the rapid decay of downstream mRNA, thereby reducing mRNA levels and translational products, is well supported by experiments. However, there remains the possibility that an indirect mechanism is responsible for the altered rates of mRNA decay and protein synthesis. Since translation itself may affect the stability of RNA (Kennell, 1986), it can be argued that RNase III reduces the frequency of translational initiation and indirectly increases mRNA turnover. If RNase III processing of the 5′ stems reduces translational initiation, then the stem must have a positive effect on initiation via either RNA structure or positive factors that interact

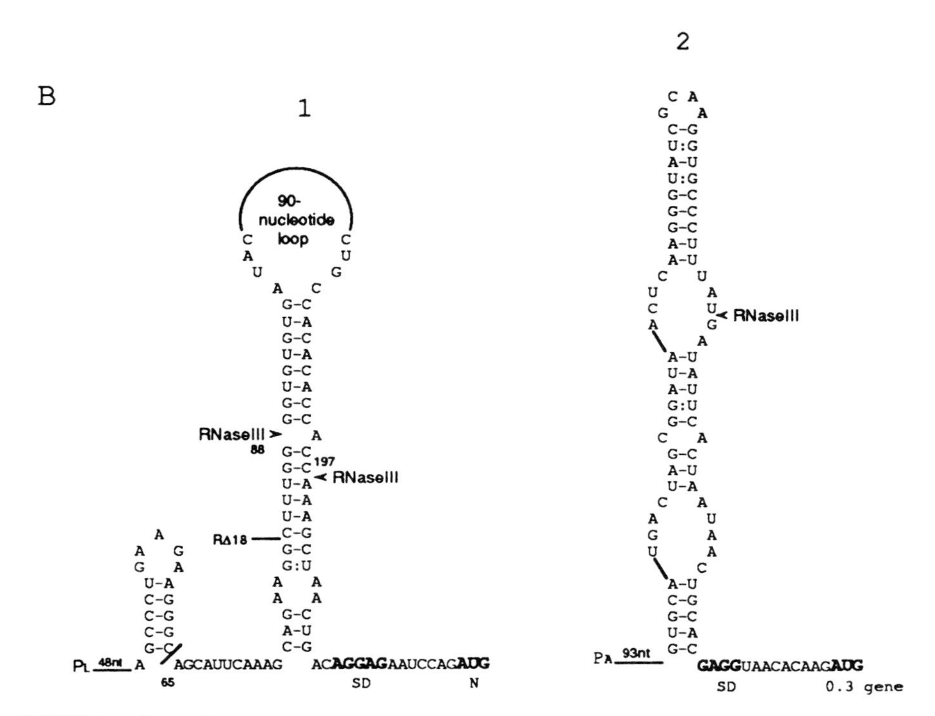

FIGURE 7 Secondary structures within 5′ noncoding leader RNAs processed by RNase III. RNase III cleavage sites are indicated in each structure, as is an RNaseE cleavage site within the *rpsO–pnp* transcript. Promoters are indicated (P), and the translation-initiation codons are in boldface. A-1 represents the *rpsO–pnp* transcript, A-2 represents the *rnc* transcript, and A-3 represents the *metY–orf15A–nusA* transcript. B-1 represents the λ *N*-leader RNA transcript, and B-2 represents the T7 *0.3* gene leader transcript. In B-1, RΔ18 indicates the endpoint of a deletion mutation that starts at base 65 (slash mark). Distances in nucleotides are indicated.

with it to influence downstream translation. Régnier and Grunberg-Manago (1990) pointed out that there would be no obvious changes in secondary structure at the initiation codon caused by processing of these transcripts and, in fact, that some of the processed transcripts appear to be translated more effectively than the unprocessed transcripts. Although this particular model for translation-dependent mRNA degradation does not explain all the effects of RNase III processing in these operon transcripts, further experiments will be required before a definite conclusion can be reached.

From analysis of *metY–orf15A–nusA–infB* transcripts, elimination of the stable stem structure between *metY* and *orf15A* decreases the stability of the *orf15A–nusA–infB* segment. There are two promoters (P1 and P2) for this operon, with P2 located between *metY* and *orf15A*. Transcripts from P2 do not include the stem structure contained in P1 transcripts. Correspondingly,

TABLE 2 Stability of Transcripts in *rnc*$^+$ and *rnc105* Hosts

Operon	Promoter	Stability (min)	
		rnc105	*rnc*$^+$
pnp	P2	15	0.8[a]
rnc era recO	Prnc	5	0.8[a]
metY orf15A nusA infB	P1	15	0.8[a]
	P2	2	2.0[b]

[a] RNaseIII processes these transcripts by removing a 5′ stem–loop structure (see Figs. 7A and 8).
[b] Not processed by RNaseIII.

P1 transcripts are sevenfold more stable than P2 transcripts in *rnc*$^-$ cells (nonprocessing conditions; Table 2). Thus, the stem structure could stabilize downstream transcripts, however, the P1 transcript also includes the *metY* gene encoding f–met–tRNA, whose structure might also affect downstream transcript stability.

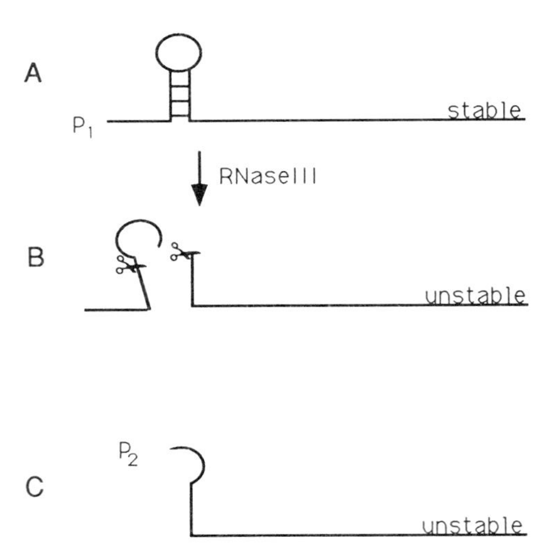

FIGURE 8 5′ RNA stems stabilize distal mRNA. Transcripts specified by the *rnc*, *rpsO–pnp*, and *metY–orf15–nusA* operons are shown. (A) The primary transcript with the structure at the 5′ end is stable. (B) When RNase III is present as indicated by scissors, the primary transcript is processed and the 5′ structure is released, causing the transcript to be unstable. (C) Primary transcripts initiated at the downstream promoter P2 in the *metY–orf15A–nusA* operon (Fig. 7A-3) do not include the proposed structure and are also unstable.

In *rnc*$^+$ cells, the P2 transcripts are unprocessed and their stability is unaffected by the presence of RNase III. However, the P1 transcripts are all processed by RNase III to generate *orf15A* transcripts containing two new 5′ ends. One transcript begins at the cleavage site on the distal side of the stem, while the other begins at the upstream cleavage site (Fig. 7A-3). The latter transcript is present because processing at the distal site is not 100% efficient. Both processed transcripts are 20- to 30-fold less stable than the primary P1 transcript (Table 2, compare *rnc105* to *rnc*$^+$), and they are also several-fold less stable than the P2 unprocessed transcripts (Table 2), even though the 5′ end of the P2 transcript is located between the two ends of the processed transcripts. The more rapid decay of the processed transcripts could be due to differences between them and the P2 transcript. First, the transcripts have different sequences that could explain the effect. Second, if RNase III exists in the cell as a complex with other nucleases, RNase III may recruit these nucleases to the processed end (Srivastava *et al.*, 1990; Miczak *et al.*, 1991). Third, as suggested by Régnier and Grunberg-Manago (1990), the difference in decay rates could result simply because the primary transcripts have a triphosphate on their 5′ end, while the processed transcripts have a monophosphate at this end. In this respect, Lin-Chao and Cohen (1991) have noticed that RNaseE cleavage of an RNA (RNA-I) responsible for ColE1 replication control generates a 5′-monophosphorylated product that is markedly less stable than an otherwise identical primary transcript with a triphosphate at the 5′ end.

Questions need to be answered with respect to the differential stabilities of the transcripts containing *orf15A* (Table 2; Fig. 8). First, is the presence and absence of the stem the key factor determining the stability of the P1 and P2 transcripts? Second, what is the deciding difference between primary P2 transcripts and processed P1 transcripts that affects their relative stability? The first question could be examined by inserting a promoter just upstream of the region encoding the stem structure and examining the stability of its mRNA in the absence of RNase III processing. The second question requires that both 5′ triphosphate and monophosphate ends be generated on transcripts with the same primary nucleotide sequence as was accomplished by Lin-Chao and Cohen (1991) for RNA-I. This could be accomplished by creating another new promoter that initiates transcription at the position of the upstream RNase III cleavage site.

B. Activation of Gene Expression by RNase III Processing within the 5′ Noncoding Region

Unlike expression of the bacterial *rnc* and *pnp* genes described in Section VI.A, the expression of the *N* gene of phage λ and the *0.3* gene of phage T7 is activated by RNase III processing within the 5′ noncoding region of their respective transcripts (Dunn and Studier, 1975; Studier *et al.*, 1979;

Kameyama *et al.*, 1991). The leader RNAs for both the *N* and *0.3* genes contain secondary structures, which are substrates for RNase III. However, unlike the stem–loop structures of the bacterial transcripts, which are greater than 30 nucleotides upstream from the ribosome binding site, those of the two phage mRNAs abut the Shine–Dalgarno sequence (compare Fig. 7A with Fig. 7B) and could interfere with ribosome entry.

1. *Activation of* λ N *Gene Expression*

The *N* gene of phage λ is the first gene transcribed from the major leftward P_L promoter following phage infection or prophage induction. The *N* gene product positively regulates the transcription of most other λ genes during phage development (Friedman and Gottesman, 1983). The AUG initiation codon of *N* is located 223 nucleotides from the start of transcription (Fig. 7B-1). This long *N*-leader RNA forms a secondary structure sensitive to endonucleolytic processing by RNase III (Lozeron *et al.*, 1976, 1977). Two specific sites of cleavage, one between nucleotides 88 and 89 and the other between nucleotides 197 and 198 of the leader, were identified both *in vivo* (Lozeron *et al.*, 1976) and *in vitro* (Steege *et al.*, 1987). These two sites are located two base pairs from each other on each side of an RNA stem structure, as is typical of most RNase III sites (Fig. 7B-1).

Long leader RNAs like that of the *N* gene often are involved in modulating the expression of the downstream gene by responding to changes in external stimuli (Landick and Yanofsky, 1987; Bardwell *et al.*, 1989). In fact, RNase III processing of the *N* leader in wild-type (rnc^+) cells enhances *N* expression threefold compared with *N* expression in cells defective for RNase III (*rnc105*) (Kameyama *et al.*, 1991). Processing at these sites has little if any effect on *N* mRNA levels (Kameyama *et al.*, 1991), and Anevski and Lozeron (1981) showed that processing in the *N*-leader region does not affect the stability of the *N* mRNA. Thus, the stimulation by RNase III is attributed to a change in the efficiency of *N* translational initiation.

The stem structure cut by RNase III includes sequences adjacent to the *N* Shine–Dalgarno sequence (Fig. 7B), which by proximity may inhibit ribosome binding. Processing at positions 88 and 197 would destabilize the lower stem structure and open the ribosome binding site. This mechanism is consistent with the observation that deletion RΔ18 (Fig. 7B-1), which prevents base pairing near the Shine–Dalgarno element, causes an increase in N synthesis in the absence of RNase III. Furthermore, *N* expression in the deletion mutant is not further stimulated by RNase III, suggesting that opening of the *N* ribosome binding site is responsible for the stimulation.

2. *Activation of T7* 0.3 *Gene Expression*

Like *N*, the *0.3* gene of T7 is the first gene to be transcribed following T7 infection, and it functions to relieve host DNA restriction (Dunn and

Studier, 1973a). It is part of the seven-kilobase precursor transcript made by the host RNA polymerase that is processed at five sites by RNase III (Dunn and Studier, 1973a, 1973b; see section II.B). The leader RNAs of *0.3* are at least 175 nucleotides long and, like the *N* leader, contain a stem structure sensitive to RNase III endonuclease. This structure is cleaved only on the distal side of the stem (Fig. 7B-2).

The processed and unprocessed transcripts of all early T7 genes are very stable. Thus, mRNA levels are not affected by processing. Instead, it appears that the translational initiation rate of the *0.3* RNA transcript is enhanced by processing of the mRNA. A processed *0.3* transcript binds ribosomes *in vitro* much better than the unprocessed transcript. *In vivo,* this is reflected in a fivefold activation of *0.3* RNA translation, dependent upon RNase III processing (Dunn and Studier, 1975; Steitz and Bryan, 1977; Studier *et al.*, 1979). The *0.4* message, which is immediately downstream of *0.3*, is contained on the processed transcripts, but its translation is unaffected *in vitro* or *in vivo* by the presence of RNase III (Studier *et al.*, 1979). Like *0.4*, expression of the other early T7 genes transcribed by the host RNA polymerase is unaffected by processing, despite the fact that some have ribosome binding sites adjacent to RNase III-processed stems. One potential explanation for why only *0.3* expression is inhibited in the unprocessed precursor RNA is that the other T7 early genes are translationally coupled to the expression of *0.4*, which has an open ribosome binding site. Ribosomes initiating at *0.4* may be able to translate through the next downstream stem structure to allow translation of the distal message. In a similar manner, translation of the remaining messages on the early transcript can be activated by this cascade of translation. Gene *0.3* is the first gene of the precursor RNA and can only be opened to translation by processing the inhibitory 5′ stem (Studier *et al.*, 1979). If this model is correct, then one would expect that, in RNase III-defective strains, translational coupling would be obligatory for full expression. Therefore, nonsense mutations, for example, in gene *0.4*, would be predicted to be translationally polar but only in an *rnc* mutant host.

Following infection by T7 or λ, *0.3* and *N* are the first phage genes expressed. These genes are ideal for amplifying and responding to external stimuli. Since RNase III has the ability to modulate expression of *0.3* and *N*, the phages may use RNase III activity in the cell to sense and respond to cell conditions. Of course, this is meaningful only if the level or activity of RNase III is altered in different growth conditions, e.g., if RNase III acts as a global control factor. Since *N* is the major early regulatory gene of λ, modulation of its expression would affect developmental pathways used by the phage. An example of another cellular mechanism used to negatively regulate N levels is the Lon protease, a component of the cellular heat-shock system (Gottesman *et al.*, 1981).

C. A Comparison of Negative and Positive Effects of Processing by RNase III in 5′ Noncoding Regions

In the examples described in Section VI (A and B), negative and positive effects on expression by RNase III processing within the 5′ noncoding region of operon transcripts are differentiated in two ways. For *rnc* and *pnp*, mRNA stability is altered by removing a 5′ structural domain in the RNA leader; for *N* and *0.3*, mRNA stability is unaffected by removal of a similar stem, but steric hindrance by the base-paired stem on translational initiation of *N* and *0.3* apparently is relieved. This steric effect is not relevant for the *rnc* and *pnp* genes, because the stem is located far upstream of the Shine–Dalgarno element. Another difference between the two sets of mRNAs is that *N* and *0.3* contain strong translational initiation signals, whereas *rnc* has a very weak Shine–Dalgarno sequence, and *pnp* uses a weak UUG start codon. Thus, protection of the *rnc* and *pnp* initiation regions by 5′ structures is likely to be more important for these mRNAs than for *N* and *0.3*. If ribosome binding limits the rate of nuclease entry into the protein-coding region of mRNA from the 5′ end, then, after processing, efficient ribosome initiation may be necessary to compete effectively against nuclease progression into the coding sequences.

D. Inhibition of Gene Expression by RNase III Processing within the 3′ Noncoding Region

Expression of four bacteriophage genes, *N* and *int* of phage λ and *1.1* and *1.2* of phage T7, is reduced by RNase III processing events within the 3′ noncoding regions of their mRNAs. The mechanism of inhibition is different for the λ and the T7 genes. For λ, processing leads to rapid decay of the *N* and *int* transcripts (Wilder and Lozeron, 1979; Guarneros *et al.*, 1982), whereas for T7, processing does not affect mRNA decay but changes RNA pairing properties of the mRNA, leading to a direct inhibition of the coordinated translation of upstream genes *1.1* and *1.2* (Saito and Richardson, 1981; see review by Gottesman *et al.*, 1982).

1. *The λ* int *and* N *Genes*

At different stages of the λ life cycle, the *int* gene is transcribed from two different promoters, P_L and P_I, but only the P_I transcript is translated. This differential expression is dependent on RNase III processing and has been reviewed in detail (Echols and Guarneros, 1983; Guarneros, 1988). RNA polymerase transcribing from P_I traverses the *int* gene and terminates at a Rho-independent terminator called T_I (Schmeissner *et al.*, 1984b), whereas the transcriptional complex from P_L is altered by the N transcriptional antiterminator function and transcribes through *int* and T_I (Schmeissner *et al.*, 1984a). The stem–loop structures formed downstream of *int* are different in the two transcripts (Fig. 9). It is this difference that is distinguished by RNase III. RNase III cleaves the read-through transcript but does not cut the T_I-terminated transcript (Fig. 10). Processing occurs in the stem–loop structure called *sib*. This processing of *sib* by RNase III prevents *int* expression from the P_L transcript (Schmeissner *et al.*, 1984a). Mutations in the *sib* site that prevent processing by RNase III increase expression of *int* from the P_L transcript (Guarneros *et al.*, 1982; Montañez *et al.*, 1986). Cells that are defective in RNase III do not process *sib* and also express *int* from P_L (Belfort, 1980; Guarneros *et al.*, 1982). Thus, there is a direct correlation between processing of the P_L *sib* transcript and reduction of Int synthesis. The reduction of Int synthesis resulting from the RNase III cleavage event is caused by a subsequent decay of *int* mRNA from the newly processed 3′ end (Guarneros *et al.*, 1982; Plunkett and Echols, 1989). This decay appears to be mediated primarily by the exonuclease, polynucleotide phosphorylase (Guarneros *et al.*, 1988) (Fig. 10).

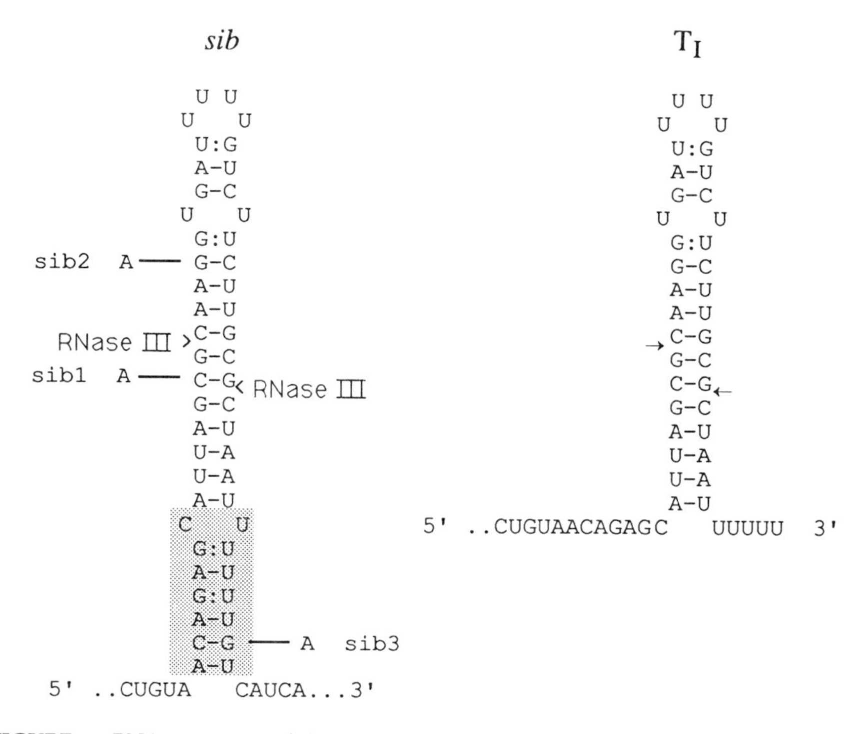

FIGURE 9 RNA structures of the cis-acting *sib* regulator and the I_I terminator of phage λ. The sequence and structure are identical in the upper stem region. The base-paired area where *sib3* is located is the additional structure of *sib* that causes it to be processed by RNase III. The *sib* mutations that prevent processing are indicated. The carets indicate the sites in *sib* where processing occurs. The small arrows represent the weak processing that occurs in the T_I structure when excess RNase III is added.

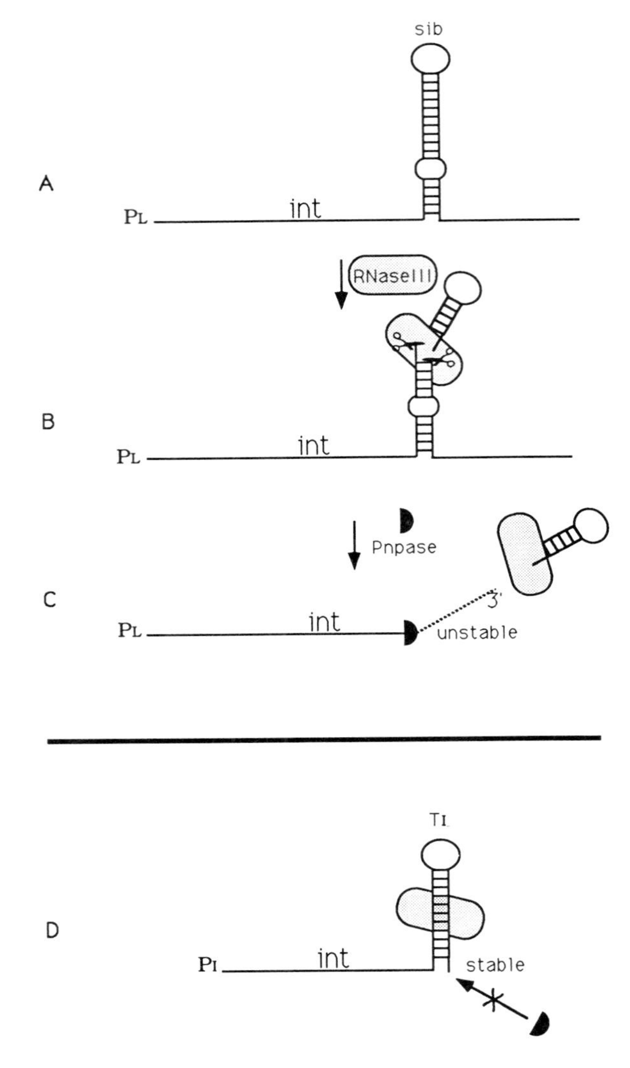

FIGURE 10 *int* retroregulation. Transcription of *int* from the P_L promoter (A, B, C) and P_I promoter (D) yields mRNAs of different stabilities. The P_I transcript terminates at T_I. The T_I structure may be bound by RNase III, but it is resistant to RNase III processing and to exonucleolytic attack by polynucleotide phosphorylase (Pnpase). The P_L transcript extends through the terminator, whereupon a different structure is formed (sib) that is sensitive to RNase III and can then be degraded by Pnpase from the cut 3′ end. RNase III is indicated by the shaded ellipse and the scissors. Pnpase is indicated by the closed half circles, and mRNA decay is indicated by the dashed line.

Since *N* mRNA decay is also caused by RNase III processing beyond the *N* coding region (Wilder and Lozeron, 1979; Hyman and Honigman, 1986), the production of N protein, like that of Int, may also be reduced. As yet, no studies have examined the effect on N protein synthesis of this *N*-distal RNase III processing event.

Endonucleolytic attack by RNase III and subsequent 3′-to-5′ exonucleolytic decay were termed negative retroregulation because these events inhibit gene expression from a site distal to the gene (Schindler and Echols, 1981; Gottesman *et al.*, 1982). Retroregulation was originally defined for *int* expression in phage λ, but since then, several other examples of retroregulation have been reported. Most of the other examples of retroregulation are generally similar in form to that described for *int*, although they may differ in the method by which the 3′ end is formed. In fact, the 3′ end may be generated by transcriptional termination (Mott *et al.*, 1985) or by processing with another species of endonuclease (Mattheakis *et al.*, 1989; Régnier and Hajnsdorf, 1991). In an example of the latter, RNase E specifically cleaves an RNA site distal to the *rpsO* gene and renders the upstream *rpsO* transcript susceptible to enhanced decay (Régnier and Hajnsdorf, 1991) (Fig. 7A-1).

2. The *T7* 1.1 *and* 1.2 *Genes*

A different form of retroregulation has been described that does not involve mRNA decay but does involve RNase III processing. In T7, the *1.1* and *1.2* messages are processed from the early precursor transcript by RNase III. The processing site (R5, Figs. 1 and 2) between genes *1.2* and *1.3* becomes the 3′ end of the *1.1–1.2* transcript. Of the five T7 early RNase III processing sites, this is the only one that is cut on both sides of the RNA stem structure (Robertson *et al.*, 1977). During T7 infections, about 40% of the R5 sites are cleaved on the proximal side (R5a) of the stem and nearly 100% are cleaved on the distal side (R5b), yielding transcripts of the *1.1* and *1.2* genes with 40% of their 3′ ends at R5a and the other 60% at R5b (Dunn, 1976; Robertson *et al.*, 1977). Saito and Richardson (1981) showed that these two transcripts with different ends are differentially expressed for proteins. The shorter transcript lacking 29 nucleotides at the 3′ end expresses both gene products efficiently, whereas the longer transcript has a much reduced expression for both proteins. This reduction is thought to result from base pairing between this extra 29-nucleotide segment and a complementary sequence that includes the *1.1* ribosome binding site at the 5′ end of the message. When translational initiation of *1.1* is blocked, translation of *1.2* is also blocked, since the two units are translationally coupled (Saito and Richardson, 1981). Thus, expression of genes *1.1* and *1.2* is dependent on the frequency of cleavage at R5a.

Saito and Richardson (1981) demonstrated that certain mutations within the R5 stem either enhanced or reduced cutting at R5a, while other changes reduced cutting at both R5a and R5b. In general, mutations that increased base pairing in the stem enhanced cleavage at R5a and enhanced expression of *1.1* and *1.2*. One exceptional mutation, HS9, reduced stem base pairing and eliminated cleavage at R5a but not R5b (Fig. 2). This mutation prevented expression of genes *1.1* and *1.2*. All other mutations that disrupted pairing in the stem eliminated cleavage at both R5a and R5b. These unprocessed transcripts expressed high levels of 1.1 and 1.2 protein, presumably because the unprocessed transcript is able to form a more stable structure at the R5 site that prevents unfolding and pairing with the 5′ end. Similarly, a strain defective for RNase III (*rnc105*) failed to process R5 and also expressed *1.1* and *1.2* efficiently (Saito and Richardson, 1981). Thus, the level or activity of RNase III in the cell might modulate expression of these two genes; i.e., more RNase III might favor cleavage at R5a and elevate expression, whereas less RNase III might at first decrease R5a cleavage and expression until a point at which processing at both R5a and R5b fails to occur.

E. Activation of Gene Expression by RNase III Processing within the 3′ Noncoding Region

Bacteriophage T7 early mRNAs are very stable, with half-lives of 10 to 20 min. These early RNAs have 3′ ends created by RNase III. Unlike most other RNase III processing sites, which are cut in both strands of the RNA stem, the T7 sites are cut on only the downstream side of the stem structure, leaving a hairpin at the 3′ end. This hairpin has been proposed to be responsible in part for the stability of these mRNAs (Dunn and Studier, 1983).

Panayotatos and Truong (1985) have cloned the T7 RNase III site R4 within a foreign RNA sequence. In this location, the R4 site was appropriately cleaved by RNase III and caused a three- to fourfold increase in the stability of the upstream RNA. RNA levels and gene expression from the upstream processed transcript were also increased several-fold. The RNA beyond the cut site remained labile like the uncut full-length RNA. Thus, processing of the T7-type RNase III sites may enhance mRNA stability upstream.

Questions remain about this particular type of positive control by RNase III. In T7, processing at these sites normally does not greatly enhance gene expression from the T7 transcripts. This may simply reflect the fact that the unprocessed T7 transcript is also very stable under the conditions used. It is clear that other stem structures can have strong stabilizing effects on upstream RNA in other systems (Mott *et al.*, 1985; Newbury *et al.*, 1987; Chen *et al.*, 1988). Another possibility, pursued in more detail in Section VI-F, is that RNase III itself may remain bound at the 3′ hairpin structure following processing and directly prevents further 3′-to-5′ exonucleolytic decay of such messages.

F. RNase III-Mediated Control of Gene Expression without Processing?

RNase III must recognize and bind to RNA before it processes. Data suggest that, at some mRNA sites, binding occurs without processing and that this binding can affect gene expression. Here, two such examples are described.

1. *Positive Control of λ* cIII *Gene Expression by RNase III*

The expression of the *c*III gene of phage λ is dependent on the presence of RNase III (Altuvia *et al.*, 1987). The requirement for RNase III is novel in that processing does not occur, but, rather, RNase III appears to stabilize one of two alternative RNA conformations at the 5′ end of the *c*III mRNA (Altuvia *et al.*, 1989, 1991; Kornitzer *et al.*, 1989) (Fig. 11). The RNA segment included in these structures extends for 10 bases upstream of the AUG

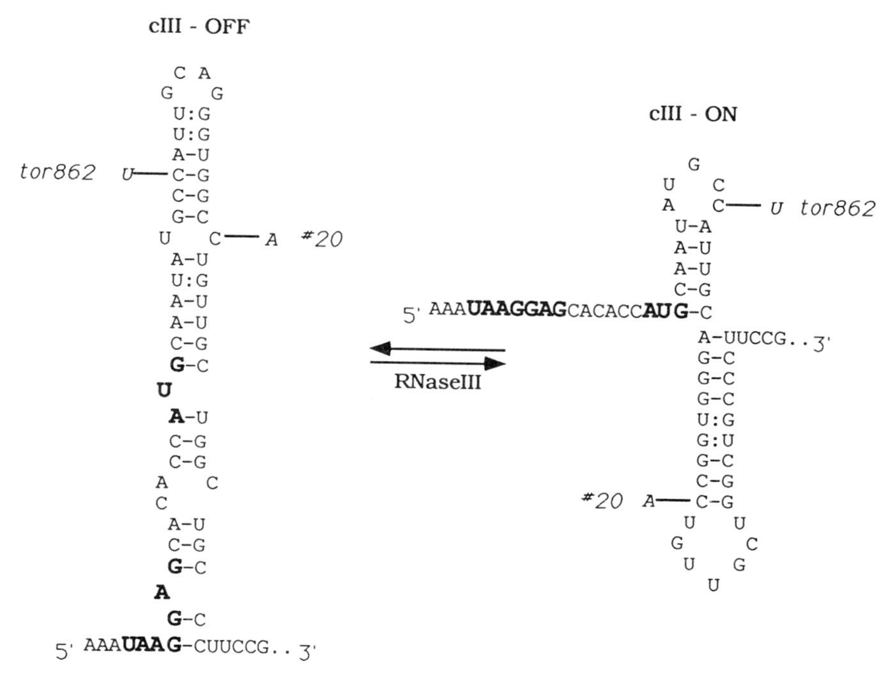

FIGURE 11 Alternative *cIII* mRNA structures. The two structures for the *c*III mRNA exist in equilibrium. The structure on the left contains the Shine–Dalgarno sequence and AUG initiation codon (indicated by the bases in bold) within the base-paired region, and *c*III translation is OFF. The structure on the right has the Shine–Dalgarno and AUG codon open for ribosome binding, and *c*III translation is ON. RNase III may bind and stabilize the ON configuration. Mutation *tor862* shifts the equilibrium to the right and favors *c*III expression, even in the absence of RNase III. Mutation No. 20 shifts the equilibrium to the left and blocks *c*III expression.

codon to 34 bases downstream. One of these conformations prevents translational initiation, while the other allows efficient translation. In the absence of RNase III, the RNA is in an equilibrium between the two structural forms but is primarily in the inhibitory conformation; RNase III, when present, is thought to bind to and stabilize the active conformation. A small percentage of this active form is cleaved *in vitro* by RNase III, but only after prolonged incubation (Altuvia *et al.*, 1991). RNase III exerts no significant effect on *c*III mRNA stability (or levels).

Specific base changes in this *c*III–RNA structure can also shift the equilibrium toward either the active or the inactive form of the RNA (Kornitzer *et al.*, 1989) (Fig. 11). Mutations that shift the equilibrium toward the active form make *c*III expression RNase III independent, whereas mutations that stabilize the inactive form are no longer fully activated by RNase III. The mutations that favor the active form also enhance ribosome binding to the *c*III translational initiation region, as determined by *in vitro* assays. Surprisingly, in this same type of assay, the presence of RNase III does not stimulate ribosome binding to the wild-type RNA (Altuvia *et al.*, 1991). For this reason, there is still concern that the RNase III effect is only indirect *in vivo* and that a second factor, dependent on RNase III, may be required to fully stabilize the active RNA form for enhanced translation of *c*III.

2. *Positive Control of* **int** *Gene Expression by RNase III*

As described in Section VI-D, λ *int* transcription occurs from two independent promoters at different stages of λ development. RNase III processes the early RNA transcripts from P_L at *sib*, revealing a 3′ end sensitive to exonucleolytic decay. This decay inactivates *int* expression and is an example of negative retroregulation by RNase III.

During later stages of λ development, *int* is transcribed from another promoter, P_I. This transcript contains only the *int* gene and terminates at the Rho-independent terminator, T_I. The terminated transcript is not processed by RNase III (Schmeissner *et al.*, 1984b) (Figs. 9 and 10). The terminal stem–loop of the terminated RNA is postulated to protect the P_I–*int* transcript from the same nuclease, polynucleotide phosphorylase, that degrades the RNase III processed P_L–*int*–*sib* transcript. Surprisingly, in the absence of RNase III (*rnc105*), *int* expression is reduced from the terminated P_I–*int*–T_I transcript (Guarneros *et al.*, 1988). The requirement for RNase III to achieve optimal expression of *int* from the terminated transcript can be explained in two ways. RNase III processing inhibits polynucleotide phosphorylase gene expression (Portier *et al.*, 1987; Takata *et al.*, 1987), and thus, in the *rnc105* mutant strain, increased phosphorylase synthesis could lead to the more rapid exonucleolytic decay of the P_I–*int*–T_I transcript. An alternative explanation is that RNase III may be able to bind to the T_I terminator stem–loop RNA without cutting it. RNase III binding to

T_I may protect the 3′ end from exonucleolytic degradation by polynucleotide phosphorylase (Fig. 10). Thus, *int* expression may be positively controlled on this transcript by RNase III binding (Guarneros, 1988). Binding of RNase III to Sp1 RNA of phage T4 has also been suggested as a mechanism to protect that RNA from degradation (Pragai and Apirion, 1981).

In vitro, the T_I terminator structure has been shown to be processed inefficiently by excess RNase III, suggesting that RNase III recognizes the T_I structure (Schmeissner *et al.*, 1984a, 1984b). The two structures, T_I and *sib*, are compared in Fig. 9. The T_I sequence, because it terminates in the run of uridines, cannot form the stable lower–stem structure that *sib* forms (see the shaded part of *sib*). Note that the *sib*3 mutation eliminates a GC base pair in this lower stem and prevents RNase III processing of *sib*. If RNase III binds to T_I, the upper stem alone may be sufficient for binding, and the extended lower stem of *sib* may provide the additional stimulus for RNase III processing. Mutations in the upper stem of *sib* have also been shown to prevent processing. Interestingly, they also reduce expression of Int from the terminated transcript, even in the presence of RNase III (Court *et al.*, 1983b; Montañez *et al.*, 1986). The results described here do not allow us to favor either of the two models that have been proposed to explain the positive effect of RNase III on *int* expression from the P_I transcript. As the *sib* mutations reduce base pairing, the weakened stem may be less resistant to attack by polynucleotide phosphorylase. Alternatively, the mutations may prevent RNase III binding to T_I and thereby cause increased sensitivity to exonuclease attack.

Biochemical as well as genetic studies can address the importance of RNase III binding in the expression of *c*III and *int*. Assays that measure binding of RNase III to the *c*III RNA structure and to the T_I terminator hairpin *in vitro* can test RNase III-binding properties independent of its processing properties. Binding can be tested directly by gel-shift assays with these particular RNAs. Alternatively, binding to the *c*III structure or to the T_I terminator hairpins may be assayed in a competition experiment with a second RNase III-sensitive substrate and limiting RNase III.

The role of RNase III at the T_I terminator hairpin can also be examined with genetic analyses. Strains with the *rnc70* mutation contain an RNase III that lacks processing activity but retains specific RNA-binding activity. Like other RNase III mutants this strain also overproduces polynucleotide phosphorylase (Table 1). By comparing *c*III and *int* expression between strains with the rnc^+, *rnc105*, or *rnc70* allele, one can determine whether the *rnc70* mutation has effects similar to those of rnc^+ or *rnc105*. If it has effects similar to those of rnc^+, then models supporting an effect on gene expression by the binding of RNase III will be strengthened. A finding that the effects of *rnc70* and *rrnc105* are similar will be less informative, since the binding properties of Rnc70 and wild-type RNase III may differ *in vivo*.

G. Involvement of RNase III in Antisense RNA Control

Some genes in bacteria and their phages are transcribed from both their sense and their antisense DNA strands. The antisense RNA molecules often act as negative regulators of translation of the sense RNA by forming duplex RNA and blocking ribosome binding or by reducing mRNA stability (see review by Simons and Kleckner, 1988). RNase III processes duplex RNA very efficiently (Robertson *et al.*, 1968). As yet, few natural antisense/sense RNA duplexes have been tested with respect to RNase III processing. The following RNAs have been tested: RNA*out*/RNA*in*, which regulate expression of the Tn10 transposase (Case *et al.*, 1990); *copA*/*copT*, which control replication function of plasmid R1 (Blomberg *et al.*, 1990); and *oop*/*c*II, which control *c*II expression in phage λ (Krinke and Wulff, 1987). All form antisense/sense RNA duplexes that are processed by RNase III.

The antisense/sense duplex includes the ribosome initiation region of the sense RNA for the Tn10 transposase system. Occlusion of the ribosome binding site by antisense RNA is found to be sufficient by itself to inhibit translational initiation of the sense RNA irrespective of RNase III processing (Case *et al.*, 1990). In this case, RNase III processes the duplex RNA, but the absence of RNase III (in *rnc105* mutants) has little, if any, effect on the expression of the polypeptide encoded by that sense RNA. Thus, RNase III cleavage and destabilization of the duplex RNA act more like a fail-safe mechanism and have little purpose other than ridding the cell of duplex RNA.

Replication of plasmid R1 is dependent on the product of *repA*. An antisense RNA called *copA* inhibits replication by pairing with *copT*, a leader RNA segment of *repA*. This duplex RNA is efficiently processed by RNase III leading to a fourfold decrease in *repA* expression compared with an RNase III, mutant (*rnc105*) control. It is not known whether processing affects *repA* mRNA decay or translational initiation (Blomberg *et al.*, 1990).

The *c*II gene product of phage λ acts during phage development as a switch that favors lysogenic development over lytic development following infection. The expression of this gene is controlled by a myriad of regulatory functions, presumably as a way to allow the phage to "detect" subtle environmental changes and adapt its life cycle accordingly (Wulff and Rosenberg, 1983). One of these controls is at the level of antisense RNA (Krinke and Wulff, 1987). Here, the antisense RNA, *oop*, hybridizes not to the translational initiation region of *c*II but rather to the 3′ terminus of the *c*II mRNA. The *oop*/*c*II duplex RNA formed is sensitive to RNase III attack, which prevents full translation of *c*II. In the absence of RNase III, there is no effect of the antisense RNA on *c*II expression. Processing by RNase III occurs within the 84-base pair *oop*/*c*II RNA duplex such that one of the cuts is within the *c*II coding sequence. The processed *c*II mRNA is then rapidly degraded (Krinke and Wulff, 1990b). Thus, RNase III processing directly inactivates *c*II expression. This inactivation is dependent on the presence of antisense *oop* RNA.

Krinke *et al.* (1991) have demonstrated that two λ phage, identical except for the ability to make antisense *oop* RNA, are indistinguishable from each other in plaque morphology, phage yield per infected cell, and lysogenization frequency. The one distinguishing difference is in phage yield following ultraviolet induction of a lysogen. Wild-type λ has a twofold larger burst size, presumably because *c*II expression is reduced by *oop* control.

H. Miscellaneous RNase III Interactions in *E. coli*

I have discussed bacterial and phage operons in which RNase III processing or binding affects gene expression. There are also examples of operons in which processing occurs, but for which no effect on expression has been observed. There are other cases in which the presence or absence of RNase III in the cell exerts an effect on gene expression, but for which processing has not been demonstrated. The first bacterial operon transcript for which RNase III processing was demonstrated was the *rplJL–rpoBC* complex operon transcript encoding ribosomal proteins and RNA polymerase subunits. Processing occurs between the ribosomal and polymerase message clusters but has essentially no effect on the translation (Barry *et al.*, 1980; Dennis, 1984; Downing and Dennis, 1987) or decay rate (Downing and Dennis, 1991) of either cut transcript. The *secE–nusG* operon transcripts are also processed by RNase III. Processing occurs in the RNA leader of the operon transcript, but no effect on the expression of either gene has been shown (Downing *et al.*, 1990).

Recently, RNase III has been shown to process the transcript of the *dicB* operon (Faubladier *et al.*, 1990). Processing generates an RNA intermediate that is converted by exonucleolytic decay to a 53-nucleotide RNA (*dicR*) that functions to inhibit cell division. Since this RNA can be made independently of RNase III processing via transcriptional termination, the presence or absence of an intact *rnc* gene has little effect upon cell division, at least under the conditions tested (Faubladier *et al.*, 1990).

Expression of another gene that encodes a polypeptide sequence similar to the ferritin H subunit of humans, but whose function in *E. coli* is unknown, is enhanced dramatically in rnc^{+} cells as compared with *rnc105* mutant cells (Izuhara *et al.*, 1991). However, Izuhara *et al.* (1991) have not determined whether RNase III directly interacts with the RNA transcript of this gene or has an indirect effect on its expression.

VII. RNase III of *Schizosaccharomyces pombe* and Other Organisms

Double-strand-specific endonucleases that cleave at specific sites have been isolated from other prokaryotic and eukaryotic organisms (Hall and Crouch, 1977; Panganiban and Whitely, 1983; Bellofatto *et al.*, 1983; Mead and Oliver, 1983; Meegan and Marcus, 1989; Iino *et al.*, 1991). Some of these almost certainly represent homologs of RNase III present in these various species; however, only the gene (*pac1*) for the endoribonuclease from the yeast *S. pombe* has been sequenced and studied genetically (Xu *et al.*, 1990; Iino *et al.*, 1991). The *pac1* gene encodes a protein of 363 amino acid residues in which the C-terminal 230 residues are 25% identical to those of RNase III from *E. coli*. It shares an identical 11-amino acid sequence, HNERLEFLGDS, with RNase III at the site of an essential glycine (G) residue implicated by the *rnc105* mutation (Fig. 4).

The *pac1* gene product was expressed in *E. coli* containing the *rnc105* mutation. Cells containing *pac1* gained a high level of double-strand-specific endonuclease activity that was lacking in the strain without *pac1*, verifying that *pac1*, in addition to having homologies to RNase III, also shares a similar activity. Expression of the *pac1* gene did not have a toxic effect on bacterial growth, nor did it complement the phenotypic defects caused by *rnc105* (Iino *et al.*, 1991). It might be noted here that a clone of the *E. coli rnc* gene was toxic when expressed in *Saccharomyces cerevesiae* (Pines *et al.*, 1988).

Unlike RNase III, which is not essential to the growth of *E. coli*, the *pac1* gene product is essential for vegetative growth of *S. pombe* cells. It can also affect mating and sporulation. The presence of multiple copies of the gene causes sterility in otherwise wild-type yeast cells. The expression of several genes normally required for mating and sporulation was shut off by the presence of extra *pac1* copies, raising the interesting question of whether mRNA processing regulates sexual development in *S. pombe*.

Retroviral reverse transcriptases are multifunctional enzymes. Recently, a double-strand RNase activity was found associated with purified HIV reverse transcriptase (Ben-Artzi *et al.*, 1992) in addition to the known RNA-dependent and DNA-dependent DNA polymerase activities and RNase H activity. This RNase activity copurified with reverse transcriptase and was found in several independent preparations from different laboratories. The enzyme specifically cleaves the HIV RNA–$tRNA^{Lys}$ double-strand complex that primes reverse transcription. Despite the evidence, there is still reason to question this activity of reverse transcriptase. The HIV enzyme studied is always expressed in and purified from *E. coli* strains that contain RNase III. As mentioned in Section II-A, RNase III tends to aggregate and may aggregate with other cellular proteins. Thus, copurification of activities is not a rigorous proof that a single enzyme is responsible.

In fact, RNase III and RNase H of *E. coli* were originally copurified and thought to be one protein with two activities (Robertson *et al.*, 1968; Crouch, 1974). Expression and purification of HIV reverse transcriptase from an *rnc* mutant strain of *E. coli* would address this objection.

VIII. Homology between RNase III and Other Double-Strand RNA-Binding Proteins

By comparing amino acid sequences of RNase III and Pac1, Iino *et al.* (1991) observed that certain segments displayed extensive homologies. When these short homologous sequences were compared individually to protein databases, one was found to have a sequence in common with a group of proteins that include known double-strand RNA-binding proteins (D. Court, unpublished). In RNase III, as well as in Pac1, this common sequence resides at the carboxy terminus of the protein (Figs. 4 and 12).

The double-stranded-RNA-dependent protein kinase of humans (Meurs *et al.*, 1990) contains this conserved segment not once but twice within its protein sequence. A protein from mice identified only as a serine/threonine-specific kinase (Icely *et al.*, 1991) has these same two conserved regions and has 62% overall identity with the human kinase (D. Court, unpublished). The human kinase, which is induced by interferon, acts as an antiviral agent by phosphorylating translational initiation factor-2, thereby interfering with translation of viral mRNA. This inhibition occurs

PROTEIN	SEQUENCE	POSITION	SIZE
Ec-RN3	SEPVVGTGSS RRKAEQAAAE QALK	221	226
Ye-RN3	KEVARAWGAN QKDAGSRAAM QALE	352	363
Va-E3L	RVFDKADGKS KRDAKNNAAK LAVD	180	190
Ro-NS34	SAEAVAKGRS KKEAKRIAAK DILD	398	402
Hu-KIN	REFPEGEGRS KKEAKNAAAK LAVE	73	550
Mu-KIN	KEFGEAKGRS KTEARNAAAK LAVD	72	519
Hu-TAR	DTSCTGQGPS KKAAKHKAAE VALK	72	345
Hu-KIN	KEYSIGTGST KQEAKQLAAK LAYL	163	550
Mu-KIN	TMYGTGSGVT KQEAKQLAAK EAYQ	158	519
Hu-TAR	RFIEIGSGTS KKLAKRNAAA KMLL	203	345

FIGURE 12 Possible protein domain for double-strand RNA binding. Ten sequences (each containing 24 amino acids) are shown from seven proteins that bind double-strand RNA. The proteins are RNase III of *E. coli* (Ec-RN3); RNase III of *S. pombe* (Y-RN3) (Iino *et al.*, 1991); E3L of vaccinia (Va- E3L) (Watson *et al.*, 1991); NS34-like protein of rotavirus (Ro-NS34) (Qian *et al.*, 1991); tar binding protein from human (Hu-TAR) (Gatignol *et al.*, 1991); serine/threonine-specific kinase of humans (Hu-KIN) (Meurs *et al.*, 1990); and serine/threonine-specific kinase of mouse (Mu-KIN) (Icely *et al.*, 1991). Note that the kinases and Hu-TAR each have two sequences represented. The shaded amino acids are conserved in all 10 sequences. The size of each protein in amino acids is indicated, and the terminal amino acid in each of the sequences shown is also indicated by the position in its respective protein.

after the kinase is activated by double-strand RNA produced during viral infections. Viruses themselves contain antikinase activities. Adenovirus produces a small double-strand RNA that actually binds to and inhibits the kinase directly (Mathews and Shenk, 1991), whereas vaccinia produces an RNA-binding protein, E3L, that binds double-strand RNA and prevents activation of the human kinase (Watson *et al.*, 1991). Both E3L and a similar rotavirus protein (Qian *et al.*, 1991) contain the conserved domain near their carboxy termini (see Fig. 2). In addition to these proteins, a protein from HeLa cells that binds to the HIV-1 TAR RNA and activates the HIV-1 long terminal repeat also contains this sequence (Gatignol *et al.*, 1991). In all of these conserved sequences, an α-helical structure in the protein can form. Such structures are also found in certain DNA binding proteins where the protein contacts the DNA.

It is interesting that RNase III, like the eukaryotic serine/threonine-specific kinases, interacts often with viral RNAs. One can imagine that this represents early attempts by bacteria to fight off phage infections (Robertson *et al.*, 1968). If this theory is correct, some of these phages, like their eukaryotic virus counterparts, have overcome the antiviral function, and these phages now use the host function for their own development.

Acknowledgments

I would like to thank S. Brown, N. Costantino, S. Dasgupta, and C. Turnbough for stimulating discussions and constructive criticism of the manuscript. C. Redmond, J. Ratliff, and A. Arthur provided typing and/or editorial assistance in preparing the manuscript. Research was sponsored by the National Cancer Institute, DHHS, under Contract No. NO1-CO-74101 with ABL. The contents of this publication do not necessarily reflect the views or policies of the Department of Health and Human Services, nor does mention of trade names, commercial products, or organizations imply endorsement by the U.S. Government.

References

Ahnn, J., March, P. E., Takiff, H. E., and Inouye, M. (1986). A GTP-binding protein of *Escherichia coli* has homology to yeast RAS proteins. *Proc. Natl. Acad. Sci. USA* **83,** 8849–8853.

Altuvia, S., Locker-Giladi, H., Koby, S., Ben-Nun, O., and Oppenheim, A. B. (1987). RNase III stimulates the translation of the *c*III gene of bacteriophage lambda. *Proc. Natl. Acad. Sci. USA* **84,** 6511–6515.

Altuvia, S., Kornitzer, D., Teff, D., and Oppenheim, A. B. (1989). Alternative mRNA structures of the cIII gene of bacteriophage lambda determine the rate of its translation. *J. Mol. Biol.* **210,** 265–280.

Altuvia, S., Kornitzer, D., Kobi, S., and Oppenheim, A. B. (1991). Functional and structural elements of the mRNA of the cIII gene of bacteriophage lambda. *J. Mol. Biol.* **218,** 723–733.

Anevski, P. J., and Lozeron, H. A. (1981). Multiple pathways of RNA processing and decay for the major leftward N-independent RNA transcript of coliphage λ. *Virology* **113,** 39–53.

Apirion, D., and Watson, N. (1975). Mapping and characterization of a mutation in *Escherichia coli* that reduces the level of ribonuclease III specific for double-stranded ribonucleic acid. *J. Bacteriol.* **124,** 317–324.

Apirion, D., and Gitelman, D. (1980). Decay of RNA in RNA processing mutants of *Escherichia coli. Mol. Gen. Genet.* **177,** 139–154.

Bardwell, J. C. A., Régnier, P., Chen, S. M., Nakamura, Y., Grunberg-Manago, M., and Court, D. (1989). Autoregulation of RNase III operon by mRNA processing. *EMBO, J.* **8,** 3401–3407.

Barry, G., Squires, C., and Squires, C. L. (1980). Attenuation and processing of RNA from the *rplJL - rpoBC* transcription unit of *Escherichia coli. Proc. Natl. Acad. Sci. USA* **77,** 3331–3335.

Belasco, J. G., Beatty, J. T., Adams, C. W., von Gabain, A., Cohen, S. N. (1985). Differential expression of photosynthesis genes in *R. capsulata* results from segmental differences in stability within the polycistronic *rxcA* transcript. *Cell* **40,** 171–181.

Belasco, J. G., and Higgins, C. F. (1988). Mechanisms of mRNA decay in bacteria: A perspective. *Gene* **72,** 15–23.

Belfort, M. (1980). The *c*II-independent expression of the phage lambda *int* gene in RNase III-defective *E. coli. Gene* **11,** 149–155.

Bellofatto, V., Amemiya, K., and Shapiro, L. (1983). Purification and characterization of an RNA processing enzyme from *Caulobacter crescentus. J. Biol. Chem.* **258,** 5467–5476.

Ben-Artzi, H., Zeelon, E., Gorecki, M., and Panet, A. (1992). Double-stranded RNA-dependent RNase activity associated with human immunodeficiency virus type 1 reverse transcriptase. *Proc. Natl. Acad. Sci. USA* **89,** 927–931.

Birenbaum, M., Schlessinger, D., and Hashimoto, S. (1978). RNase III cleavage of *Escherichia coli* rRNA precursors: Fragment release and dependence on salt concentration. *Biochemistry* **17,** 298–307.

Blomberg, P., Wagner, E. G. H., and Nordström, K. (1990). Control of replication of plasmid R1: The duplex between the antisense RNA, CopA, and its target, CopT, is processed specifically *in vivo* and *in vitro* by RNase III. *EMBO J.* **9,** 2331–2340.

Blundell, M., Craig, E., and Kennell, D. (1972). Decay rates of different mRNA in *E. coli* and models of decay. *Nature(London) New Biol* **238,** 46–49.

Bram, R. J., Young, R. A., and Steitz, J. A. (1980). The ribonuclease III site flanking 23S sequences in the 30S ribosomal precursor RNA of *Escherichia coli. Cell* **19,** 393–401.

Bremer, H., and Dennis, P. P. (1987). Modulation of chemical composition and other parameters of the cell by growth rate. *In Escherichia coli* and *Salmonella typhimurium*: Cellular and Molecular Biology (F. C. Neidhardt, J.L. Ingraham, K.B. Low, B. Magasanic, M. Schaechter, and H.E. Umbarger, Eds.), pp. 1527–1542. American Society for Microbiology, Washington, D.C.

Burgin, A. B., Parodos, K., Lane, D. J., and Pace, N. R. (1990). The excision of intervening sequences from *Salmonella* 23S ribosomal RNA. *Cell* **60,** 404–414.

Cannistraro, V. J., Subbarao, M. N., Kennell, D. (1986). Specific endonucleolytic cleavages sites for decay of *Escherichia coli* mRNA. *J. Mol. Biol.* **192,** 257–274.

Case, C. C., Simons, E. L., and Simons, R. W. (1990). The IS10 transposase mRNA is destabilized during antisense RNA control. *EMBO J.* **9,** 1259–1266.

Chelladurai, B. S., Li, H., and Nicholson, A. W. (1991). A conserved sequence element in ribonuclease III processing signals is not required for accurate *in vitro* enzymatic cleavage. *Nucleic Acids Res.* **19,** 1759–1766.

Chen, C. Y. A., Beatty, J. T., Cohen, S. N., and Belasco, J. G. (1988). An intercistronic stem–loop structure functions as an mRNA decay terminator necessary but insufficient for *puf* mRNA stability. *Cell* **52,** 609–619.

Chen, S. M., Takiff, H. E., Dubois, G. G, Barber, A. M., Bardwell, J. C. A., and Court, D. L. (1990). Expression and characterization of RNase III and Era proteins: Products of the *rnc* operon of *Escherichia coli*. *J. Biol. Chem.* **265,** 2888–2895.

Clark, M. W., and Lake, J. A. (1984). Unusual rRNA-linked complex of 50S ribosomal subunits isolated from an *Escherichia coli* RNase III mutant. *J. Bacteriol.* **157,** 971–974.

Court, D., Huang, T. F., and Oppenheim, A. B. (1983a). Deletion analysis of the retroregulatory site for lambda *int* gene. *J. Mol. Biol.* **166,** 233–240.

Court, D., Schmeissner, U., Bear, S., Rosenberg, M., Oppenheim, A. B., Montañez, C., and Guarneros, G. (1983b). Control of λ *int* gene expression by RNA processing. *In* Gene Expression (D. H. Hamer, and M. J. Rosenberg, Eds.), UCLA Symposia on Molecular and Cellular Biology New Series, Vol. 8, pp. 311–326. Alan R. Liss, New York.

Cowgill de Narvaez, C., and Schaup, H. W. (1979). *In vivo* transcriptionally coupled assembly of *Escherichia coli* ribosomal subunits. *J. Mol. Biol.* **134,** 1–22.

Crouch, R. J. (1974). Ribonuclease III does not degrade deoxyribonucleic acid-ribonucleic acid hybrids. *J. Mol. Biol. Chem.* **249,** 1314–1316.

Daniels, D. L., Subbarao, M. N., Blattner, F. R., and Lozeron, H. A. (1988). Q-mediated late gene transcription of bacteriophage λ: RNA start point and RNase III processing sites *in vivo*. *Virology* **167,** 568–577.

Dennis, P. P. (1984). Site specific deletions of regulatory sequences in a ribosomal protein-RNA polymerase operon in *Escherichia coli*: Effects on *β* and *β* gene expression. *J. Biol. Chem.* **259,** 3202–3209.

Donovan, W. P., and Kushner, S. R. (1986). Polynucleotide phosphorylase and ribonuclease II are required for cell viability and mRNA turnover in *Escherichia coli* K12. *Proc. Natl. Acad. Sci. USA* **83,** 120–124.

Downing, W. L., and Dennis, P. P. (1987). Transcription products from the *rplKAJL–rpoBC* gene cluster. *J. Mol. Biol.* **194,** 609–620.

Downing, W. L., Sullivan, S. L., Gottesman, M. E., and Dennis, P. P. (1990). Sequence and transcriptional pattern of the essential *Escherichia coli secE–nusG* operon. *J. Bacteriol* **172,** 1621–1627.

Downing, W., and Dennis, P. P. (1991). RNA polymerase activity may regulate transcription initiation and attenuation in the *rplKAJL–rpoBC* operon of *Escherichia coli*. *J. Biol. Chem.* **266,** 1304–1311.

Dunn, J. J., and Studier, F. W. (1973a). T7 early RNAs are generated by site-specific cleavages. *Proc. Natl. Acad. Sci. USA* **70,** 1559–1563.

Dunn, J. J., and Studier, F. W. (1973b). T7 early RNAs and *Escherichia coli* ribosomal RNAs are cut from large precursor RNAs *in vivo* by ribonuclease III. *Proc. Natl. Acad. Sci. USA* **70,** 3296–3300.

Dunn, J. J., and Studier, F. W. (1975). Effect of RNase III cleavage on translation of bacteriophage T7 messenger RNAs. *J. Mol. Biol.* **99,** 487–499.

Dunn, J. J. (1976). RNase III cleavage of single-stranded RNA: Effect of ionic strength on the fidelity of cleavage. *J. Mol. Biol.* **251,** 3807–3814.

Dunn, J. J., and Studier, F. W. (1983). Complete nucleotide sequence of bacteriophage T7 DNA and the locations of T7 genetic elements. *J. Mol. Biol.* **166,** 477–535.

Echols, H., and Guarneros, G. (1983). Control of integration and excision. *In* Lambda II (R. W. Hendrix, J. W. Roberts, F. W. Stahl, and R. A. Weisberg, Eds.), pp 75–92. Cold Spring Harbor Laboratory, Cold Spring Harbor, New York.

Emory, S. A., and Belasco, J. G. (1990). The *ompA* 5′ untranslated RNA segment functions in *Escherichia coli* as a growth rate regulated mRNA stabilizer whose activity is unrelated to translational efficiency. *J. Bacteriol.* **172,** 4472–4481.

Faubladier, M. Cam, K. Bouché, J. P. (1990). *Escherichia coli* cell division inhibitor *DicF* RNA of the *dicB* operon: Evidence for its generation *in vivo* by transcription termination and by RNase III and RNase E dependent processing. *J. Mol. Biol.* **212,** 461–471.

Friedman, D. I., and Gottesman, M. (1983). Lytic mode of lambda development. *In* Lambda

II (R. W. Hendrix, J. W. Roberts, F. W. Stahl, and R. A. Weisberg, Eds.), pp. 21–51. Cold Spring Harbor Laboratory, Cold Spring Harbor, New York.

Gatignol, A., Buckler-White, A., Berkhout, B., and Jeang, K-T. (1991). Characterization of a human TAR RNA-binding protein that activates the HIV-1 LTR. *Science* **251,** 1597–1600.

Gegenheimer, P., and Apirion, D. (1981). Processing of procaryotic ribonucleic acid. *Microbiol. Rev.* **45,** 502–541.

Gitelman, D. R., and Apirion, D. (1980). The synthesis of some proteins is affected in RNA processing mutants of *Escherichia coli*. *Biochem. Biophys. Res. Commun.* **96,** 1063–1070.

Gold, L. (1988). Posttranscriptional regulatory mechanisms in *Escherichia coli*. *Annu. Rev. Biochem.* **57,** 199–233.

Gotta, S. L., Miller, Jr., O. L., and French, S. L. (1991). rRNA transcription rate in *Escherichia coli*. *J. Bacteriol.* **173,** 6647–6649.

Gottesman, M., Oppenheim, A, and Court D. (1982). Retroregulation: Control of gene expression from sites distal to the gene. *Cell* **29,** 727–728.

Gottesman, S., Gottesman, M., Shaw, J. E., and Pearson, M. L. (1981). Protein degradation in *E. coli*: The lon mutation and bacteriophage lambda N and cII protein stability. *Cell* **24,** 225–233.

Gross, G., and Dunn, J. J. (1987). Structure of secondary cleavage sites of *Escherichia coli* RNase III in A3t RNA from bacteriophage T7. *Nucleic Acids Res.* **15,** 431–442.

Guarneros, G., Montañez, C., Hernandez, T., and Court, D. (1982). Posttranscriptional control of bacteriophage λ *int* gene expression from a site distal to the gene. *Proc. Natl. Acad. Sci. USA* **79,** 238–242.

Guarneros, G. (1988). Retroregulation of bacteriophage lambda *int* gene expression. *In* Current Topics in Microbiology and Immunology (A. Clarke, R. W. Compas, M. Cooper, H. Eisen, W. Goebel, H. Koprowski, F. Melchers, M. Oldstone, P. K. Vogt, H. Wagner, and I. Wilson, Eds.), pp. 1–19. Springer-Verlag, Berlin.

Guarneros, G., Kameyama, L., Orozco L., and Velasquez, F. (1988). Retroregulation of an *int–lacZ* gene fusion in a plasmid system. *Gene* **72,** 129–130.

Hall, S. H. and Crouch, R. J. (1977). Isolation and characterization of two enzymatic activities from chick embryos which degrade double-stranded RNA. *J. Biol. Chem.* **252,** 4092–4097.

Hoffman, S., and Miller, Jr., O. L. (1977). Visualization of ribosomal ribonucleic acid synthesis in a ribonuclease III-deficient strain of *Escherichia coli*. *J. Bacteriol.* **132,** 718–722.

Hyman, H. C., and Honigman, A. (1986). Transcription termination and processing sites in the bacteriophage lambda p_L operon. *J. Mol. Biol.* **189,** 131–141.

Icely, P. L., Gros, P., Bergeron, J. M., Devault, A., Afar, D. E. H., and Bell, J. C. (1991). TIK, a novel serine/threonine kinase, is recognized by antibodies directed against phosphotyrosine. *J. Biol. Chem.* **266,** 16073–16077.

Iino, Y., Sugimoto, A., and Yamamoto, M. (1991). *S. pombe pac1*, whose overexpression inhibits sexual development, encodes a ribonuclease III-like RNase. *EMBO J.* **10,** 221–226.

Inada, T., Kawakami, K., Chen, S. M., Takiff, H. E., Court, D., and Nakamura, Y. (1989). Temperature sensitive lethal mutant of Era, a G-protein in *Escherichia coli*. *J. Bacteriol.* **171,** 5017–5024.

Izuhara, M., Kazufumi, K., and Takata, R. (1991). Cloning and sequencing of an *Escherichia coli* K12 gene which encodes a polypeptide having similarity to the human ferritin H subunit. *Mol. Gen. Genet.* **225,** 510–513.

Kameyama, L. Fernandez, L., Court, D., and Guarneros, G. (1991). RNase III activation of bacteriophage lambda N synthesis. *Mol. Microbiol.* **5,** 2953–2963.

Kennell, D. (1986). The instability of messenger RNA in bacteria. *In* Maximizing Gene Expression (W. Reznikoff and L. Gold, Eds.), pp. 102–142. Butterworths, Stoneham, Massachusetts.

Kindler, P., Keil, T. V., Hofschneider, P. H. (1973). Isolation and characterization of an RNase III deficient mutant of *Escherichia coli*. *Mol. Gen. Genet.* **126,** 53–69.

King, T. C., and Schlessinger, D. (1983). S1 nuclease mapping analysis of ribosomal RNA processing in wild type and processing deficient *Escherichia coli*. *J. Biol. Chem.* **258,** 12034–12042.

King, T. C., Sirdeshmukh, R., and Schlessinger, D. (1984). RNase III cleavage is obligate for maturation but not for function of *Escherichia coli* pre-23S rRNA. *Proc. Natl. Acad. Sci. USA* **81,** 185–188.

King, C. K., Sirdeshmukh R, Schlessinger, D. (1986). Nucleolytic processing of ribonucleic acid transcripts in prokaryotes. *Microbiol. Rev.* **50,** *428–51.*

Kornitzer, D., Teff, D., Altuvia, S., and Oppenheim, A. B. (1989). Genetic analysis of bacteriophage λ *c*III gene: mRNA structural requirements for translation initiation. *J. Bacteriol.* **171,** 2563–2572.

Krinke, L., and Wulff, D. L. (1987). Oop RNA, produced from multicopy plasmids, inhibits lambda *c*II gene expression through an RNase III dependent mechanism. *Genes Dev.* **1,** 1005–1013.

Krinke, L., and Wulff, D. L. (1990a). The cleavage specificity of RNase III. *Nucleic Acids Res.* **18,** 4809–4815.

Krinke, L., and Wulff, D. L. (1990b). RNase III-dependent hydrolysis of λ *c*II-*O* gene mRNA mediated by λ OOP antisense RNA. *Genes Dev.* **4,** 2223–2233.

Krinke, L., Mahoney, M., and Wulff, D. L. (1991). The role of OOP antisense RNA in coliphage λ development. *Mol. Microbiol.* **5,** 126–127.

Landick, R., and Yanofsky, C. (1987). Transcription attenuation. *In Escherichia coli* and *Salmonella typhimurium*: Cellular and Molecular Biology (F. C. Neidhardt, J. L. Ingraham, K. B. Low, B. Magasanic, M. Schaechter, and H. E. Umbarger, Eds.), pp. 1276–1301. American Society for Microbiology, Washington, D.C.

Lin-Chao, S., and Cohen, S. N. (1991). The rate of processing and degradation of antisense RNA I regulates the replication of ColE1-type plasmids *in vitro*. *Cell* **28,** 1233–1242.

Lozeron, H. A., Dahlberg, J. E., and Szybalski, W. (1976). Processing of the major leftward mRNA of coliphage lambda. *Virology* **71,** 262–277.

Lozeron, H. A., Anevski, P. J., and Apirion, D. (1977). Antitermination and absence of processing of the leftward transcript of coliphage lambda in the RNase III-deficient host. *J. Mol. Biol.* **109,** 359–365.

Mangiarotti, G. E., Turco, E., Perlo, C., and Altruda, F. (1975). Role of precursor 16S RNA in assembly of *Escherichia coli* 30S ribosomes. *Nature* **253,** 569–571.

March, P. E., Ahnn, J., and Inouye, M. (1985). The DNA sequence of the gene (*rnc*) encoding ribonuclease III of *Escherichia coli*. *Nucleic Acids Res.* **13,** 4677–4685.

March, P. E., Lerner, C. G., Ahnn, J. A., Cui, X., and Inouye, M. (1988). The *Escherichia coli* Ras-like protein (Era) has GTPase activity and is essential for cell growth. *Oncogene* **2,** 539–544.

March, P. E., and Gonzalez, M. A. (1990). Characterization of the biochemical properties of recombinant ribonuclease III. *Nucleic Acids Res.* **18,** 3293–3298.

Mathews, M. B., and Shenk, T. (1991). Adenovirus virus-associated RNA and translational control. *J. Virol.* **65,** 5657–5662.

Mattheakis, L., Vu, L., Sor, F., and Nomura, J. (1989). Retroregulation of the synthesis of ribosomal proteins L14 and L24 by feedback repressor S8 in *Escherichia coli*. *Proc. Natl. Acad. Sci. USA* **86,** 448–452.

Mayer, J. E., and Schweiger, M. (1983). RNase III is positively regulated by T7 protein kinase. *J. Biol. Chem.* **258,** 5340–5343.

Mead, D. J., and Oliver, S. G. (1983). Purification and properties of a double-stranded ribonuclease from the yeast *Saccharomyces cerevisiae*. *Eur. J. Biochem* **137,** 501–507.

Meegan, J. M., and Marcus, P. I. (1989). Double-stranded ribonuclease coinduced with interferon. *Science* **244,** 1089–1091.

Melefors, O., and von Gabain, A. (1988). Site-specific endonucleolytic cleavages and the regulation of stability off *Escherichia coli ompA* mRNA. *Cell* **52,** 893–901.

Meurs, S. Chong, E. K., Galabru, J., Thomas, N. S., Kerr, I. M., Williams, B. R. G., and

Hovanessian, A. G. (1990). Molecular cloning and characterization of cDNA encoding human double-stranded RNA activated protein kinase induced by interferon. *Cell* **62,** 379–390.

Michalewicz, J., and Nicholson, A. W. (1992). Molecular cloning and expression of the bacteriophage T7 0.7 (protein kinase) gene. *Virology* **186,** 452–462.

Miczak, A., Srivastava, R. A. K., and Apirion, D. (1991). Location of the RNA-processing enzymes RNase III, RNaseE, and RNaseP in the *Escherichia coli* cell. *Mol. Microbiol.* **5,** 1801–1810.

Montañez, C., Bueno, J. Schmeissner, U., Court, D. L., and Guarneros, G. (1986). Mutations of bacteriophage lambda that define independent but overlapping RNA processing and transcription termination sites. *J. Mol. Biol.* **191,** 29–37.

Morgan, E. A., Ikemura, T., Lindahl, L., Fallon, A. M., and Nomura, M. (1978). Some rRNA operons in *Escherichia coli* have tRNA genes at their distal ends. *Cell* **13,** 335–344.

Morrison, P.T., Lovett, S. T., Gilson, L. E., and Kolodner, R. (1989). Molecular analysis of the *Escherichia coli recO* gene. *J. Bacteriol.* **171,** 3641–3649.

Mott, J. E., Galloway, J. L., and Platt, T. (1985). Maturation of *Escherichia coli* tryptophan operon mRNA. Evidence for 3′ exonucleolytic processing after rho-dependent termination. *EMBO J.* **4,** 1887–1891.

Nashimoto, H., Miura, A., Saito, H., and Uchida, H. (1985). Suppressors of temperature-sensitive mutations in a ribosomal protein gene, *rpsL* (S12) of *Escherichia coli* K12. *Mol. Gen. Genet.* **199,** 381–387.

Nashimoto, H., and Uchida, H. (1985). DNA sequencing of the *Escherichia coli* ribonuclease III gene and its mutations. *Mol. Gen. Genet.* **201,** 25–29.

Nasri, M., and Thomas, D. (1986). Relaxation of recognition sequence of specific endonuclease HindIII. *Nucleic Acids Res.* **14,** 811–821.

Newbury, S. F., Smith, N. H., Robinson, E. C., Hiles, I. D., and Higgins, C. F. (1987). Stabilization of translationally active mRNA by prokaryotic REP sequences. *Cell* **51,** 297–310.

Nicholson, A. W., Niebling, K. R., McOsker, P. L., and Robertson, H. D. (1988). Accurate *in vitro* cleavage by RNase III of phosphorothioate-substituted RNA processing signals in bacteriophage T7 early mRNA. *Nucleic Acids Res.* **16,** 1577–1591.

Nikolaev, N., Silengo, L., and Schlessinger, D. (1973). Synthesis of a large precursor to ribosomal RNA in a mutant of *Escherichia coli. Proc. Natl. Acad. Sci. USA* **70,** 3361–3365.

Nilsson, G., Belasco, J. G., Cohen, S. N., and von Gabain, A., (1984). Growth-rate dependent regulation of mRNA stability in *Escherichia coli. Nature (London)* **312,** 75–77.

Ono, M., and Kuwano, M. (1979). A conditional lethal mutation in an *Escherichia coli* strain with a longer chemical lifetime of messenger RNA. *J. Mol. Biol.* **129,** 343–357.

Paddock, G. V., Fukada, K., Abelson, J., Robertson, H. D. (1976). Cleavage of T4 species I ribonucleic acid by *Escherichia coli* ribonuclease III. *Nucleic Acids Res.* **5,** 1351–1371.

Panayotatos, N., and Truong, K. (1985). Cleavage within an RNase III site can control mRNA stability and protein synthesis *in vivo. Nucleic Acids Res.* **13,** 2227–2240.

Panganiban, A. T., and Whiteley, H. R. (1983). Purification and properties of a new *Bacillus subtilis* RNA processing enzyme. *J. Biol. Chem.* **258,** 12487–12493.

Pedersen, S., Rech, S., and Friesen, J. D. (1978). Functional mRNA half-lives in *Escherichia coli. Mol. Gen. Genet.* **166,** 329–336.

Pines, O., Yoon, H.-J., and Inouye, M. (1988). Expression of double-stranded-RNA-specific RNase III of *Escherichia coli* is lethal to *Saccharomyces cerevisiae. J. Bacteriol.* **170,** 2989–2993.

Plunkett III, G., and Echols, H. (1989). Retroregulation of the bacteriophage lambda *int* gene: Limited secondary degradation of the RNase III-processed transcript. *J. Bacteriol.* **171,** 588–592.

Portier, C., and Régnier, P. (1984). Expression of the *rpsO* and *pnp* genes. Structural analysis of a DNA fragment carrying their control regions. *Nucleic Acids Res.* **12,** 6091–6102.

Portier, C., Dondon, L., Grunberg-Manago, M., and Régnier, P. (1987). The first step in the functional inactivation of the *Escherichia coli* polynucleotide phosphorylase messenger is a ribonuclease III processing at the 5′ end. *EMBO J.* **6,** 2165–2170.

Pragai, B., and Apirion, D. (1981). Processing of bacteriophage T4 tRNAs: The role of RNAase III. *J. Mol. Biol.* **153,** 619–630.

Qian, Y., Jiang, B., Saif, L., Kang, S. Y., Ojeh, C. K., and Green, K. Y. (1991). Molecular analysis of the gene 6 from a porcine group C rotavirus that encodes the NS34 equivalent of group A Rotaviruses. *Virology* **184,** 752–757.

Rahmsdorf, H.-J., Pai, S.-H., Ponta, H., Herrlich, P., Roskoski, R., Jr., Schweiger, M., and Studier, F. W. (1974). Protein kinase induction in *Escherichia coli* by bacteriophage T7. *Proc. Natl. Acad. Sci. USA* **71,** 586–589.

Régnier, P., and Portier, C. (1986). Initiation, attenuation and RNase III processing of transcripts from the *Escherichia coli* operon encoding ribosomal protein S15 and polynucleotide phosphorylase. *J. Mol. Biol.* **187,** 23–32.

Régnier, P., and Grunberg-Manago, M. (1989). Cleavage by RNase III in the transcripts of the *metY-nusA-infB* operon of *Escherichia coli* releases the tRNA and initiates the decay of the downstream mRNA. *J. Mol. Biol.* **210,** 293–302.

Régnier, P., and Grunberg-Manago, M. (1990). RNase III cleavages in non-coding leaders of *Escherichia coli* transcripts control mRNA stability and genetic expression. *Biochimie* **72,** 825–834.

Régnier, P., and Hajnsdorf, E. (1991). Decay of mRNA encoding ribosomal protein S15 of *Escherichia coli* is initiated by an RNaseE-dependent endonucleolytic cleavage that removes the 3′ stabilizing stem and loop structure. *J. Mol. Biol.* **217,** 283–292.

Robertson, E. S., and Nicholson, A. W. (1990). Protein kinase of bacteriophage T7 induces the phosphorylation of only a small number of proteins in the infected cell. *Virology* **175,** 525–534.

Robertson, H. D. (1982). *Escherichia coli* ribonuclease III cleavage sites. *Cell* **30,** 669–672.

Robertson, H. D. (1977). Structure and function of RNA processing signals. *In* Nucleic Acid-Protein Recognition (H. J. Vogel, Ed.), pp. 549–568. Academic Press, New York.

Robertson, H. D. (1990). *Escherichia coli* ribonuclease III. *In* Methods in Enzymology. Vol. 181. (J. E. Dahlberg and J. N. Abelson, Eds.) pp. 189–202. Academic Press, New York.

Robertson, H. D., and Dunn, J. J. (1975). Ribonucleic acid processing activity of *Escherichia coli* ribonuclease III. *J. Biol. Chem.* **250,** 3050–3056.

Robertson, H. D., and Barany, F. (1979). Enzymes and mechanisms in RNA processing. *In* Proceedings of the 12th FEBS Congress, pp. 285–295. Pergamon Press, Oxford.

Robertson, H. D., Webster, R. E., and Zinder, N. D. (1968). Purification and properties of ribonuclease III from *Escherichia coli*. *J. Biol. Chem.* **243,** 82–91.

Robertson, H. D., Dickson, E., and Dunn, J. J. (1977). A nucleotide sequence from a ribonuclease III processing site in bacteriophage T7 RNA. *Proc. Natl. Acad. Sci. USA* **74,** 822–826.

Rosenberg, M., and Kramer, R. A. (1977). Nucleotide sequence surrounding a ribonuclease III processing site in bacteriophage T7 RNA. *Proc. Natl. Acad. Sci. USA* **74,** 984–988.

Saito, H., and Richardson, C. C. (1981). Processing of mRNA by ribonuclease III regulates expression of gene *1.2* of bacteriophage T7. *Cell* **27,** 533–542.

Schindler, D., and Echols, H. (1981). Retroregulation of the *int* gene of bacteriophage λ: Control of translation completion. *Proc. Natl. Acad. Sci. USA* **78,** 4475–4479.

Schmeissner, U., McKenney, K., Rosenberg, M., and Court, D. (1984a). Removal of a terminator structure by RNA processing regulates *int* gene expression. *J. Mol. Biol.* **176,** 39–53.

Schmeissner, U., McKenney, K. Rosenberg, M., and Court, D. (1984b). Transcription terminator involved in the expression of the *int* gene of phage lambda. *Gene* **28,** 343–50.

Schweitz, H., and Ebel, J. P. (1971). A study of the mechanism of action of *Escherichia coli* ribonuclease 3. *Biochimie* **5,** 585–593.

Simons, R. W., and Kleckner, N. (1988). Biological regulation by antisense RNA in prokaryotes. *Annu. Rev. Genet.* **22,** 567–600.

Sirdeshmukh, R., and Schlessinger, D. (1985a). Why is processing of 23S ribosomal RNA in *Escherichia coli* not obligate for its function? *J. Mol. Biol.* **186,** 669–672.

Sirdeshmukh, R., and Schlessinger, D. (1985b). Ordered processing of *Escherichia coli* 23S rRNA *in vitro*. *Nucleic Acids Res.* **13,** 5041–5054.

Srivastava, S. K., Cannistraro, V. J., and Kennell, D. (1992). Broad-specificity endoribonucleases and mRNA degradation in *Escherichia coli*. *J. Bacteriol.* **174,** 56–62.

Srivastava, A. K., and Schlessinger, D. (1988). Coregulation of processing and translation: Mature 5′ termini of *Escherichia coli* 23S ribosomal RNA form in polysomes. *Proc. Natl. Acad. Sci. USA* **85,** 7144–7148.

Srivastava, A. K., and Schlessinger, D. (1989a). Processing pathway of *Escherichia coli* 16S precursor RNA. *Nucleic Acids Res.* **17,** 1649–1663.

Srivastava, A. K., and Schlessinger, D. (1989b). *Escherichia coli* 16S rRNA 3′-end formation requires a distal transfer RNA sequence at a proper distance. *EMBO J.* **8,** 3159–3166.

Srivastava, A. K., Miczak, A., and Apirion, D. (1990). Maturation of precursor 10Sa RNA in *Escherichia coli* is a two step process: The first reaction is catalyzed by RNase III in presence of Mn^{2+}. *Biochimie* **72,** 791–80.

Stark, M. J. R., Gourse, R. L., Jemiolo D. K., and Dahlberg, A. E. (1985). A mutation in an *Escherichia coli* ribosomal RNA operon that blocks the production of precursor 23S ribosomal RNA by RNase III *in vivo* and *in vitro*. *J. Mol. Biol.* **182,** 205–216.

Steege, D. A., Cone, K. C., Queen, C., and Rosenberg, M. (1987). Bacteriophage lambda *N* gene leader RNA: RNA processing and translational initiation signals. *J. Biol. Chem.* **262,** 17651–17658.

Steitz, J. A., and Bryan, R. A. (1977). Two ribosome binding sites from the gene 0.3 messenger RNA of bacteriophage T7. *J. Mol. Biol.* **114,** 527–543.

Studier, F. W. (1975). Genetic mapping of a mutation that causes ribonuclease III deficiency in *Escherichia coli*. *J. Bacteriol.* **124,** 307–316.

Studier F.W., Dunn, J.J., and Buzash-Pollert, E. (1979). Processing of bacteriophage T7 RNAs by RNase III. *In* From Gene to Protein: Information Transfer in Normal and Abnormal Cells (T. R. Russell, K. Brew, H. Farber, and T. Schultz, Eds.), pp. 261–269. Academic Press, New York.

Takata R., Mukai, T., and Hori, K. (1985). Attenuation and processing of RNA from the *rpsO-pnp* transcription unit of *Escherichia coli*. *Nucleic Acids Res.* **13,** 7289–7297.

Takata, R., Mukai, T., and Hori, K. (1987). RNA processing by RNase III is involved in the synthesis of *Escherichia coli* polynucleotide phosphorylase. *Mol. Gen. Genet* **209,** 28–32.

Takata, R., Izuhara, M. and Hori, K. (1989). Differential degradation of the *Escherichia coli* polynucleotide phosphorylase mRNA. *Nucleic Acids Res.* **17,** 7441–7451.

Takata, R., Izuhara, M., and Akiyama, K. (1992). Processing in the 5′ region of the *pnp* transcript facilitates the site-specific endonucleolytic cleavages of mRNA. *Nucleic Acids Res.* **20,** 847–850.

Takiff, H. E., Chen, S.M., and Court, D. L. (1989). Genetic analysis of the *rnc* operon of *Escherichia coli*. *J. Bacteriol* **171,** 2581–2590.

Takiff, H. E., Baker, T., Copeland, T., Chen, S. M., and Court, D. L. (1992). Locating essential *Escherichia coli* genes by using mini-Tn10 transposons: The *pdxJ* operon. *J. Bacteriol* **174,** 1544–1553.

Talkad, V., Achord, D., and Kennell, D. (1978). Altered mRNA metabolism in ribonuclease III-deficient strains of *Escherichia coli*. *J. Bacteriol* **135,** 528–541.

Traub, P., and Nomura, M. (1969). Structure and function of *Escherichia coli* ribosomes. VI. Mechanism of assembly of 30S ribosomes studied *in vitro*. *J. Mol. Biol.* **40,** 391–413.

Watson, M., and Apirion, D. (1985). Molecular cloning of the gene for the RNA-processing enzyme RNase III of *Escherichia coli*. *Proc. Natl. Acad. Sci. USA* **82,** 849–853.

Watson, J. C., Chang, H.-W., and Jacobs, B. L. (1991). Characterization of a vaccinia virus-encoded double-stranded RNA-binding protein that may be involved in inhibition of the double-stranded RNA-dependent protein kinase. *Virology* **185,** 206–216.

Wilder, D. A., and Lozeron, H. A. (1979). Differential modes of processing and decay for the major N-dependent RNA transcript of coliphage lambda. *Virology* **99,** 241–256.

Woese, C. R. (1987). Bacterial evolution. *Microbiol. Rev.* **51,** 221–271.

Wulff, D. L., and Rosenberg, M. (1983). The establishment of repressor synthesis. *In* Lambda II (J. Hendrix, J. Roberts, F. Stahl, and R. Weisberg, Eds.), pp. 53–73. Cold Spring Harbor Laboratories, Cold Spring Harbor, New York.

Young, R. A., and Steitz, J. A. (1978). Complementary sequences 1700 nucleotides apart form a ribonuclease III cleavage site in *Escherichia coli* ribosomal precursor RNA. *Proc. Natl. Acad. Sci. USA* **75,** 3593–3597.

Xu, H.-P., Riggs, M., Rodgers, L., and Wigler, M. (1990). A gene from *S. pombe* with homology to *Escherichia coli* RNase III blocks conjugation and sporulation when overexpressed in wild type cells. *Nucleic Acids Res.* **18,** 5304.

Zillig, W., Fujiki, H., Blum, W., Janekovic, D., Schweiger, M., Rahmsdorf, H.-J., Ponta, H., and Hirsch-Kauffmann, M. (1975). *In vivo* and *in vitro* phosphorylation of DNA-dependent RNA polymerase of *Escherichia coli* by bacteriophage-T7-induced protein kinase. *Proc. Natl. Acad. Sci. USA* **72,** 2506–2510.

6

Translation and mRNA Stability in Bacteria: A Complex Relationship

CARSTEN PETERSEN

I. The Relation of Translation to mRNA Stability

Ever since the concept of messenger RNA was clearly established (Gros *et al.*, 1961; Brenner *et al.*, 1961; Jacob and Monod, 1961), it has been anticipated that ribosomes and the process of translation might play some role in determining messenger RNA stability. Here, I shall try to give an overview of the numerous studies that have been performed over the last 30 years to investigate the role of ribosomes in mRNA degradation in bacteria. As we shall see, the question of the relation of translation to mRNA stability is not one but a series of related questions, which reflect the many different ways by which ribosomes and the process of translation may influence the multiple mechanisms involved in the inactivation and degradation of messenger RNA.

A. Multiple Mechanisms of mRNA Inactivation and Degradation

Several different mechanisms contribute to the functional inactivation and degradation of mRNA in bacteria (for reviews see Petersen (1992) and the previous chapters of this volume). In *Escherichia coli* the 3′ exonucleases RNase II and polynucleotide phosphorylase are involved in degradation of mRNA, and they could also participate in the initial inactivating event for some transcripts, although most natural mRNAs are protected against this type of inactivation by stable secondary structures at their 3′ end (see Chapter 2). Considerable evidence suggests that mRNA may also be inactivated by processive nucleolytic attack proceeding in the 5′–3′ direction, although the precise mechanism and the responsible enzymatic activity have not yet been identified (see Chapter 3 and Section II,B).

Functional inactivation of some mRNAs may be accomplished by endonucleolytic cleavages, which may be directly inactivating if they occur in the ribosome binding site or within the coding sequence (e.g., see Ruckman *et al.*, 1989; Klug and Cohen, 1990). Cleavages outside of these regions may lead to inactivation indirectly by initiating processive nucleolytic processes, which eventually proceed into the essential parts of the message. Many of the cleavages known to be catalyzed by RNase III and RNase E/K in *E. coli* presumably affect mRNA stability in this way (see Chapters 4 and 5). Finally, some mRNAs may be inactivated by processes that are not nucleolytic, such as the binding of translational repressors or the formation of secondary structures that prevent initiation of translation (see Petersen (1992) and Section IV). These nonnucleolytic processes may even determine the stability of mRNA if degradation is closely coupled to functional inactivation (see below).

If an mRNA is subject to attack by several mechanisms acting independently on different target sites, the overall rate of inactivation is determined by the sum of the contributions from the individual mechanisms and will be dominated by the fastest mechanisms (Petersen, 1991; see Appendix). The sensitivity of the different decay mechanisms to modulations of the translation process may be very different, so the extent to which the translation process affects the stability of a given transcript would be expected to be strongly dependent on its specific mechanism of decay.

B. Functional Lifetime and mRNA Stability

In this chapter, the terms *functional lifetime* and *functional half-life* refer to the period of time during which an mRNA molecule serves as an active template for protein synthesis; *inactivation* is defined as a nucleolytic or nonnucleolytic event that prevents further translation and thereby limits the yield of functional polypeptides that will eventually be produced from a message. The terms *decay* and *degradation* refer to the physical destruction of an mRNA molecule; *stability* is a measure of the ability of mRNA to resist decay.

If functional inactivation of an mRNA is actually caused by nucleolytic degradative processes, the functional and the physical lifetime of the mRNA will obviously coincide. Even for an mRNA that is functionally inactivated by a nonnucleolytic event there may be a close coupling between functional inactivation and degradation, e. g., if the transcript contains very vulnerable nucleolytic target sites that are rapidly attacked as soon as the protection by ribosomes is lost. However, functional inactivation of a transcript is not necessarily followed by its immediate chemical degradation; thus, there can be considerable differences between the functional and physical half-lives of some transcripts (Schwartz *et al.*, 1970; Puga *et al.*, 1973; Yamada *et al.*, 1974; Yamamoto and Imamoto, 1975,1977; Gupta and Schlessinger, 1976; Trimble and Maley, 1976; Ono and Kuwano, 1979; Court *et al.*, 1980; Gorski *et al.*, 1985; Newbury *et al.*, 1987; Sandler and Weisblum, 1988; McCarthy *et al.*, 1991).

II. Effects of Interfering with Translation on mRNA Stability

A. An Active Role for Ribosomes in mRNA Degradation?

It has been observed repeatedly that bulk bacterial mRNA as well as specific transcripts is stabilized by antibiotics such as chloramphenicol, fusidic acid, or tetracycline, which interfere with polypeptide elongation by ribosomes (see Table 1). Paradoxically, it has been hard to determine if ribosomes are actively engaged in the degradation of mRNA or, rather, play a passive role of protecting against nucleolytic attack. Some early findings led to the suggestion that ribosomes or a special subclass of ribosomes, "killer ribosomes," might play an active role in mRNA decay via a ribosome-associated nucleolytic activity (Kuwano *et al.*, 1969) (see Fig. 1a). However, no compelling evidence for the existence of any such enzyme activity could be found (Bothwell and Apirion, 1971; Holmes and Singer, 1971).

This gave weight to the alternative interpretation, that ribosomes protect mRNA, perhaps in particular when stalled by antibiotics such as chloramphenicol (Fig. 1b). This interpretation is supported by the finding that

TABLE 1 Effect of Antibiotics on mRNA Stability in *Escherichia coli*[a]

mRNA studied	Kasugamycin	Chloramphenicol	Tetracycline	Fusidic acid	Puromycin	Reference
lacZ		+			−	Varmus *et al.*, 1971
Bulk		+		+		Craig, 1972
Bulk		+	+	+	−	Pato *et al.*, 1973
trp		+	+	+	−	Imamoto, 1973
Bulk		+			−	Cremer *et al.*, 1974
lacZ	−	+				Schneider *et al.*, 1978
bla		+				Lundberg *et al.*, 1988
ompA		+				
tetR–lacZ	0					Baumeister *et al.*, 1991
lacZ	−					

[a] In each case + indicates stabilization and − indicates destabilization, whereas no effect is indicated by a 0. The table is by no means exhaustive, it is intended only to provide a quick reference to representative studies on the subject.

a

b

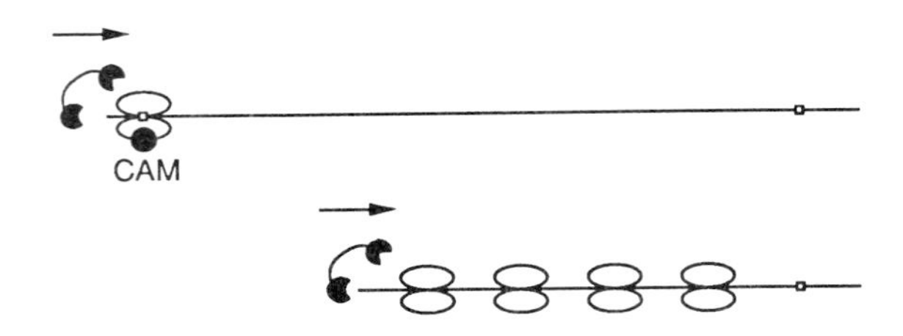

c

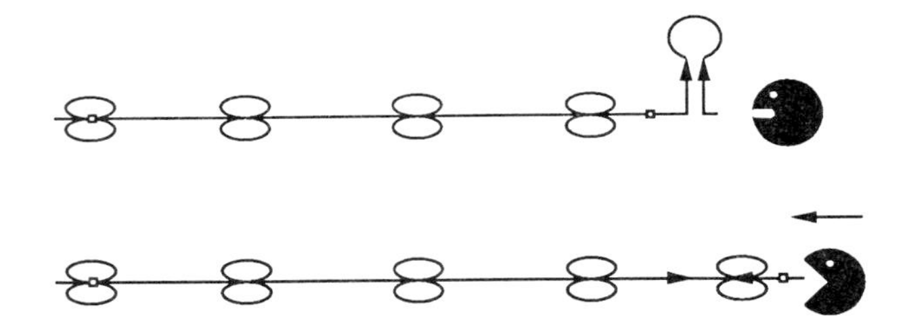

d

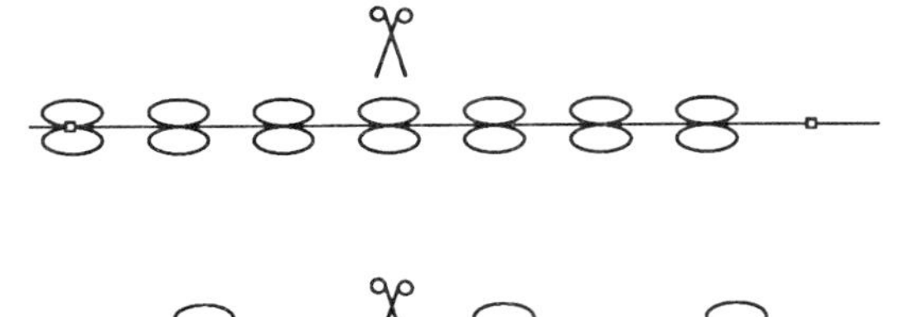

e

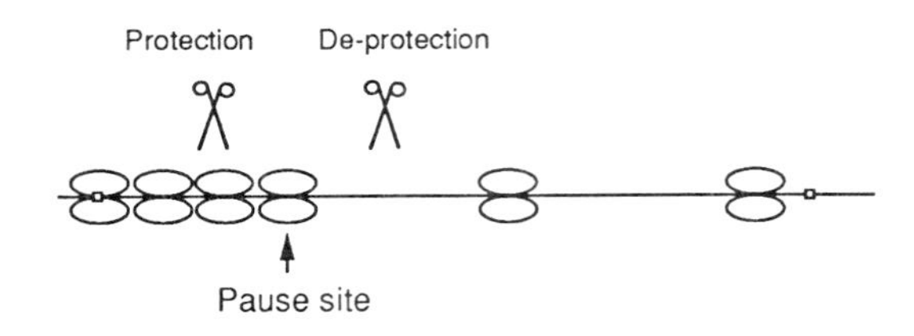

bulk mRNA and specific transcripts from the *lac* and *trp* operons are destabilized by puromycin, which causes premature release of ribosomes from the mRNA (see Table 1). The *lacZ* transcript is also destabilized by kasugamycin, which prevents initiation of translation without interfering with polypeptide elongation (Schneider *et al.*, 1978; Baumeister *et al.*, 1991). All these results suggest that ribosomes protect bacterial mRNAs against nucleolytic decay. However, one should cautiously interpret the effects of translation inhibitors because they can give rise to premature termination of transcription, "polarity," in some bacterial operons (Morse, 1971; Hansen *et al.*, 1973); prematurely terminated transcripts can be hyperlabile, perhaps owing to a lack of protective structures at their 3′ end (Morse and Yanofsky, 1969; Imamoto and Kano, 1971; Hiraga and Yanofsky, 1972).

Interestingly, there is evidence that ribosomes play an active role in the degradation of some eukaryotic transcripts. Translation of specific mRNA regions is required for the rapid degradation of several transcripts, including the MATα1 transcript in yeast, histone mRNAs, c-*fos* mRNA, and the β-tubulin transcript (see Chapters 9, 10, 12, and 13). It has been proposed that ribosomes may be directly involved in initiating the decay process by delivering a ribosome-associated nuclease to mRNA or by creating a destabilizing element recognized by the decay machinery, e.g., the nascent polypeptide. Alternatively, the passage of ribosomes may modulate the secondary structure of the transcript to expose target sites that would not otherwise be accessible to attack.

The latter possibility is reminiscent of the finding that some prokaryotic transcripts are destabilized by mutations, which allow ribosomes to read past the normal termination codon and through downstream secondary structures, which normally protect against 3′-exonucleolytic attack (Chen and Belasco, 1990; Klug and Cohen, 1991) (see Fig. 1c). Apart from these results, which probably do not reflect a physiologically important role for ribosomes in mRNA decay, there is virtually no evidence that specifically

---

FIGURE 1 Different mechanisms by which ribosomes have been proposed to affect the rate of mRNA decay. The initiation and termination codons on the mRNA are indicated as small squares. (a) A ribosome associated nucleolytic activity, a killer ribosome. (b) Protection against processive degradation in the 5′–3′direction by a ribosome stalled by chloramphenicol. The hypothetical enzyme responsible for this type of degradation is indicated to recognize the 5′ end of the transcript and cleave at a downstream position; however, this is purely speculative. [*Note added in proof*: Recent data suggest that RNase E may have this property (Bouvet and Belasco, 1992).] Normal decay in the 5′–3′direction limited by the rate of migration of the last translating ribosome is also indicated. (c) Destabilization of mRNA by ribosome-mediated unwinding of a stem–loop that protects against 3′–exonucleolytic attack. (d) Protection against endonucleolytic cleavage by dense packing of ribosomes on the mRNA caused by frequent initiations of translation. (e) Effects on mRNA stability of a translational pause site. Whether the effect is stabilizing or destabilizing depends on the position of the pause site relative to a major nucleolytic target site. See text for further details.

suggests an active role for ribosomes in prokaryotic mRNA degradation. However, in the light of the results from eukaryotic systems, it may be premature to dismiss the idea completely. Indeed it has recently been found that a specific processing cleavage of the *his* operon transcript requires the presence of a nearby ribosome (Alifano *et al.*, 1992).

B. Protection against Processive Degradation of mRNA in the 5′–3′ Direction

Inhibition of ribosome translocation by chloramphenicol or by inactivation of a temperature-sensitive elongation factor G results in stabilization of mRNA that is synthesized *after* the inhibition of translation (Cremer *et al.*, 1974; Craig, 1972). Although this newly synthesized mRNA is not translated, it is able to bind ribosomes. Release of these ribosomes by puromycin causes destabilization of the mRNA (Cremer *et al.*, 1974). These results suggest that binding at the ribosome binding site of ribosomes that are unable to translocate may be sufficient to stabilize many bacterial mRNAs, perhaps by preventing the initiation of processive degradation in the 5′–3′ direction (Fig. 1b).

Indeed, more recent studies have shown that a ribosome stalled by erythromycin during translation of the *ermA* or *ermC* leader peptides in *Bacillus subtilis* can efficiently protect dowstream mRNA sequences against degradation, even if these sequences are not actively translated (Bechhofer and Dubnau, 1987; Sandler and Weisblum, 1988, 1989; Bechhofer and Zen, 1989; Hue and Bechhofer, 1991). Apparently, the 5′ region of the λP_L transcript also has the ability to act as a barrier against 5′–3′ degradation. The chemical stability of the *trp* transcript is increased several-fold by fusion to this region, but the functional half-life is essentially unaffected (Yamamoto and Imamoto, 1975; 1977) (see Chapter 3).

Supporting evidence for directional decay of mRNA in the 5′–3′ direction comes from a study of a set of streptomycin-dependent mutant strains of *E. coli*, which differ in their characteristic rates of polypeptide elongation (Gupta and Schlessinger, 1976). In these strains the rate of decay of bulk mRNA is correlated with the rate of ribosome movement, whereas the rates of translational inactivation of the *lacZ* message and of bulk mRNA are essentially independent of changes in the rate of ribosome movement. These results are compatible with the degrading nuclease proceeding in the 5′–3′ direction behind the last translating ribosome, after the mRNA has been functionally inactivated by a mechanism that is independent of the rate of ribosome movement (Fig. 1b).

From the previous results, any physiological pertubation that inhibits or slows down the movement of ribosomes without releasing them would be expected to decrease the rate of degradation of mRNA. Indeed, bulk mRNA in *B. subtilis* and *E. coli* is stabilized during energy deprivation caused by anaerobiosis (Fan *et al.*, 1964; Nakada and Fan, 1964), and decay of the *trp* transcript is delayed during amino acid starvation (Morse *et al.*, 1969; Morse and Guertin, 1971). In contrast the rate of functional inactivation of the *lacZ* message appears to be unaffected by amino acid starvation (Kennell and Simmons, 1972; Foley *et al.*, 1981), and the same is true for the transcript encoding ornithine transcarbamylase (Hall and Gallant, 1973). It appears to be a general trend that interference with polypeptide elongation retards the degradation of mRNA, even though the functional half-life is not necessarily increased.

C. Protection against Endonucleolytic Attack

The finding that stalling of ribosomes near the 5′ end of otherwise untranslated transcripts is sufficient to prevent their degradation (Cremer *et al.*, 1974; Bechhofer and Dubnau, 1987) suggests that mRNA sequences are not generally subject to endonucleolytic attack when unprotected by ribosomes. This notion is corroborated by studies on the stability of transcripts that are deprived of ribosome protection by various means. The mRNA of the *hok* killer gene of plasmid R1 is very stable even though its translation is efficiently repressed in plasmid bearing cells to avoid killing the *E. coli* host (Gerdes *et al.*, 1990). Similarly, a *lacZ* fusion mRNA transcribed from a phage T7 promoter appears to be relatively stable, even when its translation is prevented by a mutation in the ribosome binding site (Chevrier-Miller *et al.*, 1990).

In *Rhodobacter capsulatus* the *pufA* and *pufB* messages retain a physical half-life of about 30 min even when deprived of ribosome protection by introduction of a premature stop codon a few nucleotides downstream of the initiation codon (Klug and Cohen, 1991). In *E. coli* the stability of the *bla* transcript is unaffected by insertion of premature stop codons, which leave most segments of the message unprotected by ribosomes (Nilsson *et al.*, 1987). In this case, however, very early premature stop codons located at codon 4 or codon 26 do result in some destabilization, indicating that the *bla* message requires minimal protection by ribosomes near the ribosome binding site, perhaps to prevent initiation of processive degradation in the 5′–3′ direction. In summary, it appears that ribosomes are required only for the protection of specific nucleolytic target sites where mRNA degradation may be initiated. The finding that long stretches of mRNA can be unprotected by ribosomes without being excessively unstable suggests that such sites are generally rare in bacterial coding sequences.

III. The Expected Relation between Translation Frequency and mRNA Stability

One idea that has dominated much of our thinking about the relation of translation to mRNA decay is that translation frequency and mRNA stability should be positively correlated because a more dense packing of ribosomes on mRNA would sterically protect against nucleolytic attack (Fig. 1d). However, from the previous considerations it is clear that no general correlation between translation frequency and mRNA stability should be expected when different mRNAs are compared. Some mRNAs may be intrinsically very stable in the complete absence of translation, e.g., due to protective barriers at their 5′ end. Others may be labile even when relatively frequently translated due to the presence of intrinsically vulnerable target sites (e.g., see Lundberg *et al.*, 1988). Different segments of some polycistronic mRNAs decay at essentially the same rate, although the translation frequencies of the individual cistrons differ by up to a factor of 1000 (Ray and Pearsson, 1974, 1975; Byström *et al.*, 1989). The most poorly translated cistrons in these transcripts probably lack internal targets that would require ribosome protection or contain only targets that are relatively inert even when unprotected by ribosomes. The stability of messages that are degraded by processive waves of decay initiated far upstream or downstream of the coding sequence would generally be expected to be independent of the frequency of ribosome loading.

A. Nucleolytic Inactivation at a Single Target Site

Although the requirement for ribosome protection is not universal, we have seen that some mRNAs may be efficiently protected by the presence of ribosomes near their 5′ end. It may be instructive to consider what kind of relationship one would expect between the translation frequency and mRNA stability for a hypothetical message that is inactivated exclusively by cleavage at a single nuclease target site that coincides with the ribosome binding site.

Mechanistically, translation initiation is a complex multistep process that involves the binding of a 30S ribosomal subunit to the mRNA ribosome binding site to form an initiation complex, followed by internal conformational changes in the initiation complex, binding of the 50S ribosome subunit, release of initiation factors, and translocation through the first ribosome diameter of the coding sequence to leave a cleared ribosome binding site available for the next initiation (reviewed by McCarthy and Gualerzi, 1990). For simplicity we neglect the dissociation of initiation complexes and consider the binding of a 30S subunit at the ribosome binding site as a pseudo-first order reaction characterized by a rate constant, k_b. The mean time between clearing of the ribosome binding site and binding of the next ribosome will be $t_b = 1/k_b$, whereas the mean time required for conversion of the initiation complex to an elongation competent form and for clearing of the ribosome binding site is denoted t_c. The average time between initiations is

$$t_f = t_b + t_c,$$

and the initiation frequency is

$$f = 1/(t_b + t_c).$$

The nucleolytic target site, which coincides with the ribosome binding site, is characterized by an intrinsic decay constant, k_d, corresponding to a mean-life of the target in the absence of ribosome protection of $t_d = 1/k_d$. The observable rate of inactivation of the transcript is proportional to the fraction of time that the target is exposed between initiations, F_e, given by

$$F_e = t_b/ (t_b + t_c).$$

The observed functional mean-life, t_m, is equal to t_d/F_e. Thus, we have

$$t_m = t_d (1 + t_c/t_b).$$

Usually mRNA stability is measured in terms of the half-life, $t_{1/2}$, which for monoexponential decay is related to the mean-life by $t_{1/2} = \text{In}2 \cdot t_m$. However, the mean-life is used in most of the formulas in this chapter because it gives rise to simpler expressions that are more intuitively comprehensible. In Fig. 2, I have plotted t_m as a function of the translation initiation frequency, f, when t_b is varied for two hypothetical mRNAs, which differ

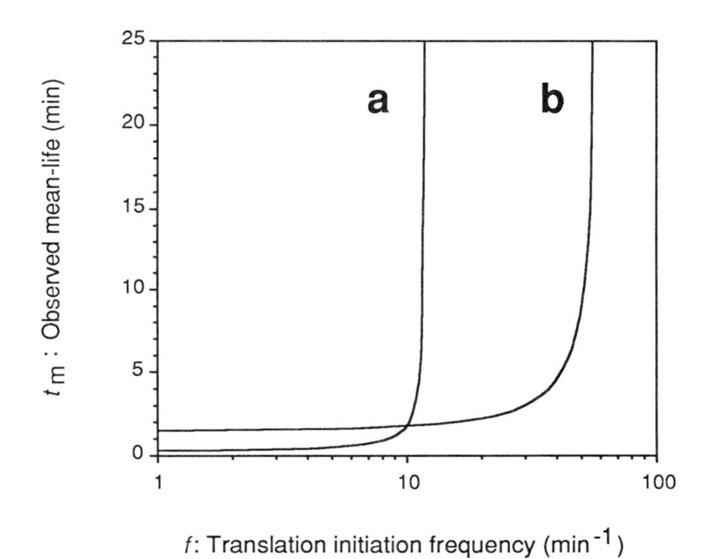

FIGURE 2 Expected relation between the translation initiation frequency (f) and the observed mean-life (t_m) for two transcripts that both contain a single nucleolytic target site, which coincides with the ribosome binding site. The intrinsic mean-life of the target site, t_d, is 0.3 min for transcript a and 1.5 min for transcript b. The average time required for clearing of the ribosome binding site, t_c, is 5 and 1 sec, respectively, for transcripts a and b. Since only relative translation frequencies have been measured in most studies, a logarithmic scale is used for f.

with respect to the intrinsic vulnerability of their target sites and the rate of clearing of their ribosome binding sites. Both mRNAs have a functional mean-life of 1.8 min at a translational frequency of 10 initiations per minute, but they respond very differently to pertubations from this value. Although these curves were derived from this particularly simple model, it should be emphasized that similar hyperbola-like relations are obtained for more complex models involving one or more target sites of specified size and location (C. Petersen, unpublished calculations).

In general we may expect three apparently different types of relation between ribosome loading frequency and mRNA stability. Near the maximally obtainable translation frequencies there will be a *steep response*, where small, perhaps undetectable, changes in translation frequency lead to drastic changes in mRNA stability. In the low range of translation frequencies we get a *flat response*, where mRNA stability is unaffected by moderate changes in the loading frequency because the target is exposed nearly all the time anyway. In between the extremes there is a *positive correlation* between initiation frequency and mRNA stability, as anticipated. Clearly, the observed response of mRNA stability to changes in the rate of ribosome loading is strongly dependent on the absolute value of the translation frequencies considered. Nevertheless, in most studies only relative translation frequencies have been measured (see Appendix). The observation of different types of response in different experimental systems (see below) may to some extent reflect differences in the range of absolute translation frequencies that were studied.

The rate of ribosome movement may vary considerably due to local variations in codon usage (e.g., see Curran and Yarus, 1989; Sørensen and Pedersen, 1991), and it has been suggested that "slow" codons in the beginning of coding sequences serve to modulate the translation frequency by delaying the clearing of the ribosome binding site (Bergmann and Lodish, 1979; Liljenström and von Heijne, 1987; Chen and Inouye, 1990). In general a reduction in the translation frequency caused by introduction of slow codons, i.e., a translation pause site, may have adverse effects on mRNA stability. If there is a ribonuclease target site downstream of the pause site, it would be more exposed, and the reduced translation would be correlated with a destabilization of the mRNA. A target upstream of the pause site, on the other hand, would enjoy increased protection, and there would be an increase in mRNA stability caused by a decreased rate of ribosome movement, i.e., an *inverse correlation* between translation frequency and mRNA stability (Fig. 1e). It is noteworthy that an increase in the clearing time in the last example would have no effect on the rate of protein synthesis, because the decrease in translation frequency would be exactly balanced by a corresponding increase in the mean-life of the mRNA, due to the increased protection of the target site. In general the regulatory potential of slow codons will be reduced if they cause increased ribosomal protection of target sites for mRNA degradation.

B. Nonnucleolytic Inactivation by Repressor Binding

Many bacterial and phage genes are subject to regulation by translational repressors, which are generally considered to regulate the frequency of ribosome loading on their target mRNAs (reviewed by McCarthy and Gualerzi, 1990). However, depending on the effective degree of reversibility of the repressor binding, some repressors may actually directly regulate the length of the period for which their target mRNA remains functional, i.e., the functional lifetime, rather than the frequency at which ribosomes are loaded during this period.

As an extreme example consider a repressor that binds so strongly to its target that it is hardly ever released again before the mRNA molecule is finally destroyed by nucleolytic attack at a target site that is not affected directly by the repressor. In Fig. 3a this situation is illustrated schematically for a hypothetical mRNA that is subject to rapid nucleolytic degradation as soon as the protection by ribosomes is lost as a consequence of repression. In this case repressor binding is made effectively irreversible by the subsequent nucleolytic decay processes acting on the repressed transcript, and so the functional and physical lifetimes of the message is determined by the time required for the repressor to bind to its target. The observable translation frequency on the other hand will be unaffected by repressor binding and will reflect the normal initiation of translation prior to repressor binding. Thus, the repression of protein synthesis will be fully accounted for by the decrease in functional mRNA lifetime.

At the other extreme consider a repressor that binds to and dissociates from its target many times during the lifetime of an individual mRNA molecule, which is eventually destroyed by a nucleolytic mechanism that is completely insensitive to changes in ribosome loading frequency (Fig. 3b). Here the functional and physical lifetimes of the mRNA will be unaffected by translational repression, and the decrease in gene expression caused by repressor binding will be fully accounted for by the decrease in the observed translation frequency.

It follows from these extreme examples that post-transcriptional regulation in general may be considered a control of the polypeptide yield per mRNA molecule, which may be defined as the product of the translation frequency and the functional mean-life of the mRNA (see Appendix). The partitioning of the repression effect between a reduction in translational frequency and a reduction in the functional mRNA lifetime will depend on the relative stability of the repressor/mRNA complex compared with the intrinsic vulnerability of the nucleolytic target(s) on the transcript. A priori any partitioning between the two extremes would be possible. In other words, the observed relation between the ribosome loading frequency and the functional mRNA lifetime could correspond to a steep response, a positive correlation, or a flat response, depending on the effective reversibility of the repressor binding. The same is true for the relation between

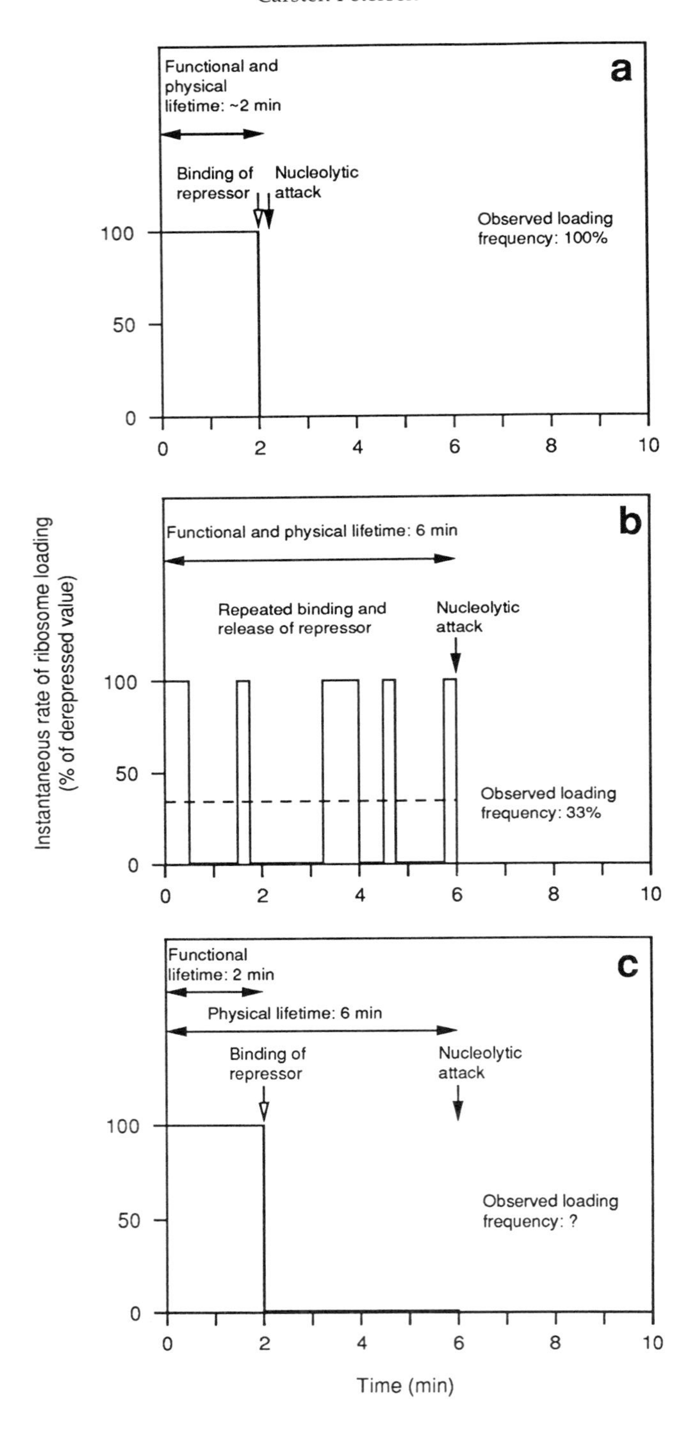
Functional and physical lifetime: ~2 min
a
Binding of repressor
Nucleolytic attack
Observed loading frequency: 100%
100
50
0
0
2
4
6
8
10
Functional and physical lifetime: 6 min
b
Repeated binding and release of repressor
Nucleolytic attack
Instantaneous rate of ribosome loading (% of derepressed value)
Observed loading frequency: 33%
100
50
0
0
2
4
6
8
10
Functional lifetime: 2 min
c
Physical lifetime: 6 min
Binding of repressor
Nucleolytic attack
Observed loading frequency: ?
100
50
0
0
2
4
6
8
10
Time (min)

translation frequency and physical mRNA stability. Indeed the physical mRNA lifetime responded to translational repression in the same way as the functional lifetime in the two extreme examples of Figs. 3a and 3b.

However, one could easily imagine that the functional and physical lifetimes of an mRNA might respond differently to repression. Consider a repressor that binds rapidly and essentially irreversibly to its target mRNA, which is eventually degraded by a slow mechanism that is completely unaffected by ribosome protection (Fig. 3c). In this case the functional lifetime of the mRNA would be determined by the process of repressor binding, whereas the physical lifetime would be determined solely by the rate of the nucleolytic process. The relation between translation frequency and mRNA stability would be strongly dependent on the method used to measure the translation frequency (cf. Appendix). If the translation frequency were defined and measured as the rate of synthesis of the encoded protein divided by the number of mRNA molecules physically present (functional or repressed), then the observed translation frequency would appear to be reduced threefold compared with the situation in the absence of repressor, whereas the physical stability of the transcript would be unaffected (a flat response). If, on the other hand, the translational frequency were defined and measured as the rate of synthesis of the encoded protein divided by the number of functional mRNA molecules present (i.e., those not bound by repressor), then the observed translation frequency would be unaffected by the repression and the decrease of gene expression would be fully accounted for by the threefold reduction of the functional mRNA lifetime (a steep response).

Although the previous examples are extreme, they show that we may expect the same types of relation between translation frequency and mRNA stability when an mRNA is subject to translational repression (nonnucleolytic inactivation) as those predicted for the simple model in Section III.A, where ribosomes protect against nucleolytic inactivation. However, the situation here is fundamentally different, because there is not a simple causal relationship between the ribosome loading frequency and the func-

---

FIGURE 3 Expected relation between the observable ribosome loading frequency and mRNA stability for three hypothetical average mRNAs that are subject to translational repression. (a) An mRNA where repressor binding is effectively irreversible due to a close coupling between functional inactivation and nucleolytic degradation. (b) An mRNA where repressor binding is effectively reversible and the rate of nucleolytic attack is unaffected by repression. (c) An mRNA where repressor binding is effectively irreversible (perhaps due to an extremely slow rate of dissociation of the repressor–mRNA complex), even though the rate of nucleolytic degradation is unaffected by repression. If it is assumed that the mRNA in a would be inactivated by nucleolytic attack with a mean-life of 6 min in the absence of repressor, the net effect of the repression in all three cases would be a threefold reduction of the rate of gene expression.

tional and physical lifetimes of mRNA. The binding of the repressor may directly affect the functional lifetime of the mRNA, and via the effect on translation it may indirectly affect the rate of nucleolytic attack on the transcript. Thus, there are two potential mechanisms of functional inactivation, and in principle both of them might contribute significantly to the observed rate of inactivation. However, in contrast to the previous example, there is not an obligatory mechanistic coupling between functional inactivation and nucleolytic attack on the transcript, and so the observed rate of functional inactivation may sometimes be faster than the rate of physical decay, as we shall see below.

IV. The Observed Relation between Translation Frequency and mRNA Stability

A. Post-Transcriptional Regulation: Control of Translation Frequency or mRNA Stability?

1. *Control of mRNA Stability*

Although the primary effect of translational repressors is to prevent initiation of translation, it appears that the resulting decrease in the rate of protein synthesis can sometimes be accounted for entirely by a reduction in mRNA concentration due to accelerated mRNA decay. This implies the seemingly paradoxical conclusion that the translational frequency is not affected by translational repression.

The *E. coli* α operon encodes the α subunit of RNA polymerase and ribosomal proteins S13, S11, S4, and L17 (Fig. 4a). Excess S4 protein represses translation of all four ribosomal protein cistrons by binding to a single operator site near the 5′ end of the transcript (Thomas *et al.*, 1987). In an *E. coli* strain with a mutation in the S4 gene, the rate of synthesis of ribosomal proteins S13, S11, and S4 is increased about 2-fold. This derepression can be fully accounted for by a concomitant 2- to 3-fold increase in the stability and steady-state amount of the mRNA, implying that the translation frequency is unchanged (Singer and Nomura, 1985). Similarly, a mutation in the translational operator site of the operon encoding the ribosomal proteins L11 and L1 results in a 10-fold increase in the rate of synthesis of these proteins accompanied by an increase of 5-fold or more in the chemical half-life of the transcript (Cole and Nomura, 1986). Presumably the repression of these transcripts is effectively irreversible, because once the repressor is bound they are rapidly attacked at extremely vulnerable nucleolytic target sites, which require frequent translation for their protection. Thus, the nonnucleolytic event of repressor binding becomes the rate-determining step in the transcripts' functional inactivation as well as their degradation, in analogy with the hypothetical mRNA of Fig. 3a.

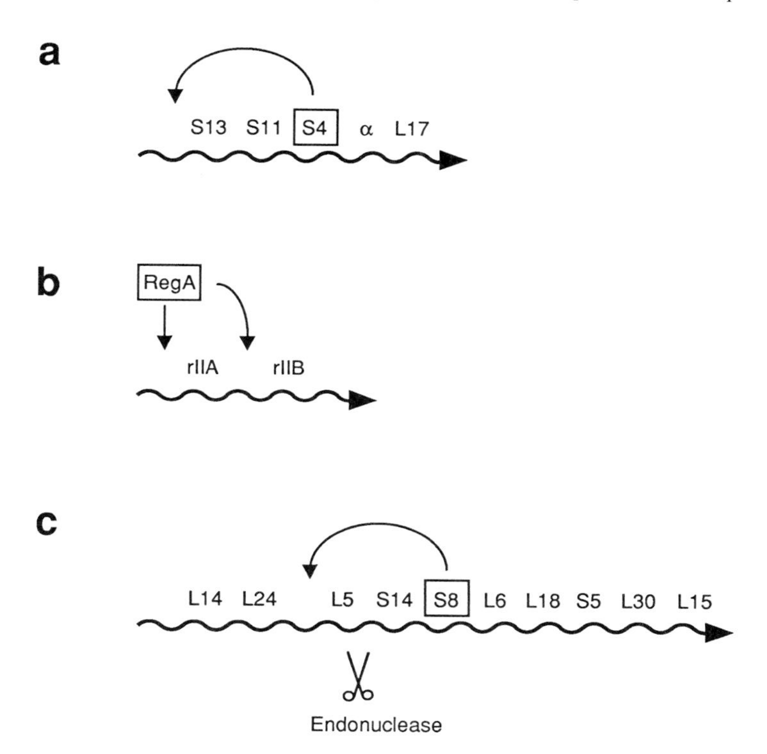

FIGURE 4 Examples of translationally regulated operons for which the effect of repression on mRNA stability has been investigated. (a) The α operon of *E. coli*. (b) The *r*II operon of bacteriophage T4. (c) The *spc* operon of *E. coli*. For each transcript the encoded proteins as well as the translational repressor (boxed) and its site of action are indicated. To my knowledge the exact location of the translational operator of the *r*IIA cistron has not been determined. However, it is tentatively indicated here to overlap the initiation codon as is the case for the known RegA binding sites on other mRNAs (Unnithan *et al.*, 1990). See text for further details.

Likewise, translational repression by anti-sense RNAs would be expected to be essentially irreversible, because the duplex formed with the mRNA often is cleaved readily by RNase III (Blomberg *et al.*, 1990; Case *et al.*, 1990; Krinke *et al.*, 1991).

Many of the "early" transcripts of bacteriophage T4 are translationally repressed by binding of the *regA* gene product (Karam and Bowles, 1974; Karam *et al.*, 1977; Hsu and Karam, 1990). For some of these, e.g., the *r*IIA message (Fig. 4b), it appears that the extent of repression can be fully accounted for by the decrease of the functional mRNA half-life, which implies that the repression is effectively irreversible (Karam *et al.*, 1977). However, this effect is apparently not due to a tight coupling between the functional inactivation of these mRNAs and their subsequent nucleolytic degradation. The *rIIA* message and five other T4 transcripts have been found to remain structurally intact in the cell for considerable periods of time after their functional inactivation by the RegA repressor, thus resembling the hypothetical mRNA of Fig. 3c (Trimble and Maley, 1976). However, for other RegA-controlled transcripts, e.g., those encoded by T4 genes 45 and 46, repression does appear to be rapidly followed by mRNA degradation, in analogy with the mRNA of Fig. 3a (Trimble and Maley, 1976; Hsu and Karam, 1990).

In the *spc* operon of *E. coli* (Fig. 4c), the stability of the mRNA segments that encode ribosomal proteins L14 and L24 is modulated by the S8 repressor protein by a mechanism termed "retroregulation" (Mattheakis *et al.*, 1989). The S8 protein binds downstream of the L14–L24 cistrons to repress translation of the distal cistrons of the operon. This repression presumably leads to exposure of an endonucleolytic target site, which upon cleavage provides an entry point for 3′ exonucleases that subsequently inactivate the upstream L14 and L24 cistrons.

Curiously, translational repression of the mRNA encoding ribosomal protein S20 in *E. coli* is accompanied by an *increase* in mRNA stability (Mackie, 1987). The reason for this remarkable inverse correlation between translation frequency and mRNA stability is unclear, but it could be that the repressor affects the stability of the transcript directly, e.g., by protecting a nucleolytic target site. In any event, it appears that the S20 message may also be protected by ribosomes. Mutations which increase the translation efficiency, e.g., by changing the suboptimal initiation codon UUG to the canonical AUG, abolish translational repression and stabilize the mRNA (Parsons *et al.*, 1988).

2. *Control of Translation Frequency*

Like the ribosomal protein genes, gene 32 of bacteriophage T4 is autogenously regulated at the translational level (reviewed by Gold, 1988). But unlike the ribosomal protein mRNAs, which even under derepressed conditions decay with half-lives of only a few minutes, the gene 32 message is extremely stable. Under repressed conditions it has a functional half-life of more than 15 min, and gene 32 mutations that abolish repression result in a 40-fold increase in the rate of protein synthesis accompanied by a 2-fold increase of the functional half-life (Russell *et al.*, 1976; Gold *et al.*, 1976). Apparently, the gene 32 message is rather insensitive to nucleolytic attack even when essentially unprotected by ribosomes, perhaps owing to protective structures near its 5′ end (see Chapter 3). Consequently, repression of this transcript is manifested predominantly in a decreased translational frequency, with only a slight effect on mRNA stability, in analogy with the hypothetical example of Fig. 3b. However, considering all the counterexamples cited above, such a partitioning of the effects of repression would appear to be the exception rather than the rule.

B. Genetic Modulation of the Ribosome Binding Site Efficiency

In several studies the relation between translation and mRNA decay has been explored by comparing the stability of transcripts that differ in translational efficiency due to genetic alterations near the ribosome binding site. The interpretation of such studies is generally difficult because it is virtually impossible to determine if such manipulations affect mRNA stability solely via their effect on the frequency of ribosome loading or whether they affect the intrinsic stability of the transcript, e.g., by introducing or destroying nucleolytic target sites or protective structures. In the latter case, any change in mRNA stability might be observed independently of the effect on translation.

Matters are further complicated if we consider that sequence alterations may affect the formation of secondary structures in the mRNA, which can efficiently prevent initiation of translation (reviewed by de Smit and van Duin, 1990). It is conceivable that the formation of such intramolecular structures in some cases might be effectively irreversible, just like the binding of trans-acting repressors. Consequently, their formation might determine the functional lifetime of the transcript directly with no measurable effect on the translation frequency, which would reflect the normal translation occurring prior to formation of the structure. As in the case of binding by translational repressors, the effect of such structures on gene expression would in general be partitioned between an effect on translation frequency and on mRNA lifetime, leading to the same classes of relation between these parameters as discussed above.

1. A Steep Response

In some studies, sequence alterations near the ribosome binding site have been found to affect gene expression and mRNA stability dramatically, with no apparent effect on the translational frequency. The functional lifetime of the *lacZ* message is severely reduced by minor sequence alterations, which increase the potential for formation of stable secondary structures in the vicinity of the ribosome binding site without significantly affecting the measurable translation frequency (Petersen, 1987; 1991) (see Fig. 5). Similar results have been reported for mutations, which generally decrease the potential for secondary structure formation upstream of the Shine–Dalgarno sequence of the *trpE* gene (Cho and Yanofsky, 1988). Such mutations give rise to an increased rate of enzyme synthesis that can be accounted for by a parallel increase in the stability of the mRNA, indicating that the ribosome loading frequency is essentially unaffected. Furthermore, formation of a long-range secondary structure that blocks the *E. coli rplL* ribosome binding site in the absence of translation of the upstream *rplJ* cistron results in a severe decrease in mRNA stability, which can essentially account for the decrease in gene expression observed for an *rplL–lacZ* gene

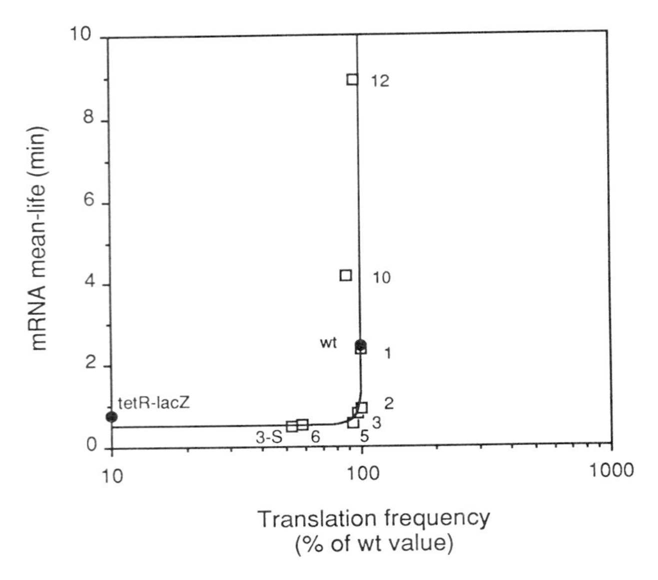

FIGURE 5 Observed relation between translation frequency and mRNA mean-life for different variants of the *lacZ* transcript. Closed circles represent chemical mean-lives calculated from the data of Baumeister *et al.* (1991) for the wild-type *lacZ* allele on λ *plac*5 (wt) and a *tetR–lacZ* gene fusion. The squares represent functional mean-lives calculated from the data of Petersen (1987, 1991) for a series of plasmid-encoded *lacZ* alleles with various sequence alterations at the ribosome binding site. The number beside each point refers to the name of the corresponding plasmid; e.g., 12 refers to pCNP12.

fusion (Petersen, 1989; and unpublished data). All these results suggest that inactivation and degradation of some mRNAs may be triggered or accelerated by the formation of secondary structures that sequester their ribosome binding site, even though the observed translation frequency is unchanged.

This conclusion is further supported by two recent studies on the relation between translation and degradation of a series of *lacZ* mRNA variants that contain different ribosome binding sites (Guillerez *et al.*, 1991; Yarchuk *et al.*, 1991). Furthermore, Yarchuk *et al.* (1991) found that the stability of some *lacZ* mRNA variants was increased in an RNase E-deficient background strain and proposed that there is competition between ribosomes and RNase E for the *lacZ* mRNA (cf. Section III.A). According to this proposal, transcripts that contain inefficient ribosome binding sites would be rapidly attacked by RNAse E, and the effect of the poor ribosome binding site would show up primarily in a reduced mRNA stability, leaving the observable translation frequency unaffected (cf. the steep part of the curves in Fig. 2). However, I recently found that the functional half-lives of the previously described *lacZ* mRNA variants (Petersen, 1987, 1991) were not increased in an RNase E-deficient strain, even though the chemical stability in some cases was dramatically increased (C. Petersen, manuscript in preparation). Thus, I believe that these *lacZ* mRNA variants are functionally inactivated by a mechanism that is independent of RNase E but sensitive to modulation of the ribosome binding site. RNase E, on the other hand, seems to be required only for the rapid degradation of the functionally inactivated transcripts. (See Chapter 4 for further discussion of the role of RNase E in mRNA degradation.)

2. *A Positive Correlation*

For some *lacZ* mRNA variants a positive correlation between the rate of ribosome loading and mRNA stability has been observed (Petersen, 1987). Sequence alterations downstream of the *lacZ* ribosome binding site that resulted in measurably reduced initiation frequencies invariably reduced the functional mRNA half-live compared with the wild-type transcript (Fig. 5). However, transcripts that were already very unstable in spite of an apparently normal translation frequency (functional mean-lives of approximately 40 sec) were not significantly destabilized by manipulations that reduce their translation frequency below the normal level (Petersen, 1987). In agreement with these results, Baumeister *et al.* (1991) found that *tetR–lacZ* fusion transcripts, which are translated at a 10-fold lower frequency than that of the *lacZ* message, have a chemical half-life that is three times shorter and that is not further reduced by a complete block of translation initiation by kasugamycin (Fig. 5). In contrast, the chemical half-life of the *lacZ* message is reduced approximately threefold to 34 sec upon inhibition with kasugamycin, in agreement with previous data obtained by Schneider *et al.* (1978). It appears that modulation of the *lacZ* ribosome binding site can give rise to any relation between translation frequency and mRNA stability, from a steep response to a positive correlation to a flat response (cf. Figs. 2 and 5).

3. *A Flat Response*

In some studies, genetic manipulation of the ribosome binding site has been found to result in dramatic changes in translational frequency with little or no effect on mRNA stability. Stanssens *et al.* (1986) compared a series of different *lacZ* mRNA variants containing a synthetic ribosome binding site with various deletions and substitutions immediately upstream of the Shine–Dalgarno sequence. These mRNAs were transcribed from the λP_L promoter so that transcriptional polarity could be eliminated by the antiterminator protein N of phage λ. Under these conditions the steady-state amount of full-length *lacZ* message visible on a Northern blot was approximately the same for all these strains, although their rates of β-galactosidase synthesis differed by up to 175-fold. It was inferred that the chemical stability of these transcripts was approximately equal, the differences of β-galactosidase synthesis being caused by differences in ribosome loading frequency. Likewise, λP_L–*lacZ* mRNA variants with ribosome binding sites from different cistrons of the *atp* operon of *E. coli* have been found to have similar chemical half-lives, in spite of great differences in their frequencies of ribosome loading (Gerstel and McCarthy, 1989; Hellmuth *et al.* 1991).

The stability of the *rnd* transcript in *E. coli* is also relatively independent of ribosome protection. The steady-state amount of the *rnd* message is unaffected, or even slightly elevated, by deletions and substitutions upstream of the Shine–Dalgarno sequence that reduce the rate of synthesis of RNase D by 2- to 10-fold (Deutscher and Zhang, 1990). The flat response of the λP_L–*lacZ* transcripts is hardly surprising in view of the previously mentioned ability of the λP_L-leader region to confer stability upon functionally inactive *trp* mRNA sequences fused to it (Yamamoto and Imamoto, 1975, 1977) (see also Chapter 3). Perhaps the *rnd* message contains a similar 5′-stabilizing region.

4. *An Inverse Correlation*

An inverse correlation between translation frequency and mRNA stability has been reported in two cases for transcripts derived from the *E. coli* *ompA* gene. The chemical half-life of the native *ompA* message is much longer during growth in rich medium than in acetate minimal medium, although the message is translated less frequently in the rich medium (Lundberg *et al.*, 1988). Furthermore, the stability of *bla* mRNA is increased when its ribosome binding site is replaced with that of the *ompA* transcript, even though its translation frequency appears to be lowered slightly (Emory and Belasco, 1990). The *ompA* leader region is known to function as an efficient mRNA stabilizer (see Chapter 3), and these results presumably reflect direct effects of this region on the intrinsic stability of mRNA rather than indirect effects mediated via changes in the ribosome loading frequency.

C. A General Correlation between Translation Frequency and mRNA Stability?

In summary, both genetic manipulation of the ribosome binding site and modulation of the loading frequency by translational repressors generally result in relations between the translation initiation frequency and mRNA stability that may be classified as a steep response, a positive correlation, or a flat response. Unfortunately the molecular mechanisms responsible for the observed relations are generally unknown. In principle all these relations may result from differences in the degree of protection of nucleolytic target sites (cf. the different regions of the response curves in Fig. 2). However, the same types of relation may be caused by differences in the effective degree of reversibility of nonnucleolytic processes that inhibit initiation of translation (cf. Section III.B).

Furthermore, manipulations that have a direct effect on the intrinsic stability of a transcript may give rise to any kind of relation between translation frequency and mRNA stability, even an inverse correlation. The same is true for manipulations that give rise to ribosome pausing or queueing phenomena. In view of these potential complications there are remarkably few reports of a clear-cut inverse correlation between translation frequency and mRNA stability.

V. Conclusion

Contrary to the situation in many eukaryotic systems there is very little evidence that ribosomes are actively involved in mRNA degradation in prokaryotic cells. On the contrary, ribosomes generally appear to protect mRNA against degradation, although such protection is not universally required. From the observed effects of translation inhibitors and premature termination codons on mRNA stability, it appears that ribosome protection is mostly required near the ribosome binding site, perhaps to protect against initiation of processive degradation in the 5′–3′ direction. Protection by ribosomes of endonucleolytic target sites within coding sequences is presumably of minor importance because such sites are scarce or relatively inert even in the complete absence of translation. In agreement with this suggestion, stability is relatively independent of ribosome protection for mRNAs which contain protective barriers at their 5′ end (e.g., *ermC*, gene 32, λP_L-fusion transcripts). However, the presence of such barriers do not necessarily increase the functional mRNA half-life, even though they protect against degradation (Yamamoto and Imamoto, 1975, 1977; Hue and Bechhofer, 1991). Perhaps this is because some mRNAs may be functionally inactivated by nonnucleolytic mechanisms in spite of the presence of protective structures at their 5′ end (see Petersen (1992) and Chapter 3 for further discussion).

There is no universal correlation between translation frequency and mRNA stability. This relation varies from gene to gene depending on the mode of mRNA inactivation, the location and intrinsic vulnerability of the nucleolytic target site(s), the range of translation initiation frequencies studied, and the method used to modulate translation frequency. The finding that translational repression frequently is effectively irreversible suggests that the functional lifetime of many mRNAs may be determined by nonnucleolytic processes, such as the binding of trans-acting repressors or the formation of secondary structures that prevent initiation of translation. However, the general role of such nonnucleolytic processes in determining mRNA stability and their relation to the various nucleolytic decay processes need to be further defined.

APPENDIX

A. Functional Lifetime, Translation Frequency, and Polypeptide Yield

Although multiple mechansims may contribute to the inactivation of a given species of mRNA, for each individual mRNA molecule there will be only one *inactivating event*, which is the event that limits the polypeptide yield from that molecule. Intuitively one might expect that an mRNA would be more likely to be inactivated at a target located near the 5′ end rather than at a target near the 3′ end, because the 5′ end has been in the cell for a considerable period of time before the 3′ end is even synthesized. Indeed a target located at the 5′ end would most likely be hit *before* an equally vulnerable target near the 3′ end. However, the *first* hit to be dealt to an mRNA molecule is not necessarily the *inactivating* one.

Consider a hypothetical long message, e.g., a *lacZ* transcript, the 5′ end of which is synthesized at time $t = 0$. At $t = 15$ sec, when three ribosomes have initiated translation, further initiation is prevented by an endonucleolytic cleavage at the ribosome binding site. At $t = 90$ sec, the 3′ end of the message is synthesized and 4 sec later the first ribosome terminates translation, but then, at $t = 95$ sec, the transcript is attacked by a 3′ exonuclease that removes the distal part of the coding sequence. The two ribosomes still engaged in polypeptide elongation will produce only truncated nonfunctional proteins. Thus, in this case the inactivating event is *not* the cleavage occurring at $t = 15$ sec, but rather the exonucleolytic attack occurring at $t = 95$ sec, which limits the yield of the transcript to one functional polypeptide. If the rate of ribosome movement is considered to be relatively constant, it follows that the *functional lifetime* of a given message is defined by the minimal time span occurring between the synthesis and the inactivation of any essential segment of that message, which in the present example is approximately 5 sec. Thus the probability that a message will be inactivated at a specific target is dependent only on its relative vulnerability compared with the other targets on the same molecule, not on its position.

The *yield of complete polypeptides* that is produced from an mRNA molecule is given by the product of its functional lifetime and the frequency at which ribosomes pass the target site before it is inactivated. Accordingly, the average polypeptide yield for the message (Y) will be equal to the product of the functional mean-life (t_m) and the translation frequency (f) and will be independent of the location of the target site(s):

$$Y = ft_m.$$

The observed translation frequency is equal to the frequency of ribosome loading only if premature termination of translation is negligible, which is not always the case (Manley, 1978; Jørgensen and Kurland, 1990). In the steady state, the rate of protein synthesis (R_p) is given by

$$R_p = R_m Y,$$

where R_m is the rate of synthesis of completed transcripts, which is equal to the frequency of transcription initiation if there is no premature termination of transcription (transcriptional polarity).

B. Measurements of Functional Half-Lives

A functional mRNA molecule may be considered one that is still discharging ribosomes at its 3′ end and releasing complete polypeptides. The steady-state amount of functional mRNA (M) is given by

$$M = R_m t_m.$$

In order to measure the functional half-life of an mRNA species, one would ideally want to measure the number of functional message molecules (M) at different times after a complete inhibition of their synthesis. Unfortunately, the number of functional mRNA molecules cannot be measured directly. What can be measured is the rate of protein synthesis (R_p) that these molecules give rise to:

$$R_p = Mf.$$

In measurements of functional mRNA stability it is generally assumed that the translation frequency is constant, so that the number of functional mRNA molecules may be considered to be proportional to the observed rate of protein synthesis. This assumption is not necessarily justified when functional half-lives are measured during severely perturbed physiological conditions, such as those imposed by amino acid starvation or by addition of antibiotics. Under such conditions the kinetics of protein synthesis, and thus the observed functional half-life, will reflect changes in the number of functional mRNA molecules as well as changes in their translation frequency. Unfortunately, it is not trivial to separate these effects, due to the inherent difficulty of separately measuring the translation frequency without knowledge of the number of functional mRNA molecules, as we shall see below.

C. Experimental Determination of Relative and Absolute Translation Frequencies

Translation frequencies are generally estimated from a measurement of the steady-state rate of protein synthesis (Rp) and some estimate of the number of functional mRNA molecules (M), using the relation $f = R_p/M$. Usually M is inferred directly from some kind of hybridization measurement (Kennell and Riezman, 1977; Mackie, 1987; Lundberg *et al.*, 1988; Emory and Belasco, 1990). Alternatively, by making use of the relationship $M = R_m t_m$, the number of functional mRNA molecules may be estimated by measuring t_m and estimating R_m from data obtained with an operon fusion to a downstream reporter gene (Petersen, 1987). Both of these methods have their drawbacks. In principle, the number of functional molecules present cannot be deduced from a hybridization measurement unless the precise pathway for functional inactivation and degradation of the mRNA is known (Kennell and Riezman, 1977). This is rarely the case, and so M may be overestimated due to the presence of relatively stable but nonfunctional mRNA molecules, which may also give rise to differences between the functional and the physical half-lives. On the other hand, measurements of R_m by operon fusion may be compromised if the rate of expression of the reporter gene is not a true measure of the rate of synthesis of the upstream message (Petersen, 1987, 1991).

In most studies only relative translation frequencies have been estimated, with one exception. The absolute rate of ribosome loading on the *lacZ* message was estimated to be one initiation per 3.2 sec by Kennell and Riezman (1977). From hybridization measurements of the steady-state amount of *lacZ* message, the average rate of transcription initiation on the *lac* promoter was estimated to be one initiation per 3.3 sec. This is fourfold higher than more recent estimates based on comparisons of different promoters in a system specifically designed to measure promoter strength (Deuschle *et al.*, 1986). Thus, it would appear that the amount of *lacZ* message present was overestimated several-fold by Kennell and Riezman. However, their estimate of the cellular rate of synthesis of β-galactosidase molecules [20 monomers per second per cell, corresponding to 10% of total cellular protein synthesis as calculated from the average composition of an *E. coli* cell (Neidhardt, 1987)] was also overestimated several-fold compared with measurements based on radioactive labeling and gel electrophoresis (Dalbow and Bremer, 1975). Consequently, their estimated translation frequency of the *lacZ* message is probably in the correct range, but given the uncertainties in the determination of the components that went into its calculation, the precise value is questionable. Determination of ribosome loading frequencies is a difficult business.

Acknowledgments

I thank Dr. O. Karlström for helpful comments on the manuscript. This work was supported by the Danish Center for Microbiology.

References

Alifano, P., Piscitelli, C., Blasi, V., Rivellini, F., Nappo, A. G., Bruni, C. B., and Carlomagno, M. S. (1992). Processing of a polycistronic mRNA requires a 5′*cis* element and active translation. *Mol. Microbiol.* **6,** 787–798.

Baumeister, R., Flache, P., Melefors, Ö., von Gabain, A., and Hillen, W. (1991). Lack of a 5′ non-coding region in Tn1721 encoded *tetR* mRNA is associated with a low efficiency of translation and a short half-life in *Escherichia coli. Nucleic Acids Res.* **19,** 4595–4600.

Bechhofer, D. H., and Dubnau, D. (1987). Induced mRNA stability in *Bacillus subtilis. Proc. Natl. Acad. Sci. U.S.A.* **84,** 498–502.

Bechhofer, D. H., and Zen, K. H. (1989). Mechanism of erythromycin-induced *ermC* mRNA stability in *Bacillus subtilis. J. Bacteriol.* **171,** 5803–5811.

Bergmann, J. E., and Lodish, H. F. (1979). A kinetic model of protein synthesis. *J. Biol. Chem.* **254,** 11927–11937.

Blomberg, P., Wagner, E. G. H., and Nordström, K. (1990). Control of replication of plasmid R1: The duplex between the antisense RNA, CopA, and its target, CopT, is processed specifically *in vivo* and *in vitro* by RNase III. *EMBO J.* **9,** 2331–2340.

Bothwell, A. L. M., and Apirion, D. (1971). Is RNase V a manifestation of RNase II? *Biochem. Biophys. Res. Comm.* **44,** 844–851.

Bouvet, P., and Belasco, J. G. (1992). Control of RNase E-mediated RNA degradation by 5′-terminal base pairing in *E. coli. Nature (London)* **360,** 488–491.

Brenner, S., Jacob, F., Meselson, M. (1961). An unstable intermediate carrying information from genes to ribosomes for protein synthesis. *Nature (London)* **190,** 576–581.

Byström, A. S., von Gabain, A., and Björk, G. R. (1989). Differentially expressed *trmD* ribosomal protein operon of *Escherichia coli* is transcribed as a single polycistronic mRNA species. *J. Mol. Biol.* **208,** 575–586.

Case, C. C., Simons, E. L., and Simons, R. W. (1990). The IS*10* transposase mRNA is destabilized during antisense RNA control. *EMBO J.* **9,** 1259–1266.

Chen, C. Y. A., and Belasco, J. G. (1990). Degradation of *pufLMX* mRNA in *Rhodobacter capsulatus* is initiated by nonrandom endonucleolytic cleavage. *J. Bacteriol.* **172,** 4578–4586.

Chen, G. F. T., and Inouye, M. (1990). Suppression of the negative effect of minor arginine codons on gene expression. Preferential usage of minor codons within the first 25 codons of the *Escherichia coli* genes. *Nucleic Acids Res.* **18,** 1465–1473.

Chevrier-Miller, M., Jaques, N., Raibaud, O., and Dreyfus, M. (1990). Transcription of single-copy hybrid *lacZ* genes by T7 RNA polymerase in *Escherichia coli*: mRNA synthesis and degradation can be uncoupled from translation. *Nucleic Acids Res.* **18,** 5787–5792.

Cho, K. O., and Yanofsky, C. (1988). Sequence changes preceding a Shine–Dalgarno region influence *trpE* mRNA translation and decay. *J. Mol. Biol.* **204,** 51–60.

Cole, J. R., and Nomura, M. (1986). Changes in the half-life of ribosomal protein messenger RNA caused by translational repression. *J. Mol. Biol.* **188,** 383–392.

Court, D., de Crombrugghe, B., Adhya, S., and Gottesman, M. (1980). Bacteriophage lambda

Hin function. II. Enhanced stability of lambda messenger RNA. *J. Mol. Biol.* **138,** 731–743.

Craig, E. (1972). Synthesis of specific, stabilized messenger RNA when translocation is blocked in *Escherichia coli*. *Genetics* **70,** 331–336.

Cremer, K. J., Silengo, L., and Schlessinger, D. (1974). Polypeptide formation and polyribosomes in *Escherichia coli* treated with chloramphenicol. *J. Bacteriol.* **118,** 582–589.

Curran, J. F., and Yarus, M. (1989). Rates of aminoacyl-tRNA selection at 29 sense codons *in vivo*. *J. Mol. Biol.* **209,** 65–77.

Dalbow, D. G., and Bremer, H. (1975). Metabolic regulation of β-galactosidase synthesis in *Escherichia coli*. *Biochem. J.* **150,** 1–8.

de Smit, M. H., and van Duin, J. (1990). Control of prokaryotic translational initiation by mRNA secondary structure. *Prog. Nucleic Acid Res. Mol. Biol.* **38,** 1–35.

Deuschle, U., Kammerer, W., Gentz, R., and Bujard, H. (1986). Promoters of *Escherichia coli*. A hierarchy of *in vivo* strength indicates alternate structures. *EMBO J.* **5,** 2987–2994.

Deutscher, M. P., and Zhang, J. (1990). Ribonucleases: diversity and regulation. *In* Post-Transcriptional Control of Gene Expression (J. E. G. McCarthy, and M. F. Tuite, Eds.), NATO ASI Series Vol. H 49, pp.1–11. Springer-Verlag, Berlin/Heidelberg.

Emory, S. A., and Belasco, J. G. (1990). The *ompA* 5′ untranslated RNA segment functions in *Escherichia coli* as a growth-rate-regulated mRNA stabilizer whose activity is unrelated to translational efficiency. *J. Bacteriol.* **172,** 4472–4481.

Fan, D. P., Higa, A., and Levinthal, C. (1964). Messenger RNA decay and protection. *J. Mol. Biol.* **8,** 210–222.

Foley, D., Dennis, P., and Gallant, J. (1981). Mechanism of the *rel* defect in beta-galactosidase synthesis. *J. Bacteriol.* **145,** 641–643.

Gerdes, K., Thisted, T., and Martinussen, J. (1990). Mechanism of post-segregational killing by the *hok/sok* system of plasmid R1: *sok* antisense RNA regulates formation of a *hok* mRNA species correlated with killing of plasmid-free cells. *Mol. Microbiol.* **4,** 1807–1818.

Gerstel, B., and McCarthy, J. E. G. (1989). Independent and coupled translational initiation of *atp* genes in *Escherichia coli*: Experiments using chromosomal and plasmid-borne *lacZ* fusions. *Mol. Microbiol.* **3,** 851–859.

Gold, L. (1988). Post-transcriptional regulatory mechanisms in *Escherichia coli*. *Annu. Rev. Biochem.* **57,** 199–233.

Gold, L., O'Farrell, P. Z., and Russel, M. (1976). Regulation of gene 32 expression during bacteriophage T4 infection of *Escherichia coli*. *J. Biol. Chem.* **251,** 7251–7262.

Gorski, K., Roch, J. M., Prentki, P., and Krisch, H. M. (1985). The stability of bacteriophage T4 gene 32 mRNA: A 5′ leader sequence that can stabilize mRNA transcripts. *Cell* **43,** 461–469.

Gros, F., Hiatt, H., Gilbert, W., Kurland, C. G., Risebrough, R. W., and Watson, J. D. (1961). Unstable ribonucleic acid revealed by pulse labelling of *Escherichia coli*. *Nature* (*London*) **190,** 581–585.

Guillerez, J., Gazeau, M., and Dreyfus, M. (1991). In the *Escherichia coli lacZ* gene the spacing between the translating ribosomes is insensitive to the efficiency of translation initiation. *Nucleic Acids Res.* **19,** 6743–6750.

Gupta, R. S., and Schlessinger, D. (1976). Coupling of rates of transcription, translation, and messenger ribonucleic acid degradation in streptomycin-dependent mutants of *Escherichia coli*. *J. Bacteriol.* **125,** 84–93.

Hall, B. G., and Gallant, J. A. (1973). On the rate of messenger decay during amino acid starvation. *J. Mol. Biol.* **73,** 121–124.

Hansen, M. T., Bennett, P. M., and von Meyenburg, K. (1973). Intracistronic polarity during dissociation of translation from transcription in *Escherichia coli*. *J. Mol. Biol.* **77,** 589–604.

Hellmuth, K., Rex, G., Surin, B., Zinck, R., and McCarthy, J. E. G. (1991). Translational coupling varying in efficiency between different pairs of genes in the central region of the *atp* operon of *Escherichia coli*. *Mol. Microbiol.* **5,** 813–824.

Hiraga, S., and Yanofsky, C. (1972). Hyper-labile messenger RNA in polar mutants of the tryptophan operon of *Escherichia coli. J. Mol. Biol.* **72,** 103–110.

Holmes, R., and Singer, M. (1971). Inability to detect RNase V in *Escherichia coli* and comparison of other ribonucleases before and after infection with coliphage T7. *Biochem. Biophys. Res. Commun.* **44,** 837–843.

Hsu, T., and Karam, J. D. (1990). Transcriptional mapping of a DNA replication gene cluster in bacteriophage T4. *J. Biol. Chem.* **265,** 5303–5316.

Hue, K. K., and Bechhofer, D. H. (1991). Effect of *ermC* leader region mutations on induced mRNA stability. *J. Bacteriol.* **173,** 3732–3740.

Imamoto, F. (1973). Diversity of regulation of genetic transcription. I. Effect of antibiotics which inhibit the process of translation on mRNA metabolism in *Escherichia coli. J. Mol. Biol.* **74,** 113–136.

Imamoto, F., and Kano, Y. (1971). Inhibition of transcription of the tryptophan operon in *Escherichia coli* by a block in initiation of translation. *Nature New Biol.* **232,** 169–173.

Jacob, F., and Monod, J. (1961). Genetic regulatory mechanisms in the synthesis of proteins. *J. Mol. Biol.* **3,** 318–356.

Jørgensen, F., and Kurland, C. G. (1990). Processivity errors of gene expression in *Escherichia coli. J. Mol. Biol.* **215,** 511–521.

Karam, J. D., and Bowles, M. G. (1974). Mutation to overproduction of bacteriophage T4 gene products. *J. Virol.* **13,** 428–438.

Karam, J., McCulley, C., and Leach, M. (1977). Genetic control of mRNA decay in T4 phage-infected *Escherichia coli. Virology* **76,** 685–700.

Kennell, D., and Riezman, H. (1977). Transcription and translation initiation frequencies of the *Escherichia coli lac* operon. *J. Mol. Biol.* **114,** 1–21.

Kennell, D. and Simmons, C. (1972). Synthesis and decay of messenger ribonucleic acid from the lactose operon of *Escherichia coli* during amino-acid starvation. *J. Mol. Biol.* **70,** 451–464.

Klug, G., and Cohen, S. N. (1990). Combined actions of multiple hairpin loop structures and sites of rate-limiting endonucleolytic cleavage determine differential degradation rates of individual segments within polycistronic *puf* operon mRNA. *J. Bacteriol.* **172,** 5140–5146.

Klug, G., and Cohen, S. N. (1991). Effects of translation on degradation of mRNA segments transcribed from the polycistronic *puf* operon of *Rhodobacter capsulatus. J. Bacteriol.* **173,** 1478–1484.

Krinke, L., Mahoney, M., and Wulff, D. L. (1991). The role of the OOP antisense RNA in coliphage λ development. *Mol. Microbiol.* **5,** 1265–1272.

Kuwano, M., Kwan, C. N., Apirion, D., and Schlessinger, D. (1969).Ribonuclease V of *Escherichia coli.* I. Dependence on ribosomes and translocation. *Proc. Natl. Acad. Sci. USA* **64,** 693–700.

Liljenström, H., and von Heijne, G. (1987). Translation rate modification by preferential codon usage: intragenic position effects. *J. Theor. Biol.* **124,** 43–55.

Lundberg, U., Nilsson, G., and von Gabain, A. (1988). The differential stability of the *Echerichia coli ompA* and *bla* mRNA at various growth rates is not correlated to the efficiency of translation. *Gene* **72,** 141–149.

Mackie, G. A. (1987). Posttranscriptional regulation of ribosomal protein S20 and stability of the S20 mRNA species. *J. Bacteriol.* **169,** 2697–2701.

Mattheakis, L., Vu, L., Sor, F., and Nomura, M. (1989). Retroregulation of the synthesis of ribosomal proteins L14 and L24 by feedback repressor S8 in *Escherichia coli. Proc. Natl. Acad. Sci. U.S.A* **86,** 448–452.

McCarthy, J. E. G., Gerstel, B., Surin, B., Wiedemann, U., and Ziemke, P. (1991). Differential gene expression from the *Escherichia coli atp* operon mediated by segmental differences in mRNA stability. *Mol. Microbiol.* **5,** 2447–2458.

McCarthy, J. E. G., and Gualerzi, C. (1990). Translational control of prokaryotic gene expression. *Trends Genet.* **6,** 78–85.

Manley, J. L. (1978). Synthesis and degradation of termination and premature-termination fragments of β-galactosidase *in vitro* and *in vivo*. *J. Mol. Biol.* **125,** 407–432.

Morse, D. E. (1971) Polarity induced by chloramphenicol and relief by *suA*. *J. Mol. Biol.* **55,** 113–118.

Morse, D. E., and Guertin, M. (1971). Regulation of mRNA utilization and degradation by amino-acid starvation. *Nature New Biol.* **232,** 165–169.

Morse, D. E., Mosteller, R. D., and Yanofsky, C. (1969). Dynamics of synthesis, translation, and degradation of *trp* operon messenger RNA in *E. coli*. *Cold Spring Harbor Symp. Quant. Biol.* **34,** 725–739.

Morse, D. E., and Yanofsky, C. (1969). Polarity and the degradation of mRNA. *Nature* (*London*) **224,** 329–331.

Nakada, D., and Fan, D. P. (1964). Protection of β-galactosidase messenger RNA from decay during anaerobiosis in *Escherichia coli*. *J. Mol. Biol.* **8,** 223–230.

Neidhardt, F. C. (1987). Chemical composition of *Escherichia coli*. *In Escherichia coli* and *Salmonella Typhimurium*: Cellular and Molecular Biology (F. C. Neidhardt, J. L. Ingraham, K. B. Low, B. Magasanik, M. Schaechter, and H. E. Umbarger, Eds.), Vol. 1, pp. 3–6. American Society for Microbiology, Washington, D.C.

Newbury, S. F., Smith, N. H., Robinson, E. C., Hiles, I. D., and Higgins, C. F. (1987). Stabilization of translationally active mRNA by prokaryotic REP sequences. *Cell* **48,** 297–310.

Nilsson, G., Belasco, J. G., Cohen, S. N., and von Gabain, A. (1987). Effect of premature termination of translation on mRNA stability depends on the site of ribosome release. *Proc. Natl. Acad. Sci. U.S.A.* **84,** 4890–4894.

Ono, M., and Kuwano, M. (1979). A conditional lethal mutation in an *Escherichia coli* strain with a longer chemical lifetime of messenger RNA. *J. Mol. Biol.* **129,** 343–357.

Parsons, G. D., Donly, B. C., and Mackie, G. A. (1988). Mutations in the leader sequence and initiation codon of the gene for ribosomal protein S20 (*rpsT*) affect both translational efficiency and autoregulation. *J. Bacteriol.* **170,** 2485–2492.

Pato, M. L., Bennett, P. M., and von Meyenburg, K. (1973). Messenger ribonucleic acid synthesis and degradation in *Escherichia coli* during inhibition of translation. *J. Bacteriol.* **116,** 710–718.

Petersen, C. (1987). The functional stability of the *lacZ* transcript is sensitive towards sequence alterations immediately downstream of the ribosome binding site. *Mol. Gen. Genet.* **209,** 179–187.

Petersen, C. (1989). Long-range translational coupling in the *rplJL-rpoBC* operon of *Escherichia coli*. *J. Mol. Biol.* **206,** 323–332.

Petersen, C. (1991) Multiple determinants of functional mRNA stability: Sequence alterations at either end of the *lacZ* gene affect the rate of mRNA inactivation. *J. Bacteriol.* **173,** 2167–2172.

Petersen, C. (1992). Control of functional mRNA stability in bacteria: Multiple mechanisms of nucleolytic and non-nucleolytic inactivation. *Mol. Microbiol.* **6,** 277–282.

Puga, A., Borrás, M. T., Tessman, E. S., Tessman, I. (1973). Difference between functional and structural integrity of messenger RNA. *Proc. Natl. Acad. Sci. U.S.A* **70,** 2171–2175.

Ray, P. N., and Pearsson, M. L. (1974). Evidence for post-transcriptional control of the morphogenetic genes of bacteriophage lambda. *J. Mol. Biol.* **85,** 163–175.

Ray, P. N., and Pearsson, M. L. (1975). Functional inactivation of bacteriophage λ morphogenetic gene mRNA. *Nature* (*London*) **253,** 647–650.

Ruckman, J., Parma, D., Tuerk, C., Hall, D. H., and Gold, L. (1989). Identification of a T4 gene required for bacteriophage mRNA processing. *New Biologist* **1,** 54–65.

Russel, M., Gold, L., Morrisett, H., and O'Farrell, P. Z. (1976). Translational, autogenous regulation of gene 32 expression during bacteriophage T4 infection. *J. Biol. Chem.* **251,** 7263–7270.

Sandler, P., and Weisblum, B. (1988). Erythromycin-induced stabilization of *ermA* messenger RNA in *Staphylococcus aureus* and *Bacillus subtilis*. *J. Mol. Biol.* **203,** 905–915.

Sandler, P., and Weisblum, B. (1989). Erythromycin-induced ribosome stall in the *ermA* leader: A barricade to 5′-to-3′ nucleolytic cleavage of the *ermA* transcript. *J. Bacteriol.* **171,** 6680–6688.

Schneider, E., Blundell, M., and Kennell, D. (1978). Translation and mRNA decay. *Mol. Gen. Genet.* **160,** 121–129.

Schwartz, T., Craig, E., and Kennell, D. (1970). Inactivation and degradation of messenger ribonucleic acid from the lactose operon of *Escherichia coli*. *J. Mol. Biol.* **54,** 299–311.

Singer, P., and Nomura, M. (1985). Stability of ribosomal protein mRNA and translational feedback regulation in *Escherichia coli*. *Mol. Gen. Genet.* **199,** 543–546.

Stanssens, P., Remaut, E., and Fiers, W. (1986). Inefficient translation initiation causes premature transcription termination in the *lacZ* gene. *Cell* **44,** 711–718.

Sørensen, M. A., and Pedersen, S. (1991). Absolute *in vivo* translation rates of individual codons in *Escherichia coli*. *J. Mol. Biol.* **222,** 265–280.

Thomas, M. S., Bedwell, D. M., and Nomura, M. (1987). Regulation of α operon gene expression in *Escherichia coli:* A novel form of translational coupling. *J. Mol. Biol.* **196,** 333–345.

Trimble, R. B., and Maley, F. (1976). Level of specific prereplicative mRNAs during bacteriophage T4 *regA*-, 43- and T4 43- infection of *Escherichia coli* B. *J. Virol.* **17,** 538–549.

Unnithan, S., Green, L., Morrissey, L., Binkley, J., Singer, B., Karam, J., and Gold, L. (1990). Binding of the bacteriophage T4 regA protein to mRNA targets: An initiator AUG is required. *Nucleic Acids Res.* **18,** 7083–7092.

Varmus, H. E., Perlman, R. L., and Pastan, I. (1971). Regulation of *lac* transcription in antibiotic-treated *E. coli*. *Nature New Biol.* **230,** 41–44.

Yamada, Y., Whitaker, P. A., and Nakada, D. (1974). Early to late switch in bacteriophage T7 development: Functional decay of T7 early messenger RNA. *J. Mol. Biol.* **89,** 293–303.

Yamamoto, T., and Imamoto, F. (1975). Differential stability of *trp* messenger RNA synthesized originating at the *trp* promoter and P_L promoter of lambda *trp* phage. *J. Mol. Biol.* **92,** 289–309.

Yamamoto, T., and Imamoto, F. (1977). Function of the *tof* gene product in modifying chemical stability of *trp* messenger RNA synthesized from the P_L promoter of λ *trp* phage. *Mol. Gen. Genet.* **155,** 131–138.

Yarchuk, O., Iost, I., and Dreyfus, M. (1991). The relation between translation and mRNA degradation in the *lacZ* gene. *Biochimie* **73,** 1533–1541.

PART II

EUKARYOTES

7

mRNA Degradation in Eukaryotic Cells: An Overview

GEORGE BRAWERMAN

I. Importance in Control of Gene Expression

The process of mRNA decay is generally recognized by now as a major control point for gene expression. The potential for regulation is obvious in eukaryotes, where a wide diversity in decay rates is seen. The high degree of metabolic stability of some mRNA species, such as the globin mRNAs, no doubt contributes to their accumulation to high steady-state levels. Highly unstable species with half-lives of about 10–15 min, such as the lymphokine and protooncogene mRNAs, are normally present at low steady-state levels. Although low mRNA abundance could be achieved perhaps more easily through a low rate of transcription, the high rate of turnover permits rapid cessation of the production of a protein when it is no longer needed and also its more rapid induction (see Chapter 9). This is of particular importance in the case of proteins that play critical regulatory roles for brief periods during a developmental process.

The fact that the decay of many mRNAs is subject to regulation provides a powerful means for controlling gene expression during physiological transitions. A wide variety of stimuli and cellular signals have been found to affect the decay of selected mRNA species (see Chapter 8). Notably, steroid hormones tend to affect the stability of the target mRNAs as well as the transcription of the target genes. Such combined actions can lead to more rapid and more marked shifts in the expression a particular genes. Sometimes, the regulation of expression of a particular protein can be attributed primarily or solely to a shift in the rate of decay of the corresponding mRNA. Such is the case for the synthesis of histones, which takes place only during the S phase of the cell cycle (see Chapter 12), the synthesis of transferrin receptor, which is turned off when the cell is exposed to excess iron (see Chapter 11), and the synthesis of tubulin, which is regulated by the degree of depolymerization of microtubules (see Chapter 10). The transient induction of proteins encoded in some of the highly unstable mRNAs may involve mRNA stabilization in addition to activation of gene transcription (see Chapter 9).

Some developmental processes, such as erythroid differentiation, require massive reorganization of the pattern of gene expression. In the latter case, induction of globin gene transcription is coupled with destabilization of a large number of cellular mRNAs (Volloch and Housman, 1981; Krowczynska *et al.*, 1985). This permits a rapid shift from a relatively undifferentiated cell to one devoted primarily to the production of hemoglobin. Shifts in mRNA stability are also known to contribute to *Xenopus* oocyte and early embryo development (see Chapter 8).

mRNA destabilization is also one of the various strategies used by viruses to take over the biosynthetic machinery of host cells. For instance, the Herpes virion has been found to carry a protein that promotes rapid degradation of cellular mRNA (see Chapter 16).

II. Basis for Selectivity in mRNA Degradation

A. Sources of Diversity in Decay Rates

The diversity in mRNA decay rates implies the existence of specific endonucleolytic targets on individual mRNA species. This diversity could be based on interactions with a battery of endonucleases, each recognizing a particular type of sequence or structure. Such an arrangement could provide a means for independent regulation of groups of mRNAs. Discrimination could also be achieved by having less specific targets embedded to varying degrees within complex mRNA configurations. Binding of protein factors could provide additional complexity to the masking structures. Regulation could then be achieved by a shift in RNA conformation and/or by activation or release of a masking protein (see Chapter 8). Regulation by an RNA-binding factor has been clearly demonstrated in the case of the transferrin receptor mRNA (see Chapter 11). The regulation of histone mRNA stability also appears to operate by binding of a protein to its 3′-terminal stem–loop structure (see Chapter 12).

Another means for achieving diversity is suggested by the finding that mammalian mRNA chains carry an endonuclease activity as part of their native nucleoprotein structure (see Chapters 8 and 17). This activity is normally repressed by a nuclease inhibitor, and it is conceivable that nuclease-inhibitor interactions may be modulated by features of the secondary structure of individual mRNA species (Bandyopadhyay *et al.*, 1990).

The apparent link between translation and mRNA decay provides the potential for an additional element of complexity, due to the fact that many mRNA species are subject to translational control. Thus, the processes that control translation initiation could affect mRNA degradation. Also, events during the translation elongation process, such as ribosome stalling or frame-shifting, could affect the stability of individual mRNAs (see Chapter 13).

B. Special Sequences Promote mRNA Destabilization

It is quite well established by now that special RNA sequences can determine the metabolic lifetime of mRNA. The known "stability" determinants seem to function by promoting rapid decay of species in which they reside. This feature indicates that eukaryotic mRNA chains are intrinsically stable in the absence of special features that promote rapid degradation. Such is the case of the long AU-rich segment in the 3′ UTR of the highly unstable mRNAs for cytokines, protooncogenes, and interferons (see Chapter 9). The 3′ stem–loop of histone mRNAs also serves as a signal for rapid degradation at the end of the phase of DNA synthesis (see Chapter 12). A sequence element in the 3′ UTR of transferrin receptor mRNA may function in the same fashion, under the control of an iron-responsive element (see Chapter 11). Messenger RNA instability can also be conferred by coding region sequences (see Chapters 9 and 13).

C. Protecting Elements at the mRNA Termini

The high degree of metabolic stability of mRNAs that lack destabilizing elements implies that these long-lived mRNAs are normally protected from the action of cellular RNases. In prokaryotes, and possibly also in the mitochondria and chloroplasts of eukaryotes, a stem–loop structure at the 3′ end protects against the action of 3′ exonucleases (see Chapters 2 and 14). A similar protection mechanism has been postulated in eukaryotes, with the poly(A) tail in conjunction with the poly(A)-binding protein (PABP) taking on the role of the prokaryotic stem–loop structure. Although there is some correlation between rapid poly(A) shortening and rapid mRNA decay, clear evidence for a direct protecting effect of this 3′-terminal sequence *in vivo* is scarce. Some evidence suggests that the poly(A) tail may be involved in the translation process (see Chapter 15). Thus the possibility must be considered that mRNA chains may become destabilized as a result of a defect in translation initiation caused by the absence of the poly(A)–protein complex. Such a possibility is supported by indications that AU-rich destabilizing sequences can affect mRNA translation efficiency (Tonouchi *et al.*, 1989; Kruys *et al.*, 1989). A better understanding of the translation initiation process and of the possible role of the poly(A)–PABP complex in this process will be required to resolve this problem.

The cap structure present at the 5′ terminus of eukaryotic mRNA has also been implicated as a protecting element. In this case, the evidence, although indirect, seems more compelling. It was originally observed that addition of a cap to RNA transcripts renders them more resistant to degradation in cell extracts (Furuichi *et al.*, 1977). More significantly, perhaps, the same effect was observed when capped and uncapped transcripts were injected into *Xenopus* oocytes. These findings suggested that mRNA chains need to be protected against 5′ exonucleolytic attack in the cell and that the cap structure functions as a protecting barrier. The fact that exoribonucleases that cleave in a 5′-to-3′ direction are present in eukaryotic cells (see Chapter 17) suggests that a protecting structure at the 5′ end is necessary. These findings point clearly to the need for such protection in the nucleus, since most of the 5′ exonucleases identified so far have been obtained from nuclear extracts. Studies with cytoplasmic extracts of mouse ascites cells indicate that this kind of protection also operates in the cytoplasm (Coutts *et al.*, 1993).

In order to understand the significance of protection by 5′ blocked termini, it must be determined whether this kind of protection is absolute or whether 5′ exonucleolytic attack can be initiated under certain circumstances. An enzyme that can hydrolyze the pyrophosphate linkage in the cap structure on mRNA has been indentified in yeast, and such a cleavage would leave the mRNA vulnerable to the action of 5′ exonucleases (see Chapter 17). It is conceivable that a decapping event, possibly modulated by events during the translation initiation process, could serve as an initial step in mRNA degradation in yeast (see model in Fig. 1). Although cap hydrolysis is also seen in mammalian cytoplasmic extracts, the enzyme involved can act only on very short capped oligonucleotides (Nuss *et al.*, 1975). Thus it seems unlikely that 5′ exonucleolytic attack in mammalian cells can be triggered by prior removal of the cap structure from intact mRNA.

D. Protection against Endonucleases

In view of the widespread occurrence of intracellular endoribonucleases, there may be a need for protection of mRNA against random endonucleolytic cleavage. It is possible that this kind of protection is due primarily to the segregation of cellular nucleases in compartments such as lysosomes, but some endonucleases may be present in the cytoplasm in a diffusible state, allowing them access to mRNA molecules. However, cleavage by these enzymes is impeded by RNase inhibitors. Such inhibitors have been identified in a variety of mammalian cells (see Chapter 16) and may be ubiquitous. Thus it is possible that interactions between endonucleases

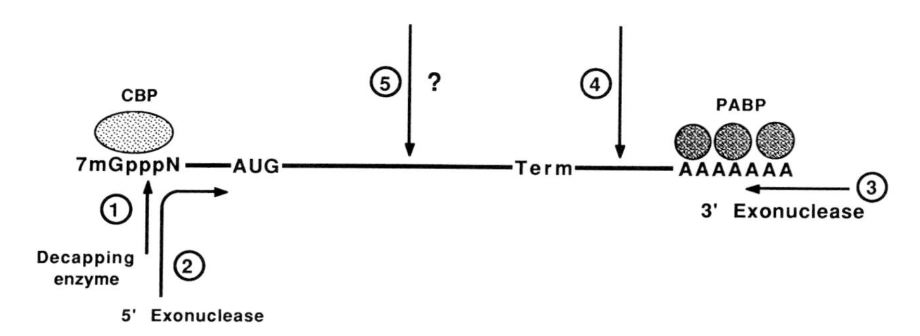

FIGURE 1 Exonucleolytic and endonucleolytic attack of eukaryotic mRNA. mRNA depicted with cap-binding protein (CBP) bound to the cap structure and poly(A)-binding proteins (PABP) bound to the poly(A) sequence. In yeast, decapping enzyme can remove 7mGpp from the 5′ end (1), thus leaving mRNA vulnerable to digestion by 5′ exonuclease (2). Reaction (1) does not take place in mammals, but 5′ exonuclease that can attack capped RNA could carry out reaction (2) in these cells. The poly(A) sequence at the 3′ end can be degraded in exonucleolytic fashion (3) by poly(A)-specific 3′ exonuclease (in mammals) or by PABP-dependent poly(A) nuclease (in yeast). Endonucleolytic cleavage in the 3′ UTR (4) causes loss of the poly(A) tail and of part of UTR, leaving the 5′ UTR and the coding region still intact. The possibility of endonucleolytic cleavage within the coding region (5) is suggested by the occurrence of coding-region instability determinants, but such cleavages have not been detected so far.

and RNase inhibitors play a major role in protecting mRNA against random degradation. Such interactions, if properly modulated, could also contribute to controlling mRNA decay in the cell (Bandyopadhyay *et al.*, 1990).

III. Nucleases Involved in mRNA Decay, and Their Targets

A. Nature of Target Sites for Endonuclease

It has been difficult to identify cleavages in individual mRNAs that could be attributed to an initial inactivating step. Decay intermediates are usually not detectable, presumably because of their rapid destruction in the cell. In the case of histone mRNAs, truncated molecules produced by cleavages in the 3′-terminal stem–loop structure have been observed. The fact that these fragments accumulate to higher levels when the mRNAs are destabilized suggests that they are true intermediates in the regulated decay process (see Chapters 12 and 16). It remains to be established whether this fragmentation is due to processive 3′-exonucleolytic attack or to specific endonucleolytic cleavages (see Chapter 12). Fragmentation products have been identified in the cases of 9E3 and *gro* mRNAs (Stoeckle and Hanafusa, 1989). In these two instances, the enzyme involved is clearly an endonuclease, since both the upstream and the downstream cleavage fragments were seen. Truncated globin and VLDL apoprotein mRNAs that may have been generated *in vivo* by fragmentation at AU dinucleotide sequences have also been indentified (see Chapter 8).

The relation between the destabilizing sequences identified so far and the nucleolytic targets controlled by these sequences is rather obscure. In the case of the histone mRNAs, it seems that the 3′ stem–loop, which represents the structure that responds to the destabilizing signal, is also the site of initial cleavage (see Chapter 12). In the transferrin receptor mRNA, the structure involved in regulation appears to be distinct from the destabilizing sequence. The initial cleavage sites are not known in this case, although a fragment corresponding to cleavage in the vicinity of the control region has been seen (see Chapter 11). The AU-rich destabilizing sequence of many short-lived mRNAs does not appear to serve as target for nucleolytic cleavage. In the case of 9E3 mRNA, which carries such a sequence, an initial endonucleolytic cleavage takes place in the 3′ UTR at a site quite distant from the AU-rich element (Stoeckle and Hanafusa, 1989). In the *c-myc* and *c-fos* mRNAs, the presence of the AU-rich element leads to rapid degradation of the poly(A) tail (see Chapters 9 and 15).

B. Identification of Nucleases

The identification and characterization of nucleases and other protein factors that carry out the rate-determining steps in mRNA decay could go a long way toward elucidating the mechanisms by which the cell can discriminate between individual mRNA species and thereby achieve selective and regulatable mRNA degradation. Although the genetic approach would seem like the rational way to deal with this problem, this approach is available only to the study of lower eukaryotes such as yeast (see Chapters 13, 15, and 17) and has had limited success so far.

For higher eukaryotes, the only approach currently available is the isolation of nucleases that appear to function in a selective fashion in cell extracts. Some cell-free systems display remarkable selectivity in the degradation of individual mRNA species (see Chapter 16) and offer great promise as starting points for the isolation of mRNA-specific nucleases. It is no easy task, however, to obtain objective evidence that any nuclease identified in this fashion is in fact involved in mRNA turnover in intact cells. The histone mRNA decay process provides a potentially useful experimental system for the identification of an mRNA-degrading enzyme. Cleavage products that appear to represent the initial stage of the *in vivo* decay process have been detected, and these cleavages can be reproduced in a cell-free system (see Chapter 16). Isolation and characterization of the enzyme responsible for these cleavages should resolve questions concerning its mode of action, the basis for its specificity, and the manner in which it permits regulation of histone mRNA decay. Several other examples of highly selective mRNA degradation *in vitro* have been described (see Chapter 16), thus making the prospect of isolating relevant nucleases and accessory factors seem like a realistic and exciting goal. It may seem surprising, therefore, that no meaningful progress has been achieved so far in this direction.

The search for exonucleases possibly involved in the mRNA decay process has been somewhat more rewarding. Exonucleases that cleave in a 5′-to-3′ direction have been identified in yeast and in mammalian cells (see Chapter 17). Many of these enzymes were obtained from nuclear extracts, but it is possible that the yeast exonucleases function in the cytoplasm. Recent studies have led to the identification of a 5′ exonuclease in cytoplasmic extracts of murine sarcoma-180 ascites cells (Coutts and Brawerman, 1993). Exoribonucleases seem to be unaffected by the RNase inhibitors that block many endonucleases. Thus, these enzymes are likely to be functional in the intracellular environment and to have the potential to attack mRNA. However, it remains to be demonstrated unambiguously that the mRNA in the cytoplasm is accessible to these enzymes. Since most 5′ exonucleases cannot attack RNA chains terminated with a cap structure, 5′ exonucleolytic attack of mRNA would require prior removal of this structure. A decapping enzyme has been identified in yeast, and it is possible that mRNA decay in this organism could be initiated by decapping followed by 5′ exonucleolytic attack (see Chapter 17). The cytoplasmic 5′ exonuclease identified in mouse ascites cells is capable of attacking capped RNA molecules (Coutts and Brawerman, 1993), and in this case the protection of capped transcripts seen in cytoplasmic extracts may involve interaction of the 5′ terminus with proteins. Since mammalian cells appear to be unable to hydrolyze the cap on intact mRNA chains (Nuss *et al.*, 1975), 5′ exonucleolytic attack in these cells would require such an enzyme capable of bypassing the cap structure. Proteins bound to the 5′-terminal region of eukaryotic mRNA could be involved in the control of decay, either by blocking cap hydrolysis (in yeast) or by interfering with the exonucleolytic attack itself (in mammals). Further characterization of the enzymes (exonuclease and cap hydrolase) and of their potential targets at the 5′ end of the native mRNA–protein complexes could help to define an important mRNA-decay process.

Exonucleolytic degradation in the 3′–5′ direction has also been implicated in the mRNA decay process (see Chapter 16). A 3′ exonuclease could easily account for the apparent protective effect of the poly(A) tail at the 3′ end of mRNA, as well as for some other features of the degradation of certain mRNA species. Although a crude preparation derived from polyribosomes has been shown to display characteristics expected of a 3′ exonuclease (Ross *et al.*, 1987), no such eukaryotic enzyme capable of digesting the entire RNA chain has been purified and characterized so far. A 3′ exonuclease specific for the poly(A) tail, but unable to continue the degradation process beyond the poly(A) junction, has been identified in mammalian cell extracts (Astrom *et al.*, 1991). This enzyme could account for the poly(A) shortening process in these cells. The yeast poly(A) nuclease that requires the presence of the poly(A)-binding protein also appears to be a 3′ exonuclease active only on the poly(A) tail (see Chapter 15).

IV. Relation of mRNA Decay to the Translation Process

A. Inhibitors of Protein Synthesis Prevent the Rapid Decay of Unstable mRNAs

There is much evidence for a link between translation and mRNA decay. First, inhibitors of protein synthesis tend to stabilize unstable mRNAs (see Chapters 9–12). The effect of cycloheximide, an inhibitor of polypeptide chain elongation, has been attributed to physical protection by the ribosomes that accumulate on the mRNA. However, puromycin, which causes premature release of ribosomes from the mRNA and thus leaves much of the RNA chain without this kind of protection, also leads to mRNA stabilization. Inhibitors of translation initiation have the same effect. The one common feature shared by these different inhibitors is that they interrupt the supply of protein molecules to the cell. This has led to the suggestion that a nuclease or other protein factor involved in mRNA decay is itself unstable and needs to be continuously replenished. This explanation could easily account for many of the aspects of the relation between translation and decay. However, the very rapid response to these inhibitors and the equally rapid resumption of normal decay upon their removal would require an extremely high rate of turnover of the hypothetical factor, corresponding to a half-life of a few minutes (Rahmsdorf *et al.*, 1987; Wilson and Treisman, 1988). In support of the model is the fact that a soluble factor required for the rapid *in vitro* decay of *c-myc* mRNA is inactive when the extract is derived from cycloheximide-treated cells (Brewer and Ross, 1989).

B. A Linkage between Translation and Decay of Individual mRNAs

There are indications of a more intimate relation between the translating ribosome and the degradation of the mRNA chain being translated. In histone mRNA, introduction of termination codons that lead to release of the ribosomes upstream of the normal translation termination site results in stabilization of the mRNA (see Chapter 12). The critical feature appears to be the need for the ribosome to reach a site sufficiently close to the 3′ stem–loop, which is presumed to be the recipient of the initial cleavage in the decay process. This has led to the suggestion that the ribosome itself may carry the nuclease involved in this cleavage. The same explanation has been offered to account for the fact that decay of tubulin mRNA also requires translation beyond a certain point in the RNA coding region (see Chapter 10). In the latter case, however, the critical instability determinant has been traced to the N-terminal tetrapeptide of the nascent polypeptide chain, and the requirement for ribosome progression could be due simply to the need for the N terminus to protrude from the ribosome structure so that it can be accessible to other components of the mRNA decay machinery. In the case of transferrin receptor mRNA, rapid decay apparently does not require the action of a translating ribosome. This is indicated by the fact that preventing translation initiation on this mRNA does not abolish its rapid decay in response to iron abundance (see Chapter 11).

C. Does the Ribosome Carry a Nuclease Capable of Degrading mRNA?

The effects of translation on the decay of histone and tubulin mRNA, as well as the fact that the coding region itself can carry instability determinants, can be explained by the action of a ribosome-bound nuclease that is normally inactive and that can be activated in the course of translation when it reaches a certain site on the mRNA. In support of this model is the finding of a nuclease activity that is loosely bound to polyribosomes (see Chapter 16). However, it remains to be demonstrated that this nuclease does not become associated with the polyribosome fraction under conditions of cell disruption in low salt. The search for an enzyme that is tightly bound to polyribosomes has led to the finding of a nuclease activity that is associated with the mRNA itself and not with either of the ribosomal subunits (Bandyopadhyay *et al.*, 1990). In the absence of a well-characterized nuclease that clearly is bound to ribosomal particles, the question of whether translating ribosomes carry an mRNA-degrading activity must be left unanswered at present.

Other explanations have been offered for the apparent involvement of ribosomes in the decay of some mRNA species. To account for the occurrence of coding-region instability determinants, it has been proposed that ribosome stalling at certain sites could activate an mRNA-degrading process (see Chapter 13). It has also been suggested that ribosome progression causes mRNA conformation changes that unmask nuclease target sites.

D. A Model That Links mRNA Decay to the Translation Initiation Process

It is not possible at present to formulate a model that can account for all the features of the relation between translation and decay. Moreover, it is quite likely that different decay processes operate on different mRNA species and even on the same mRNA. However, the presence of a poly(A) tail at the 3′ end of almost all mRNAs suggests a general mechanism for mRNA degradation, if one accepts the proposition that the poly(A) functions in both translation and mRNA protection. Current evidence indicates that the poly(A) tail, in conjunction with the PABP, facilitates translation initiation (Grossi de Sa *et al.*, 1988; Munroe and Jacobson, 1990). The poly-(A)–PABP complex appears to be involved in the joining of the 60S ribosomal subunit to the 40S subunit already associated with the mRNA (Sachs and Davis, 1989). It could be that faulty or delayed interaction between the subunits at this point would lead to occasional nucleolytic cleavages in the 5′ UTR, possibly triggered by a 5′ exonuclease (see Fig. 2). In this manner, the poly(A) tail would exert its protective effect indirectly, by preventing mRNA cleavage at the 5′ end. The rate-determining step in mRNA decay would then be either degradation of the poly(A) tail (as in the case of the *c-fos* and *c-myc* mRNAs) or cleavage in the body of the mRNA, which would also lead to loss of the poly(A). The 3′ stem–loop of histone mRNAs could take on the role of the poly(A) tail, by also facilitating the translation initiation step (see Chapter 12). Verification of this model will require more precise knowledge of the events taking place during initiation of protein synthesis.

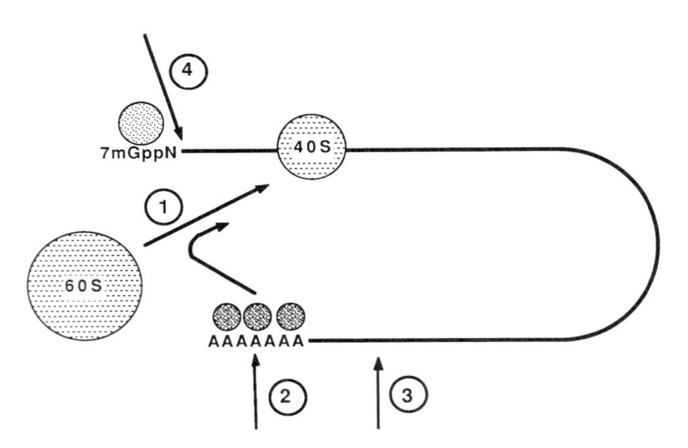

FIGURE 2 A model for linkage between translation initiation and mRNA decay. The poly(A) tail together with the poly(A)-binding proteins is shown to participate in the translation initiation step, by promoting the binding of the 60S ribosomal subunit to the 40S subunit attached to initiation site on mRNA (1). Loss of the poly(A) tail, by exonucleolytic degradation (2) or by endonucleolytic cleavage in 3′ UTR (3), causes loss of the facilitating effect of the poly(A)–PABP complex on step (1). The resulting defect in initiation leads to 5′ exonucleolytic attack on mRNA (4).

References

Astrom, J., Astrom, A., and Virtanen, A. (1991). *In vitro* deadenylation of mammalian mRNA by a HeLa cell 3′ exonuclease. *EMBO J.* **10,** 3067–3071.

Bandyopadhyay, R., Coutts, M., Krowczynska, A., and Brawerman, G. (1990). Nuclease activity associated with mammalian mRNA in its native state: Possible basis for selectivity in mRNA decay. *Mol. Cell. Biol.* **10,** 2060–2069.

Brewer, G., and Ross, J. (1989). Regulation of *c-myc* mRNA stability *in vitro* by a labile destabilizer with an essential nucleic acid component. *Mol. Cell. Biol.* **9,** 1996–2006.

Coutts, M., and Brawerman, G. (1993). A 5′ exoribonuclease from cytoplasmic extracts of mouse sarcoma 180 ascites cells. *Biochim. Biophys Acta* (in press).

Coutts, M., Krowczynska, A., and Brawerman, G. (1993). Protection of mRNA against nucleases in cytoplasmic extracts of mouse sarcoma ascites cells. *Biochim. Biophys. Acta* (in press).

Furuichi, Y., LaFiandra, A., and Shatkin, A. J. (1977). 5′-terminal structure and mRNA stability. *Nature (London)* **266,** 235–239.

Grossi de Sa, M. F., Standart, N., Martins de Sa, C., Akhayat. O., Huesca, M., and Scherrer, K. (1988). The poly(A)-binding protein facilitates *in vitro* translation of poly(A)-rich mRNA. *Eur. J. Biochem.* **176,** 521–526.

Krowczynska, A., Yenofsky, R., and Brawerman, G. (1985). Regulation of messenger RNA stability in mouse erythroleukemia cells. *J. Mol. Biol.* **181,** 231–239.

Kruys, V., Marinx, O., Shaw, G., Deschamps, J., and Huez, G. (1989). Translational blockade imposed by cytokine-derived UA-rich sequences. *Science* **245,** 852–855.

Munroe, D., and Jacobson, A. (1990). mRNA poly(A) tail, a 3′ enhancer of translation initiation. *Mol. Cell. Biol.* **10,** 3441–3455.

Nuss, D. L., Furuichi, Y., Koch, G., and Shatkin, A. J. (1975). Detection in HeLa cell extracts of a 7-methyl guanosine specific enzyme activity that cleaves m7GpppNm. *Cell* **6,** 21–27.

Rahmsdorf, H. J., Schonthal, A., Angel, P., Liftin, M., Ruther U., and Herrlich, P. (1987). Posttranscriptional regulation of c-fos mRNA expression. *Nucleic Acids Res.* **15,** 1643–1659.

Ross, J., Kobs, G., Brewer G., and Peltz, S. W. (1987). Properties of theexonuclease activity that degrades H4 histone mRNA. *J. Biol. Chem.* **262,** 9374–9381.

Sachs, A. B., and Davis, R. W. (1989). The poly(A)-binding protein is required for poly(A) shortening and 60S ribosomal subunit-dependent translation initiation. *Cell* **58,** 858–867.

Stoeckle, M. Y., and Hanafusa, H. (1989). Processing of 9E3 mRNA and regulation of its stability in normal and Rous Sarcomavirus-transformed cells. *Mol. Cell. Biol.* **9,** 4738–4745.

Tonouchi, N., Miwa, K., Karasuyama, H., and Matsui, H. (1989). Deletion of 3′ untranslated region of human BSF-2 mRNA causes stabilization of the mRNA and high-level expression in mouse NIH3T3 cells. *Biochem. Biophys. Res. Commun.* **163,** 1056–1062.

Volloch, V., and Housman, D. (1981). Relative stability of globin and non-globin mRNA in terminally differentiating Friend cells. *In* Organization and Expression of Globin Genes (G. Stamatoyannopoulos and A. N. Nienhuis, Eds.), pp. 251–257. Alan R. Liss, New York.

Wilson, T., and Treisman, R. (1988). Removal of poly(A) and the consequent degradation of *c-fos* mRNA facilitated by 3′ AU-rich sequences. *Nature (London)* **336,** 396–399.

8

Hormonal and Developmental Regulation of mRNA Turnover

DAVID L. WILLIAMS, MARTHA SENSEL,
MONICA McTIGUE, AND ROBERTA BINDER

I. Introduction

In addition to regulating gene expression at the level of transcription, hormonal and developmental signals also modulate gene expression by altering mRNA stability or mRNA degradation. The regulation of mRNA stability during development may be stage-specific and apply to large classes of mRNA transcripts. During the developmental cycle of *Dictyostelium discoideum*, for example, cell aggregation results in the stabilization of a class of mRNAs while disaggregation destabilizes these aggregation-specific mRNAs (Mangiarotti *et al.*, 1983). In contrast, regulation of mRNA turnover by steroid hormones is typically specific for one or a few mRNAs while most cellular mRNAs are unaffected. The estrogen-induced alterations in stability of apolipoprotein II (apoII) and vitellogenin (VTG) mRNAs in chicken liver, for example, do not alter the turnover of other hepatocyte mRNAs (Gordon *et al.*, 1988). These observations suggest that nature has evolved strategies for regulating mRNA turnover that may be global or specific and must include pathways for transducing the hormonal or developmental signal to the cellular mechanisms responsible for mRNA degradation. After a brief historical introduction and a survey of some more extensively studied systems, this chapter will focus on recent studies that ask about the mechanisms underlying the process of regulated mRNA turnover.

Control of Messenger RNA Stability

II. History

Regulated mRNA turnover has been studied over the past 25 years and was recognized early on as an important means of regulating gene expression. From studies on the expression of specialized protein products in terminally differentiated cells, it was recognized that a major committment of the cell to the synthesis of a specific protein could be realized without a major specialization in the transcription of the corresponding gene if the specific mRNA were very stable (Kafatos, 1972). Another important facet of mRNA turnover was the recognition that the steady-state levels of mRNAs with rapid turnover rates will respond quickly to signals that alter either transcription or mRNA degradation compared with mRNAs with slow turnover rates. Extrapolation from the formulations of Gaddum (1944) on drug delivery and of Berlin and Schimke (1965) on protein turnover illustrated that the steady-state level of an mRNA is a function of both synthesis and degradation rates, but the time course of changing the mRNA level in response to a stimulus is determined solely by the degradation rate of the mRNA (Kafatos, 1972). Thus a regulatory protein that must be increased and decreased rapidly in a cell and is controlled at the pretranslational level will likely be encoded by an mRNA with a rapid turnover. This is the case for mRNAs encoding cytokines and transcriptional regulators like the protooncogene mRNAs, c-*myc* and c-*fos,* which increase rapidly in cells in response to transcriptional activation and decrease rapidly when transcription is reduced (Kelly *et al.*, 1983; Greenberg and Ziff, 1984; Shaw and Kamen, 1986). The acute response characteristics of these mRNAs are due to rapid turnover rates.

Regulation of mRNA turnover by steroid hormones was first documented in studies with the estrogen-stimulated chick oviduct. In this system, hormone withdrawal caused translatable ovalbumin mRNA to decay 5- to 10-fold faster than predicted from the turnover rate estimated in the presence of estrogen (Palmiter and Carey, 1974). Subsequent studies with hybridization probes confirmed this finding and extended it to the family of estrogen-regulated mRNAs that encode egg white proteins in tubular gland cells (Cox, 1977; Hynes *et al.*, 1979). Similar experiments with chick liver showed that the mRNAs for the estrogen-induced yolk proteins VTG and apoII decayed 7- to 8-fold more rapidly following hormone withdrawal than predicted from the $t_{1/2}$ estimated by approach-to-steady-state methods in the presence of estrogen (Wiskocil *et al.*, 1980). This finding was also extended to VTG mRNA in *Xenopus* liver, in which case it was possible to show directly through pulse–chase analysis in organ culture that VTG mRNA decayed more slowly in the presence of hormone than after hormone withdrawal (Brock and Shapiro, 1983). Readdition of hormone to cultures withdrawn from hormone also led to resumption of the slow decay mode for VTG mRNA, suggesting that the effects of the hormone on mRNA stability are reversible and not the result of permanent covalent modifications of the mRNA.

In the initial studies of Palmiter and Carey (1974), it was noted that the more rapid decay of ovalbumin mRNA following hormone withdrawal was consistent with two possibilities: estrogen either stabilized the mRNA or regulated a degradative capacity. They noted, for example, that ovalbumin mRNA decayed more rapidly during hormone withdrawal following 4 days of estrogen treatment compared with 1 day of estrogen treatment. Recent studies of apoII mRNA and VTG mRNA turnover in chick liver bear on this point. Gordon *et al.* (1988) found that apoII mRNA had the same turnover rate ($t_{1/2}$=13 hr) in the presence of estrogen or after hormone withdrawal when the prior period of estrogen treatment was 1–3 days. However, after prolonged estrogen treatment (5–14 days), hormone withdrawal led to the rapid decay of apoII and VTG mRNAs ($t_{1/2}$=1.5 hr) (Fig. 1A) compared with the slow decay after short-term estrogen treatment (Fig. 1B). These results suggest that estrogen induced an activity responsible for the rapid destabilization of apoII mRNA and VTG mRNA following hormone withdrawal. The destabilization activity appears to be selective for specific mRNAs in that, following hormone withdrawal, the steady-state levels of at least two other mRNAs and the overall profile of polyribosomes were unchanged. In addition, the destabilization activity appears to be quiescent in the presence of estrogen and is activated only following hormone withdrawal (Gordon *et al.*, 1988). These data in chick liver and the studies with chick oviduct noted above suggest that estrogen induces a selective destabilization activity that rapidly removes these mRNAs when hormone is withdrawn. The idea that estrogen regulates a destabilization activity is conceptually and mechanistically different from the idea that estrogen causes a selective stabilization of specific mRNAs.

Recent studies by Schoenberg and colleagues (Schoenberg *et al.*, 1989; Pastori *et al.*, 1990, 1991a) with *Xenopus* liver have shown that estrogen causes the coordinate cytoplasmic destabilization of serum albumin mRNA and other mRNAs encoding secreted plasma proteins but not mRNAs encoding cytoplasmic proteins. This response to estrogen is in contrast to the effect on VTG mRNA, which is stable in the presence of estrogen and destabilized following hormone withdrawal (Brock and Shapiro, 1983). Thus, estrogen appears to selectively destabilize some mRNAs, have no effect on most mRNAs, and stabilize yet another mRNA.

Other steroid hormones and thyroid hormone are known to alter mRNA turnover both by stabilizing and by destabilizing specific mRNAs (Table 1). In addition, protein hormones, growth factors, biogenic amines, and developmental stimuli also alter mRNA turnover (Tables 2 and 3). The pattern of regulation may be cell type-specific and mRNA-specific, and may exhibit opposing effects on different mRNAs in the same cell at the same time in response to the same agent. These examples also illustrate the

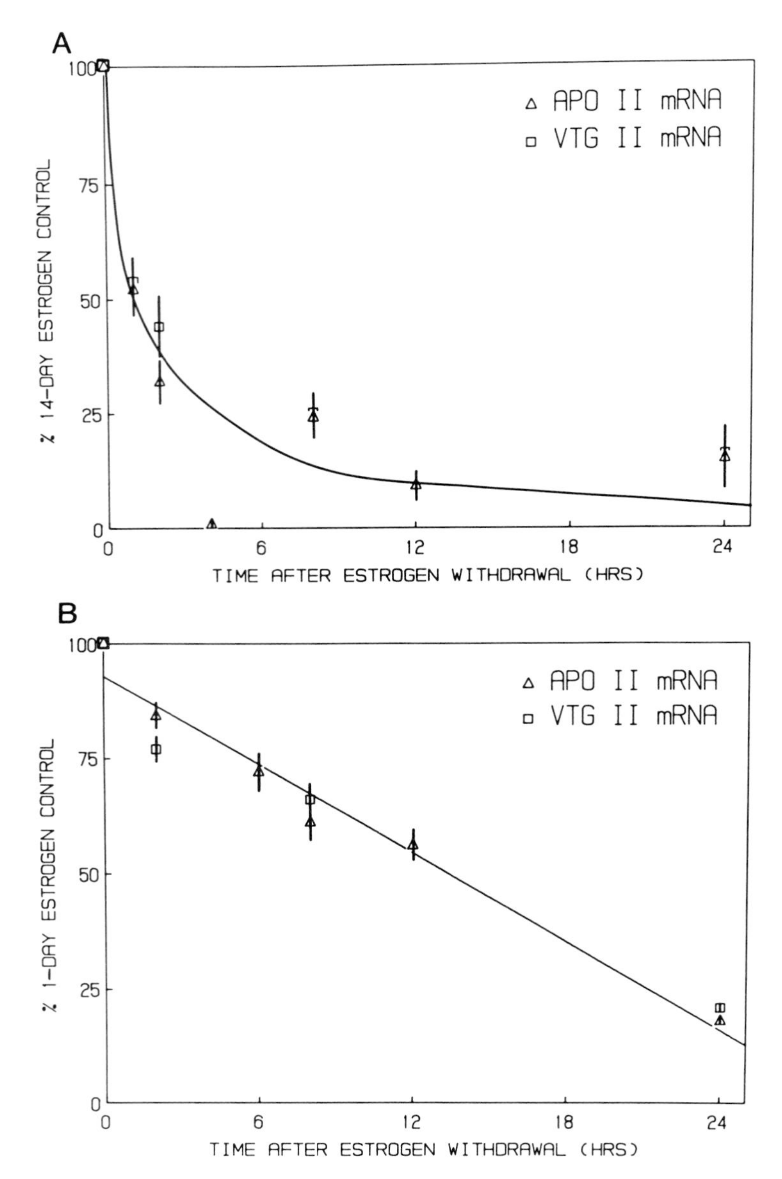

FIGURE 1 Induction of the rapid decay mode for apoII and VTG II mRNAs. Chickens were treated with estrogen for 14 days (A) or 1 day (B) prior to withdrawal of hormone. Data points show the amount of apoII or VTG II mRNA remaining at the indicated times. The half-lives for decay of these mRNAs is 1.5 hr with the 14-day estrogen treatment (A) and 12 hr with the 1-day treatment (B). [Adapted from Gordon *et al.* (1988).]

TABLE 1 Effects of Steroid and Thyroid Hormones on mRNA Turnover[a]

Effector	Cell	mRNA	Effect	Comment	Reference
Estrogen	Chick liver	apoII, VTG	ID	Slow induction	Gordon *et al.* (1988)
Estrogen	*Xenopus* liver	VTG	S	NBIPS	Brock and Shapiro (1983)
Estrogen	Chick oviduct	Ovalbumin, conalbumin	S or ID	Slow induction	Palmiter and Carey (1974); Cox (1977); Hynes *et al.* (1979).
Estrogen	*Xenopus* liver	Albumin, γ-fibrinogen, transferrin, trypsin inhibitor, 12B mRNA	D	NBIPS[b]	Schoenberg *et al.* (1989); Pastori *et al.* (1990); Pastori *et al.* (1991a); Moskaitis *et al.* (1991)
Estrogen	MCF-7 Mammary carcinoma	*c-myc*	S		Santos *et al.* (1988)
Estrogen, Progesterone	Chick oviduct	Hsp 108	S		Baez *et al.* (1987)
Progesterone, RU486	MCF-7 Mammary Carcinoma	Fatty acid synthetase	S		Chalbos *et al.* (1991)
Glucocorticoid	U937 Promonocyte	IL-1β	D		Lee *et al.* (1988)
Glucocorticoid	L929	IFN-β	D	NBIPS	Peppel *et al.* (1991)
Glucocorticoid	H4IIE Hepatoma	PEPCK	S	Slow induction	Petersen *et al.* (1989)
Glucocorticoid	Fibroblast	Growth hormone	S	BIPS; slow induction; poly (A) tail lengthening	Paek and Axel (1987)
Glucocorticoid	Rat liver	HMG-CoA reductase	D		Simonet and Ness (1989)
Glucocorticoid	Rat liver	S-11 mRNA	S	BIPS; slow induction	Wong and Oppenheimer (1986)

continues

TABLE 1 *continued*

Effector	Cell	mRNA	Effect	Comment	Reference
Glucocorticoid	COMMA D cells	β-Casein	S[c]		Poyet *et al.* (1989)
Glucocorticoid	HT-1080 Fibrosarcoma	Fibronectin	S	BIPS	Dean *et al.* (1988)
Thyroid hormone	Rat liver	HMG-CoA reductase	S	BIPS	Simonet and Ness (1988)
Thyroid hormone	Rat liver	S-11 mRNA	S	BIPS; slow induction	Wong and Oppenheimer (1986)
Thyroid hormone	Chick hepatocyte	Malic enzyme	S	BIPS	Back *et al.* (1986)
Thyroid hormone	Rat liver	S-14 mRNA	S	Rapid increase in nuclear pre-mRNA	Narayan and Towle (1985)
Thyroid hormone	Rat pituitary cells	Thyrotropin β-subunit	D	Poly (A) shortening	Krane *et al.* (1991)
Androgen	Rat prostate	Androgen-binding proteins	S	Increase in nuclear pre-mRNA	Page and Parker (1982)

[a] *Abbreviations:* VTG, vitellogenin; ID, induced destabilization; S, stabilized; D, destabilized; NBIPS, not blocked by inhibitors of protein synthesis; BIPS, blocked by inhibitors of protein synthesis; IL-1β, interleukin-1β; IFN-β, interferon-β; PEPCK, phosphoenolpyruvate carboxykinase.

[b] Refers only to albumin destabilization.

[c] In the presence of prolactin and insulin.

TABLE 2 Effects on mRNA Turnover of Agents Acting through PKC, TK, and Ca^{2+}

Effector	Cell	mRNA	Effect	Pathway	Comment	Reference
Phorbol ester	MCF-7 mammary carcinoma	Estrogen receptor	D	PKC	NBIPS	Saceda *et al.* (1991)
Phorbol ester	T lymphoblast	GM-CSF	S	PKC		Shaw and Kamen (1986)
CD-28 T-cell receptor activation	T cell	GM-CSF, IL-2, IFN-γ, TNF-α	S	TK, PKC		Lindsten *et al.* (1989)
Phorbol ester	U937 Promonocyte	TGF-β1	S	PKC	Slow effect	Wager and Assoian (1990)
Phorbol ester, TNF-α	WI38 Fibroblast	GM-CSF, G-CSF	S	PKC		Koeffler *et al.* (1988)
Phorbol ester, differentiation	HL-60 Promonocytes	*c-fms*	S	PKC		Weber *et al.* (1989)
Phorbol ester	Monocyte	*c-fms*	D	PKC		Weber *et al.* (1989)
Phorbol ester	EL-4 thymoma	GM-CSF	S	PKC		Bickel *et al.* (1990)
INF-2, INF-β	Daudi lymphoblast	*c-myc*	D			Dani *et al.* (1985) Knight *et al.* (1985)
Ca^{2+} ionophore	PB-3c mast cells	IL-3	S	Ca^{2+}	NBIPS	Wodnar-Filpowicz and Moroni (1990)

Abbreviations: S, stabilized; D, destabilized; NBIPS, not blocked by inhibitors of protein synthesis; BIPS, blocked by inhibitors of protein synthesis; PKC, protein kinase C; TK, tyrosine kinase; GM-CSF, granulocyte–macrophage colony-stimulating factor; IL-2, interleukin 2; IFN-γ, interferon; TNF-α, tumor necrosis factor; TGF-β1, transforming growth factor β1; G-CSF, granulocyte colony-stimulating factor; IL-3, interleukin 3.

TABLE 3 Effects on mRNA Turnover of Miscellaneous Agents and Agents that Act Through PKA[a]

Effector	Cell	mRNA	Effect	Pathway	Comments	Reference
Cell Aggregation, cAMP	*Dictyostelium discoideum*	Many	S	cAMP receptor		Mangiarotti *et al.* (1983)
cAMP	*Dictyostelium discoideum*	117 mRNA	D	cAMP receptor	BIPS	Juliani *et al.* (1990)
Isoproterenol	Vas deferens cell line	β-Adrenergic receptor	D	PKA		Hadcock *et al.* (1989)
Isoproterenol	C6 glioma	LDH-A	S	PKA		Jungman *et al.* (1983)
Glucagon	Hepatocytes	Malic enzyme	D	PKA	Rapid effect	Back *et al.* (1986)
cAMP	H-4 hepatoma	TAT	D	PKA		Smith and Liu (1988)
cAMP	FTO-2B hepatoma	PEPCK	S	PKA	NBIPS[b]	Hod and Hanson (1988)
Insulin, -thrombin	CCL39 fibroblast	*c-myc*	S			Blanchard *et al.* (1985)
TGF-β1	3T3	1(I) collagen	S[c]			Penttinen *et al.* (1988)
TGF-β1	Hep3B	Albumin, apo AI	D			Morrone *et al.* (1989)
Lipopolysaccharide	Macrophage	GM-CSF	S		BIPS	Thorens *et al.* (1987)
Prolactin	Mammary gland	β-Casein	S[d]			Guyete *et al.* (1979)

[a] *Abbreviations:* S, stabilized; D, destabilized; NBIPS, not blocked by inhibitors of protein synthesis; BIPS, blocked by inhibitors of protein synthesis; PKA, protein kinase A; LDH-A, lactate dehydrogenase A; TAT, tyrosine aminotransferase; PEPCK, phosphoenolpyruvate carboxykinase; TGF-β1, transforming growth factor β1.

[b] Y. Hod, personal communication.

[c] Observed only in confluent cells.

[d] In the presence of insulin and hydrocortisone.

wide variety of mRNAs that are regulated in this manner. Although some examples of regulated mRNA turnover have been known for 20 years, it is clear from Tables 1–3 that most examples have been uncovered in the very recent past. This recent and rapid increase in our knowledge of specific mRNAs and agents that regulate their turnover suggests that we are only beginning to appreciate the widespread nature and importance of mRNA turnover in the regulation of gene expression.

III. Mechanistic Aspects of Regulated mRNA Turnover

The mechanisms through which mRNA turnover is altered are not understood in detail for any hormone or developmental signal. Nevertheless, a variety of recent studies have uncovered parts of the puzzle, and we can use this information to outline a basic model of regulated turnover that is informative as to the state of the art and useful in formulating hypotheses to be tested experimentally. Figure 2 shows a model that includes a number of features that have been identified or are likely to occur. At the level of mRNA structure, for example, target sites (TS) that act as cleavage sites for either endonuclease (EN) or exonuclease (EX) activities must occur in the mRNA. Turnover factors (TF) modulate the susceptibility of TS to nuclease attack by either increasing susceptibility (mRNA destabilization) or decreasing susceptibility (mRNA stabilization). As illustrated in the model (Fig. 2), a TF may modulate nuclease attack by binding independently to mRNA to alter susceptibility to a freely diffusible nuclease (see 1, Fig. 2), by interacting with a freely diffusible nuclease to localize the nuclease to the mRNA (see 2, Fig. 2), or by activating a nuclease that is normally bound to the mRNA in an inactive state (see 3, Fig. 2). A TF might also act by targeting the mRNA to a cellular locale or organelle where degradation occurs. TF is shown to interact with stability or degradation elements in the mRNA that, for simplicity, are termed turnover elements (TE). In the model the activity of the TF is modulated by signal transduction pathways. This could occur directly, for example, by changes in phosphorylation states mediated via intracellular signaling pathways or indirectly by regulating the level of TF gene expression. Signal transduction pathways that alter the activation state or quantity of TF are in turn coupled to developmental or hormonal signals by appropriate cell surface or intracellular receptors.

Since components of the model have not been completely defined for any single mRNA, one could argue that a simpler model would suffice. In a simpler model, for example, the nuclease could itself be the turnover factor that is linked to signal transduction pathways and the nuclease target site could be the turnover element in the mRNA. In the simplest case, nucleolytic cleavage is an mRNA self-catalyzed reaction that is modulated by interaction with a turnover factor. While it is possible that specific

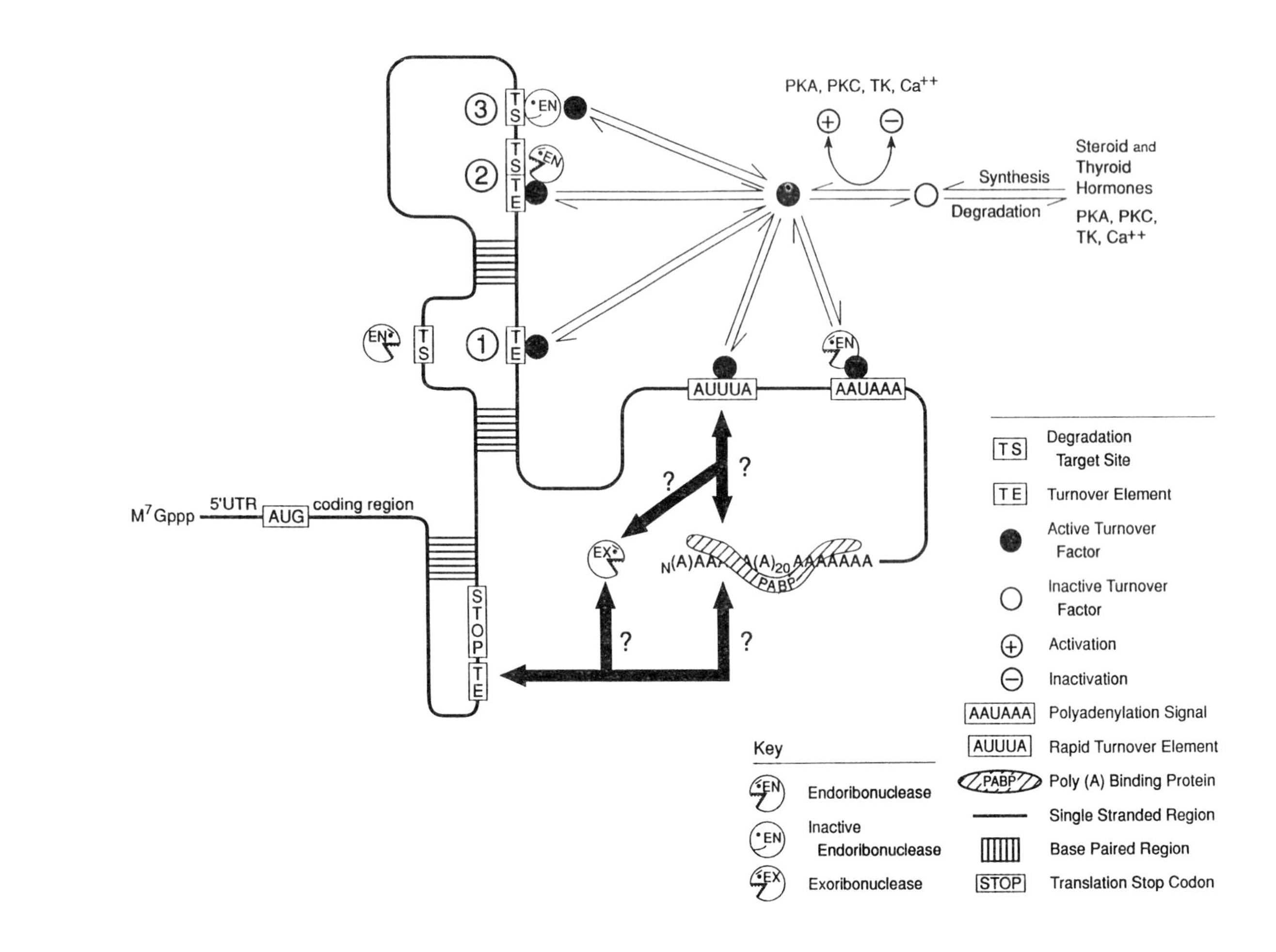

PKA, PKC, TK, Ca++
Steroid and Thyroid Hormones
Synthesis
Degradation
PKA, PKC, TK, Ca++
M7Gppp
5'UTR
AUG
coding region
TS
TE
STOP
AUUUA
AAUAAA
N(A)AA A(A)20 AAAAAAA
PABP
?
Key
Endoribonuclease
Inactive Endoribonuclease
Exoribonuclease
Degradation Target Site
Turnover Element
Active Turnover Factor
Inactive Turnover Factor
Activation
Inactivation
Polyadenylation Signal
Rapid Turnover Element
Poly (A) Binding Protein
Single Stranded Region
Base Paired Region
Translation Stop Codon

nucleases or self-catalysis is responsible for the turnover of some mRNAs, circumstantial evidence to support most aspects of the model is available. Furthermore, by analogy with other regulated processes in mRNA metabolism, mRNA turnover is likely to occur by multiple pathways and to employ an array of factors that explain inherent differences in turnover rates among mRNAs, that provide for specificity in differential regulation of mRNA turnover by hormones and other signaling agents, and that explain cell-type specificity in turnover rates of specific mRNAs. With these points in mind we can examine the evidence for the components shown in the model in Fig. 2.

IV. Linkage to Signal Transduction Pathways

The model in Fig. 2 shows signal transduction pathways interacting with TF to alter mRNA turnover. Although the actual biochemical steps linking a signal transduction pathway to turnover of a specific mRNA have not been established, the examples of regulated mRNA turnover listed in Tables 1–3 are categorized, for convenience, according to the signal transduction pathways through which the effector or stimulus is generally thought to act. In Table 2, for example, T cell activation via the T cell receptor is believed to involve both protein kinase C and tyrosine kinases (Helman *et al.*, 1991; Klausner and Samelson, 1991). Phosphorylation events mediated by one or both kinases *might* be the linkage between receptor activation and mRNA turnover.

It is interesting to note that in each of the listed categories (Tables 1–3), there are examples of regulated mRNA turnover that show characteristics expected for linkage via activation or inactivation of preexisting TF and examples that show characteristics expected for linkage via alterations in

---

FIGURE 2 Model of regulated mRNA turnover. The model shows a composite of features that have been identified in analyses of mRNA turnover in many systems. Symbols and RNA elements are defined in the key except for those explained below. As shown, active TFs alter susceptibility to nuclease by interacting with nuclease or with TEs in the mRNA. TEs in the model are shown, for simplicity, only in single stranded domains but could also occur in base-paired regions. The AUUUA element represents the rapid turnover element identified in numerous cytokine, lymphokine, and protooncogene mRNAs. The polyadenylation signal AAUAAA is shown as a potential nuclease target site (Binder *et al.*, 1989) to suggest that this may be a bifunctional element that participates in cytoplasmic mRNA turnover as well as in nuclear polyadenylation. The thick arrows linking the AUUUA element and the coding region TE with exonuclease and the poly(A) binding protein/poly (A) tail is meant to indicate communication between these elements and the processes of poly(A) tail shortening and mRNA degradation. Such communication could arise through protein–protein interactions, RNA–protein interactions, base pairing interactions with the RNA component of a TF, or long-range base pairing interactions within the mRNA.

TF gene expression. The former is typified by responses that occur quickly and are not blocked by inhibitors of protein synthesis, while the latter occur slowly and are blocked by inhibitors of protein synthesis. Where available, this information is noted in Tables 1–3. Note that the effects of protein synthesis inhibitors on mRNA turnover are complex and not well understood. Inhibition of protein synthesis stabilizes some mRNAs, destabilizes some mRNAs, and has no effect on other mRNAs. Because one cannot evaluate the effect of protein synthesis inhibition on the induction process if the inhibitor itself alters the turnover of the mRNA, comments on protein synthesis inhibition are given in Tables 1–3 only for those mRNAs whose turnover is insensitive to the inhibitors. Although not intended to be com-

TABLE 4 Effects of Protein Synthesis Inhibitors on mRNA Turnover[a]

mRNA	Inhibitor	Inhibitor effect	Reference
ApoII	CHX	Superinduces	Sensel and Williams, unpublished data
	PUR	Superinduces	
9E3	CHX	Superinduces	Sugano *et al.* (1987)
Estrogen receptor	CHX	No effect	Saceda *et al.* (1991)
GM-CSF	CHX	No effect/slight induction	Thorens *et al.* (1987)
IL-3	CHX	No effect	Wodnar-Filipowicz (1990)
Vitellogenin	CHX	No effect	Blume and Shapiro (1989)
Albumin	CHX	No effect	Moskaitis *et al.* (1991)
	MDMP	Decreases	
IL-1β	CHX	Superinduces	Lee *et al.* (1988)
IFN-β	CHX	Superinduces, does not prevent poly(A) shortening	Peppel and Baglioni (1991)
Transferrin receptor	CHX	Stabilizes	Koeller *et al.* (1991)
	PUR	Stabilizes	
1(I) collagen	CHX	No effect	Penttinen *et al.* (1988)
Creatine phosphokinase	CHX	Stabilizes	Pontecorvi *et al.* (1988)
Malic enzyme	PUR	No effect	Back *et al.* (1986)
Thyrotropin β-subunit	CHX	Induces poly(A) shortening	Krane *et al.* (1991)
β-casein	CHX	Decreases	Poyet *et al.* (1989)
PEPCK	PACT	No effect	Y. Hod, personal communication
	CHX	Stabilizes	
c-fms (monocytes)	CHX	Destabilizes	Weber *et al.* (1989)

[a] *Abbreviations:* CHX, cycloheximide; MDMP, 2-(4-methyl-2,6-dinitroanilino)-N-methylpropionamide; PUR, puromycin; PACT, pactamycin; GM-CSF, granulocyte–macrophage colony-stimulating factor; IL-3, interleukin 3; IL-1β, interleukin 1β; IFN-β, interferon β; PEPCK, phosphoenolpyruvate carboxykinase.

prehensive, we have also listed in Table 4 examples of mRNAs that are unaffected, stabilized, or destabilized by specifc inhibitors.

A. Steroid and Thyroid Hormones

The effects of steroid and thyroid hormones on gene expression are generally believed to occur via interaction with specific receptor proteins that act as hormone-dependent transcriptional activators (Carson-Jurica *et al.*, 1990). Are effects on mRNA turnover mediated by these receptor proteins? Findings such as the ability of receptor antagonists to block effects on mRNA turnover (Schoenberg *et al.*, 1989; Peppel *et al.*, 1991) provide circumstantial evidence for receptor involvement. While suggestive, such results alone cannot establish this point. A recent study illustrates the ambiguity inherent in experiments with antagonists. Both progesterone and the anti-progestin RU486 stabilize fatty acid synthetase mRNA, but at the same time RU486 inhibits transcription of the gene, whereas progesterone stimulates its transcription (Chalbos *et al.*, 1991). One could interpret these results as suggesting that RU486 alters mRNA stability by interacting with a protein other than the nuclear progesterone receptor. An equally plausable explanation is that RU486 can activate the progesterone receptor adequately to elicit mRNA stabilization but not adequately to activate transcription of the fatty acid synthetase gene. The best evidence that classical receptors are required for altered mRNA stability is provided by comparing cells with and without the receptor. RU486 alters fatty acid synthetase mRNA stability in MCF-7 cells, which contain progesterone receptor, but not in MDA-MB231 cells, which lack progesterone receptor (Chalbos *et al.*, 1991). Similarly, Nielsen and Shapiro (1990) have used transient transfections to study the effect of estrogen on the expression of a minivitellogenin gene that is driven by a thymidine kinase promoter and also lacks most of the vitellogenin coding region. They interpret the estrogen-dependent accumulation of minivitellogenin mRNA as evidence for mRNA stabilization and show that this effect requires estrogen receptor. While this evidence for receptor involvement is clearly suggestive, additional studies with mutant or inactivated receptors will be required to answer this question definitively.

If steroid receptors are required for regulated mRNA turnover, what is their role in this process? At present, there is no evidence that steroid receptors act directly in the cytoplasm to alter mRNA stability. Estrogen, progesterone, and androgen receptors are predominantly nuclear proteins (see Carson-Jurica *et al.* (1990) for review). Unliganded glucocorticoid receptors may be localized in the cytoplasm, but they localize to the nucleus upon hormone binding. Nuclear estrogen receptors in chicken and *Xenopus* liver are present in very low abundance compared with the number of regulated mRNA molecules in the cytoplasm (Lazier, 1978; Hayward *et al.*, 1980). Thus, it is unlikely that the receptor itself interacts with cytoplasmic mRNA in a stoichiometric way to alter stability. More likely is the possibility that receptors indirectly participate in regulated mRNA turnover because they are required for the transcriptional regulation of a turnover factor. Consistent with this point is the prolonged period of estrogen treatment (3–5 days) required to elicit the rapid decay mode for apoII and VTG mRNAs in chicken liver (Fig. 1) (Gordon *et al.*, 1988). Similarly, stabilization of PEPCK-CAT chimeric mRNA in rat hepatoma cells occurred after only 8 hr of glucocorticoid treatment (Petersen *et al.*, 1989), and the stabilization of growth hormone mRNA by glucocorticoid in fibroblast cell lines required 12–24 hr (Paek and Axel, 1987). These results are consistent with an indirect role of receptor in the induction of a TF responsible for altered mRNA stability. On the other hand, more rapid and direct effects of steroids on mRNA turnover also occur. For example, destabilization of interferon-β mRNA by glucocorticoid occurs within 2 hr and is not blocked by inhibitors of protein synthesis (Peppel *et al.*, 1991). Similarly, the destabilization of albumin mRNA by estrogen in *Xenopus* liver is not blocked by inhibitors of protein synthesis (Moskaitis *et al.*, 1991). These results suggest that steroid hormones can alter mRNA stability independently of newly synthesized protein, perhaps by activating a preexisting signaling pathway. *In vitro* studies, for example, suggest a rapid activation of protein kinase by estradiol in rat uterine nuclei (Thampan, 1988). This observation might reflect a phosphorylation-based signaling pathway that rapidly communicates the hormonal signal to the cytoplasm and may or may not include the classical receptor protein.

Effects of thyroid hormone on mRNA stability in liver cells can occur rapidly (20 min) as observed with S14 mRNA (Narayan and Towle, 1985) or after a substantial lag period (6 hr) as observed with S11 mRNA (Wong and Oppenheimer, 1986). Effects of thyroid hormones on mRNA turnover are typically blocked by inhibitors of protein synthesis, suggesting the possible involvement of a labile protein factor (Wong and Oppenheimer, 1986; Back *et al.*, 1986; Simonet and Ness, 1988). Receptors for thyroid hormones are also predominantly nuclear and unlikely to be directly involved in regulating cytoplasmic mRNA stability. Interestingly, the rapid stabilization of the rat liver S14 mRNA in response to thyroid hormone is seen with the nuclear mRNA precursor before an effect is evident with cytoplasmic S14 mRNA (Narayan and Towle, 1985). A similar result was seen upon androgen treatment of rat prostate gland; nuclear precursors to androgen binding protein mRNAs increased much more than the level of transcription (Page and Parker, 1982). These results raise the possibility that thyroid hormone receptor or androgen receptor might interact directly with primary transcripts to alter mRNA stability or processing in the nucleus. This area has not received much attention but might be a fruitful avenue for study given the availability of expression vectors to synthesize receptor proteins and pre-mRNAs *in vitro*.

B. Agents That Act through Protein Kinase C

As indicated in Table 2, a wide variety of mRNAs are regulated by agents that activate the protein kinase C (PKC) pathway either directly, such as phorbol esters, or indirectly, such as via receptor activation. Phorbol esters have been used extensively to study the stabilization of cytokine and lymphokine mRNAs that frequently accompanies cell differentiation. Available evidence indicates that PKC activation may lead to mRNA stabilization or destabilization, and these effects may or may not be blocked by inhibitors of protein synthesis. Upon activation of cell surface receptors that are coupled to phospholipase C, hydrolysis of phosphotidylinositol activates PKC directly via diacylglycerol and indirectly via inositol*tris*phosphate-mediated Ca^{2+} release. In addition, a recent study suggests that intracellular Ca^{2+} may regulate mRNA turnover independently of PKC. In this study, interleukin 3 (IL-3) mRNA was markedly stabilized in the mast cell line PB-3c by the calcium ionophore A23187 but not by phorbol ester (Wodnar-Filipowicz and Moroni, 1990). IL-3 and GM-CSF mRNAs were also induced by A23187 in cells in which PKC was down-regulated. The stabilization of IL-3 mRNA by A23187 was not affected by inhibition of protein synthesis. These results suggest the occurrence of a Ca^{2+}-dependent signaling pathway that alters IL-3 mRNA turnover independently of PKC via preexisting TF.

Studies on the stabilization of lymphokine mRNAs upon T cell receptor activation also suggest that tyrosine kinases may play a role in altering mRNA turnover. Activation of the CD-28 receptor in human T cells, for example, stabilizes GM-CSF, IL-2, INF-γ, and TNF-α mRNAs (Lindsten *et al.*, 1989). Although the linkage between mRNA turnover and the T cell receptor is not known, T cell activation is believed to involve tyrosine-kinase-mediated steps prior to PKC activation (Klausner and Samelson, 1991; Helman *et al.*, 1991).

C. Agents That Act through Protein Kinase A

A wide variety of agents are known to act through cell surface receptors and G proteins to activate adenylate cyclase and the protein kinase A (PKA) signaling pathway. Table 3 shows a number of examples in which mRNA turnover is altered by agents that act in this fashion. Activation of this pathway by isoproterenol stabilizes LDH-A mRNA in C6 glioma cells (Jungmann *et al.*, 1983), and activation by cAMP stabilizes phosphoenolpyruvate carboxykinase (PEPCK) mRNA in rat hepatoma cells (Hod and Hanson, 1988). In contrast, isoproteranol destabilizes β-receptor mRNA in a vas deferens cell line (Hadcock *et al.*, 1989), and cAMP and glucagon destabilize malic enzyme mRNA in chick hepatocytes (Back *et al.*, 1986). The effect of glucagon on malic enzyme mRNA degradation is very rapid and possibly a direct result of activating the protein kinase A pathway. An interesting effect occurs in *Dictyostelium discoideum*, in which low levels of cAMP stabilize aggregation-specific mRNAs (Mangiarotti *et al.*, 1983) while high levels of cAMP selectively destabilize the mRNA for 117 antigen (Juliani *et al.*, 1990). Different effects of high and low cAMP concentrations may indicate two different cAMP-dependent signaling pathways acting on mRNA turnover.

D. Miscellaneous Agents and Developmental Transitions

Table 3 also lists examples in which mRNA turnover is regulated by agents that act via signaling pathways that are not well understood and examples in which regulated mRNA turnover accompanies developmental transitions for which no specific biochemical signal is known. In combination with glucocorticoid, prolactin has a dramatic effect on the turnover of casein mRNA in mammary gland explants, stabilizing this mRNA by as much as 20-fold (Guyette *et al.*, 1979; Eisenstein and Rosen, 1988). Developmental changes in mRNA stability have been shown to play an important role in terminally differentiated cells that are specialized for the synthesis of one or a few proteins. The predominance of globin among newly synthesized reticulocyte proteins is attributed to stabilization of globin mRNA and destabilization of non-globin mRNAs during red cell development (Lodish and Small, 1976; Aviv *et al.*, 1976; Volloch and Housman, 1981a, 1981b). Similarly, studies with cell lines reflecting different stages of B cell development indicate that increased stability of immunoglobulin mRNA plays a major role in the marked accumulation of this mRNA in differentiated B cells (Cox and Emtrage, 1989; Genovese and Milcarek, 1990; Genovese *et al.*, 1991)

Changes in mRNA stability are also known to occur during oocyte maturation in both frog and mouse oocytes. In both systems a substantial fraction of maternal mRNA is stored in a stable and translationally inactive form (De Leon *et al.*, 1983; Richter, 1991). During meiotic maturation, specific mRNAs are recruited for translation from the inactive pool, in a process that is accompanied by increased degradation of these mRNAs. Recruitment of the mRNA to the translationally active state occurs simultaneously with cytoplasmic elongation of the poly(A) tail (McGrew *et al.*, 1989; Huarte *et al.*, 1987). This process requires specific sequence elements in the 3′ untranslated region that act to overcome a default deadenylation mechanism (Varnum and Wormington, 1990; Fox and Wickens, 1990). The relationships among the increase in poly(A) tail length, translational recruitment, and subsequent mRNA degradation have been difficult to determine.

In mouse oocytes, the increase in poly(A) tail length and subsequent degradation of tissue plasminogen activator mRNA are mimicked by injected fragments of the 3′ untranslated region, suggesting that polyadenylation and degradation do not require translation (Vassalli *et al.*, 1989). It is not known whether degradation of tissue plasminogen activator mRNA requires elongation of the poly(A) tail per se or whether it is recruitment from the translationally inactive mRNA pool that permits mRNA degradation.

V. Turnover Elements and trans-Acting Turnover Factors

A. mRNA Turnover Elements

Although a wide variety of mRNAs are regulated by changes in turnover, in only a few instances have TEs responsible for hormonal or developmental regulation been identified. In the case of glucocorticoids, Petersen *et al.* (1989) have shown that the 3′ untranslated region (UTR) of PEPCK mRNA is necessary for mRNA stabilization by the hormone. The 3′ UTR of PEPCK mRNA conferred glucocorticoid-dependent stabilization upon a heterologous mRNA when stably transfected into a rat hepatoma cell line. Interestingly, the accumulation kinetics of the heterologous mRNA showed a lag period of 6–8 hr before mRNA stabilization was apparent. The authors suggest that stabilization mediated by the 3′ UTR may require a glucocorticoid-induced TF. The AU-rich element in the 3′ UTR of IFN-β also has been shown to mediate mRNA destabilization by glucocorticoids (Peppel *et al.*, 1991). When IFN-β mRNAs with deletions in the 3′ UTR were stably expressed in L929 cells, glucocorticoid destabilized intact IFN-β mRNA but did not effect IFN-β mRNA lacking a specific AU-rich domain. Glucocorticoid-mediated destabilization of IFN-β mRNA was not affected by cycloheximide, suggesting that induction of a new protein factor was not necessary.

Phorbol ester treatment of a T lymphoblast cell line stabilized the normally labile mRNA for GM-CSF (Shaw and Kamen, 1986). These investigators showed that an AU-rich motif containing multiple copies of the element AUUUA was responsible for the rapid turnover of GM-CSF mRNA (see Chapter 9). This AU-rich motif is present in the 3′ UTR of many cytokine, lymphokine, and protooncogene mRNAs (Caput *et al.*, 1986), and acts to destabilize globin mRNA when inserted into the 3′ UTR of this normally stable mRNA (Shaw and Kamen, 1986). In an attempt to identify specific turnover elements, Bickel *et al.* (1990) have studied the stabilization of GMCSF mRNA in EL-4 thymoma cells. They found that the 3′ UTR of GM-CSF confers rapid mRNA turnover and phorbol-ester-mediated stabilization when present in an heterologous mRNA encoding chloramphenicol acetyl transferase (CAT) (Iwai *et al.*, 1991). Their experiments yielded several interesting results. First, the 3′ UTR and the AU-rich motif of GM-CSF conferred rapid turnover of CAT mRNA in EL-4 thymoma cells and in NIH 3T3 cells, suggesting that rapid turnover mediated by AU-rich motifs occurs in diverse cell types. Second, phorbol-ester-induced stabilization of CAT mRNA due to the GM-CSF 3′ UTR was seen in EL-4 but not NIH 3T3 cells, suggesting a cell-type-specific mechanism. Third, concanavalin-A-stimulated stabilization of GM-CSF mRNA was not conferred by the 3′UTR, suggesting that separate mRNA elements mediate stabilization induced by phorbol esters and mitogens. Fourth, the AU-rich motif alone did not confer phorbol-ester-mediated stabilization to the heterologous mRNA. This result argues that stabilization of GM-CSF mRNA does not result simply from blocking the AU-rich motif or by competition between positive and negative TFs interacting with this motif. This interpretation is supported by deletion–substitution mutants that identified a 60-nucleotide region 5′ to the AU-rich motif that was necessary for phorbol-ester-induced stabilization (Iwai *et al.*, 1991). Further analysis shows that mRNA stabilization induced by the Ca^{2+} ionophore A23187 is conferred in part by the AU-rich motif and in part by the 60-nucleotide phorbol-ester-responsive element (D. H. Pluznik, personal communication).

In relation to the model in Fig. 2, these results with GM-CSF mRNA suggest that this mRNA contains multiple TEs within the 3′ UTR that can act together or independently to alter stability depending upon the input received from intracellular signaling pathways. In the case of mRNAs containing AU-rich motifs, these elements may serve to overcome the normally rapid turnover conferred by the AU-rich motif. These results and the studies with glucocorticoids noted above illustrate the importance of the 3′ UTR in constitutive and hormonally regulated turnover. Further, since mitogen-stimulated stabilization of the mRNA was not conferred by the 3′ UTR of GM-CSF mRNA, a mitogen-responsive TE may occur in the coding region or 5′ UTR. A coding region TE which mediates rapid deadenylation and mRNA degradation has been identified in c-*fos* mRNA (Shyu *et al.*, 1989, 1991). A coding region TE is illustrated in Fig. 2. Similarly, mRNA destabilization may involve recognition of a nascent protein as occurs with tubulin mRNA (Yen *et al.*, 1988).

B. Turnover Factors

A number of studies have identified putative turnover factors that bind to specific domains and elements within mRNAs that are regulated by hormones or developmental stimuli. In the case of mRNAs regulated by steroid hormones, two chicken liver cytosolic proteins of 60 and 34 kDa bind to the 3′ UTR of apoII mRNA but not to the coding region or 5′ UTR (Ratnasabapathy *et al.*, 1990). Figure 3 shows the binding domains for these factors illustrated as shaded regions on the secondary structure model of

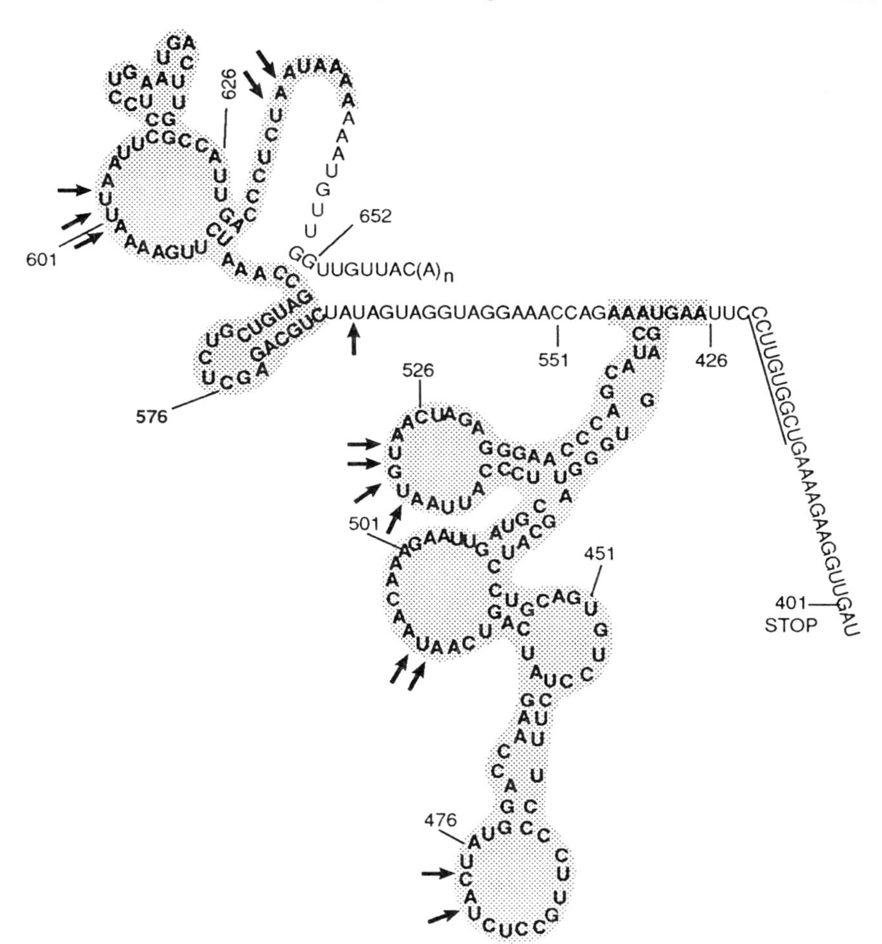

FIGURE 3 Protein-binding domains and degradation target sites in the 3′ untranslated region of apolipoprotein II mRNA. The model shows the proposed secondary structure of the 3′ untranslated region of apoII mRNA as determined from nuclease and dimethyl sulfate reactivities (Hwang *et al.*, 1989). The arrows show the endonuclease cleavage sites as determined by primer extension analysis (Binder *et al.*, 1989). The underlined nucleotides are base paired within the coding region in the complete model from which the region shown here was taken. The shaded areas represent the binding domains for the 34-kDa (upstream region) and the 60-kDa (downstream region) cytosolic proteins from chicken liver. [Reproduced from Ratnasabapathy *et al.* (1990).]

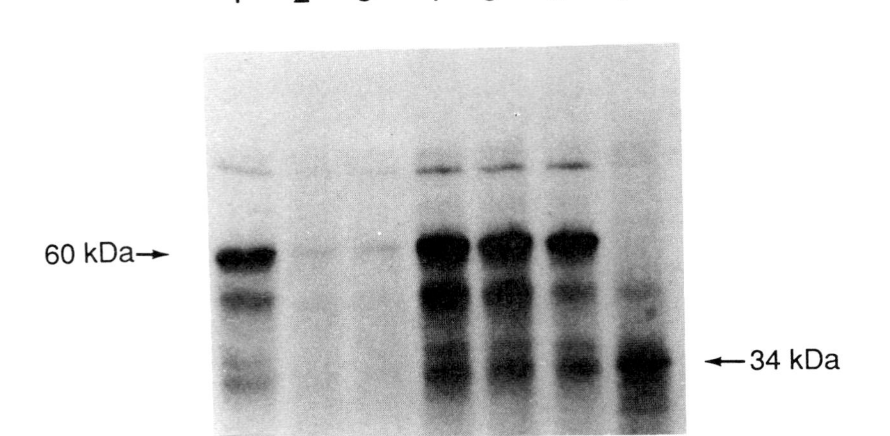

FIGURE 4 UV crosslinking-label transfer reaction with apoII mRNA fragment (nucleotides 568-643) with and without competitors. ApoII mRNA fragment (568–643) (see Fig. 3) labeled with [α-^{32}P]UTP was incubated with 40 μg of cytosolic protein in the absence (lane 1) or presence (lanes 2–6) of competitor mRNA fragments. After incubation and UV cross-linking, samples were digested with ribonucleases, and the ^{32}P proteins were analyzed by SDS–polyacrylamide gel electrophoresis and autoradiography. Unlabeled competitor mRNA fragments were lanes 2 and 3, 12.5- and 25-fold molar excess of fragment 568–643; lanes 4 and 5, 12.5- and 25-fold molar excess of fragment 400–547; and lane 6, 12.5-fold molar excess of coding region fragment 90–380. Lane 7 shows the proteins labeled by ^{32}P-labeled mRNA fragment 400–547. The positions of the 60-kDa and 34-kDa proteins are shown. [Reproduced from Ratnasabapathy *et al.* (1990).]

the 3′ UTR of apoII mRNA (Hwang *et al.*, 1989). Gel shift and UV cross-linking experiments indicate that the factors bind independently to the 3′ UTR. The UV cross-linking experiment shown in Fig. 4, for example, shows that the downstream domain (nucleotide 568–643) identifies a 60-kDa protein (lane 1) while the upstream domain identifies a 34-kDa protein (lane 7). Competition of the labeled downstream domain with the unlabeled downstream domain effectively prevents the interaction (lanes 2 and 3), whereas little competition is shown by the upstream domain (lanes 4 and 5) or the coding region (lane 6). Similar competition experiments show specificity in the interaction of the 34-kDa factor with the upstream domain (Ratnasabapathy *et al.*, 1990). As noted in Fig. 3, these binding domains contain 5′–AAU–3′/5′–UAA–3′ target sites (arrows) for apoII mRNA degradation (Binder *et al.*, 1989). The binding of these proteins to domains that contain target sites may indicate that these proteins participate in apoII mRNA degradation, although this remains to be tested. Liang and Jost (1991) have identified a 66-kDa protein that binds to an oligonucleotide containing the 12 nucleotide 5′ UTR of chicken VTG mRNA. This protein is found in polyribosomes from hen liver and is induced by estrogen treatment of roosters. When the 66-kDa protein was added to a cell-free extract, a synthetic RNA containing three copies of the VTG oligonucleotide was partially protected from degradation. The authors suggest that the 66-kDa protein may act to stabilize VTG mRNA.

Bohjanen *et al.* (1991) took advantage of the observation that lymphokine and c-*myc* mRNA stabilities can be differentially regulated in human T cells (Lindsten *et al.*, 1989). They reasoned that T cells must contain factors that can distinguish between the 3′ UTRs of lymphokine mRNAs and those of c-*myc* mRNA even though both mRNAs show rapid turnover and contain AU-rich motifs. Their results showed that resting T lymphocytes contain a factor, AU-A, that binds to the AU-rich motifs of *c-myc* mRNA and the lymphokine mRNAs GM-CSF, IL-2, and TNF-α (Bohjanen *et al.*, 1991). This 34-kDa protein has many similarities to the AU binding protein identified by Malter (1989) in Jurkat cells and by Vakalopoulou *et al.* (1991) in HeLa cells, although some differences in binding specificities and molecular weights were reported. When T cells were activated by stimulation of the T cell receptor–CD3 complex, the amount of AU-A binding did not change, but a distinct 30-kDa factor, AU-B, was transiently induced. AU-B recognized the AU-rich motifs of lymphokine mRNAs but not c-*myc* mRNA, thereby providing a possible explanation for the discrimination in regulating mRNA turnover. Induction of AU-B activity was blocked by inhibiting protein synthesis. When anti-CD-3-stimulated cells were given phorbol ester to stabilize GM-CSF mRNA, the binding of AU-B was diminished while AU-A was unchanged. Further studies with different AU-rich oligonucleotides showed that AU-A and AU-B have different specificities and that AU-B has an affinity 10-fold higher than that of AU-A for the repeating AUUUA motif in GM-CSF mRNA (Bohjanen *et al.*, 1992). These results are consistent with the idea that AU-A and AU-B may regulate GM-CSF mRNA turnover by competing for binding to the AU-rich motif. The pattern of AU-B induction and the decrease in AU-B binding after phorbol ester treatment (Bohjanen *et al.*, 1991) are consistent with AU-B promoting mRNA degradation and AU-A promoting stabilization, although the activities of these factors remain to be tested.

Malter and Hong (1991) have also identified an inducible binding activity that recognizes the AU-rich motif and correlates with stabilization of GM-CSF mRNA in peripheral blood mononuclear cells. Treatment of cells with phorbol ester or the Ca^{2+} ionophore A23187, or both, rapidly induced the AU-binding factor in a process that was not blocked by inhibitors of protein and RNA synthesis. The binding activity of this factor was eliminated by potato acid phosphatase. This suggests that the factor has an essential phosphate group and raises the possibility that activation occurs via PKC-mediated phosphorylation.

VI. Degradation Target Sites and Nucleases

The actual mechanisms through which mRNA degradation occurs are poorly understood. To some extent this reflects the difficulty of detecting low levels of mRNA degradation intermediates *in vivo*. In addition, there is often uncertainty in assigning relevance to intermediates generated in cell-free extracts, which often contain a mixture of nucleases that are released from sequestered intracellular sites when cells are broken. Nevertheless, putative degradation intermediates have been identified for several hormone-regulated and constitutively expressed mRNAs in animal cells. These data suggest that degradation occurs by at least two pathways for polyadenylated mRNAs. A third pathway applicable to nonpolyadenylated histone mRNAs will not be considered here.

One pathway for mRNA degradation appears to involve endonucleolytic cleavage without prior removal or shortening of the poly(A) tail. No change in poly(A) tail length was seen during either slow or rapid decay of apoII mRNA in chick liver (Binder *et al.*, 1989). Poly(A) tail lengths ranged from 40 to 180 nucleotides with no substantial change in the minimum length or heterogeneity with mRNA half-lives of 13 or 1.5 hr. Serum albumin mRNA in *Xenopus* liver has a very short poly(A) tail of only 17 nucleotides both when the mRNA is stable and after destabilization by estrogen treatment (Schoenberg *et al.*, 1989). Similarly, the rapid degradation of mutant globin mRNAs containing premature termination codons is not accompanied by rapid or complete deadenylation (Shyu *et al.*, 1991).

Evidence for endonuclease-initiated mRNA turnover is provided by analysis of truncated mRNAs presumed to be degradation intermediates. mRNAs truncated at specific sites in the 3′ UTR have been identified with several relatively high abundance mRNAs. Albrecht *et al.* (1984) used S1 nuclease protection analysis to show that globin mRNA is cleaved at AU-rich sites in the 3′ UTR when isolated directly from reticulocytes. Globin mRNA purified after polyribosome isolation contains more truncated molecules, and incubation of polyribosomes *in vitro* generated further cleavages at AU-rich sites in the 3′ UTR (Bandyopadhyay *et al.*, 1990). This degradation was relatively insensitive to dilution, suggesting that cleavages might be due to nuclease already bound to the 3′ UTR. A similar result was seen with the estrogen-regulated apoII mRNA in chick liver, in which cleavages at AU-rich sites in the 3′ UTR occurred during polyribosome isolation or during incubation of liver homogenates (Shelness *et al.*, 1987; Hwang *et al.*, 1989; MacDonald and Williams, 1992). Cleavage at these sites may reflect activation of a tightly bound nuclease since cleavage was insensitive to dilution and was not inhibited by exogneous RNA, heparin, or RNasin.

Low levels of cleavage at AU-rich sites were also seen with apoII mRNA isolated directly from liver, presumably reflecting *in vivo* degradation intermediates (Bakker *et al.*, 1988; Binder *et al.*, 1989). Cleavage sites were detected with both S1 nuclease protection and primer extension analyses, indicating that degradation occurred through endonucleolytic cleavage. These sites as mapped by primer extension analysis are shown (arrows) in Fig. 3. Cleavage sites occurred exclusively in the 3′ UTR and primarily at 5′–AAU–3′ and 5′–UAA–3′ trinucleotides that were localized to single-stranded loop domains in a secondary structure model (Hwang *et al.*, 1989). These cleavage sites were shown to be single-stranded and accessible both in naked mRNA and in polyribosomal mRNP, as judged by reactivity with chemical and enzymatic probes of RNA structure (Binder *et al.*, 1989; Hwang *et al.*, 1989). Interestingly, not all AU-rich sites in single-stranded domains were cleavage target sites, and AAU or UAA trinucleotides that were base-paired were not target sites. These data indicate that the typical endonucleolytic cleavage site is a 5′–AAU–3′ or 5′–UAA–3′ that is present in an accessible single-stranded loop domain. Neither an accessible domain alone nor an UAA or AAU trinucleotide alone is sufficient for endonucleolytic cleavage. Specificity may be provided by the localization of these trinucleotides to accessible single-stranded domains. In addition, the same apoII mRNA degradation intermediates were seen in chick liver irrespective of whether the mRNA was undergoing slow ($t_{1/2}$ = 13 hr) or rapid ($t_{1/2}$ = 1.5 hr) decay. This result suggests that estrogen-induced destabilization does not involve a different mechanism but, more likely, reflects increased targeting of the mRNA for degradation by a preexisting pathway. Interestingly, one of the degradation target sites in apoII mRNA is the AAUAAA polyadenylation signal itself (Binder *et al.*, 1989). This finding raises the possiblility that the polyadenylation signal may be a bifunctional element that also serves as a target site for cytoplasmic mRNA degradation.

Chick embryo fibroblasts that are serum stimulated or infected with Rous sarcoma virus express 9E3 mRNA via increased transcription and mRNA stabilization (Stoeckle and Hanafusa, 1989). Truncated forms of 9E3 mRNA arise by endonucleolytic cleavage in the 3′ UTR, but not within the several copies of the AUUUA element present in this mRNA. Instead, cleavage occurs at a UC-rich sequence similar to the self-cleavage site of hepatitis delta virus (Sharmeen *et al.*, 1988). The mRNA for another inflammatory mediator termed MONAP also displays truncated molecules (Kowalski and Denhardt, 1989). MONAP mRNA is cleaved at several AU-rich sites located primarily within the 3′UTR; 60% of the cleavage sites are within 50 nucleotides of an AUUUA element, but none correspond to the element itself. Endonucleolytic cleavage within the 3′ UTR also occurs in the *Xlhbox2B* mRNA in *Xenopus* oocytes (Brown and Harland, 1990). Using a cell-free system, Brewer and Ross (1988) also detected cleavage sites in the 3′ UTR of c-*myc* mRNA that were near but not immediately adjacent to or within the AUUUA element. In this case it was not determined whether the truncated molecules arose by endo- or exonuclease cleavage.

The above examples suggest that endonucleolytic cleavage within the

TABLE 5 Examples of mRNA Cleavage Sites

RNA/enzyme	Mechanism	Location	Sequence/structure	Reference
β-globin/ mRNP-associated	Endonuclease or exonuclease?	3′ UTR	No consensus AU-rich	Bandyopadhyay et al., 1990; Albrecht et al., 1984
Histone H4/ polysome-associated	Endonuclease or exonuclease?	3′ UTR	Intermediates located at the -5 and -12 positions, relative to the 3′ terminus 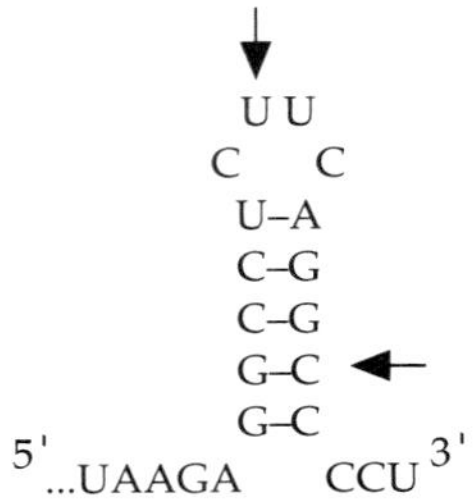 (stem-loop: UU / C C / U–A / C–G / C–G / G–C ← / G–C; 5′...UAAGA CCU 3′; ↓ at loop)	Ross et al., 1986
c-*myc*/ polysome-associated; accelerated by a cytoplasmic factor	Endonuclease or exonuclease?	3′ UTR	Intermediates mapped to within ±3 nucleotides 1 UUUUUAUUUAAGUA...GAUUUUUUUCUAUUGU (41 ↑) (AUUUA consensus) UUUAGAAAAAAUAAAAUAACUGGCAAAUAUA (55 ↑, 64 ↑, 73 ↑)	Brewer and Ross, 1988
MONAP	Endonuclease or exonuclease?	3′ UTR	No consensus AU-rich 60% of intermediates are within 50 nucleotides of an AUUUA consensus	Kowalski and Denhardt, 1989
9E3	Endonuclease	3′ UTR	Sequence in region of cleavage is homologous to self-cleavage site in hepatitus delta virus 5′ 3′ UUCCUCCUGCUCCCCUGG —9E3 UUCCUCUUCGGG —Delta virus (↑ cleavage)	Stoeckle and Hanafusa, 1989

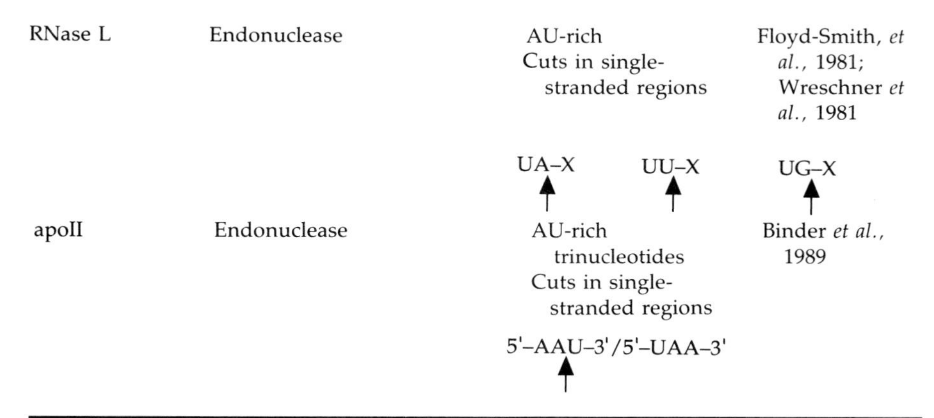

RNase L	Endonuclease	AU-rich Cuts in single-stranded regions UA↑–X UU↑–X UG↑–X	Floyd-Smith, *et al.*, 1981; Wreschner *et al.*, 1981
apoII	Endonuclease	AU-rich trinucleotides Cuts in single-stranded regions 5'–AA↑U–3'/5'–UAA–3'	Binder *et al.*, 1989

Note The table describes cleavage site characteristics for five polyadenylated mRNAs and histone H4 mRNA based on the analysis of truncated mRNA species believed to be degradation intermediates. Also shown are the specificity characteristics of endoribonuclease RNase L.

3′ UTR occurs frequently in the degradation of eukaryotic mRNAs. Table 5 summarizes cleavage sites that have been identified in a number of polyadenylated mRNAs as well as in the nonpolyadenylated histone H4 mRNA. As indicated in Table 5, in some cases it is not known if the cleavages result from endonuclease or exonuclease activity. Cleavages are often within AU-rich regions, and, in these cases, it may be important for these target sites to be in single-stranded domains. In the cases examined thus far, the AUUUA elements that confer rapid mRNA degradation in cytokine and protooncogene mRNAs do not appear to be actual target sites for degradation. A simple scheme suggested by these data is that endonucleolytic cleavage within the 3′ UTR is the rate-limiting targeting event and is followed by rapid exonucleolytic removal of the resultant fragments. This targeting event would provide a key site for regulation by turnover factors that may perturb RNA structure or serve to recruit nuclease to target sites.

Endonucleolytic cleavage also occurs frequently in the processing and turnover of prokaryotic mRNAs by RNase K (Lundberg *et al.*, 1990), RNase M (Cannistraro and Kennel, 1989; Cannistraro *et al.*, 1986), RNase E (Mudd *et al.*, 1988, 1990; Carpousis *et al.*, 1989), and RNase III (Gegenheimer and Apirion, 1981). RNases E and M appear to cleave within single-stranded AU-rich regions, suggesting that this overall endonucleolytic targeting strategy is well conserved through evolution. However, it is clear from analyses of prokaryotic mRNA turnover as well as in studies noted above (Bandyopadhyay *et al.*, 1990) that endonucleolytic cleavages in mRNA also occur at sites other than AU-rich elements. Thus, multiple endonucleases of differing specificity will likely be important for mRNA turnover in eucaryotic cells as well.

A second pathway for mRNA degradation appears to involve rapid shortening of the poly(A) tail before degradation of the main body of the mRNA. Poly(A) tail shortening occurs during rapid turnover of protooncogene and cytokine mRNAs such as c-*fos* (Wilson and Treisman, 1988), c-*myc* (Brewer and Ross, 1988), and IFN-β (Peppel and Baglioni, 1991). The poly(A) tail is not completely removed in this process but is shortened to approximately 30–40 residues, which may still bind one molecule of the poly(A) binding protein. Shyu *et al.* (1989) identified two turnover elements in c-*fos* mRNA that mediate rapid mRNA decay. One element is the AU-rich motif in the 3′ UTR containing three copies of the pentanucleotide AUUUA, whereas the other is located in the coding region and is not AU-rich. Further studies showed that when inserted into the coding region of β-globin mRNA, the c-*fos* coding region TE conferred rapid deadenylation and rapid turnover (Shyu *et al.*, 1991). This result was also seen when the AU-rich TE was inserted into the 3′ UTR of globin mRNA, indicating that two distinct TEs in c-*fos* mRNA act to facilitate poly(A) tail shortening and rapid mRNA decay. These investigators also showed that single point mutations within the three AUUUA elements in the AU-rich motif had little effect on the rate or extent of deadenylation but markedly slowed degradation of the mRNA (Shyu *et al.*, 1991). Therefore, mRNA turnover by this pathway may require poly(A) tail shortening, but in the absence of intact AUUUA TEs, neither the process of poly(A) tail shortening nor a short poly(A) tail per se is sufficient to initiate rapid turnover. This also is an interesting result in light of the findings of Bohjanen *et al.* (1992) that point mutations within the AUUUA element can shift the binding specifity of the AU-rich motif between two putative T cell TFs that may have opposing effects on mRNA turnover.

Steroid and thyroid hormones may also influence the same or related pathways that alter the length of the poly(A) tail. In mouse fibroblasts, in which glucocorticoids increase the stability of transfected human growth hormone mRNA, poly(A) tail length was increased (Paek and Axel, 1987). Wu and Miller (1991) have shown that progesterone treatment of ovine pituitary cells, which reduces luteinizing hormone mRNA levels, promotes poly(A) tail shortening for mRNAs encoding the α and β chains of luteinizing hormone. It is not known whether the shortened poly(A) tail is associated with altered mRNA turnover. Krane *et al.* (1991) showed that increased turnover of rat thyrotropin β-subunit mRNA in response to thyroid hormone was accompanied by poly(A) tail shortening. Interestingly, cycloheximide treatment alone promoted poly(A) tail shortening for this mRNA.

The endonucleases , exonucleases, and poly(A)-shortening nucleases responsible for the various degradation steps noted above have yet to be identified and characterized. It is likely that these enzymes will be identified in the near future as more studies are carried out with cell-free systems and as genetic analysis is applied to the problem, particularly in yeast. As a harbinger of information to come, several recent studies have mimicked key aspects of mRNA turnover in cell-free extracts. Brewer and Ross (1988) showed that poly(A) shortening and degradation of c-*myc* mRNA occur in extracts prepared from erythroleukemia cells, and Pei and Calame (1988) showed the importance of the 3′ UTR for c-*myc* mRNA degradation *in vitro*. A cytosolic factor that binds to the c-*myc* AU-rich motif and destabilizes c-*myc* mRNA *in vitro* has been described and partially purified (Brewer, 1991). Pastori *et al.* (1991) recently described a polyribosomal endonuclease activity in *Xenopus* liver that is activated during estrogen-induced destabilization of albumin mRNA and selectively degrades albumin mRNA *in vitro*. A ribonuclease activity in extracts of U937 cells selectively degrades TGF-β mRNA, and this activity is reduced in extracts from phorbol-ester-treated cells in which TGF-β mRNA is stabilized (Wager and Assoian, 1990). These initial studies suggest that cell-free systems can be used for the identification of components required for basal and regulated mRNA turnover and for the analysis of the molecular mechanisms of mRNA degradation.

One nuclease that remains a candidate for an important role in cellular mRNA metabolism is RNase L, an endoribonuclease with a specificity for UA and UU dinucleotides in single-stranded domains (Lengeyl, 1982). RNase L is normally latent, but is activated by 2′–5′oligoadenylate (2–5A), which in turn is synthesized by 2–5A synthetase when the enzyme is activated by binding to double-stranded RNA. The best known role of this nuclease system is as part of an antiviral response in which interferon-induced 2–5A synthetase is activated by viral RNA, resulting in RNase L activation and degradation of viral RNA. Fully activated levels of the RNase L system also lead to degradation of cellular mRNA and even ribosomal RNA, and it has been this seemingly nonspecific nucleolytic action that has discouraged consideration of this system as important in normal mRNA turnover. Nevertheless, 2–5A synthetase and RNase L are expressed in a wide variety of tissues and cells independent of the interferon response. It has been shown that 2–5A synthetase is induced by glucocorticoids (Krishnan and Baglioni, 1980), platelet-derived growth factor (Garcia-Blanco *et al.*,1989), estrogen withdrawal from chick oviduct (Stark *et al.*, 1979), anti-estrogen treatment of breast cancer cell lines (Viano *et al.*, 1990), and retinoic acid (Bourgeade and Besancon, 1984). These findings suggest a broader role for this system in RNA metabolism. The problem of apparent nonspecificity was solved in theory over a decade ago by Nilsen and Baglioni (1979), who showed that selective degradation of a specific RNA could occur when 2–5A synthetase was localized to a single-stranded RNA by virtue of a covalently linked double-stranded tail. These results imply that selectivity is due to the colocalization of 2–5A synthetase and RNase L to the target RNA or that the 2–5A concentration is high enough to transiently activate unassociated RNase L only in the immediate microenvironment of the target mRNA. Support for this idea is provided by a recent study in which the partially base-paired trans-acting response element (TAR) of human immunodeficiency virus (HIV) was shown to bind to and activate 2–5A synthetase when the TAR element was present in the 5′ region of a chimeric CAT mRNA (Schroder *et al.*, 1990). The Tat protein, which normally activates HIV expression by interaction with the TAR element, competed with 2–5A synthetase for binding to TAR and prevented activation of 2–5A synthetase. In interferon-treated HeLa cells, transient expression of CAT mRNA was reduced by the cis-linked TAR element, and this effect was blocked by coexpression of Tat protein. These results are suggestive that the 2–5A synthetase–RNase L system can be activated in such a way as to discriminate among mRNAs *in vivo*. In Fig. 2, 2–5A synthetase can be viewed as a TF that selectively interacts with a double stranded domain in the mRNA. In addition, as shown in Table 5, the specificity of RNase L is consistent with a role in the cleavage of mRNA at AU-rich sites.

VII. Summary and Perspective

The model in Fig. 2 is based on the premise that mRNA turnover is regulated primarily through the interaction of TFs with specific elements in the mRNA. TFs modulate the susceptibility of the mRNA to either endoribonuclease or exoribonuclease by interacting with TEs or with nuclease that is bound to the mRNA. The activity and/or the concentration of TFs is regulated by a variety of intracellular signal transduction pathways that are responsive to hormones, growth factors, biogenic amines, and uncharacterized developmental stimuli.

TEs have been identified in several mRNAs to date. In the case of the AUUUA element, which specifies rapid turnover of cytokine and protooncogene mRNAs, the functional element has been delimited to a relatively small sequence. Analyses of c-*fos* mRNA and GM-CSF mRNA indicate that regulated mRNA turnover results from the activity of multiple TEs within the mRNA. In c-*fos* mRNA, a TE in the coding region and a distinct element in the 3′ UTR act independently to alter the rate of poly(A) tail shortening and the overall degradation rate of the mRNA. Differences in sensitivity to transcription inhibitors suggest that the mechanisms through which these elements act are somewhat different (Shyu *et al.*, 1989). c-*myc* mRNA also contains a coding region TE that is necessary for mRNA stabilization by cycloheximide (Wisdom and Lee, 1991). GM-CSF contains two TEs in the 3′ UTR that respond independently to different intracellular signals. While our understanding of turnover elements is still quite fragmentary, the picture emerging involves an interplay between multiple elements located in the 3′ UTR and other regions of the mRNA. This is remniscent of transcriptional regulation, which in many cases results from a mosaic of DNA regulatory elements that interact and communicate via different transcription and enhancer factors. A major focus of research in the immediate future will be the further characterization of these TEs and the identification of other TEs that mediate altered mRNA turnover in response to a wide variety of agents.

Putative TFs that recognize specific elements or domains within mRNAs have been identified in a number of systems. Proteins that recognize the AUUUA rapid turnover element have been identified and characterized in extracts of HeLa cells, Jurkat cells, T cells, and peripheral blood mononuclear cells. We can anticipate the purification, cloning, and analysis of these and a variety of newly identified putative TFs in the next few years. Analyses of these factors via expression studies and in cell-free extracts will be required to confirm their function as TFs and to explore how these factors participate in regulated mRNA turnover. Cell-free systems that faithfully duplicate the pathways and specificity of differential mRNA turnover *in vivo* will be essential for these studies. Appropriate cell-free systems also will be central to the identification and purification of relevant ribonucleases. As described above, several cell-free systems with these properties have been described.

The model in Fig. 2 shows two additional features not considered above that are likely to be important for mRNA turnover; that is, long range interactions within the mRNA and mRNA secondary (or higher-order) structure (see Fig. 2, legend). The finding that AUUUA elements in c-*fos* mRNA mediate exonucleolytic shortening of the poly(A) tail implies communication between the element and the nucleolytic process. This could occur via interaction of the poly(A) tail itself with the 3′ UTR (Wilson and Treisman, 1988), interaction between the poly(A) binding protein and AU-rich motifs in the 3′UTR (Brewer and Ross, 1988), or protein–protein interactions between bound TFs and the poly(A) binding protein or nucleases. Such interactions within the 3′ UTR are easy to conceptualize in the context of an mRNP particle in which distant regions (along the mRNA) are brought into proximity through highly ordered secondary or higher-order structure. More difficult to conceptualize is how TEs within the coding region are involved in long-range interactions since ribosome transit would likely disrupt and displace bound TFs. This phenomenon is similar to the problem of transcriptional regulatory elements and enhancers within the transcribed portion of a gene, except that, in the case of transcription, a protein factor may associate with the noncoding strand during polymerase transit.

The model in Fig. 2 also shows regions of secondary structure within the 3′ UTR as well as base pairing between the 3′ UTR and the coding region. As judged by studies with prokaryotic mRNAs (Gegenheimer and Apirion, 1981; Belasco *et al.*, 1985; Klug and Cohen, 1990), histone mRNA (Graves *et al.*, 1987), and the iron response element in the 3′ UTR of transferrin receptor mRNA (Casey *et al.*, 1988), RNA secondary structure is likely to play an important role in mRNA turnover. Base pairing between the coding region and the 3′ UTR is one possible way for coding region TEs to communicate with the 3′ UTR without the problem of displacing a TF during ribosome transit. A recent study with chicken apoII mRNA, for example, showed the presence of stable base pairing (T_m = 50°C) between the 3′ UTR and the coding region (Hwang *et al.*, 1989). The region of the 3′ UTR involved in this interaction is underlined in Fig. 3. This structure would likely occur only transiently in polyribosomal mRNA due to ribosome transit. This raises the possibility that alternate structures may occur in the 3′ UTR as a result of ribosome transit. Perhaps such alternate structures play a role in mRNA turnover by transiently exposing a degradation target site. These points are speculative, but they illustrate that long-range communication and long-range base pairing interactions occur in mRNA. Understanding how such intramolecular communication occurs will be important for understanding the mechanisms of regulated mRNA turnover.

Acknowledgments

This work was supported by NIH Grant DK 18171. M.S. was supported by a fellowship from the American Heart Association, New York State Affiliate. M.M. was supported by NIH Training Grant GM 07518. For helpful discussions and for providing information prior to publication, we thank Daniel Schoenberg, Craig Thompson, David Shapiro, Dov Pluznik, and Yaacov Hod.

References

Albrecht, G., Krowczynska, A., and Brawerman, G. (1984). Configuration of β-globin messenger RNA in rabbit reticulocytes. *J. Mol. Biol.* **178,** 881–896.

Aviv, H., Voloch, Z., Bastos, R., and Levy, S. (1976). Biosynthesis and stability of globin mRNA in cultured erythroleukemic friend cells. *Cell* **8,** 495–503.

Back, D. W., Wilson, S. B., Morris, S. M., Jr., and Goodridge, A. G. (1986). Hormonal regulation of lipogenic enzymes in chick embryo hepatocytes in culture. *J. Biol. Chem.* **261,** 12555–12561.

Baez, M., Sargan, D. R., Elbrecht, A., Kulomaa, M. S., Zarucki-Schulz, T., Tsai, M.-J., and O'Malley, B. W. (1987). Steroid hormone regulation of the gene encoding the chicken heat shock protein hsp 108. *J. Biol. Chem.* **262,** 6582–6588.

Bakker, O., Arnberg, A. C., Noteborn, M. H. M., Winter, A. J., and AB, G. (1988). Turnover products of the apo very low density lipoprotein II messenger RNA from chicken liver. *Nucleic Acids Res.* **16,** 10109–10119.

Bandyopadhyay, R., Coutts, M., Krowczynska, A., and Brawerman, G. (1990). Nuclease activity associated with mammalian mRNA in its native state: Possible basis for selectivity in mRNA decay. *Mol. Cell. Biol.* **10,** 2060–2069.

Belasco, J. G., Beatly, J. T., Adams, C. W., von Gabain, A., and Cohen, S. N. (1985). Differential expression of photosynthesis genes in *R. capsulata* results from segmental differences in stability within the polycistronic rxcA transcript. *Cell* **40,** 171–181.

Berlin, C. M., and Schimke, R. T. (1965). Influence of turnover rates on the responses of enzymes to cortisone. *Mol. Pharmacol.* **1,** 149–156.

Bickel, M., Cohen, R. B., and Pluznik, D. H. (1990). Post-transcriptional regulation of granulocyte-macrophage colony-stimulating factor synthesis in murine T cells. *J. Immunol.* **145,** 840–845.

Binder R., Hwang, S.-P. L., Ratnasabapathy, R., and Williams, D. L. (1989). Degradation of apolipoprotein II mRNA occurs via endonucleolytic cleavage at 5′-AAU-3′/5′-UAA-3′ elements in single-stranded loop domains of the 3′-noncoding region. *J. Biol. Chem.* **264,** 16910–16918.

Blanchard, J.-M., Piechaczyk, M., Dani, C., Chambard, J.-C., Franchi, A., Pouyssegur, J., and Jeanteur, P. (1985). c-*myc* gene is transcribed at high rate in G_0-arrested fibroblasts and is post-transcriptionally regulated in response to growth factors. *Nature* (*London*) **317,** 443–445.

Blume, J. E. and Shapiro, D. J. (1989). Ribosome loading, but not protein synthesis, is required for estrogen stabilization of *Xenopus laevis* vitellogenin mRNA. *Nucleic Acids Res.* **17,** 9003–9014.

Bohjanen, P. R., Petryniak, B., June, C. H., Thompson, C. B., and Lindsten, T. (1991). An inducible cytoplasmic factor (AU-B) binds selectively to AUUUA multimers in the 3′ untranslated region of lymphokine mRNA. *Mol. Cell. Biol.* **11,** 3288–3295.

Bohjanen, P. R., Petryniak, B., June, C. H., Thompson, C. B., and Lindsten, T. (1992). AU RNA-binding factors differ in their binding specificities and affinities. *J. Biol. Chem.* **267,** 6302–6309.

Bourgeade, M. F., and Besancon, F. (1984). Induction of 2′,5′-oligoadenylate synthetase by retinoic acid in two transformed human cell lines. *Cancer Res.* **44,** 5355–5360.

Brewer G. (1991). An A + U-rich element RNA-binding factor regulates c-*myc* mRNA stability in vitro. *Mol. Cell. Biol.* **11,** 2460–2466.

Brewer, G., and Ross, J. (1988). Poly(A) shortening and degradation of the 3′ A + U-Rich sequences of human c-*myc* mRNA in a cell-free system. *Mol. Cell. Biol.* **8,** 1697–1708.

Brock, M. L., and Shapiro, D. J. (1983). Estrogen stabilizes vitellogenin mRNA against cytoplasmic degradation. *Cell* **34,** 207–214.

Brown, B. D., and Harland, R. M. (1990). Endonucleolytic cleavage of a maternal homeo box mRNA in *Xenopus* oocytes. *Genes Dev.* **4,** 1925–1935.

Cannistraro, V. J., Subbarao, M. N., and Kennel, D (1986). Specific endonucleolytic cleavage sites for decay of *Escherichia coli* mRNA. *J. Mol. Biol.* **192,** 257–274.

Cannistraro, V. J. and Kennel, D (1989). Purification and characterization of ribonuclease M and mRNA degradation in *Escherichia coli*. *Eur. J. Biochem.* **181,** 363–370.

Caput, D., Beutler, B., Hartog, K., Thayer, R., Brown-Shimer, S., and Cerami, A. (1986). Identification of a common nucleotide sequence in the 3′ untranslated region of mRNA molecules specifying inflammatory mediators. *Proc. Natl. Acad. Sci. USA* **83,** 1670–1674.

Carpousis, A. J., Mudd, E. A., and Krisch, H. M. (1989). Transcription and mRNA processing upstream of bacteriophage T4 gene 32. *Mol. Gen. Genet.* **219,** 39–48.

Carson-Jurica, M. A., Schrader, W. T., and O'Malley, B. W. (1990). Steroid receptor family: Structure and functions. *Endocrine Rev.* **11,** 201–220.

Casey, J. L., Hentze, M. W., Koeller, D. M., Caughman, S. W., Rouault, T. A., Klausner, R. D., and Harford, J. B. (1988). Iron-responsive elements: Regulator control mRNA levels and translation. *Science* **240,** 924–928.

Chalbos, D., Galtier, F., Emiliani, S., Rochefort, H. (1991). The anti-progestin RU486 stabilizes the progestin-induced fatty acid synthetase mRNA but does not stimulate its transcription. *J. Biol. Chem.* **266,** 8220–8224.

Cox, A., and Emtage, J. S. (1989). A 6-fold difference in the half-life of immunoglobulin μ heavy chain mRNA in cell lines representing two stages of B cell differentiation. *Nucleic Acids Res.* **17,** 10439–10454.

Cox, R. F. (1977). Estrogen withdrawal in chick oviduct. Selective loss of high abundance classes of polyadenylated messenger RNA. *Biochemistry* **16,** 3433–3442.

Dani, C., Mechti, N., Piechaczyk, M., Lebleu, B., Jeanteur, Ph., and Blanchard, J. M. (1985). Increased rate of degradation of c-*myc* in interferon-treated Daudi cells. *Proc. Natl. Acad. Sci. USA* **82,** 4896–4899.

Dean, D. C., Newby, R. F., and Bourgeois, S. (1988). Regulation of fibronectin biosynthesis by dexamethasone, transforming growth factor β, and cAMP in human cell lines. *J. Cell Biol.* **106,** 2159–2170.

De Leon, V., Johnson, A., and Bachvarova, R. (1983). Half-lives and relative amounts of stored and polysomal ribosomes and poly(A)$^+$ mRNA in mouse oocytes. *Dev. Biol.* **98,** 400–408.

Eisenstein, R. S. and Rosen, J. M. (1988). Both cell substratum regulation and hormonal regulation o milk protein gene expression are exerted at the posttranscriptional level. *Mol. Cell. Biol.* **8,** 3183–3190.

Floyd-Smith, G., Slattery, E., and Lengyel, P. (1981). Interferon action: RNA cleavage pattern of a (2′–5′) oligoadenylate-dependent endonuclease. *Science* **212,** 1030–1031.

Fox, C. A., and Wickens, M. (1990). Poly (A) removal during oocyte maturation: A default reaction selectively prevented by specific sequences in the 3′ UTR of certain maternal mRNAs. *Genes Dev.* **4,** 2287–2298.

Gaddum, J. H. (1944). Repeated doses of drugs. *Nature (London)* **153,** 494.

Garcia-Blanco, M. A., Lengyel, P., Morrison, E., Brownlee, C., Stiles, C. D., Rutherford, M., Hannigan, G., and Williams, B. R. G. (1989). Regulation of 2′,5′-oligoadenylate synthetase gene expression by interferons and platelet-derived growth factor. *Mol. Cell. Biol.* **9,** 1060–1068.

Gegenheimer, P., and Apirion, D. (1981). Processing of procaryotic ribonucleic acid. *Microbiol. Rev.* **45,** 502–541.

Genovese, C., and Milcarek, C. (1990). Increased half-life of μ immunoglobulin mRNA during mouse B cell development increases its abundancy. *Mol. Immunol.* **27,** 733–743

Genovese, C., Harrold, S., and Milcarek, C. (1991). Diferential mRNA stabilities affect mRNA levels in mutant mouse myeloma cells. *Somatic Cell Mol. Genet.* **17,** 69–81.

Gordon, D. A., Shelness, G. S., Nicosia, M., and Williams, D. L. (1988). Estrogen-induced destabilization of yolk precursor protein mRNAs in avian liver. *J. Biol. Chem.* **263,** 2625–2631.

Graves, R. A., Pandey, N. B., Chodchoy, N., and Marzluff, W. F. (1987). Translation is required for regulation of histone mRNA degradation. *Cell* **48,** 615–626.

Greenberg, M. E., and Ziff, E. B. (1984). Stimulation of mouse 3T3 cells induces transcription of the c-*fos* oncogene. *Nature (London)* **311,** 433–438.

Guyette, W. A., Matusik, R. J., and Rosen, J. M. (1979). Prolactin-mediated transcriptional and post-transcriptional control of casein gene expression. *Cell* **17,** 1013–1023.

Hadcock, J. R., Wang, H.-Y., and Malbon, C. C. (1989). Agonist-induced destabilization of β-adrenergic receptor mRNA. *J. Biol. Chem.* **264,** 19928–19933.

Hayward, M. A., Mitchell, T. A., and Shapiro, D. J. (1980). Induction of estrogen receptor and reversal of the nuclear/cytoplasmic receptor ratio during vitellogenin synthesis and withdrawal in *Xenopus laevis*. *J. Biol. Chem.* **255,** 11308–11312.

Helman Finkel, T., Kubo, R. T., and Cambier, J. C. (1991). T-cell development and transmembrane signaling: Changing biological responses through an unchanging receptor. *Immunol. Today* **12,** 79–85.

Hod, Y. and Hanson, R. W. (1988). Cyclic AMP stabilizes the mRNA for phosphoenolpyruvate carboxykinase (GTP) against degradation. *J. Biol. Chem.* **263,** 7747–7752.

Hwang, S.-P. L., Eisenberg, M., Binder, R., Shelness, G. S., and Williams, D. L. (1989). Predicted structures of apolipoprotein II mRNA constrained by nuclease and dimethyl sulfate reactivity: Stable secondary structures occur predominantly in local domains via intraexonic base pairing. *J. Biol. Chem.* **264,** 8410–8418.

Hynes, N. E., Groner, B., Sippel, A. E., Jeep, S., Wurtz, T., Nguyen-Huu, M. C., Giesecke, K., and Schtz, G. (1979). Control of cellular content of chicken egg white protein specific RNA during estrogen administration and withdrawal. *Biochemistry* **18,** 616–624.

Iwai, Y., Bickel, M., Pluznik, D. H., and Cohen, R. B. (1991). Identification of sequences within the murine granulocyte-macrophage colony-stimulating factor mRNA 3′-untranslated region that mediate mRNA stabilization induced by mitogen treatment of EL-4 thymoma cells. *J. Biol. Chem.* **266,** 17959–17965.

Juliani, M. H., Souza, G. M., Klein, C. (1990). cAMP stimulation of *Dictyostelium discoideum* destabilizes the mRNA for 117 antigen. *J. Biol. Chem.* **265,** 9077–9082.

Jungmann, R. A., Kelley, D. C., Miles, M. F., and Milkowski, D. M. (1983). Cyclic AMP regulation of lactate dehydrogenase. *J. Biol. Chem.* **258,** 5312–5318.

Kafatos, F. C. (1972). mRNA stability and cellular differentiation. *In* Karolinska Symposia on Research Methods in Reproductive Endocrinology (E. Diczfalusy, Ed.). Fifth Symposium: Gene Transcription in Reproductive Tissue, May 29–31, 1972, Stockholm, Sweden.

Kelly, K., Cochran, B. H., Stiles, C. D., and Leder, P. (1983). Cell-specific regulation of the c-*myc* gene by lymphocyte mitogens and platelet-derived growth factor. *Cell* **35,** 603–610.

Klausner, R. D., and Samelson, L. E. (1991). T cell antigen receptor activation pathways: The tyrosine kinase connection. *Cell* **64,** 875–878.

Klug, G., and Cohen, S. N. (1990). Combined actions of multiple hairpin loop structures and sites of rate-limiting endonucleolytic cleavage determine differential degradation rates of individual segments within polycistronic *puf* operon mRNA. *J. Bacteriol.* **172,** 5140–5146.

Knight, E., Jr., Anton, E. D., Fahey, D., Friedland, B. K., and Jonak, G. J. (1985). Interferon regulates c-*myc* gene expression in Daudi cells at the post-transcriptional level. *Proc. Natl. Acad. Sci. USA* **82,** 1151–1154.

Koeffler, H. P., Gasson, J., and Tobler, A. (1988). Transcriptional and posttranscriptional modulation of myeloid colony stimulating factor expression by tumor necrosis factor and other agents. *Mol. Cell. Biol.* **8,** 3432–3438.

Koeller, D. M., Horowitz, J. A., Casey, J. L., Klausner, R. D., and Harford, J. B. (1991). Translation and the stablity of mRNAs encoding the transferrin receptor and c-*fos*. *Proc. Natl. Acad. Sci. USA* **88,** 7778–7782.

Kowalski, J., and Denhardt, D. T. (1989). Regulation of the mRNA for monocyte-derived neutrophil-activating peptide in differentiating HL60 promyelocytes. *Mol. Cell. Biol.* **9,** 1946–1957.

Krane, I. M., Spindel, E. R., and Chin, W. W. (1991). Thyroid hormone decreases the stability and the poly(A) tract length of rat thyrotropin β-subunit messenger RNA. *Mol. Endocrinol.* **5,** 469–475.

Krishnan, I., and Baglioni, C. (1980). Increased levels of (2′–5′) oligo(A) polymerase activity in human lymphoblastoid cells treated with glucocorticoids. *Proc. Natl. Acad. Sci. USA* **77,** 6506–6510.

Lazier, C. B. (1978). Ontogeny of the vitellogenic response to oestradiol and of the soluble nuclear oestrogen receptor in embryonic-chick liver. *Biochem. J.* **174,** 143–152.

Lee, S. W., Tsou, A.-P., Chan, H., Thomas, J., Petrie, K., Eugui, E. M., and Allison, A. C. (1988). Glucocorticoids selectively inhibit the transcription of the interleukin 1β gene and decrease the stability of interleukin 1β mRNA. *Proc. Natl. Acad. Sci. USA* **85,** 1204–1208.

Lengyel, P. (1982). Biochemistry of interferons and their actions. *Annu. Rev. Biochem.* **51,** 251–282.

Liang, H., and Jost, J.-P. (1991). An estrogen-dependent polysomal protein binds to the 5′ untranslated region of the chicken vitellogenin mRNA. *Nucleic Acids Res.* **19,** 2289–2294.

Lindsten, T., June, C. H., Ledbetter, J. A., Stella, G., and Thompson, C. B. (1989). Regulation of lymphokine messenger RNA stability by a surface-mediated T cell activation pathway. *Science* **244,** 339–343.

Lodish, H. F., and Small, B. (1976). Different lifetimes of reticulocyte messenger RNA. *Cell* **7,** 59–65.

Lundberg, U., von Gabain, A., Melefors, O. (1990). Cleavages in the 5′ region of the omp A and bla mRNA control stability: Studies with an *E. Coli* mutant altering mRNA stability and a novel endoribonuclease. *EMBO J.* **9,** 2731–2741.

MacDonald, C. C. and Williams, D. L. (1992). Proteins associated with the messenger ribonucleoprotein particle for the estrogen-regulated apolipoprotein II mRNA. *Biochemistry* **31,** 1742–1748.

Malter, J. S. (1989). Identification of an AUUUA-specific messenger RNA binding protein. *Science* **246,** 664–666.

Malter, J. S. and Hong, Y. (1991). A redox switch and phosphorylation are involved in the posttranslational up-regulation of the adenosine-uridine binding factor by phorbol ester and ionophore. *J. Biol. Chem.* **266,** 3167–3171.

Mangiarotti, G., Ceccarelli, A., and Lodish, H. F. (1983). Cyclic AMP stabilizes a class of developmentally regulated *Dictyostelium discoideum* mRNAs. *Nature* (*London*) **301,** 616–618.

McGrew, L. L., Dworkin-Rastl, E., Dworkin, M. B., and Richter, J. D. (1989). Poly(A) elongation during *Xenopus* oocyte maturation is required for translational recruitment and is mediated by a short sequence element. *Genes Dev.* **3,** 803–815.

Morrone, G., Cortese, R., and Sorrentino, V. (1989). Post-transcriptional control of negative acute phase genes by transforming growth factor beta. *EMBO J* **8,** 3767–3771.

Moskaitis, J. E., Buzek, S. W., Pastori, R. L., and Schoenberg, D. R. (1991). The estrogen-regulated destabilizaton of *Xenopus* albumin mRNA is independent of translation. *Biochim. Biophys. Res. Commun.* **174,** 825–830.

Mudd, E. A., Prentki, P., Belin, D., and Krisch, H. M. (1988). Processing of unstable bacteriophage T4 gene 32 mRNAs into a stable species requires *Escherichia coli* ribonuclease E. *EMBO J.* **7,** 3601–3607.

Mudd, E. A., Carpousis, A. J., and Krisch, H. M. (1990). *Escherichia coli* RNAse E has a role in the decay of bacteriophage T4 mRNA. *Genes Dev.* **4,** 873–881.

Narayan, P. and Towle, H. C. (1985). Stabilization of a specific nuclear mRNA precursor by thyroid hormone. *Mol. Cell. Biol.* **5,** 2642–2646.

Newbury, S. F., Smith, N. H., Robinson, E. C., Hiles, I. D., and Higgins, C. F. (1987). Stabilization of translationally active mRNA by procaryotic REP sequences. *Cell* **48,** 297–310.

Nielsen, D. A. and Shapiro, D. J. (1990). Estradiol and estrogen receptor-dependent stabilization of a minivitellogenin mRNA lacking 5,100 nucleotides of coding sequence. *Mol. Cell. Biol.* **10,** 371–376.

Nilsen, T. W., and Baglioni, C. (1979). Mechanism for discrimination between viral and host mRNA in interferon-treated cells. *Proc. Natl. Acad. Sci. USA* **76,** 2600–2604.

Paek, I. and Axel, R. (1987). Glucocorticoids enhance stability of human growth hormone mRNA. *Mol. Cell. Biol.* **7,** 1496–1507.

Page, M. J. and Parker, M. G. (1982). Effect of androgen on the transcription of rat prostatic binding protein genes. *Mol. Cell Endocrinol.* **27,** 343–355.

Palmiter, R. D. and Carey, N. H. (1974). Rapid inactivation of ovalbumin messenger ribonucleicacid after acute withdrawal of estrogen. *Proc. Natl. Acad. Sci. USA* **71,** 2357–2361.

Pastori, R. L., Moskaitis, J. E., Smith, L. H., and Schoenberg, D. R. (1990). Estrogen regulation of *Xenopus laevis* γ-fibrinogen gene expression. *Biochemistry* **29,** 2599–2605.

Pastori, R. L., Moskaitis, J. E., Buzek, S. W., and Schoenberg, D. R. (1991a). Coordinate estrogen-regulated instability of serum protein-coding messenger RNAs in *Xenopus laevis*. *Mol. Endocrinol.* **5,** 461–468.

Pastori, R. L., Moskaitis, J. E., and Schoenberg, D. R. (1991b). Estrogen-induced ribonuclease activity in *Xenopus* liver. *Biochemistry* **30,** 10490–10498.

Pei, R. and Calame, K. (1988). Differential stability of c-*myc* mRNAs in a cell-free system. *Mol. Cell. Biol.* **8,** 2860–2868.

Penttinen, R. P., Kobayashi, S., and Bornstein, P. (1988). Transforming growth factor β increases mRNA for matrix proteins both in the presence and in the absence of changes in mRNA stability. *Proc. Natl. Acad. Sci. USA* **85,** 1105–1108.

Peppel, K., Vinci, J. M., and Baglioni, C. (1991). The AU-rich sequences in the 3′ untranslated region mediate the increased turnover of interferon mRNA induced by glucocorticoids. *J. Exp. Med.* **173,** 349–355.

Peppel, K. and Baglioni, C. (1991). Deadenylation and turnover of interferon-β mRNA. *J. Biol. Chem.* **266,** 6663–6666.

Petersen, D. D., Koch, S. R., and Granner, D. K. (1989). 3′ noncoding region of phosphoenolpyruvate carboxykinase mRNA contains a glucocorticoid-responsive mRNA-stabilizing element. *Proc. Natl. Acad. Sci. USA* **86,** 7800–7804.

Pontecorvi, A., Tata, J. R., Phyillaier, M., and Robbins, J. (1988). Selective degradation of mRNA: the role of short-lived proteins in differential destabilization of insulin-induced creatine phosphokinase and myosin heavy chain mRNAs during rat skeletal muscle L_6 cell differentiation. *EMBO J* **7,** 1489–1495.

Poyet, P., Henning, S., and Rosen, J. M. (1989). Hormone-dependent β-casein mRNA stabilization requires ongoing protein synthesis. *Mol. Endocrinol.* **3,** 1961–1968.

Ratnasabapathy, R., Hwang, S.-P. L., and Williams, D. L. (1990). The 3′ untranslated region of apolipoprotein II mRNA contains two independent domains that bind distinct cytosolic factors. *J. Biol. Chem.* **265,** 14050–14055.

Ross, J., Peltz, S. W., and Brewer, G. (1986) Histone mRNA degradation *in vivo*: The first detactable step occurs at or near the 3′ terminus. *Mol. Cell Biol.* **6,** 4362–4371.

Richter, J. D. (1991) Translational control during early development. *Bioessays* **13,** 179–183.

Saceda, M., Knabbe, C., Dickson, R. B., Lippman, M. E., Bronzert, D., Lindsey, R. K., Gottardis, M. M., and Martin, M. B. (1991). Post-transcriptional destabilization of estrogen receptor mRNA in MCF-7 cells by 12-*O*-tetradecanoylphorbol-13-acetate. *J. Biol. Chem.* **266,** 17809–1

Santos, G. F., Scott, G. K., Lee, W. M. F., Liu, E., and Benz, C. (1988). Estrogen-induced post-transcriptional modulation of c-*myc* proto-oncogene expression in human breast cancer cells. *J. Biol. Chem.* **263,** 9565–9568.

Schoenberg, D. R., Moskaitis, J. E., Smith, L. H., Jr., and Pastori, R. L. (1989). Extranuclear estrogen-regulated destabilization of *Xenopus laevis* serum albumin mRNA. *Mol. Endocrinol.* **3,** 805–814.

Schröder, H. C., Ugarkovic, D., Wenger, R., Reuter, P., Okamoto, T., and Mller, W. E. G. (1990). Binding of TAT protein to TAR region of human immunodeficiency virus type 1 blocks TAR-mediated activation of (2′–5′)oligoadenylate synthetase. *AIDS Res. Hum. Retroviruses* **6,** 659–672.

Sharmeen, L., Kuo, M. Y. P., Dinter-Gottlieb, G., and Taylor, J. (1988). Antigenomic RNA f human hepatitis delta virus can undergo self-cleavage. *J. Virol.* **62,** 2674–2679.

Shaw, G. and Kamen R. (1986). A conserved AU sequence from the 3′ untranslated region of GM-CSF mRNA mediates selective mRNA degradation. *Cell* **46,** 659–667.

Shelness, G. S., Binder, R., Hwang, S.-P. L., MacDonald, C., Gordon, D. A., and Williams, D. L. (1987). Sequence and structural elements associated with the degradation of apolipoprotein II mRNA. *In* Molecular Biology of RNA: New Perspectives (M. Inouye and B. Dudock, Eds.), pp. 381–399. Academic Press, New York.

Shyu, A.-B., Greenberg, M. E., and Belasco, J. G. (1989). The c-*fos* transcript is targeted for rapid decay by two distinct mRNA degradation pathways. *Genes Dev.* **3,** 60–72.

Shyu, A.-B., Belasco, J. G., and Greenberg, M. E. (1991). Two distinct destabilizing elements in the c-*fos* message trigger deadenylation as a first step in rapid mRNA decay. *Genes Dev.* **5,** 221–231.

Simonet, W. S. and Ness, G. C. (1988). Transcriptional and posttranscriptional regulation of rat hepatic 3-hydroxyl-3-methylglutaryl-coenzyme A reductase by thyroid hormones. *J. Biol. Chem.* **263,** 12448–12453.

Simonet, W. S. and Ness, G. C. (1989). Post-transcriptional regulation of 3-hydroxy-3-methylglutaryl-CoA reductase mRNA in rat liver. *J. Biol. Chem.* **264,** 569–573.

Smith, J. D. anf Liu, A. Y.-C. (1988). Increased turnover of the messenger RNA encoding tyrosine aminotransferase can account for the desensitization and de-induction of tyrosine aminotransferase by 8-bromo-cyclic AMP treatment and removal. *EMBO J.* **7,** 3711–3716.

Stark, G. R., Dower, W. J., Schimke, R. T., Brown, R. E., and Kerr, I. M. (1979). 2-5A synthetase: Assay, distribution, and variation with growth or hormone status. *Nature (London)* **278,** 471–473.

Stoeckle, M. Y. and Hanafusa, H. (1989). Processing of 9E3 mRNA and regulation of its stability in normal and rous sarcoma virus-transformed cells. *Mol. Cell. Biol.* **9,** 4738–4745.

Sugano, S., Stoeckle, M. Y., and Hanafusa, H. (1987). Transformation by rous sarcoma virus induces a novel gene with homology to a mitogenic platelet protein. *Cell* **49,** 321–328.

Thampan, R. V. (1988). Estradiol-stimulated nuclear ribonucleoprotein transport in the rat uterus: A molecular basis. *Biochemistry* **27,** 5019–5026.

Thorens, B., Mermod, J.-J., and Vassalli, P. (1987). Phagocytosis and inflammatory stimuli induce GM-CSF mRNA in macrophages through posttranscriptional regulation. *Cell* **48,** 671–679.

Vakalopoulou, E., Schaack, J., and Shenk, T. (1991). A 32-kilodalton protein binds to AU-rich domains in the 3′ untranslated regions of rapidly degraded mRNAs. *Mol. Cell. Biol.* **11,** 3355–3364.

Varnum, S. M., and Wormington, W. M. (1990). Deadenylation of maternal mRNAs during *Xenopus* oocyte maturation does not require specific cis-sequences: A default mechanism for translational control. *Genes Dev.* **4,** 2278–2286.

Vassalli, J. D., Huarte, J., Belin, D., Gubler, P., Vassalli, A., O'Connell, M. L., Parton, L. A., Rickles, R. J., and Strickland, S. (1989). Regulated polyadenylation controls mRNA translation during meiotic maturation of mouse oocytes. *Genes Dev.* **3,** 2163–2171.

Viano, I., Silvestro, L., Giubertoni, M., Diazani, C., Genazzani, E., Di Carlo, F. (1990). Induction of 2′–5′ oligoadenylate synthetase and activation of ribonuclease in tamoxifen treated human breast cancer cell lines. *J. Biol. Regul. Homeostatic Agents* **3,** 167–174.

Volloch, V., and Housman, D. (1981a). Relative stability of globin and non-globin mRNA in terminally differentiating Friend cells. *In* Organization and Expression of Globin Genes (G. Stamatoyannopoulos and A. W. Nienhuis, Eds.), pp. 251–257. Alan R. Liss, New York.

Volloch, V., and Housman, D. (1981b). Regulation of stability of non-globin mRNA in Friend cells. *In* Organization and Expression of Globin Genes (G. Stamatoyannopoulos and A. W. Nienhuis, Eds.), pp. 259–266. Alan R. Liss, New York.

Wager, R. E., and Assoian, R. K. (1990). A phorbol ester-regulated ribonuclease system controlling transforming growth factor β1 gene expression in hematopoietic cells. *Mol. Cell. Biol.* **10,** 5983–5990.

Weber, B., Horiguchi, J., Luebbers, R., Sherman, M., and Kufe, D. (1989). Posttranscriptional stabilization of c-*fms* mRNA by a labile protein during human monocytic differentiation. *Mol. Cell. Biol.* **9,** 769–775.

Wilson, T., and Treisman, R. (1988). Removal of poly(A) and consequent degradation of c-*fos* mRNA facilitated by 3′ AU-rich sequences. *Nature* (*London*) **336,** 396–399.

Wisdom, R., and Lee, W. (1991). The protein-coding region of c-*myc* mRNA contains a sequence that specifies rapid mRNA turnover and induction by protein synthesis inhibitors. *Genes Dev.* **5,** 232–243.

Wiskocil, R., Bensky, P., Dower, W., Goldberger, R. F., Gordon, J. I., and Deeley, R. G. (1980). Coordinate regulation of two estrogen-dependent genes in avian liver. *Proc. Natl. Acad. Sci. USA* **77,** 4474–4478.

Wodnar-Filipowicz, A., and Moroni, C. (1990). Regulation of interleukin 3 mRNA expression in mast cells occurs at the posttranscriptional level and is mediated by calcium ions. *Proc. Natl. Acad. Sci. USA* **87,** 777–781.

Wong, N. C. W., and Oppenheimer, J. H. (1986). Multihormonal regulation and kinetics of induction of a hepatic mRNA sequence which is slowly responsive to triiodothyronine. *J. Biol. Chem.* **261,** 10387–10393.

Wreschner, D. H., McCauley, J. W., Skehel, J. J., and Kerr, I. M. (1981) Interferon action - sequence specificity of the ppp(A2′p)-dependent ribonuclease. *Nature* (*London*) **289,** 414–417.

Wu, J. C., and Miller, W. L. (1991). Progesterone shortens poly(A) tails of the mRNAs for alpha and beta subunits of ovine luteinizing hormone. *Biol. Reprod.* **45,** 215–220.

Yen, T. J., Machlin, P. S., and Cleveland, D. W. (1988). Autoregulated instability of β-tubulin mRNAs by recognition of the nascent amino terminus of β-tubulin. *Nature* (*London*) **334,** 580–585.

9

Control of the Decay of Labile Protooncogene and Cytokine mRNAs

MICHAEL E. GREENBERG AND JOEL G. BELASCO

Among the most short-lived mammalian mRNAs are those encoded by the early response gene family (reviewed in Rivera and Greenberg, 1990; Herschman, 1991). The early response genes (ERGs, which have also been termed immediate early genes) are a class of over 100 genes whose transcription is rapidly and transiently activated within minutes after cells are exposed to a wide range of extracellular stimuli including growth factors, cytokines, and neurotransmitters. The expression of ERG mRNAs lasts for only a few hours because these mRNAs are synthesized for only a brief period of time after cell stimulation and because, once they are processed and transported to the cytoplasm, ERG mRNAs are rapidly degraded. The 10- to 30-min half-lives of ERG mRNAs are unusually short compared with those of most mammalian mRNAs, which typically survive for hours or days. This chapter will focus on studies of the cellular mechanisms that allow for the selective rapid degradation of ERG mRNAs.

I. Introduction

The ERG mRNAs can be divided into several groups based on both their function and their mode of decay. The first ERGs to be characterized were members of the protooncogene family and include c-*fos*, c-*myc*, and c-*jun* (reviewed in Sheng and Greenberg, 1990; Herschman 1991). Aberrant expression or mutation of any one of these genes can lead to oncogenesis. Interestingly, one mechanism by which these protooncogenes can be rendered oncogenic is by mutations that lead to the stabilization of their mRNA

Control of Messenger RNA Stability

(Lee *et al.*, 1988; Ruther *et al.*, 1987; Rabbits *et al.*, 1985; Piechaczyk *et al.*, 1985).

The c-*fos* and c-*myc* genes are the best characterized of the early response genes. Prior to growth factor stimulation of cells, both the c-*fos* and the c-*myc* mRNAs are virtually undetectable. Upon growth factor stimulation of a variety of cell types, c-*fos* trancription is transiently induced within 5 min (Greenberg and Ziff, 1984). Transcription of c-*myc* is activated somewhat later, around 30 min after growth factor addition (Kelly *et al.*, 1983; Greenberg and Ziff, 1984). Within 30–60 min after the initial stimulation event, the cytoplasmic level of c-*fos* and c-*myc* mRNA peaks and begins to decline (Fig. 1). The rapid disappearance of these messages is due not only to a shutoff of ERG transcription but also to the short half-lives of these mRNAs. The observation that the c-*fos* and c-*myc* protooncogenes are rapidly and transiently activated by extracellular stimuli and that their expression is tightly controlled at multiple levels raised the possibility that they might encode key regulators of cell growth and differentiation. This idea has been substantiated by a variety of experiments demonstrating that both c-*fos* and c-*myc* encode transcription factors that are important regulators of gene expression (reviewed by Curran and Franza, 1988; Lüscher and Eisenman, 1990). Understanding the mechanisms controlling the rapid decay of this family of mRNAs is therefore likely to give important insights into the processes of normal cell growth and differentiation and how the deregulation of these events can lead to oncogenesis.

Other members of the ERG family include genes that are related to c-*fos*, c-*jun*, and c-*myc*, as well as a number of other structurally distinct transcription factors. To date, a common characteristic of the ERGs that encode transcription factors is that their mRNAs are all very short lived. Therefore, induction of transient transcription of these ERGs results in only a very brief period of gene expression. A distinct subset of ERGs encode cytokines, interferons, and mediators of inflamation. The expression of members of this class of ERGs is transiently induced by a variety of extracel-

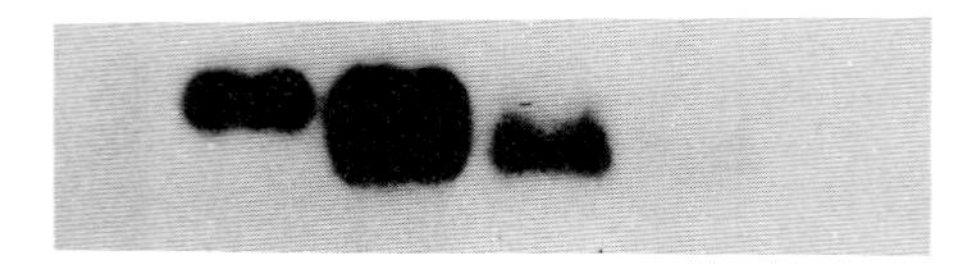

FIGURE 1 Transient induction of c-*fos* mRNA in growth-factor-stimulated cells. Total cytoplasmic RNA was isolated at time intervals after serum stimulation of quiescent NIH3T3 fibroblasts. The relative concentration of c-*fos* mRNA in each RNA sample was determined by Northern blot analysis. Times indicate minutes after serum stimulation.

lular stimuli, but in contrast to members of the *fos*, *jun*, and *myc* families discussed above, the increase in mRNA levels is due in large part to mRNA stabilization (Thorens *et al.*, 1987; Lindsten *et al.*, 1989). Well-characterized examples of this group of ERGs are the genes that encode cytokines such as granulocyte–monocyte colony stimulating factor (GM-CSF) (Shaw and Kamen, 1986), γ-interferon (Lindsten *et al.*, 1989), and the *gro* α and 9E3 gene products (Stoeckle and Hanafusa, 1989; Stoeckle, 1992). In this chapter we will focus our discussion of cytokine mRNAs on the GM-CSF, *gro* α, and 9E3 messages.

GM-CSF stimulates the growth, differentiation, and activation of hematopoietic cells of myeloid lineage and also functions as a growth factor for early multipotential cells and erythroid precursors (reviewed in Metcalf, 1985). It is produced in a variety of cell types, including B cells, T cells, monocytes, endothelial cells, and fibroblasts. Synthesis of GM-CSF in T cells is induced by treatment with phorbol esters, antigens, lectins, interleukins 2 and 3, and γ-interferon (Shaw and Kamen, 1986; Lindsten *et al.*, 1989). The *gro* α gene product is a small secreted protein that appears to be a mediator of inflammation as well as a potent neutrophil chemoattractant and an autocrine growth factor (reviewed in Stoeckle and Barker, 1990). Both this protein and the related product of the chicken 9E3 gene are members of the platelet factor 4 family. The expression of *gro* α mRNA is induced in fibroblasts, synovial cells, and endothelial cells in response to inflammatory stimuli, including interleukin 1, tumor necrosis factor, and bacterial lipopolysaccharide. The level of both *gro* α and GM-CSF mRNA is low in most cells prior to physiological stimulation; however, the abundance of these mRNAs increases dramatically within minutes after activation of the appropriate cell type. Nuclear run-on transcription analyses and mRNA half-life measurements indicate that the increase in GM-CSF and *gro* α mRNA induced by extracellular stimulation is not due just to enhanced levels of transcription (Shaw and Kamen, 1986; Stoeckle, 1992). Instead, the increase in GM-CSF and *gro* α mRNA expression is in large part a consequence of a significant increase in the stability of these mRNAs, which are normally very labile.

The identification of growth factor inducible mRNAs that are extremely unstable (e.g., c-*fos* and c-*myc*) and labile cytokine mRNAs that are stabilized by extracellular stimuli (e.g., GM-CSF and *gro* α) raises an important question. What are the cellular mechanisms that control the differential stability of these short-lived mRNAs? Efforts to understand these mRNA decay pathways have focused first on identifying the mRNA sequence determinants that target the messages for decay. The characterization of these mRNA decay determinants has facilitated the development of both *in vitro* and *in vivo* approaches for defining the cellular mechanisms controlling the decay of ERG mRNAs.

II. The Sequence Determinants Controlling ERG mRNA Decay

The first hint that the 3′ untranslated regions (UTRs) of the ERG mRNAs might contain important determinants of instability was an experiment by Treisman (1985). He found that replacement of the 3′ half of c-*fos* mRNA with the corresponding region of the highly stable β-globin message resulted in significant stabilization in NIH3T3 mouse fibroblasts. Comparative analysis of the nucleotide sequence of the c-*fos* 3′ untranslated region revealed that it is well conserved through evolution. In addition, the 3′ UTRs of c-*fos* and many other ERG mRNAs are notable in that they contain long segments that are rich in adenine and uridine nucleotides when compared with the 3′ UTRs of typical mammalian mRNAs (Figs. 2 and 3) (Shaw and Kamen, 1986). Taken together, these results suggested that the AU-rich regions of the ERG mRNAs might function as critical determinants of mRNA instability (Caput *et al.*, 1986; Shaw and Kamen, 1986). This hypothesis has received strong support from a variety of studies (see below).

Shaw and Kamen (1986) were the first to demonstrate that a conserved

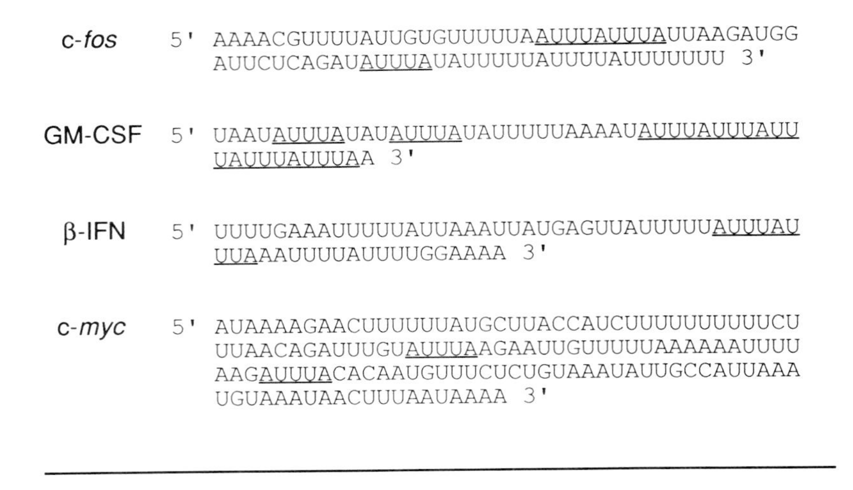

FIGURE 2 AU-rich destabilizing elements. Sequences are listed for four AREs (from the human c-*fos*, GM-CSF, β-interferon, and c-*myc* mRNAs) that have been shown to function as mRNA destabilizing elements in mammalian cells. Also shown for comparison is the 3′ UTR of rabbit β-globin mRNA, a long-lived message that lacks an ARE. AUUUA pentanucleotides are underlined.

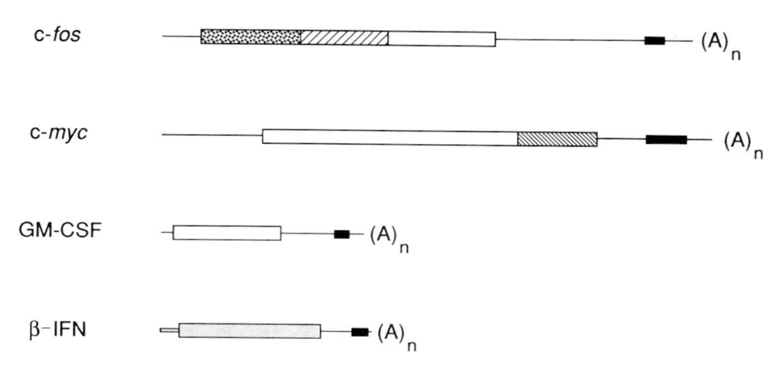

FIGURE 3 Destabilizing elements in ERG mRNAs. The locations of known mRNA destabilizing elements in the human c-*fos*, c-*myc*, GM-CSF, and β-interferon mRNAs are indicated. Thin lines, UTRs. Open rectangles, protein-coding regions. Hatched and stippled areas, coding-region destabilizing elements. Black rectangles, AREs. For each destabilizing element, the minimal functional unit may well be smaller than indicated. The non-ARE destabilizing element in β-interferon mRNA may reside in the coding region, the 5′ UTR, or both.

51-nucleotide AU-rich sequence present in the 3′ UTR of human GM-CSF mRNA is a potent mRNA destabilizing sequence (Figs. 2 and 3). In their experiments, rabbit β-globin mRNA and various derivatives thereof were expressed in mouse fibroblast cells from a constitutive promoter. β-Globin mRNA levels were then measured by RNase-protection analysis at various times after transcription was inhibited with actinomycin D. Insertion of the GM-CSF AU-rich element (ARE) into the 3′ UTR of rabbit β-globin mRNA reduced the half-life of this message to less than 30 min, whereas the insertion of a variant sequence containing a number of interspersed G and C residues resulted in a chimeric mRNA that was stable for more than 2 hr. Nuclear run-on transcription analysis revealed that insertion of the ARE had no detectable effect on the level of β-globin gene transcription. Taken together these experiments suggested that the 51-nucleotide AU-rich sequence functions primarily as an instability element. These findings were particularly provocative because the authors noted the presence of a similar AU-rich sequence in the 3′ UTRs of a wide range of labile protooncogene and cytokine mRNAs. This suggested that ERG mRNAs might be degraded by a common pathway involving cellular factors that somehow recognize the AREs. As most of the ERG AREs contain multiple copies of the sequence motif AUUUA, it was speculated that this sequence might be critical for recognition of AU-rich elements by the cellular mRNA decay machinery.

More recent experiments have further substantiated the hypothesis that the AU-rich sequences in the 3′ UTRs of ERG mRNAs function as destabilizing determinants (Jones and Cole, 1987; Bonnieu *et al.*, 1988; Wilson and Treisman, 1988; Shyu *et al.*, 1989). For example, the c-*myc* 3′ UTR contains two long AU-rich sequences. Evidence that at least one of these c-*myc* AREs functions as a destabilizing element has been obtained by analysis of the decay of neomycin–*myc* (*neo–myc*) hybrid mRNAs bearing the c-*myc* 3′ UTR. These experiments indicate that fusion to the c-*myc* 3′ UTR accelerates degradation of *neo* mRNA and that deletion of a 140-nucleotide AU-rich sequence preceding the upstream c-*myc* poly(A) addition site (see Figs. 2 and 3) impedes decay mediated by the c-*myc* 3′ UTR (Jones and Cole, 1987).

Interestingly, several studies indicate that at least two classes of AREs exist that function by somewhat distinct mechanisms. The AU-rich sequences present in cytokine mRNAs, such as GM-CSF, function as destabilizing elements whose activity is regulated by extracellular stimuli (Shaw and Kamen, 1986; Lindsten *et al.*, 1989). For example, stimulation of human T lymphoblasts with the phorbol ester TPA leads to stabilization of GM-CSF mRNA. The stabilizing effect of TPA on mRNA bearing the GM-CSF 3′ UTR requires both the ARE and an additional 3′ UTR element located 100–160 nucleotides upstream of the ARE (Iwai *et al.*, 1991). A variety of other examples in which the stability of a cytokine message containing an ARE increases upon treatment of cells with a hormone or cytokine have also been described (Shaw and Kamen, 1986; Lindsten *et al.*, 1989). These observations have prompted the hypothesis that the 3′ UTR associates with a protein factor whose binding activity can be induced by certain extracellular stimuli. The interaction of the binding factor with the ARE or another 3′ UTR element might then protect the mRNA from degradation. Conversely, the decay rate of ERG mRNAs can also accelerate upon hormone treatment. For example, in L929 cells treated with the synthetic glucocorticoid dexamethasone, the lifetime of β-interferon mRNA is reduced; this hormone-induced destabilization is ARE-dependent (Peppel *et al.*, 1991).

In contrast to the AU-rich element present in GM-CSF mRNA, the ARE in the 3′ UTR of the c-*fos* protooncogene transcript functions as a potent destabilizing element whose activity appears not to be affected by growth factor or cytokine stimulation (Shyu *et al.*, 1989; Lindsten *et al.*, 1989). When the 75-nucleotide c-*fos* ARE (Figs. 2 and 3) was inserted into the 3′ UTR of rabbit β-globin mRNA (to generate BBB+ARE mRNA), it decreased the half-life of this normally stable message from >8 hr to about 30 min (1989) (Fig. 4). The chimeric BBB+ARE message is very labile, regardless of whether it is transiently synthesized from a c-*fos* promoter in growth-factor-stimulated fibroblasts or constitutively expressed from a strong viral enhancer in continuously growing cells. The differential regulation of the stability of mRNAs containing the c-*fos* and GM-CSF AREs suggests that, while these two elements are both AU-rich, there must be some important sequence difference that accounts for their distinct behavior. One difference between these two classes of ARE is that the AU-rich sequences present in GM-CSF-like mRNAs contain three or more contiguous AUUUA motifs,

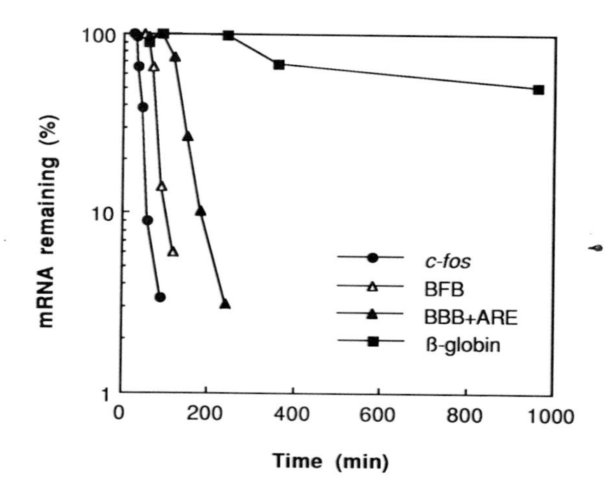

FIGURE 4 mRNA degradation mediated by the c-*fos* ARE and protein-coding region. Total cytoplasmic RNA was isolated at time intervals after serum stimulation of NIH3T3 cells that had been transiently transfected with plasmid pBBB, pBBB + ARE, or pBFB. Degradation of β-globin mRNA (BBB), BBB + ARE mRNA (β-globin mRNA with a copy of the c-*fos* ARE inserted into its 3′ UTR), BFB mRNA (the c-*fos* protein-coding region flanked by β-globin 5′ and 3′ UTRs), and the endogenous c-*fos* transcript was monitored by RNase protection analysis. The percentage of mRNA remaining is plotted as a function of time after induction of transient transcription. See Shyu *et al.* (1989) for experimental details.

while those in the protooncogene messages are generally scattered throughout the element (Fig. 2) (Shaw and Kamen, 1986).

A direct comparison of the ability of the c-*fos*, c-*myc*, and GM-CSF AREs to function as destabilizing elements lends support to the idea that there are at least two classes of ARE. Schuler and Cole (1988) have found that GM-CSF and protooncogene mRNA 3′ UTRs can be independently regulated in the same cell type. Fusion of the 3′ UTR of c-*fos* or c-*myc* to a *neo* reporter gene yielded mRNAs with very short half-lives in all cell lines tested. However, the 3′ UTR of GM-CSF mRNA, when fused to the *neo* message, had a destabilizing effect in some but not all cell lines. These results suggest the existence of trans-acting factors that differentially recognize the two classes of ARE, and support the idea that these two classes of ARE have distinct functions.

The characterization of the mechanism by which these distinct classes of ARE function should be facilitated by a careful *in vivo* analysis of the structural features of AREs that are essential for their activity as destabilizing determinants. The importance for rapid mRNA decay of the AUUUA motif within the 3′ UTRs of ERG mRNAs has been addressed in several recent studies. An alteration of all three AUUUA pentanucleotides in the c-*fos* ARE by single U-to-A mutations significantly impaired the c-*fos* ARE's ability to function as a destabilizing element when inserted into β-globin mRNA (Shyu *et al.*, 1991). This result suggests that the AUUUA pentanucleotide or an overlapping sequence is important for rapid ARE-mediated mRNA decay. Nevertheless, it is important to emphasize that the simple presence of any AU-rich sequence or of an isolated AUUUA pentanucleotide is not enough to guarantee that an mRNA will be unstable. For example, the highly stable rabbit β-globin mRNA has an AUUUA pentanucleotide in its 3′ UTR (Fig. 2). Fine mutagenesis of AREs will be necessary to establish the precise nucleotide sequences that specify these elements as mRNA decay determinants.

A more detailed analysis of the instability determinants present in ERG mRNAs indicates that a number of these messages contain potent destabilizing elements in addition to the ARE (Fig. 3). These other determinants of instability have been best characterized for the c-*fos* and c-*myc* mRNAs (Kabnick and Housman, 1988; Shyu *et al.*, 1989, 1991; Wisdom and Lee, 1991). When human c-*fos* mRNA is expressed from its own promoter in fibroblasts stimulated by growth factors, its decay can be monitored in the absence of transcriptional inhibitors because c-*fos* transcription is rapidly shut off within minutes after the initial induction event (Greenberg and Ziff, 1984). Under these conditions Shyu *et al.* (1989) found that deletion of the c-*fos* ARE or replacement of the entire c-*fos* 3′ UTR with that of a stable mRNA has little effect on the decay rate of the c-*fos* message. While the c-*fos* 3′ UTR does contain a potent ARE that has a dramatic destabilizing effect when inserted into the 3′ UTR of a normally stable message, the deletion of this element from the c-*fos* message has only a small effect on the mRNA decay rate because of the presence of additional destabilizing elements elsewhere in the message. Replacement of various regions of the stable β-globin mRNA (5′ UTR, coding region, 3′ UTR) with the corresponding region of the c-*fos* message led to the identification of the c-*fos* coding region as a second effective destabilizing domain (Shyu *et al.*, 1989). Replacement of the coding region of β-globin mRNA with that of c-*fos* (to generate BFB mRNA) resulted in a dramatic reduction in the mRNA half-life from >8 hr to <30 min (Fig. 4). More recent studies have indicated that the c-*fos* coding region can be dissected into at least two distinct instability elements (S. Schiavi, C. Wellington, A.-B. Shyu, C.-Y. Chen, M. E. Greenberg, and J. G. Belasco, unpublished observations) (Fig. 3). In-frame insertion of each individual coding-region determinant into the coding region of β-globin mRNA destabilizes the latter message, although to a lesser degree than caused by the entire c-*fos* coding region.

In addition to c-*fos* mRNA, a coding region determinant of instability has also been identified within c-*myc* mRNA (Wisdom and Lee, 1991) (Fig. 3). The presence of this additional destabilizing element in the c-*myc* coding region provides a simple explanation for why deletion of the entire AU-rich portion of the c-*myc* 3′ UTR has no effect on the lability of c-*myc* mRNA

(Laird-Offringa *et al.*, 1991). β-Interferon mRNA also possesses a second instability determinant in addition to its ARE; this second element has been localized to a segment comprising the 5′ UTR and coding region of the interferon message (Whittemore and Maniatis, 1990) (Fig. 3). The instability determinants in *gro α* and 9E3 mRNA have not been mapped by mutational analysis; degradation of these mRNAs appears to involve endonucleolytic cleavage at a 3′ UTR site adjacent to a shared sequence motif: UCCCPy-UGGA (Stoeckle and Hanafusa, 1989; Stoeckle, 1992) (Fig. 5).

III. Mechanisms of mRNA Decay: The Importance of Protein Synthesis

Several features of the *in vivo* decay process have given important insights into the cellular mechanisms controlling the decay of ERG mRNAs. Early studies of c-*fos* mRNA decay revealed that the degradation process is dependent on protein synthesis (Cochran *et al.*, 1984; Greenberg *et al.*, 1986). A wide range of translational inhibitors that block protein synthesis by different mechanisms were found to lead to a signficant stabilization of ERG mRNAs. Since these inhibitors of protein synthesis also block the shutoff of transcription that occurs following growth factor stimulation of ERG transcription, it was necessary in these experiments to monitor mRNA decay in the presence of actinomycin D to block further transcription. Under these conditions, inhibition of protein synthesis with cycloheximide increases the half-life of c-*fos* mRNA from ~15 min to several hours (Fort *et al.*, 1987; Wilson and Treisman, 1988).

The requirement of protein synthesis for rapid ERG mRNA decay might be explained in several ways, which are not mutually exclusive. One possibility is that mRNA decay occurs by a mechanism that requires that the ERG message be actively translated. For example, a ribonuclease that is critical for degrading the mRNA might be associated with the translating ribosomes. An alternative explanation of the requirement for protein synthesis is that ERG mRNA degradation may require a labile protein that decays away in the presence of protein synthesis inhibitors. Arguing against

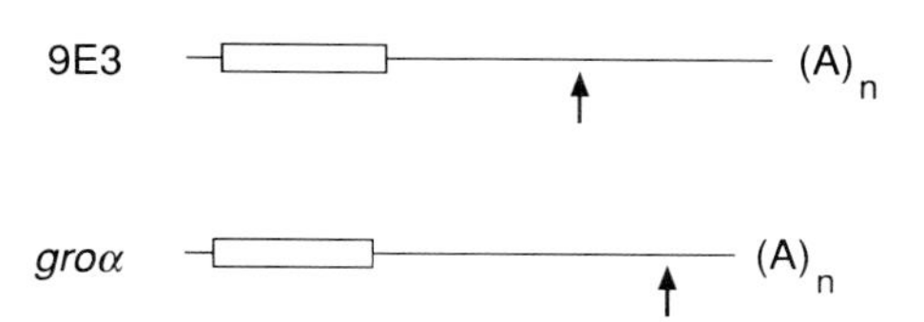

FIGURE 5 Endonucleolytic cleavage sites in the 3′ UTR of 9E3 and *gro α* mRNA. Thin lines, UTRs. Rectangles, protein-coding regions. Arrows, apparent sites of endonucleolytic cleavage.

this latter possibility is the finding that c-*fos* mRNA decay is arrested immediately upon treatment of cells with cycloheximide (Wilson and Treisman, 1988). To explain this swift impact of translation inhibition by invoking a labile protein would require a putative polypeptide of extraordinary instability. On the other hand, a role for a labile protein in ERG mRNA decay is supported by the observation that ERG mRNAs are stabilized when cells are treated with any of several different inhibitors (e.g., cycloheximide, anisomycin, puromycin) that block protein synthesis at distinct steps in translation. Presumably, if a particular aspect of ERG mRNA translation was required for proper mRNA decay, different protein synthesis inhibitors might be expected to affect mRNA stability to different extents. In contrast, if a labile protein was involved in the decay process, all translational inhibitors would be expected to have a similar effect on mRNA half-life. At present, a complete explanation for the effect of protein synthesis inhibitors on ERG mRNA decay is not available.

Several recent studies have addressed the importance of mRNA translation for rapid mRNA decay by introducing mutations within specific mRNAs to alter their translatability (Koeller *et al.*, 1991; Cole and Mango, 1990). The effect of these mutations on mRNA stability was then analyzed. The results of these studies are provocative, but are somewhat difficult to interpret. A mutation that replaced one of the c-*myc* initiation codons with a stop codon resulted in a dramatic stabilization of this normally labile mRNA (Cole and Mango, 1990). The mutant *myc* mRNA was found to be associated with only one ribosome per message, suggesting that the mRNA was not being translated efficiently. However, it is unclear whether this stabilization of c-*myc* mRNA occurs because, once the mRNA is no longer associated with multiple ribosomes, it adopts a secondary structure that is resistant to degradation by cellular ribonucleases or because the failure of the mRNA to associate with polysomes renders it inaccessible to attack by a ribosome-associated nuclease. In another study, the importance of translation for rapid mRNA degradation was analyzed by inserting into the 5′ UTR an iron response element (IRE) that allows translation of the mRNA to be regulated by the abundance of iron (Koeller *et al.*, 1991) (see Chapter 11). When iron levels in the surrounding medium are low due to treatment with the iron chelator desferrioxamine, translation of mRNAs containing an IRE in the 5′ UTR is inhibited. As iron levels increase, the translation of the IRE-containing mRNA is potentiated. When translation of an unstable chimeric mRNA containing an IRE in its 5′ UTR and the c-*fos* ARE in its 3′ UTR was inhibited by treating cells with desferrioxamine, degradation was hardly affected. This finding suggested that the c-*fos* ARE can function effectively as an mRNA instability determinant whether or not the mRNA that contains it is translated. This conclusion is consistent with the model, discussed above, that a labile protein is required for the proper decay of c-*fos* and other ERG mRNAs. However, the results of this experiment might also be interpreted in another way. Even though the message is rapidly degraded whether or not it is being translated, it remains possible that the pathway of decay differs depending on whether the mRNA is associated with translating ribosomes. For example, ARE-mediated decay may be translation dependent, but once the mRNA is no longer associated with ribosomes, it may be rapidly degraded by an alternative pathway that is ARE-independent.

Whether or not mRNA translation is necessary for the function of the destabilizing elements in the c-*fos* and c-*myc* coding regions is of particular interest. It is not yet clear for these mRNAs whether it is the mRNA sequence itself or the nascent polypeptide that it encodes that mediates rapid mRNA degradation. There is precedent for the participation of a nascent polypeptide in mRNA decay, as the amino terminal residues of nascent β-tubulin can direct accelerated degradation of β-tubulin mRNA (Yen *et al.*, 1988) (see Chapter 10). If the determinants of instability in the coding region of c-*fos* or c-*myc* could function in the absence of protein synthesis, this would suggest that the corresponding protein domain encoded by the instability determinant is not required for its function as an mRNA destabilizer. However, preliminary experiments with translation inhibitors and modified c-*fos* mRNAs suggest that translation of the c-*fos* coding region is required for its determinants of instability to function properly (S. Schiavi, C. Wellington, M. E. Greenberg, and J. G. Belasco, unpublished observations). This leaves open the intriguing possibility that the nascent c-*fos* and/or c-*myc* proteins may themselves play a role in the process of ERG mRNA decay.

Clearly, additional experiments are required to determine which determinants of mRNA instability require mRNA translation in order to function effectively and which instability elements can carry out their destabilizing function in the absence of mRNA translation. Together, the available evidence suggests that, while a labile protein may be involved in the decay process, the translation of ERG mRNAs is also important for proper functioning of at least some of their mRNA decay determinants.

IV. Deadenylation as the First Step in ERG mRNA Decay

In addition to demonstrating a requirement for protein synthesis in the rapid degradation of c-*fos* mRNA, early studies of c-*fos* mRNA decay also revealed that this message shortens significantly prior to its decay in growth-factor-stimulated cells (Treisman, 1985; Wilson and Treisman, 1988). When the decay of c-*fos* mRNA was analyzed by Northern blotting, the length of this message was found to decrease swiftly by about 200 nucleotides before the message was rapidly degraded (Treisman, 1985). Subsequent studies have revealed that this shortening of c-*fos* mRNA occurs at its 3′ end and is due to the removal of its poly(A) tail (Wilson and Treisman, 1988; Shyu *et al.*, 1991) (Fig. 6). The deadenylation of ERG mRNAs prior to their decay may be a mechanism by which many of these messages are degraded, as other ERG mRNAs, including c-*myc* and β-interferon mRNA, also have been shown to undergo deadenylation prior to decay (Laird-Offringa, 1990; Peppel and Baglioni, 1991). A careful analysis of the length of the c-*fos* mRNA poly(A) tail at various times after serum stimulation of NIH 3T3 fibroblasts indicates that the transcribed portion of the c-*fos* message is stable until the poly(A) is completely removed, whereupon rapid degradation of the mRNA body ensues (Shyu *et al.*, 1991). This finding suggests that deadenylation of c-*fos* and other ERG mRNAs is the first step in their decay.

Most mRNAs are not rapidly deadenylated in growth-factor-stimulated cells. For example, β-globin mRNA expressed from the c-*fos* promoter undergoes slow and incomplete shortening of its poly(A) tail (Fig. 6). Remarkably, the insertion into β-globin mRNA of any of the c-*fos* mRNA instability determinants, from either the c-*fos* coding region or the 3′ UTR, markedly increases the rate and extent of degradation of the β-globin message (Shyu *et al.*, 1991; S. Schiavi, M. E. Greenberg, and J. G. Belasco, unpublished observations) (Fig. 6). Moreover, as observed for the wild-type c-*fos* message, decay of the transcribed portion of β-globin variants containing the c-

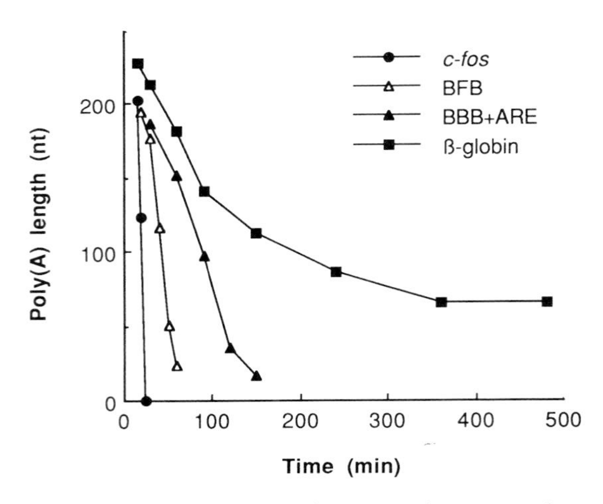

FIGURE 6 Deadenylation mediated by the c-*fos* ARE and protein-coding region. Total cytoplasmic RNA was isolated at time intervals after serum stimulation of untransfected NIH 3T3 cells or NIH3T3 cells that had been stably transfected with plasmid pBFB or transiently transfected with plasmid pBBB + ARE or pBBB. Deadenylation of c-*fos* , BFB, BBB + ARE, and β-globin (BBB) mRNA was monitored by Northern blot analysis. Poly(A) tail lengths are plotted as a function of time after induction of transient transcription. [See Fig. 4 for a description of BFB and BBB + ARE mRNA and Shyu *et al.* (1991) for experimental details.]

fos ARE or coding region does not commence until deadenylation is virtually complete. These findings suggest that at least one function of the c-*fos* instability elements is to promote rapid deadenylation of mRNA as a first step in the decay process.

The mechanism by which the various c-*fos* destabilizing elements direct deadenylation is not yet known. *In vivo,* the poly(A) tails of all eukaryotic mRNAs are bound by poly(A) binding protein (PABP) (reviewed in Sachs, 1990; Bernstein and Ross, 1989) (see Chapter 15). One possibility is that the instability determinants in the c-*fos* coding region and 3′ UTR might each act to strip this PABP from the poly(A) tail and expose the tail to ribonuclease digestion. On the other hand, the discovery by Sachs of a PABP-dependent ribonuclease (PAN) in yeast that specifically degrades poly(A) tails bound by PABP (A. Sachs, personal communication) (see Chapter 15) suggests by analogy that the presence of bound PABP may instead be necessary for deadenylation of ERG mRNAs. Normally, digestion by yeast PAN stops after poly(A) has been shortened to a length of 15 to 25 nucleotides. However, a 3′ UTR element present in at least one labile yeast message, *MFA2*, directs PAN to remove the entire poly(A) tail. It remains to be established whether this *MFA2* element is a yeast analog of the AREs found in many mammalian ERG mRNAs.

The finding that poly(A) tail removal is followed immediately by degradation of the transcribed portion of the c-*fos*, c-*myc*, and β-interferon messages suggests that, directly or indirectly, poly(A) protects these messages from ribonuclease digestion. One simple explanation for how poly(A) could accomplish this is suggested by the discovery in mammalian cell extracts of a ribonuclease activity that degrades RNAs not protected at the 3′ end by PABP (Bernstein *et al.*, 1989). *In vivo,* such an enzyme might rapidly degrade ERG mRNAs once their poly(A) tails have been shortened so severely as to prevent further binding by PABP (Fig. 7). In contrast, long-lived transcripts like β-globin mRNA, whose poly(A) tail is never shortened to less than 60 nucleotides in transfected fibroblasts, would retain indefinitely their ability to bind PABP at the 3′ end and would therefore be resistant to degradation by this ribonuclease. The absence of detectable c-*fos* or c-*myc* mRNA degradation intermediates that might correspond to the products of endonuclease cleavage in the transcribed portion of these messages suggests that these mRNAs may be degraded by a 3′ exonuclease after they are deadenylated. Of course other, more complex explanations for how deadenylation facilitates ERG mRNA decay are possible given the role of poly(A) in other processes, such as translation. It remains to be established whether deadenylation is an early step in the decay of most ERG mRNAs, although the widespread presence of AU-rich sequences in these messages suggests that deadenylation may trigger the decay of many ERG transcripts.

In contrast to the c-*fos* message, discrete fragments of 9E3 and *gro* α mRNA are observed that apparently correspond to the 5′ and 3′ products

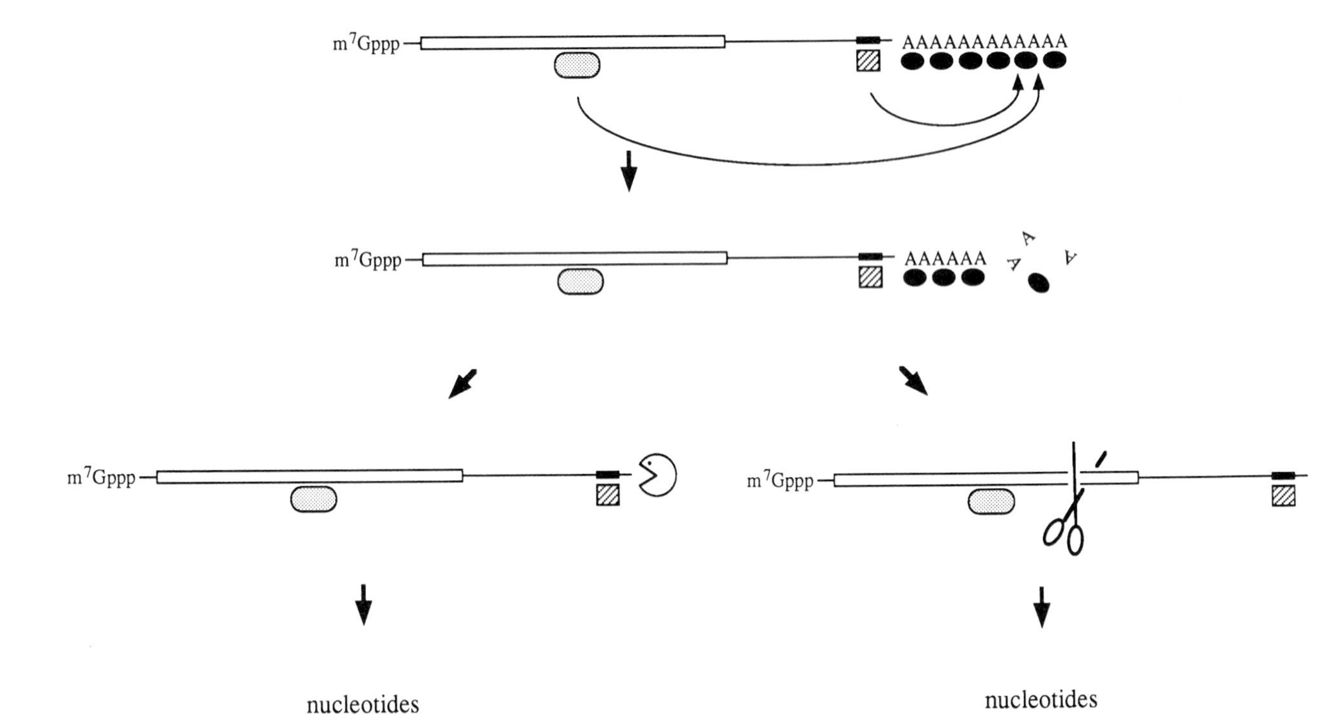

FIGURE 7 Mechanism of c-*fos* mRNA degradation. Cytoplasmic factors (stippled or hatched shapes) that bind to the c-*fos* ARE or to one of the instability determinants in the c-*fos*-coding region are postulated to interact with the poly(A) tail (shown as a string of As, each of which represents about 30 adenosine residues), which exists as a complex with poly(A) binding protein (closed ovals). This interaction leads to rapid and complete digestion of the poly(A) tail by a deadenylating enzyme. Deadenylation then exposes the c-*fos* message to swift degradation, perhaps by a 3′ exonuclease (voracious head) or an endonuclease (scissors).

of endonuclease cleavage at a conserved site within the 3′ UTR of these two messages (Stoeckle and Hanafusa, 1989; Stoeckle, 1992). Available evidence suggests that these cleavage products are likely to represent important intermediates in 9E3 and *gro* α mRNA decay and that the 9E3 cleavage event may not require prior deadenylation (Fig. 8). These findings imply that the 9E3 and *gro* α mRNAs are degraded by a mechanism that is at least somewhat distinct from the decay mechanism of c-*fos* and c-*myc* mRNA, although it remains possible that these two protooncogene transcripts are cleaved endonucleolytically following deadenylation to generate intermediates that are too unstable to be detected (Fig. 7).

V. The Cellular Factors That Control Rapid ERG mRNA Decay

A more complete understanding of the mechanism by which AREs and coding region instability determinants target ERG mRNAs for rapid decay requires the identification of cellular factors that recognize these elements. Recent studies from several laboratories have employed RNA binding assays to identify proteins that specifically interact with AU-rich sequences

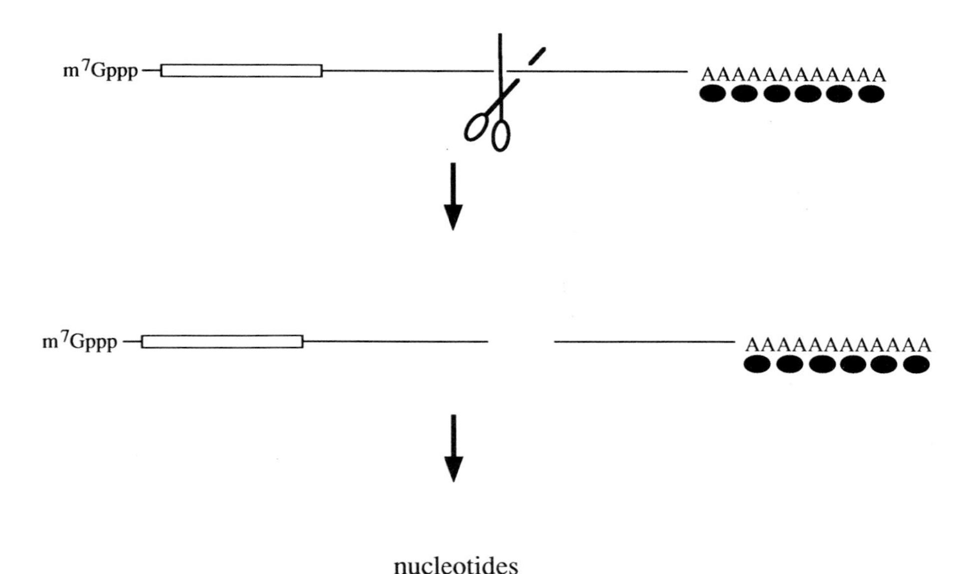

FIGURE 8 Mechanism of 9E3 mRNA degradation. Endonucleolytic cleavage (scissors) within the 3′ UTR of 9E3 mRNA is postulated to trigger overall degradation of the message. This endonucleolytic cleavage apparently does not require prior removal of the poly(A) tail (represented by a string of As), which forms a complex with poly(A) binding protein (closed ovals).

within the ERG mRNAs. So far, at least three distinct protein factors have been characterized that bind *in vitro* to AU-rich instability determinants (Bohjanen *et al.*, 1991, 1992; Brewer, 1991; Gillis and Malter, 1991; Vakalopoulou *et al.*, 1991). One of these ARE binding factors, AU-B (and its close relative AU-C), is a cytoplasmic T cell protein that displays high affinity for the AU-rich region of lymphokine mRNAs, such as GM-CSF, interleukin 2, and tumor necrosis factor α (TNF-α) mRNA, but that does not bind appreciably to an AU-rich sequence derived from the c-*myc* 3′ UTR (Bohjanen *et al.*, 1991, 1992). AU-B and AU-C preferentially interact with (AUUU) tandem multimers, and their selective binding to the 3′ UTR of cytokine mRNAs might account, at least in part, for the differential regulation of cytokine and protooncogene mRNA decay. Additional support for the possibility that AU-B and AU-C play a role in the control of cytokine mRNA degradation comes from the finding that their activity in T cells parallels the induction of cytokine mRNAs following cell stimulation via the T cell receptor. However, there is as yet no direct evidence that AU-B or AU-C functions as an mRNA degradation or stabilization factor *in vivo*.

Although AU-B and AU-C do not bind to a c-*myc* AU-rich sequence, other protein factors have been characterized that do interact with the AREs of protooncogene mRNAs. One of these factors, Auf, which has been purified from the cytosol of K562 erythroleukemia cells, appears to contain two distinct polypeptides 37 and 40 kDa in size. Auf binds to the AREs of both c-*myc* and GM-CSF mRNA (Brewer, 1991). Moreover, Auf stimulates the decay of c-*myc* mRNA in an *in vitro* cell-free system developed for studying mRNA degradation (see Chapter 16). The observation that Auf can accelerate the process of mRNA decay *in vitro* lends support to the idea that it may be involved in controlling mRNA degradation *in vivo*.

Another cellular factor that can interact with AU-rich elements *in vitro* is a 34-kDa protein present in both stimulated and unstimulated T cells (Bohjanen *et al.*, 1991). This factor, named AUBF or AU-A, binds *in vitro* to AU-rich sequences in the 3′ UTR of GM-CSF, γ-interferon, interleukin 3, TNF-α, c-*fos*, and v-*myc* mRNA (Bohjanen *et al.*, 1991; Gillis *et al.*, 1991). Another ARE binding factor has been identified in HeLa cell extracts and may be the same as or closely related to AUBF and AU-A (Vakalopoulou *et al.*, 1991). This 32-kDa HeLa cell protein has been shown to bind to the AREs of GM-CSF, c-*fos*, and c-*myc* mRNA. Multiple U-to-G mutations in a synthetic ARE that significantly reduce its effectiveness as an mRNA instability determinant *in vivo* also prevent binding of the HeLa factor *in vitro*. An interesting property of AUBF/AU-A and the HeLa cell factor is that they are localized primarily in the nucleus rather than in the cytoplasm, where the degradation of mature mRNAs normally occurs. The presence of ARE binding factors in the nucleus suggests that these proteins may complex with ERG mRNAs prior to their transport to the cytoplasm. This raises the possibility that, in addition to their function as degradation determinants,

AREs might also play some role in the nucleus or in the transport of ERG mRNAs from the nucleus to the cytoplasm.

The existence of multiple ARE binding factors, sometimes coexisting within a single cell, suggests an intriguing degree of complexity for ARE function. Additional studies will be required to establish the importance of these various factors for the decay or stabilization of ERG mRNAs *in vivo*.

VI. Conclusions

Substantial progress has been made toward identifying the sequence determinants and protein factors that control the rapid decay of ERG mRNAs. Nevertheless, the cellular mechanisms by which these mRNAs are targeted for decay are still poorly understood. Deadenylation appears frequently to be a first step in the process of decay, but it is not yet known how ERG mRNA instability determinants interact with the poly(A) tail to facilitate its removal. The mechanism(s) by which the transcribed regions of ERG mRNAs are degraded following deadenylation is also unclear. Nor is it known whether the destabilizing elements that mediate deadenylation have an additional role to play in this later stage of the decay process. The development of effective *in vitro* degradation assays coupled with further *in vivo* analysis should shed light on these issues. These approaches should facilitate the further characterization of the cellular factors that carry out the process of ERG mRNA degradation. Ultimately, a full understanding of the mechanisms controlling protooncogene and cytokine mRNA decay will require that the steps in the decay pathway(s) be defined precisely *in vivo* and then reconstituted *in vitro* using purified components.

Acknowledgments

We gratefully acknowledge the support of the National Institutes of Health (Grant CA43855 to M.E.G. and Grant GM42720 to J.G.B.) and the American Cancer Society (Faculty Research Award FRA379 to M.E.G.).

References

Bernstein, P., Peltz, S. W., and Ross, J. R. (1989). The poly(A)-binding protein complex is a major determinant of mRNA stability *in vitro*. *Mol. Cell. Biol.* **9,** 659–670.

Bernstein, P., and Ross, J. (1989). Poly(A), poly(A)-binding protein and the regulation of mRNA stability. *TIBS* **14,** 373–377.

Blobel, G. A., and Hanafusa, H. (1991). The v-*src* inducible gene 9E3/pCEF4 is regulated by both its promoter upstream sequence and its 3′ untranslated region. *Proc. Natl. Acad. Sci. USA* **88,** 1162–1166.

Bohjanen, P. R., Petryniak, B., June, C. H., and Thompson, C. B. (1991). An inducible cytoplasmic factor (AU-B) binds selectively to AUUUA multimers in the 3′ untranslated region of lymphokine mRNA. *Mol. Cell. Biol.* **11,** 3288–3295.

Bohjanen, P. R., Petryniak, B., June, C. H.,Thompson, C. B., and Lindsten, T. (1992). AU RNA-binding factors differ in their binding specificities and affinities. *J. Biol. Chem.* **267,** 6302–6309.

Bonnieu, A., Piechaczyk, M., Marty, L., Cuny, M., Blanchard, J. M., Fort, P., and Jeanteur, P. (1988). Sequence determinants of c-*myc* mRNA turn-over: Influence of 3′ and 5′ non-coding regions. *Oncogene Res.* **3,** 155–166.

Bonnieu, A., Rech, J., Jeanteur, P., and Fort, P. (1989). Requirements for c-*fos* mRNA down regulation in growth stimulated murine cells. *Oncogene* **4,** 881–888.

Bonnieu, A., Roux, P., Marty, L., Jeanteur, P., and Piechaczyk, M. (1990). AUUUA motifs are dispensable for rapid degradation of the mouse c-*myc* RNA. *Oncogene* **5,** 1585–1588.

Brewer, G. (1991). An A + U-rich element RNA-binding factor regulates c-*myc* mRNA stability in vitro. *Mol. Cell. Biol.* **11,** 2460–2466.

Brewer, G., and Ross, J. (1988). Poly(A) shortening and degradation of the 3′ A + U-rich sequences of human c-*myc* mRNA in a cell-free system. *Mol. Cell. Biol.* **8,** 1697–1708.

Brewer, G., and Ross, J. (1989). Regulation of c-*myc* mRNA stability *in vitro* by a labile destabilizer with an essential nucleic acid component. *Mol. Cell. Biol.* **9,** 1996–2006.

Caput, D., Beutler, B., Hartog, K., Thayer, R., Brown-Shimer, S., and Cerami, A. (1986). Identification of a common nucleotide sequence in the 3′-untranslated region of mRNA molecules specifying inflammatory mediators. *Proc. Natl. Acad. Sci. USA* **83,** 1670–1674.

Cochran, B. H., Zullo, J., Verma, I. M., and Stiles, C. D. (1984). Expression of the c-*fos* gene and of a *fos*-related gene is stimulated by platelet-derived growth factor. *Science* **226,** 1080–1082.

Cole, M. D., and Mango, S. E. (1990). *cis*-Acting determinant of c-*myc* mRNA stability. *Enzyme* **44,** 167–180.

Curren, T., and Franza, B. R. (1988). Fos and jun: The AP-1 connection. *Cell* **55,** 395–397.

Fort, P., Rech, J., Vie, A., Piechaczyk, M., Bonnieu, A., Jeanteur, P., and Blanchard, J. M. (1987). Regulation of c-*fos* gene expression in hamster fibroblasts: Initiation and elongation of transccription and mRNA degradation. *Nucleic Acids Res.* **15,** 5657–5667.

Gillis, P., and Malter, J. S. (1991). The adenosine–uridine binding factor recognizes the AU-rich elements of cytokine, lymphokine, and oncogene mRNAs. *J. Biol. Chem.* **266,** 3172–3177.

Greenberg, M. E., Hermanowski, A. L., and Ziff, E. B. (1986). Effect of protein synthesis inhibitors on growth factor activation of c-*fos*, c-*myc*, and actin gene transcription. *Mol. Cell. Biol.* **6,** 1050–1057.

Greenberg, M. E., and Ziff, E. B. (1984). Stimulation of 3T3 cells induces transcription of the c-*fos* proto–oncogene. *Nature (London)* **311,** 433–438.

Herschman, H. R. (1991). Primary response genes induced by growth factors and tumor promoters. *Annu. Rev. Biochem.* **60,** 281–319.

Iwai, Y., Bickel, M., Pluznik, D. H., and Cohen, R. B. (1991). Identification of sequences within the murine granulocyte-colony-stimulating factor mRNA 3′-untranslated region that mediate mRNA stabilization induced by mitogen treatment of EL-4 thymoma cells. *J. Biol. Chem.* **266,** 17959–17965.

Jones, T. R., and Cole, M. D. (1987). Rapid cytoplasmic turnover of c-*myc* mRNA: Requirement of 3′ untranslated sequences. *Mol. Cell. Biol.* **7,** 4513–4521.

Kabnick, K. S., and Housman, D. E. (1988). Determinants that contribute to cytoplasmic stability of human c-*fos* and β-globin mRNAs are located at several sites in each mRNA. *Mol. Cell. Biol.* **8,** 3244–3250.

Kelly, K., Cochran, B. H., Stiles, C. D., and Leder, P. (1984). Cell-specific regulation of the c-*myc* gene by lymphocyte mitogens and platelet-derived growth factor. *Cell* **35,** 603–610.

Koeller, D. M., Horowitz, J. A., Casey, J. L., Klausner, R. D., and Harford, J. B. (1991).

Translation and the stability of mRNAs encoding the transferrin receptor and c-*fos*. *Proc. Natl. Acad. Sci. USA* **88,** 7778–7782.

Laird-Offringa, I. A., De Wit, C. L., Elfferich, P., and Van Der Eb, A. J. (1990). Poly(A) tail shortening is the translation-dependent step in c-*myc* mRNA degradation. *Mol. Cell. Biol.* **10,** 6132-6140.

Laird-Offringa, I. A.,Elfferich, P., and Van der Eb, A. J. (1991). Rapid c-*myc* mRNA degradation does not require (A + U)-rich sequences or complete translation of the mRNA. *Nucleic Acids Res.* **19,** 2387–2394.

Laird-Offringa, I. A. (1992). What determines the instability of *c-myc* proto-oncogene mRNA? *BioEssays* **14,** 119–124.

Lee, W. M., Lin, C., and Curran, T. (1988). Activation of the transforming potential of the human *fos* proto-oncogene requires message stabilization and results in increased amounts of partially modified *fos* protein. *Mol. Cell. Biol.* **8,** 5521–5527.

Lindsten, T., June, C. H., Ledbetter, J. A.,Stella, G., and Thompson, C. B. (1989). Regulation of lymphokine messenger RNA stability by a surface-mediated T cell activation pathway. *Science* **244,** 339–343.

Lüscher, B., and Eisenman, R. N. (1990). New light on Myc and Myb. I. Myc. *Genes Dev.* **4,** 2025–2035.

Malter, J. S. (1989). Identification of an AUUUA-specific messenger RNA binding protein. *Science* **246,** 664–666.

Meijlink, F., Curran, T., Miller, A. D., and Verma, I. M. (1985). Removal of a 67-base-pair sequence in the noncoding region of protooncogene *fos* converts it to a transforming gene. *Proc. Natl. Acad. Sci. USA* **82,** 4987–4991.

Metcalf, D. (1985). The granulocyte macrophage colony-stimulating factors. *Science* **229,** 16–22.

Peppel, K., and Baglioni, C. (1991). Deadenylation and turnover of interferon-β mRNA. *J. Biol. Chem.* **266,** 6663–6666.

Peppel, K., Vinci, J., and Baglioni, C. (1991). The AU-rich sequences in the 3′ untranslated region mediate the increased turnover of interferon mRNA induced by glucocorticoids. *J. Exp. Med.* **173,** 349–355.

Piechaczyk, M., Yang, J.-Q., Blanchard, J. M., Jeanteur, P., and Marcu, K. B. (1985). Posttranscriptional mechanisms are responsible for accumulation of truncated c-*myc* mRNAs in murine plasma cell tumors. *Cell* **42,** 589–597.

Rabbitts, P. H., Forster, A., Stinson, M. A., and Rabbitts, T. H. (1985). Truncation of exon 1 from the c-*myc* gene results in prolonged c-*myc* mRNA stability. *EMBO J.* **4,** 3727–3733.

Rivera, V. M., and Greenberg, M. E. (1990). Growth factor-induced gene expression: The ups and downs of c-*fos* regulation. *New Biol.* **2,** 751–758.

Ruther, U., Garber, C., Komitowski, D., Müller, R., Müller, E. G., and Wagner, E. F. (1987). Deregulated c-*fos* expression interferes with normal bone development in transgenic mice. *Nature (London)* **325,** 412–416.

Sachs, A. B. (1990). The role of poly(A) in the translation and stability of mRNA. *Curr. Opinions Cell Biol.* **2,** 1092-1098.

Schuler, G. D., and Cole, M. D. (1988). GM-CSF and oncogene mRNA stabilities are independently regulated in trans in a mouse monocytic tumor. *Cell* **55,** 1115–1122.

Shaw, G., and Kamen, R. (1986). A conserved AU sequence from the 3′ untranslated region of GM-CSF mRNA mediates selective mRNA degradation. *Cell* **46,** 659–667.

Shyu, A.-B., Belasco, J. G., and Greenberg, M. E. (1991). Two distinct destabilizing elements in the c-*fos* message trigger deadenylation as a first step in rapid mRNA decay. *Genes Dev.* **5,** 221–231.

Shyu, A.-B., Greenberg, M. E., and Belasco, J. G. (1989). The c-*fos* transcript is targeted for rapid decay by two distinct mRNA degradation pathways. *Genes Dev.* **3,** 60–72.

Stoeckle, M. Y. (1992). Removal of a 3′ non-coding sequence is an initial step in degradation of groα mRNA and is regulated by interleukin-1. *Nucleic Acids. Res.* **20,** 1123–1127.

Stoeckle, M. Y., and Barker, K. A. (1990). Two burgeoning families of platelet factor 4-related proteins: Mediators of the inflammatory response. *New Biol.* **2,** 313–323.

Stoeckle, M. Y., and Hanafusa, H. (1989). Processing of 9E3 mRNA and regulation of its stability in normal and *Rous Sarcoma* virus-transformed cells. *Mol. Cell. Biol.* **9,** 4738–4745.

Thorens, B., Mermod, J.-J., and Vassalli, P. (1987). Phagocytosis and inflammatory stimuli induce GM-CSF mRNA in macrophages through posttranscriptional regulation. *Cell* **48,** 671–679.

Treisman, R. (1985). Transient accumulation of c-*fos* RNA following serum stimulation requires a conserved 5′ element and c-*fos* 3′ sequences. *Cell* **42,** 889–902.

Vakalopoulou, E., Schaack, J., and Shenk, T. (1991). A 32-kilodalton protein binds to AU-rich domains in the 3′ untranslated regions of rapidly degraded mRNAs. *Mol. Cell. Biol.* **11,** 3355–3364.

Whittemore, L.-A., and Maniatis, T. (1990). Postinduction turnoff of beta-interferon gene expression. *Mol. Cell. Biol.* **64,** 1329–1337.

Wilson, T., and Treisman, R. (1988). Removal of poly(A) and consequent degradation of c-*fos* mRNA facilitated by 3′ AU-rich sequences. *Nature* (*London*) **336,** 396–399.

Wisdom, R., and Lee, W. (1991). The protein-coding region of c-*myc* mRNA contains a sequence that specifies rapid mRNA turnover and induction by protein synthesis inhibitors. *Genes Dev.* **5,** 232–243.

Yen, T. J., Machlin, P. S., and Cleveland, D. W. (1988). Autoregulated instability of β-tubulin mRNAs by recognition of the nascent amino terminus of β-tubulin. *Nature* (*London*) **334,** 580–585.

10

Translationally Coupled Degradation of Tubulin mRNA

NICHOLAS G. THEODORAKIS AND DON W. CLEVELAND[1]

I. Introduction

The organization of the cytoplasm of the eukaryotic cell is largely determined by a network of three filament systems: the 40-Å microfilaments, the 80- to 100-Å intermediate filaments, and the 250-Å microtubules, which collectively compose the cytoskeleton. These filament systems are responsible for cell motility, intracellular trafficking, cell division, and the variety of shapes that a eukaryotic cell can assume. To cite an extreme example, a neuron can extend an axonal process more than 1 m long, the formation and maintenance of which are dependent on the cytoskeleton.

The cytoskeletal filaments are protein polymers that are assembled from monomeric or dimeric protein subunits. The monomers that comprise the microfilaments are actin; the intermediate filaments are formed by the assembly of any of a family of related proteins called, appropriately enough, the intermediate filament proteins. The subunit component of the microtubules is a heterodimer of two related proteins, α- and β-tubulin.

To appreciate more fully the mechanisms by which the synthesis of the cytoskeletal subunits is controlled, it is useful to reflect on the physical nature of the cytoskeleton and its components. Although the cytoskeleton tends to be visualized as a scaffold, this metaphor is misleading because it implies that these filament systems are static structures. Instead, it is now clear that the cytoskeleton is highly dynamic. For example, in mammalian

[1] To whom correspondence should be addressed.

cells the interphase microtubule arrays nucleated by the nuclear-bound centrosome are completely disassembled prior to mitosis and a new array comprising at least 10 times as many microtubules is assembled to form the mitotic spindle. Beyond this cell-cycle-dependent rearrangement, the microtubules are inherently in a state of dynamic flux with unassembled tubulin subunits such that at any given time, individual microtubules are either growing slowly or shrinking rapidly, with approximately half of the total tubulin in the cell in the form of subunits. This property, called dynamic instability (Mitchison and Kirschner, 1984), has important consequences for the regulation and assembly of microtubules. As Mitchison and Kirschner (1987) showed, under such conditions if the number of microtubule nucleating sites and the cell volume remain constant, then the steady-state level of unassembled tubulin subunits *increases* as the level of total tubulin increases. In contrast, without dynamic instability the level of unassembled subunits is self-buffered by the polymer; assembly would continue until the level of unassembled subunits was depleted to a unique value (called the critical concentration) (Oosawa and Asakura, 1975) determined by the equilibrium constant for subunit dissociation. Dynamic instability makes the concept of a critical concentration of unassembled subunits always at equilibrium with the polymer no longer meaningful. Rather, the concentration of unassembled tubulin subunits at steady state is a variable determined by the number of microtubule ends and the total concentration of tubulin (Mitchison and Kirschner, 1987). In this light, if a cellular mechanism existed to fix one tubulin subunit concentration (independent of assembly) and to regulate the number of microtubules (e.g., by controlling the nucleation capacity of centrosomes), then both the total tubulin concentration and the stability of individual microtubules would be uniquely determined.

While a discussion of microtubule nucleation is beyond the scope of our effort here, what is now clear is the existence of a regulatory pathway that does control tubulin subunit synthesis by measuring the unassembled tubulin subunit concentration. As we shall see, such regulated changes in tubulin content are the result of changes in synthetic rates that are specified by selective, cotranslational degradation of cytoplasmic tubulin mRNA.

II. Tubulin Synthesis Is Autoregulated

Evidence that first demonstrated the presence of a mechanism for regulating tubulin synthesis according to the concentration of unassembled tubulin subunits arose from using a variety of drugs that alter the equilibrium between polymer and subunits. For example, colchicine and related drugs (colcemid, nocodazole) bind to the tubulin subunit, thus preventing the assembly of polymer. As a result of the combination of this block in polymerization and dynamic instability, cells treated with colchicine rapidly lose their microtubules and most tubulin in the cell is converted into the subunit form. This increase in unassembled tubulin is accompanied by a very specific change in the pattern of gene expression: of all the proteins whose synthesis is detectable by two-dimensional gel electrophoresis, the only change that consistently occurs when microtubules are depolymerized by colchicine is the decrease in synthesis of the subunits of microtubules, α- and β-tubulin (Ben Ze'ev *et al.*, 1979; Cleveland *et al.* 1981) (see Fig. 1). Similarly, when cells are treated with taxol, a drug that induces assembly and thereby decreases the concentration of tubulin subunits, the cells respond with a selective increase in the level of tubulin synthesis (Cleveland *et al.*, 1981, 1988).

How does the cell monitor the distribution of tubulin between the subunit and the polymer? The two obvious possibilities are that the cell senses the level of unassembled tubulin and that the cell can sense the amount of microtubule polymer. Two experiments support the former interpretation. The first of these used the properties of the vinca alkaloid vinblastine, which causes tubulin to precipitate into insoluble "paracrystals," thereby decreasing polymer mass without increasing the level of soluble unassembled tubulin. If cells could sense a decrease in the level of unassembled tubulin (as caused by vinblastine or taxol), then the level of tubulin synthesis should not decrease during vinblastine treatment; however, if cells sense a decrease in polymer (as caused by vinblastine or colchicine), then the level of tubulin synthesis should decrease when cells are treated with vinblastine. In fact, cells treated with vinblastine respond by increasing the synthesis of tubulin, supporting the view that the cell senses changes in the level of monomer, rather than the level of polymer (Ben Ze'ev *et al.*, 1979; Cleveland *et al.*, 1981). A second, more direct approach that does not rely on drugs was to elevate the level of tubulin subunits by microinjection of unassembled tubulin into cells; this resulted in a specific decrease in the synthesis of α- and β-tubulin (Cleveland *et al.*, 1983). Together, these results strongly support the notion that the cell monitors the level of unassembled tubulin. The changes in tubulin synthetic rates could result from changes in the efficiency of translation of tubulin mRNAs or from changes in the abundance of those mRNAs, thus leading to the measured changes in synthesis. Discrimination between these two possibilities emerged from measurement of tubulin mRNA levels. This revealed that changes in the level of tubulin synthesis are not due to altered translation of tubulin mRNAs, but rather to specific changes in the abundance of tubulin mRNAs (Cleveland *et al.*, 1981). As discussed above, this "autoregulation" of tubulin synthesis based on tubulin subunit concentration makes sense from a physiological point of view: for dynamic microtubules, specifying the unassembled tubulin level determines the total tubulin content.

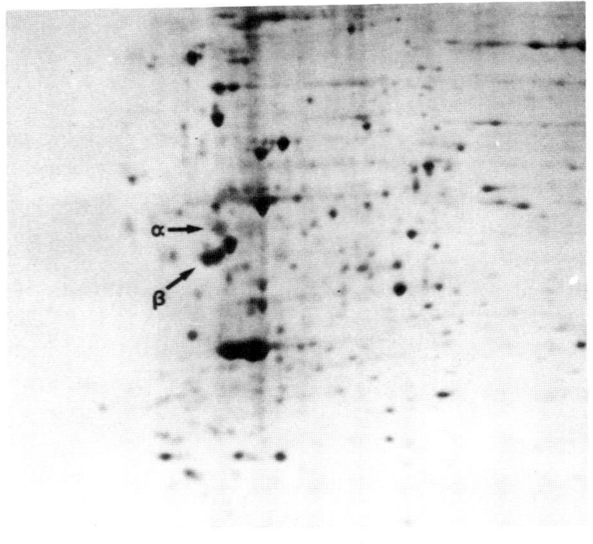

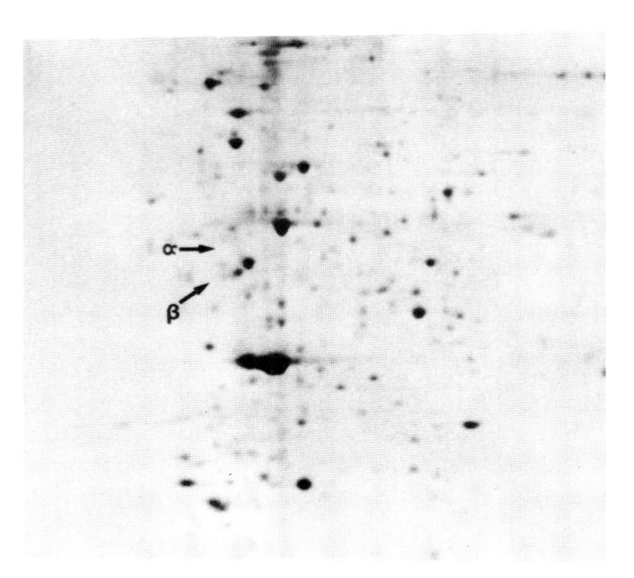

FIGURE 1 Selective inhibition of α- and β-tubulin synthesis in response to elevation of tubulin subunit concentration following colchicine-induced microtubule disassembly. Duplicate dishes of CHO cells were incubated in the absence or presence of colchicine for 4 hr. The cells were then metabolically labeled with [^{35}S]methionine and the proteins were analyzed by two-dimensional gel electrophoresis and fluorography. The positions of α- and β-tubulin are indicated by the arrows.

III. Tubulin Synthesis Is Regulated by Changes in the Stability of Cytoplasmic Tubulin mRNAs

The autoregulated changes in tubulin mRNA levels seen after altering tubulin subunit levels could derive from three general pathways: transcriptional regulation, changes in RNA processing or transport, or regulation of cytoplasmic mRNA stability. Evidence that transcription does not play a role came from an experiment performed by Cleveland and Havercroft (1983). These researchers measured the apparent rate of tubulin gene transcription using the nuclear "run-on" transcription assay and found that there were no differences in the transcription rates of α- and β-tubulin genes between nuclei isolated from colchicine-treated and those isolated from untreated cells. However, one potential caveat in interpretation of this experiment is that there are many tubulin pseudogenes (e.g., Lee *et al.*, 1983); if these are transcriptionally active but unregulatable, the results of the run-on assay may not be meaningful.

The most convincing evidence that tubulin synthesis is regulated post-transcriptionally came from studies on enucleated cells (Caron *et al.*, 1985; Pittenger and Cleveland, 1985). Such cytoplasts were prepared by fragmenting actin filaments with cytochalasin, followed by centrifugation to remove the nuclei while the cytoplasm remained attached to the substratum. These studies demonstrated that the regulatory mechanism that controls tubulin mRNA levels must be cytoplasmic, since the cytoplasts were still able to respond to colchicine by decreasing the synthesis of β-tubulin. This finding rules out a requirement for nuclear events as control points for the regulation of tubulin synthesis. Therefore, the decrease in β-tubulin mRNA levels in response to elevation of tubulin subunit levels must be due to a decrease in the stability of cytoplasmic tubulin mRNA.

A third piece of evidence that tubulin synthesis is controlled post-transcriptionally is that the abundance of RNAs encoded by hybrid gene constructs containing only a small portion of the β-tubulin coding region (the first exon) fused to the herpes simplex virus thymidine kinase gene was still subject to changes in the level of tubulin subunits, even though these constructs contained no sequences from the tubulin promoter, introns, or downstream sequences (Gay *et al.*, 1987).

IV. The Minimal Regulatory Sequence That Confers the Selective Instability of β-Tubulin mRNA Is the First Four Translated Codons

If tubulin mRNA stability is selectively altered by changes in the level of unassembled tubulin subunits, it is reasonable to suppose that specific sequences in tubulin mRNAs might identify those mRNAs as substrates for regulated instability. To address this, hybrid genes bearing parts of the β-tubulin gene were transfected into cultured fibroblasts, and the levels of the resulting mRNAs were examined before and after altering the tubulin subunit levels with drugs that interfere with tubulin assembly. Sequences from β-tubulin mRNA that can confer onto a heterologous mRNA the ability to respond to changes in the level of tubulin subunits were initially localized to a small domain in the first exon that contained protein coding sequences; sequences in the 5′ untranslated region or other exons were not required (Gay *et al.*, 1987; Yen *et al.*, 1988a). A series of constructs that fused progressively smaller regions of the β-tubulin coding sequence to a test gene (herpes simplex virus thymidine kinase) showed that the autoregulatory domain was contained within the first 13 translated nucleotides of β-tubulin mRNA, which encode the first four translated codons, Met–Arg–Glu–Ile (MREI); in contrast, the stability of a comparable RNA that carried only the first 7 translated nucleotides of β-tubulin mRNA was not responsive to tubulin subunit concentration (Yen *et al.*, 1988a). That this 13-nucleotide domain was necessary for selective instability was further shown by deletion of either the second codon or the third and fourth codons of an otherwise authentic β-tubulin gene. This yielded an mRNA that was not subject to autoregulation. Combined with the earlier transfection results, these experiments demonstrated that this short, translated sequence was both necessary and sufficient for the autoregulated control of β-tubulin gene expression.

V. Degradation of β-Tubulin mRNA Is Mediated by Cotranslational Binding of a Cellular Factor to the β-Tubulin Nascent Peptide

Additional clues to the nature of the mechanism that regulates β-tubulin mRNA stability came from studies that showed that β-tubulin mRNA degradation is linked to translation (Pachter *et al.*, 1987). Protein synthesis inhibitors that disrupt polysomes, such as pactamycin, which inhibits translational initiation, or puromycin, which causes premature peptide chain release, prevent the autoregulated degradation of β-tubulin mRNA. In contrast, a 90% inhibition of protein synthesis caused by treatment with a low concentration of cycloheximide, which slows ribosome translocation but leaves polysomes intact, actually enhanced autoregulated instability. These findings suggested that only β-tubulin mRNAs that are attached to polysomes are substrates for regulated degradation. This was confirmed by examining β-tubulin RNAs into which signals for premature translational termination were introduced at codon 27 (Pachter *et al.*, 1987) or at codon 41 (Yen *et al.*, 1988a); neither of these mRNAs were substrates for regulated instability. Together, these data demonstrate that β-tubulin mRNA must be located on polysomes in order to be degraded. Interestingly, several other mRNAs have been shown also to be cotranslationally degraded. For example, cell-cycle-dependent histone mRNA degradation, which accelerates upon the cessation of DNA synthesis, requires protein synthesis. Furthermore, histone mRNA must be translated to within 300 nucleotides of the 3′ end of the mRNA in order for it to be a substrate for this selective degradation at the end of DNA synthesis (Graves *et al.*, 1987) (see Chapter 12).

The finding of a short domain (13 nucleotides encoding the first four amino acids of β-tubulin) that conferred the ability of an mRNA to respond to changes in the level of tubulin subunits suggested two models for the recognition of tubulin mRNA. In one model, unassembled tubulin subunits directly bind to the autoregulatory sequence in the tubulin mRNA, somehow triggering its degradation. In the second model tubulin subunits (or some other cellular factor) bind cotranslationally not to the mRNA, but rather to the nascent β-tubulin polypeptide as it emerges from the ribosome; this model accounts for the findings that the autoregulatory domain lies in a translated region of the mRNA and that only polysomal mRNAs are degraded. These two models make very different predictions as to the nucleotide sequences necessary for specifying autoregulation. In particular, if the second model is correct, then all mutations in the mRNA sequence that maintain the same peptide sequence must result in an mRNA that is still subject to regulation. An abundance of evidence has now proven that the polypeptide recognition model is correct.

The first evidence that supports the polypeptide-recognition model came from the observation that yeast, human, and chicken β-tubulin amino terminal sequences all were functional in specifying autoregulation in mouse cells (Yen *et al.*, 1988a). Consistent with the polypeptide model, these genes encode the same amino-terminal peptide, but the mRNA sequences have diverged. A more direct approach was to create a series of point mutations in the second and third codons of a human β-tubulin mRNA (Yen *et al.*, 1988b). Substitution of any of the 6 arginine codons at the second codon resulted in mRNAs that were still substrates for autoregulated instability (Fig. 2), whereas RNAs containing any of 11 mutated second codons that specified other amino acids were not. More recent work in which the arginine codon at the second position was altered to a codon for either of the two other basic amino acids, lysine or histidine, demonstrated that only arginine codons at the second position yielded mRNAs that were subject to autoregulated instability (C. J. Bachurski and D. W. Cleveland, unpublished results). A similar mutational analysis at the third codon (which encodes glutamate in all known β-tubulin mRNAs) revealed that only mutants that retained codons for glutamate (or the other acidic amino acid, aspartate) remain selectively regulated by tubulin subunit levels; mutations to codons for other amino acids were not.

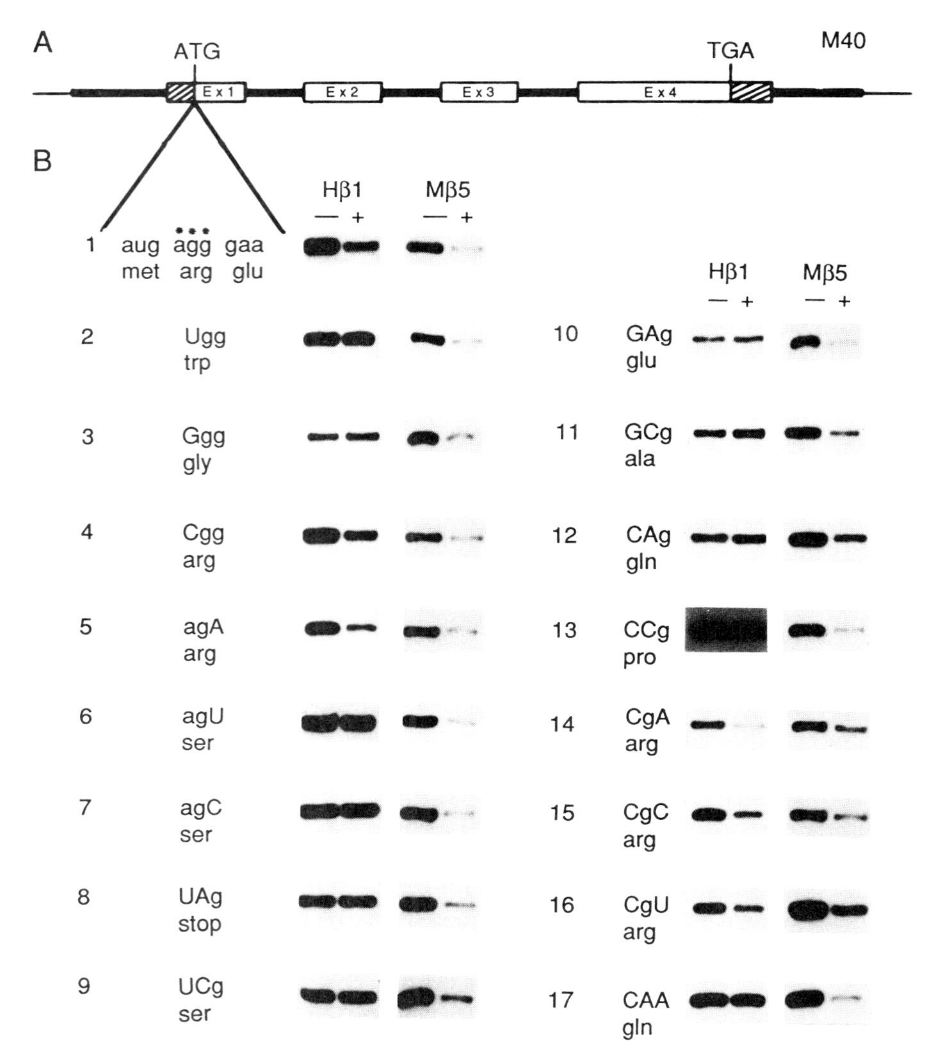

FIGURE 2 Autoregulated instability of β-tubulin mRNAs with mutations at codon 2. (A) Schematic diagram of the human β-tubulin gene used for mutagenesis at codon 2, which encodes arginine in the wild-type gene. The gene sequences are indicated by the bold line and the boxes, which represent exons. The hatched areas in the boxes indicate regions of the gene that are transcribed but not translated. (B) Tubulin RNA levels in transfected cells. Mouse L cells were transfected with various mutant human tubulin genes; the mutation in the second codon and the corresponding amino acid encoded by the mutated codon are indicated to the left of each panel. Duplicate dishes from each transfection were incubated in the absence (−) or presence (+) of colchicine for 5 hr. Cytoplasmic RNA was isolated and analyzed by S1 nuclease analysis using probes that detect either RNA from the transfected human gene (Hβ1) or the endogenous mouse β-tubulin mRNA (mβ5). [The data are reprinted with permission from Yen *et al.* (1988b).]

Other evidence that the regulatory sequence is the nascent peptide sequence rather than the mRNA sequence came from shifting the amino-terminal peptide sequence out of its normal translated frame, while leaving its position in the RNA nearly unchanged (Yen *et al.*, 1988b). A mutation that shifted the reading frame of the autoregulatory sequence after the first codon resulted in a mRNA that could not be regulated, although a similar mutation that shifted the reading frame after the fifth codon (thus preserving the autoregulatory sequence MREI) could still be autoregulated, as long the encoded peptide was greater than 40 amino acids long. These data, together with the observation that the mRNA must be translated for more than 40 codons (presumably to allow the nascent polypeptide to emerge from the ribosome), led to the model that an increase in the level of tubulin subunits allowed the cotranslational binding of a cellular factor, perhaps tubulin itself, to the nascent β-tubulin polypeptide. Presumably, this binding event alters the polysomes to allow the binding or activation of a nuclease that can degrade tubulin mRNA.

Buttressing this evidence from molecular genetic methods, direct physical support for a required interaction of a cellular factor with the nascent β-tubulin polypeptide came from microinjecting into cultured cells monoclonal antibodies that bind to the amino-terminal β-tubulin domain (Theodorakis and Cleveland, 1992). As expected, cells that were injected with this amino-terminal β-tubulin antibody could still suppress the synthesis of α-tubulin in response to elevation of tubulin subunit levels after colchicine-induced microtubule disassembly; however, the presence of the antibody completely prevented measurable changes in β-tubulin synthesis (Fig. 3). Microtubule dynamics were not affected by injection of the antibody, indicating that the disruption of β-tubulin autoregulation was not due to the inability of microtubules to disassemble. Furthermore, that α-tubulin autoregulation was *not* affected indicated that the stabilization of β-tubulin mRNA could not be explained by an antibody-induced depletion of the pool of unassembled tubulin subunits. Rather, protection of β-tubulin mRNAs from regulated degradation must be due to a specific interaction of the antibody with the nascent β-tubulin polypeptide. Presumably, the binding of the antibody to the nascent polypeptide prevents the cotranslational binding of some cellular factor(s) to that polypeptide, thereby disrupting the normal autoregulatory pathway. When combined with the transfection experiments, the evidence is overwhelming that the regulatory target for β-tubulin synthesis is the recognition of the nascent polypeptide, rather than of the mRNA itself (see Fig. 4).

It may seem surprising that the sequence that specifies autoregulation of tubulin mRNA stability in response to changes in the level of tubulin subunits can be as short as 4 amino acids. However, two observations strengthen the claim that this domain can confer the required specificity. The first is that this regulatory sequence must lie at the amino terminus.

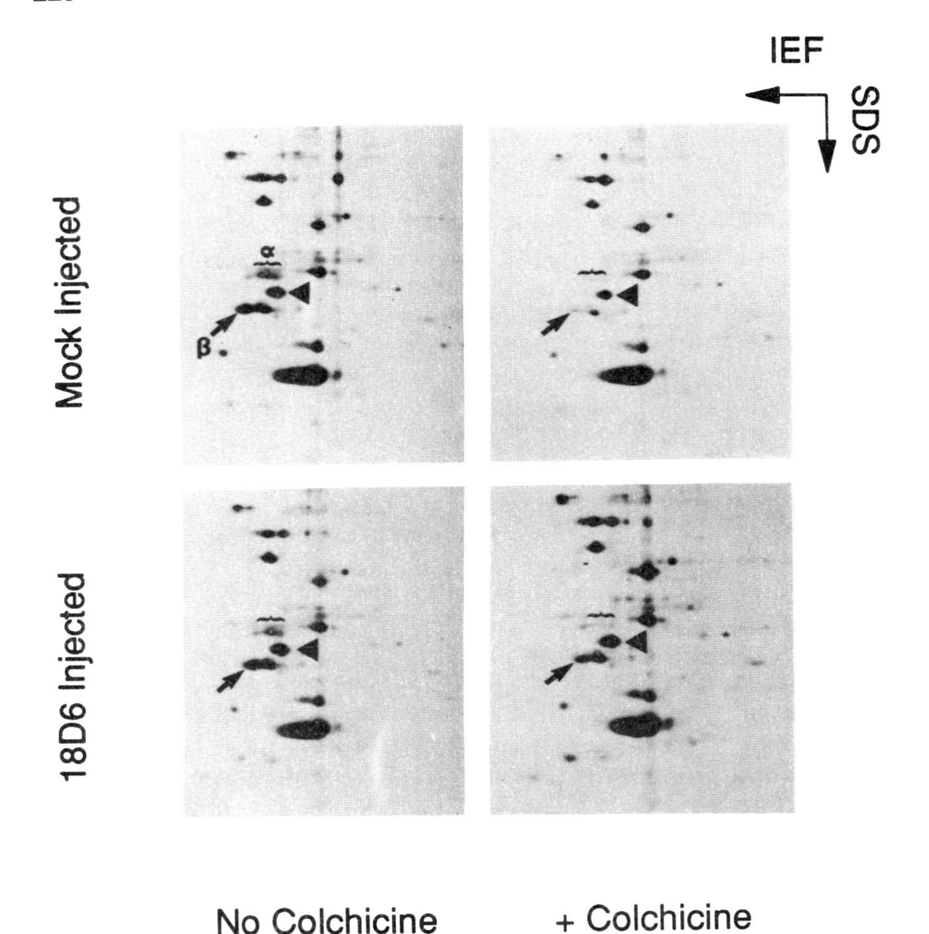

FIGURE 3 Microinjection of antibodies to the amino terminus of β-tubulin disrupts the autoregulation of β-tubulin synthesis. CHO cells growing on glass chips were either injected with buffer alone (Mock) or injected with buffer containing 15 mg/ml mouse monoclonal antibody to the amino terminus of β-tubulin (18D6). The cells were then incubated in the presence or absence of colchicine for 4 hr and metabolically labeled with [^{35}S]methionine, and the labeled proteins were analyzed by two-dimensional electrophoresis and fluorography. The positions of α- and β-tubulin are indicated by the bracket and the arrow, respectively. The position of a protein that migrates near the tubulins (probably vimentin) is indicated by the arrowhead.

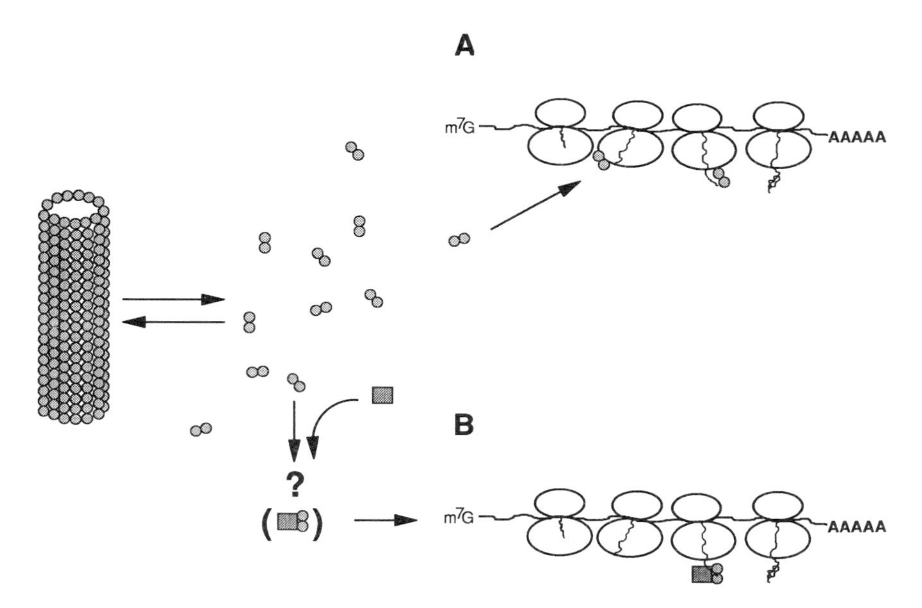

FIGURE 4 A model for autoregulated β-tubulin mRNA instability. (A) Unassembled tubulin subunits (small circles) in the context of the ribosome (large ovals) bind to the nascent β-tubulin polypeptide as it emerges from the ribosome. (B) Alternatively, another cellular factor (such as a molecular chaperone; small square) assists in the binding of tubulin to the β-tubulin nascent peptide. This complex activates a nuclease that then degrades the polysome-bound mRNA.

This was initially shown by translocation of a sequence encoding the 12 amino-terminal amino acids of β-tubulin into the interior of a test gene; this resulted in an mRNA that was not responsive to changes in tubulin subunit levels (Yen *et al.*, 1988b). In fact, later work has shown that insertion of as few as two codons (Met–Gly) as the first two codons in an otherwise authentic β-tubulin mRNA abolishes the ability of that mRNA to be autoregulated (Bachurski *et al.*, manuscript in preparation). Second, a search of current protein databases reveals that no other known eukaryotic protein has MREI at the amino terminus and furthermore that all β-tubulins (save one) begin with MREI. [The one possible exception to this last feature is *Achlya klebsiana* β-tubulin, which is reported to begin with MREL. Interestingly, this unicellular protoctist is resistant to many microtubule depolymerizing drugs (Cameron *et al.*, 1990).]

VI. What Binds to the Nascent β-Tubulin Peptide?

One simple prediction of the autoregulatory model for controlling β-tubulin synthesis is that the cellular factor that interacts with the nascent β-tubulin polypeptide is the tubulin subunit itself. This idea is intellectually appealing for two reasons. First, this would represent a very simple way to couple the increase of tubulin subunits to the binding of the nascent peptide. Second, protein–protein interactions between tubulin subunits are involved in tubulin heterodimer formation and microtubule assembly; it therefore requires no great stretch of imagination to suppose that similar interactions could also be used for cotranslational interactions between the tubulin subunit and the nascent peptide. However, theoretical considerations suggest that, if there is a direct interaction between tubulin subunits and the nascent β-tubulin polypeptide, it must be of low affinity. The concentration of unassembled tubulin in cells with a normal complement of microtubules is about 10 μM (Hiller and Weber, 1978). Therefore, if the affinity were much greater than this ($K_d < 10\ \mu M$), then all of the nascent peptide would be bound by tubulin subunits even before microtubules are depolymerized with colchicine.

Theodorakis and Cleveland (1992) looked for such an interaction between tubulin and the amino-terminal β-tubulin peptide by using two equilibrium methods. Using either gel filtration or equilibrium dialysis, no meaningful (i.e., physiologically relevant) binding of purified tubulin subunits to a synthetic peptide that corresponds to the first 12 amino acids of β-tubulin could be detected. Therefore, the supposition that there exists a simple, direct interaction between tubulin and the nascent β-tubulin polypeptide is probably not correct.

If not tubulin itself, what might be the regulatory factor that binds to the nascent β-tubulin polypeptide? One possibility that comes to mind is that the tubulin–peptide interaction may require the presence of additional factors. Inasmuch as the degradation of β-tubulin mRNA occurs while the RNA is attached to ribosomes, one obvious possibility is that the productive complex is composed of tubulin, the nascent polypeptide, and the ribosome.[2] This possibility could be tested by using *in vitro* translation systems to which tubulin subunits are added.

Alternatively, one might imagine the existence of another factor that monitors the level of free tubulin in the cell and that can bind to the nascent peptide, either alone or in a complex with tubulin. Although there is no direct experimental evidence that supports or refutes this notion, if such a factor exists it must be ubiquitous, since autoregulation of tubulin synthesis is found in all animals examined (Cleveland *et al.*, 1981; Lau *et al.*, 1985); furthermore, it must be abundant, since tubulin is an abundant protein, and it must be able to bind both to tubulin subunits and to the nascent polypeptide.

[2] In this context, it is curious that ribosomes have been found in association with microtubules assembled from sea urchin oocytes (Suprenant *et al.*, 1989).

One class of proteins that comes to mind are the molecular chaperones; these are proteins that can bind to incompletely folded or assembled proteins, thereby assisting in protein folding and/or assembly (Ellis, 1990). There are two classes of chaperone proteins that are particularly ubiquitous and abundant: the hsp70 family of proteins and the chaperonin (hsp60 or groEL) class of proteins. The hsp70 family of proteins have recently been shown to be peptide binding proteins (Flynn *et al.*, 1989) that may interact with nascent peptides on polysomes (Beckmann *et al.*, 1990). Moreover, a cell line selected for resistance to microtubule inhibitors has alterations in at least one of these proteins, suggesting that it may interact with tubulin (Ahmad *et al.*, 1990). Another provocative observation is that this cell line also has a mutation in another protein, one of the family of the hsp60 or groEL class (chaperonin) of molecular chaperones (Ahmad *et al.*, 1990; Gupta, 1990). That tubulin subunit folding is facilitated by interactions with such chaperones is consistent with the finding that newly synthesized tubulin is associated with a large complex that has many characteristics of chaperonins (Yaffe *et al.*, 1992). Nevertheless, since chaperones of the hsp70 and/or hsp60/groEL class are believed to interact with a wide variety of substrates, it is difficult to imagine how they would specifically target tubulin mRNAs for degradation. Rather, we think it is more likely that their involvement in the more general phenomenon of folding and assembly of newly synthesized proteins may be required for productive interaction of mature tubulin subunits with the nascent polypeptide.

In any event, the identification of factors beyond the tubulin subunits themselves that are required for regulated tubulin mRNA stability will likely need to await an *in vitro* system in which tubulin mRNA stability can be selectively regulated.

VII. Regulation of α-Tubulin Synthesis

The regulation of α-tubulin synthesis, like β-tubulin synthesis, is apparently controlled by selective changes in mRNA stability in response to changes in the level of unassembled tubulin subunits (Ben Ze'ev *et al.*, 1978; Cleveland *et al.*, 1979) (see Fig. 1). It is thus tempting to speculate that the mechanisms that regulate the two are similar. Indeed, like β-tubulin mRNA, only polysomal α-tubulin mRNAs are preferentially lost when tubulin subunit concentrations are elevated by treatment with colchicine (Fig. 5). Additionally, linkage of α-tubulin RNA stability with tubulin subunit levels is prevented when polysomes are disrupted by protein synthesis inhibitors (Pachter *et al.*, 1987; Gong and Brandhorst, 1988).

Inspection of the amino terminal sequences of α-tubulin polypeptides reveals that all vertebrate α-tubulins begin with tetrapeptide MREC, while α-tubulins from some simpler eukaryotes begin with MREV or MREI. This

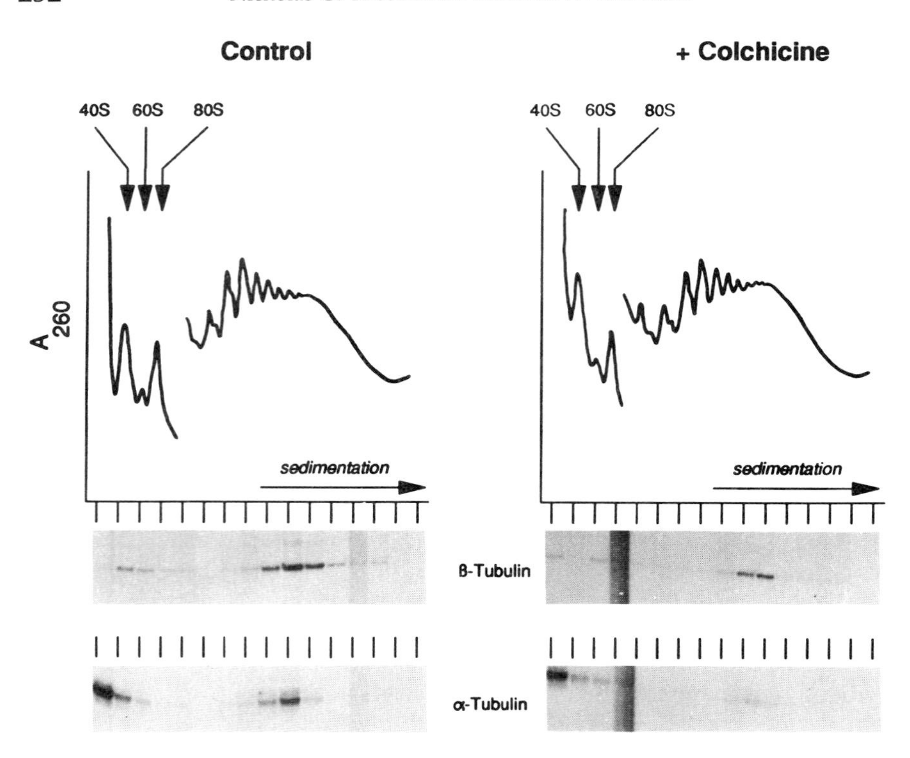

FIGURE 5 Colchicine-induced elevation of unassembled tubulin subunit levels causes loss of polysomal α- and β-tubulin mRNAs. Mouse L cells were incubated in the absence or presence of colchicine for 4 hr. Cytoplasmic extracts were prepared and sedimented on sucrose gradients to analyze the distribution of mRNAs on polysomes. The absorbance at 260 nm was monitored during collection; 16 fractions of 0.75 ml each were collected from each gradient. RNA was extracted from each fraction and analyzed for the presence of α- and β-tubulin mRNA by S1 nuclease protection. The top panels show the absorbance profile from each gradient. The direction of sedimentation is from left to right. The break in the absorbance profile after the 80S peak is due to a twofold change in the sensitivity of the spectrophotometer. The bottom panel shows the distribution of α- and β-tubulin mRNAs in each gradient.

is intriguingly similar to the MREI autoregulatory domain identified for β-tubulin; indeed, the data already described above do not rule out the possibility that the minimal domain conferring autoregulation for β-tubulin might be MRE. However, introduction of mutations in the fourth codon of β-tubulin to produce all possible amino acids at the fourth position reveals that neither the α-tubulin amino terminus MREC nor the remaining 17 amino acids are efficient substrates for autoregulation; the only RNA that is regulated encodes the initial MREI-encoding β-tubulin mRNA (Bachurski *et al.*, manuscript in preparation). Therefore, the minimal β-tubulin sequence necessary for autoregulation is the tetrapeptide MREI, and the regulatory domain controlling α-tubulin mRNA decay cannot be the analogous amino-terminal domain.

Despite the similarities to β-tubulin, the domains on α-tubulin that confer autoregulated instability have proven to be difficult to identify. Many hybrid gene constructs that encode portions of α-tubulin mRNA fused to heterologous test RNAs have been examined by transfection of cells with DNA, but for reasons yet to be explained, all were poorly translated (Sisodia, Yen, Theodorakis, and Cleveland, unpublished observations). The stability of each was not altered by changes in the tubulin subunit concentration, even though the wild-type mRNAs in the host cells were autoregulated as expected. Constructs containing full-length α-tubulin protein coding regions transcribed from heterologous promoters and linked to heterologous 5′ and 3′ untranslated sequences have also been examined. Again, all were translated poorly and their levels were not regulated by tubulin subunit levels (Coulson and Cleveland, unpublished observations). Constructs that contained the full length α-tubulin protein sequences with most of the α-tubulin 5′ and 3′ untranslated regions (but under the control of a heterologous promoter) were translated as efficiently as the endogenous α-tubulin mRNAs, but were not regulated by changes in the level of tubulin subunits (Bachurski *et al.*, manuscript in preparation). Finally, transfection of a complete wild-type rat α-tubulin gene into hamster cells yielded mRNAs that are identical to the authentic rat α-tubulin mRNA. Despite their efficient translation, even these wild-type mRNAs were not regulated by tubulin levels (Bachurski *et al.*, manuscript in preparation). Possibly, there are species-specific, tissue-specific, or isotype-specific regulatory domains in α-tubulin. Nevertheless, since there is no such specificity in the regulation of β-tubulin mRNA, it seems reasonable to conclude that the mechanism that controls α-tubulin mRNA levels is different than that for β-tubulin mRNA.

VIII. Other Mechanisms Regulating Tubulin Expression

Might there be multiple levels of regulation of tubulin expression? This seems very likely to be the case. First of all, tubulin genes are encoded by multigene families; many of the isotypes have differential tissue-specific expression (see, e.g., Raff, 1984; Cleveland, 1987; Joshi and Cleveland, 1990). It thus appears that the first level of control is transcriptional: the establishment of the transcriptional activation of particular tubulin isotypes in a given tissue. The second level of control is the selective destabilization of tubulin mRNAs in response to changes in the concentration of tubulin subunits, as discussed in this chapter.

Could there be a third (post-translational) level of control? A relevant example comes from studies using *Drosophila* tubulin mutants. In a *Drosophila* mutant that produces an unstable testis-specific β-tubulin subunit, the excess (wild-type) α-tubulin is degraded (Kemphues *et al.*, 1982). Similarly, in mammalian cells the expression of an unstable β-tubulin also leads to degradation of the excess α-tubulin (Boggs and Cabral, 1987). Additionally, overexpression of normal β-tubulin leads to rapid degradation of the excess β-tubulin (Sisodia *et al.*, 1990; Coulson and Cleveland, unpublished results). Excess β-tubulin in mammalian cells thus is likely to be unstable (see also Sawada and Cabral, 1989). However, despite many attempts, no cell lines that constitutively overexpress α-tubulin to a great degree have been isolated, perhaps indicating that α-tubulin overexpression is lethal to mammalian cells (Sisodia, Coulson, and Cleveland, unpublished results).

In budding yeast, however, overexpression of β-tubulin relative to α-tubulin is lethal, whereas overexpression of α-tubulin is not, instead resulting in a rapid loss of the excess accumulated α-tubulin protein (Burke *et al.*, 1989; Katz *et al.*, 1990; Weinstein and Solomon, 1990). In yeast, therefore, it would seem that the cell need only control the level of β-tubulin, since excess α-tubulin does not accumulate. However, the relevance of these observations to animal cell tubulins is uncertain, given the much lower concentration of tubulin in budding yeast. Indeed, there is no evidence in budding yeast for the autoregulated mRNA destabilization mechanism described here. This is not surprising, since the properties of yeast tubulin are very different from animal tubulins; most of the tubulin in yeast does not appear to be cytoplasmic, but instead is used to form an intranuclear mitotic spindle. Therefore it is not surprising that budding yeast has not been observed to retain a *cytoplasmic* mechanism for regulating the level of tubulin. Together with the results obtained from *Drosophila*, these results indicate that, beyond the RNA degradation mechanism discussed here, there is also a post-translational mechanism for balancing the level of α- and β-tubulin levels.

IX. A Model for Cotranslational Tubulin RNA Degradation: Parallels with Other Examples of Cotranslational mRNA Decay

The sum of the present evidence suggests the following model for the autoregulated degradation of β-tubulin mRNA mediated by the concentration of unassembled subunits. As diagrammed in Fig. 4, free tubulin subunits (either alone or in association with another protein, such as a chaperonin) bind to the ribosome in such a way as to detect the presence of emerging polypeptides bearing the autoregulatory sequence. When this sequence is bound by the complex, a conformational change is triggered that alters the ribosomes translocation activity. This alteration either exposes or activates a latent nuclease that is present on the ribosome, allowing the degradation of the polysome-bound mRNA. Evidence that the ribosome is integrally involved in this β-tubulin mRNA degradation (apart from its obvious role of synthesizing the polypeptide recognition sequence) comes from studies using translation inhibitors (Pachter *et al.*, 1987; Gay *et al.*, 1989). Low concentrations of cycloheximide or anisomycin slow ribosome translocation and accelerate tubulin mRNA degradation. However, when either a single drug at a higher concentration or a combination of drugs added at low concentrations is used to abolish greater than 99% of protein synthesis and freeze ribosomes on the mRNA, autoregulated degradation of tubulin mRNA is prevented (Gay *et al.*, 1989). Emetine, another translation elongation inhibitor that freezes ribosomes on mRNA, also disrupts the autoregulated degradation of tubulin mRNA. The simplest interpretation of these data is that translational elongation is required for the degradation of tubulin mRNA.

Another unsolved problem is the identity of the nuclease that degrades tubulin mRNA. We think it is likely that this nuclease either is associated with the ribosome or is an integral part of the ribosome. Support for this view comes from the observation that nucleases can be identified and purified in polysome fractions from the cell (Ross and Kobs, 1986). Association of the nuclease with polysomes also makes practical sense, since it would facilitate the localization of the nuclease to its substrate, ribosome-bound mRNA.

The demonstration that the binding of the same cellular component to the nascent β-tubulin polypeptide must occur for the regulated degradation of β-tubulin mRNA raises the question of how this binding event is transduced into degradation of the mRNA. Obviously, the binding of the nascent polypeptide per se cannot be a sufficient stimulus for degradation of the polysome-bound mRNA because injection of an antibody that binds the nascent polypeptide does not cause degradation of the mRNA, but rather inhibits its decay (see Fig. 3). Moreover, the signal sequence of the nascent polypeptides translated on the rough endoplasmic reticulum is transiently bound by the signal recognition particle; this binding event does not cause those mRNAs to be degraded, although translation is transiently arrested (Walter and Blobel, 1981; Wolin and Walter, 1989). Another example of a nascent polypeptide influencing gene expression is provided by the bacteriophage T4 gene *60* (Weiss *et al.*, 1990). During translation of this mRNA, 50 bases of the RNA are translationally bypassed by the ribosome; this event is inferred to require an interaction of a factor with the nascent polypeptide. A common feature of these three examples is that the binding of a nascent polypeptide somehow influences events that occur during translation; however, only in the β-tubulin example is this binding linked to RNA degradation.

Future examination of the predictions of this model for tubulin mRNA decay, including the possible association of tubulin subunits with ribosomes or other cellular factors, the possible interaction of tubulin subunits with polysome-bound nascent polypeptides, and the possible ribosome association of the nuclease that is responsible for the regulated degradation of tubulin mRNA, now awaits a successful *in vitro* system in which regulated tubulin mRNA stability can be reproduced.

References

Ahmad, S., Ahuja, R., Venner, R., and Gupta, R. S. (1990). Identification of a protein altered in mutants resistant to microtubule inhibitors as a member of the major heat shock protein (hsp70) family. *Mol. Cell. Biol.* **10,** 5160–5165.

Beckmann, R. P., Mizzens, L. E., and Welch, W. J. (1990). Interaction of hsp70 with newly synthesized proteins: Implications for protein folding and assembly. *Science* **246,** 850–854.

Ben Ze'ev, A., Farmer, S. R., and Penman, S. (1979). Mechanisms of regulating tubulin synthesis in cultured mammalian cells. *Cell* **17,** 319–325.

Boggs, B., and Cabral, F. (1987). Mutations affecting assembly and stability of tubulin: Evidence for a nonessential β-tubulin in CHO cells. *Mol. Cell. Biol.* **7,** 2700–2702.

Burke, D., Gasdaska, P., and Hartwell, L. (1989). Dominant effects of tubulin overexpression in *Saccharomyces cerevisiae*. *Mol. Cell. Biol.* **9,** 1049–1059.

Caron, J. M., Jones, A. L., Ball, L. B., and Kirschner, M. W. (1985). Autoregulation of tubulin synthesis in enucleated cells. *Nature (London)* **317,** 648–651.

Cleveland, D. W. (1987). The multitubulin hypothesis revisited: What have we learned? *J. Cell Biol.* **104,** 381–383.

Cleveland, D. W. (1988). Autoregulated instability of tubulin mRNAs: A novel eukaryotic regulatory mechanism. *Trends Biochem. Sci.* **13,** 339–343.

Cleveland, D. W., and Havercroft, J. C. (1983). Is apparent autoregulatory control of tubulin synthesis nontranscriptionally controlled? *J. Cell Biol.* **97,** 919–924.

Cleveland, D. W., Lopata, M. A., Sherline, P., and Kirschner, M. W. (1981). Unpolymerized tubulin modulates the level of tubulin mRNAs. *Cell* **25,** 537–546.

Cleveland, D. W., Pittenger, M. F., and Feramisco, J. R. (1983). Elevation of tubulin levels by microinjection suppresses new tubulin synthesis. *Nature (London)* **305,** 738–740.

Ellis, R. J. (1990). Molecular chaperones: The plant connection. *Science* **250,** 954–959.

Flynn, G. C., Chappell, T. G., and Rothman, J. E. (1989). Peptide binding and release by proteins implicated as catalysts of protein assembly. *Science* **245,** 385–390.

Gay, D. A., Yen, T. J., Lau, J. T. Y, and Cleveland, D. W. (1987). Sequences that confer β-tubulin autoregulation through modulated mRNA stability reside within exon 1 of a β-tubulin mRNA. *Cell* **50,** 671–679.

Gay, D. A., Sisodia, S. S., and Cleveland, D. W. (1989). Autoregulatory control of β-tubulin mRNA stability is linked to translational elongation. *Proc. Natl. Acad. Sci. USA* **86,** 5763–5767.

Gong, Z., and Brandhorst, B. P. (1988). Stabilization of tubulin mRNA by inhibitors of protein synthesis in sea urchin embryos. *Mol. Cell. Biol.* **8,** 3518–3525.

Gupta, R. S. (1990). Mitochondria, molecular chaperone proteins, and the *in vivo* assembly of microtubules. *Trends Biochem. Sci.* **15,** 415–418.

Graves, R. A., Pandey, N. B., Chodchoy, N., and Marzluff, W. F. (1987). Translation is required for regulation of histone mRNA degradation. *Cell* **48,** 615–626.

Hiller, G., and Weber, K. (1978). Radioimmunoassay for tubulin: A quantitative comparison of the tubulin content of different established tissue culture cells and tissues. *Cell* **14,** 795–804.

Joshi, H. C., and Cleveland, D. W. (1990) Diversity among tubulin subunits: Toward what functional end? *Cell Motil. Cytoskel.* **16,** 159–163.

Katz, W., Weinstein, B., and Solomon, F. (1990). Regulation of tubulin levels and microtubule assembly in *Saccharomyces cerevisiae*: Consequences of altered tubulin gene copy number. *Mol. Cell. Biol.* **10,** 5286–5294.

Kemphues, K. J., Kaufman, T. C., Raff, R. A., and Raff, E. C. (1982). The testis-specific β-tubulin subunit of *Drosophila melanogaster* has multiple functions in spermatogenesis. *Cell* **31,** 655–670.

Lau, J. T., Pittenger, M. F., and Cleveland, D. W. (1985). Reconstruction of appropriate tubulin and actin gene regulation after transient transfection of cloned β-tubulin and β-actin genes. *Mol. Cell. Biol.* **5,** 1611–1620.

Lee, M. G., Kewis, S. A., Wilde, C. D., and Cowan, N. J. (1983). Evolutionary history of a multigene family: An expressed human β-tubulin gene and three processed pseudogenes. *Cell* **33,** 477–487.

Mitchison, T., and Kirschner, M. (1984). Dynamic instability of microtubule growth. *Nature (London)* **312,** 237–242.

Mitchison, T. J., and Kirschner, M. W. (1987). Some thoughts on the partitioning of tubulin between monomer and polymer under conditions of dynamic instability. *Cell Biophys.* **11,** 35–55.

Oosawa, R., and Asakura, S. (1975). Thermodynamics of the Polymerization of Protein. Academic Press, New York.

Pachter, J. S., Yen, T. J., and Cleveland, D. W. (1987). Autoregulation of tubulin expression is achieved through specific degradation of polysomal tubulin mRNAs. *Cell* **51,** 283–292.

Pittenger, M. F., and Cleveland, D. W. (1985). Retention of autoregulatory control of tubulin synthesis in cytoplasts: Demonstration of a cytoplasmic mechanism that regulates the level of tubulin expression. *J. Cell Biol.* **101,** 1941–1952.

Raff, E. C. (1984). Genetics of microtubule systems. *J. Cell Biol.* **99,** 1–10.

Ross, J., and Kobs, G. (1986). H4 histone mRNA decay in cell-free extracts initiates at or near the 3′ terminus and proceeds 3′ to 5′. *J. Mol. Biol.* **188,** 579–593.

Sawada, T., and Cabral, F. (1989). Expression and function β-tubulin isotypes in Chinese hamster ovary cells. *J. Biol. Chem.* **264,** 3103–3020.

Sisodia, S. S., Gay, D. A., and Cleveland, D. W. (1990). In vivo discrimination among β-tubulin isotypes: Selective degradation of a type IV β-tubulin isotype following overexpression in cultured animal cells. *New Biol.* **2,** 66–76.

Suprenant, K. A., Tempero, L. B., and Hammer, L. E. (1989). Association of ribosomes with in vitro assembled microtubules. *Cell Motil. Cytoskel.* **14,** 401–415.

Theodorakis, N. G., and Cleveland, D. W. (1992). Physical evidence for cotranslational regulation of β-tubulin mRNA degradation. *Mol. Cell Biol.* **12,** 791–799.

Walter, P., and Blobel, G. (1981). Translocation of proteins across the endoplasmic reticulum III. Signal recognition particle causes signal sequence-dependent and site-specific arrest of chain elongation that is released by microsomal membranes. *J. Cell Biol.* **91,** 557–561.

Weinstein, B., and Solomon, F. (1990). Phenotypic consequences of tubulin overproduction in *Saccharomyces cerevisiae*: Differences between alpha-tubulin and beta-tubulin. *Mol. Cell. Biol.* **10,** 5295–5304.

Weiss, R. B., Huang, W. M., and Dunn, D. M. (1990). A nascent peptide is required for ribosomal bypass of the coding gap in bacteriophage T4 gene *60*. *Cell* **62,** 117–126.

Wolin, S. L., and Walter, P. (1989) Signal recognition particle mediates a transient elongation arrest of preprolactin in reticulocyte lysate. *J. Cell Biol.* **109,** 2617–2622.

Yaffe, M. B., Farr, G. W., Miklos, D., Horwich, A. L., Sternlicht, M. L., and Sternlicht, H. (1992). TCP1 complex is a molecular chaperone in tubulin biogenesis. *Nature* (*London*) **358,** 245–248.

Yen, T. J., Gay, D. A., Pachter, J. S., and Cleveland, D. W. (1988a). Autoregulated changes in stability of polyribosome-bound β-tubulin mRNAs are specified by the first thirteen translated nucleotides. *Mol. Cell. Biol.* **8,** 1224–1235.

Yen, T. J., Machlin, P. S., and Cleveland, D. W. (1988b). Autoregulated instability of β-tubulin mRNAs by recognition of the nascent amino terminus of β-tubulin. *Nature* (*London*) **334,** 580–585.

11

Iron Regulation of Transferrin Receptor mRNA Stability

JOE B. HARFORD

I. Overview of Cellular Iron Homeostasis

With the documented exceptions of certain *Lactobacillus* and *Bacillus* species (Neilands, 1974; Stubbe, 1990), all organisms require a nearly continuous source of environmental iron for growth and the maintenance of a wide range of metabolic pathways. Iron's flexible coordination chemistry and favorable redox potential are responsible for the fact that the metal is utilized as a cofactor for a variety of both structural and metabolic functions. In association with proteins, iron can exist as part of a heme prosthetic group or in one of several characteristic iron–sulfur complexes. In addition, there exists a heterogeneous collection of enzymes that contain neither heme nor an iron–sulfur complex but that contain iron or require it for activity (Crichton, 1991).

Although iron is the second most abundant metal in the Earth's crust, iron metabolism presents some unique biological challenges. In an oxygen-containing environment, iron is readily oxidized from the ferrous to the ferric state. This conversion has two notable consequences for cells and organisms. First, ferric iron is less biochemically available owing to its low solubility (about 10^{-17} *M* at neutral pH). Second, iron in conjunction with oxygen becomes a dangerous generator of hydroxyl radicals, which have a variety of toxic effects on cells (Gutteridge, 1989). Thus, much of the current efforts to understand cellular iron metabolism are aimed at understanding how cells obtain iron from the environment, how they detoxify excess iron, and how they regulate these processes in a way that balances critical metabolic needs with the extreme toxicity of iron.

II. Iron Acquisition

In the dividing cells of higher eukaryotes, iron is acquired via the transferrin cycle involving the iron-carrying serum protein transferrin (Tf) and the membrane-associated transferrin receptor (TfR). The TfR is among the most studied of cellular receptors, and there have appeared several reviews that have dealt with various aspects of the biochemistry and cell biology of the TfR and with its involvement in cellular iron metabolism (Hanover and Dickson, 1985; Huebers and Finch, 1987; Crichton and Charloteaux-Wauters, 1987; Kühn, 1989; Harford *et al.*, 1990). Here, attention will be focused on the regulation of the expression of the TfR and in particular the regulation of TfR mRNA stability that occurs in response to alterations in iron availability.

III. Iron Sequestration

Protection from the harmful effects of iron plus oxygen is in large measure accomplished through the process of iron sequestration. The best characterized protein that subserves this function is ferritin, a highly conserved protein found in all vertebrates (Munro and Linder, 1978; Theil, 1987). Ferritin-like sequestration compounds have been reported in bacteria as well as in simple eukaryotes. Vertebrate ferritin subunits assemble into a well-characterized structure containing a total of 24 subunits that form a spherical shell into which as many as 4500 atoms of iron can be sequestered and detoxified as a ferric oxyhydroxide micelle. Whether and to what extent iron that has entered ferritin can be remobilized for acute metabolic needs remain unclear. A detailed description of ferritin structure and function is beyond the scope of this chapter.

IV. Regulation of Cellular Iron Homeostasis

It has been known since the 1940s that ferritin levels vary directly with the amount of iron given to an individual (Granick, 1946). In the mid-1970s, Munro and colleagues presented evidence indicating that this regulation took place at the translational level since actinomycin D did not prevent the increase in ferritin observed upon the injection of iron into rat liver (Zahringer *et al.*, 1976; Munro and Linder, 1978) or treatment of hepatoma cells with iron (Rogers and Munro, 1987).

The cloning and characterization of the full gene encoding human *H* chain ferritin (Hentze *et al.*, 1986) allowed identification of the cis-acting element responsible for translational regulation of ferritin synthesis by iron. Deletion analysis localized the region responsible for translational control to about 30 nucleotides of the 212-nucleotide 5′ untranslated region (UTR) of human *H* chain ferritin (Hentze *et al.*, 1987b). Examination of these 30 nucleotides revealed that they could be folded into a moderately stable stem–loop structure (see Fig. 1) that was similar to potential stem–loop structures present in the 5′UTR of the other ferritin genes that had been cloned at the time (Hentze *et al.*, 1988). This sequence element, when placed in the 5′UTR of heterologous mRNAs, was sufficient to transfer iron-dependent translational control to these transcripts. The sequence element that was both necessary and sufficient for iron-dependent translational control was termed an iron-responsive element, or IRE (Hentze *et al.*, 1987a). At about the same time, Munro and colleagues (Aziz and Munro, 1987; Leibold and Munro, 1987, 1988) demonstrated that the corresponding sequence element from rat ferritin mRNA also functioned as a translational control element. The importance of the IRE sequence for translational control was supported by the finding that point mutations could completely ablate the ability of the sequence to control translation ; when the first C of the IRE loop was removed, the IRE failed to function (Hentze *et al.*, 1987a). Similarly, when mutations were made that disrupted the upper stem, IRE function was lost (Hentze *et al.*, 1987a; Leibold *et al.*, 1990). Further characterization of the key features of the IRE of rat ferritin *L* mRNA have been reported in the detailed study of Barton *et al.* (1990).

It has been known for some time that the expression of the TfR is also regulated. Within populations of proliferating cells, TfR synthesis is modulated by iron availability in a manner suggestive of feedback regulation, as fewer receptors are expressed when iron is abundant and more receptors are expressed when iron is scarce (Ward *et al.*, 1982; Pelicci *et al.*, 1982; Mattia *et al.*, 1984). Quantitative immunoprecipitation of TfR translated *in vitro* from RNA isolated from iron-treated and iron-chelator-treated (iron-starved) K562 cells indicated that the iron-dependent alterations in TfR biosynthesis were a reflection of alterations in the level of TfR mRNA (Mattia *et al.*, 1984; Rao *et al.*, 1985).

The availability of a molecular probe for the TfR mRNA in the form of the cDNA clone pcD-TR1 (McClelland *et al.*, 1984) allowed the direct quantitation of TfR mRNA. The steady-state level of TfR mRNA was found to increase after cells were treated with an iron chelator and to decrease after exogenous iron was supplied (Rao *et al.*, 1986). The TfR cDNA clone also provided the means to assess, by nuclear run-on assays, the effect of cellular iron levels on transcription of the TfR gene. Nuclei from cells treated with the iron chelator desferrioxamine incorporated more [α-^{32}P]UTP into TfR RNA than did nuclei from cells treated with hemin, an iron source (Rao *et al.*, 1986). These results suggested that at least a portion of the iron regulation of TfR mRNA levels was occurring at the level of gene transcription. The transcription rate of the TfR gene has also been assessed by nuclear run-on experiments in activated T cells (Kronke *et al.*, 1985) and in

Consensus IRE

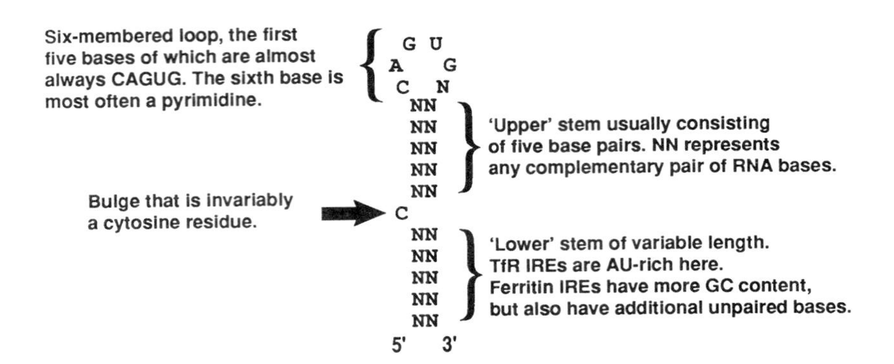
Six-membered loop, the first five bases of which are almost always CAGUG. The sixth base is most often a pyrimidine.
'Upper' stem usually consisting of five base pairs. NN represents any complementary pair of RNA bases.
Bulge that is invariably a cytosine residue.
'Lower' stem of variable length. TfR IREs are AU-rich here. Ferritin IREs have more GC content, but also have additional unpaired bases.
5'
3'

Examples of IREs

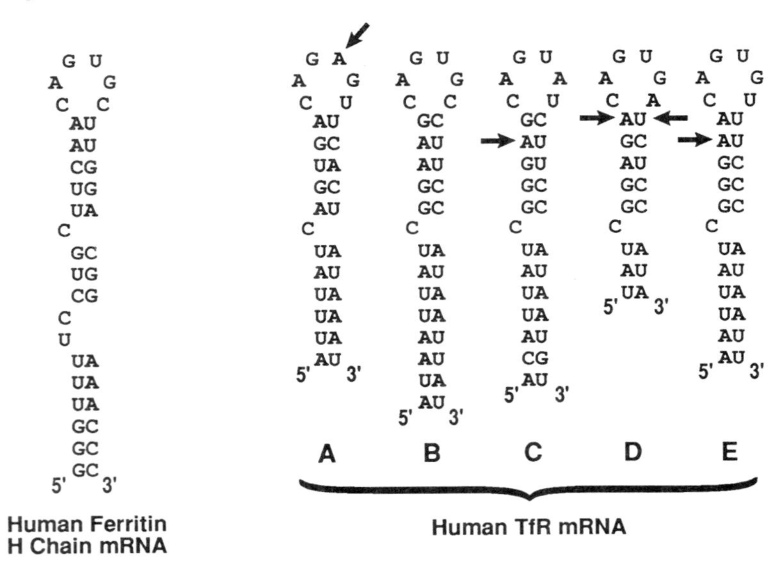
A
B
C
D
E
Human Ferritin H Chain mRNA
Human TfR mRNA

FIGURE 1 The sequence and proposed structures of IREs. Based on the sequences of the IREs of all known ferritin mRNAs and those of all known TfR mRNAs, a consensus IRE motif is depicted. Features of the IRE consensus are indicated. In the lower portion of the figure are shown examples of IREs from the human ferritin H chain mRNA and from the human TfR mRNA (labeled A–E as described by Casey *et al.* (1988a). Nucleotides indicated by arrows in the TfR mRNA IREs are bases that are different in the corresponding motifs of the chicken TfR mRNA (Koeller *et al.*, 1989; Chan *et al.*, 1989).

HL60 cells treated with dibutryl cAMP (Trepel *et al.*, 1987). Activation of T cells leads to increased TfR expression, whereas cAMP treatment of HL60 results in decreased TfR expression. The corresponding changes in nuclear run-on data that were observed in both instances are consistent with a transcriptional component for these types of regulation of TfR gene expression.

However, the predominant locus of iron regulation of TfR expression lies not in the promoter but in sequences corresponding to the 3′UTR of the TfR mRNA (Owen and Kühn, 1987; Casey *et al.*, 1988a, 1988b). Fusion to a 2-kb fragment of the TfR cDNA corresponding to the 3′UTR of the TfR mRNA was shown to be sufficient to confer iron regulation onto the expression of a chimeric gene encoding human growth hormone. Thus it appears that two genetic loci participate in regulation of TfR expression by iron (Casey *et al.*, 1988b). One is located in the region 5′ of the transcription start site, and the other corresponds to the 3′UTR of the TfR transcript.

The region within the 3′UTR of the TfR mRNA that is the predominant locus of iron regulation has been defined as being within a fragment of 678 nucleotides (Casey *et al.*, 1988a). Deletion of this fragment ablated iron regulation of TfR transcript levels, whereas a deletion of essentially all of the 3′UTR except the fragment of 678 nucleotides was without effect. Interestingly, two independent and nonoverlapping deletions (separated by greater than 200 nucleotides) within this 678-nucleotide TfR mRNA fragment each eliminated nearly all regulation by iron, suggesting that more than one sequence element in this region participates in iron regulation. Examination of the sequence of the critical 678-nucleotide fragment revealed that a high degree of RNA secondary structure is possible within this region (see Fig. 2). A very similar secondary structure can be formed by corresponding sequences within the 3′UTR of the chicken TfR mRNA (Koeller *et al.*, 1989; Chan *et al.*, 1989).

One of the most exciting discoveries made upon inspection of the possible secondary structure of the regulatory region of the TfR mRNA was the presence there of five stem–loop structures that bore striking similarity to the IRE found within the 5′UTR of mRNAs encoding ferritin (Casey *et al.*, 1988a). Although the IREs of ferritin and TfR mRNAs can be fit to a consensus motif, the IREs contained in the TfR mRNA have sequences distinct from those of known ferritin mRNAs, particularly in the bases that make up the stems of the IRE stem–loops (see Fig. 1).

V. The Iron-Responsive Element-Binding Protein as the Cellular "Ferrostat"

The presence of IREs in the transcripts for ferritin and the TfR suggested that iron regulation of the fates of these mRNAs might involve a trans-acting factor that could bind to these elements. Such a binding factor could

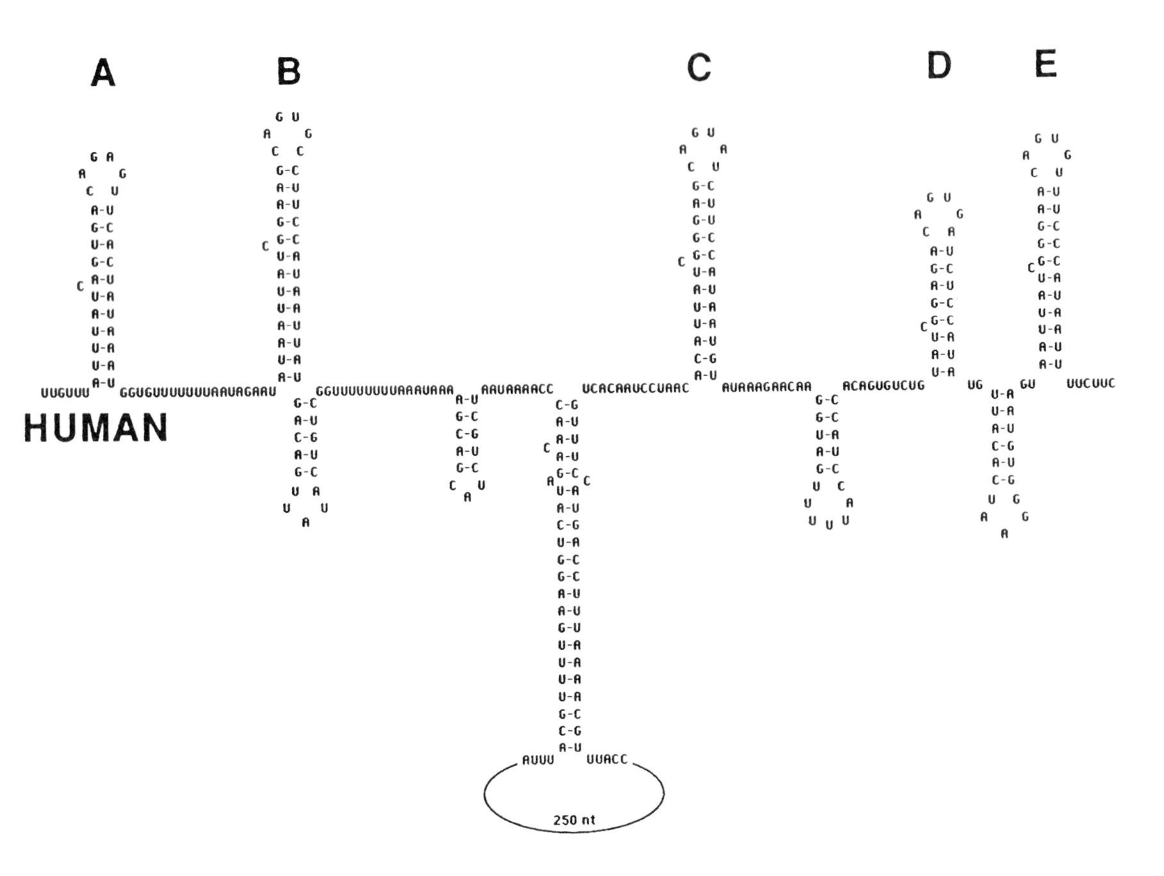
A
B
C
D
E
HUMAN
250 nt

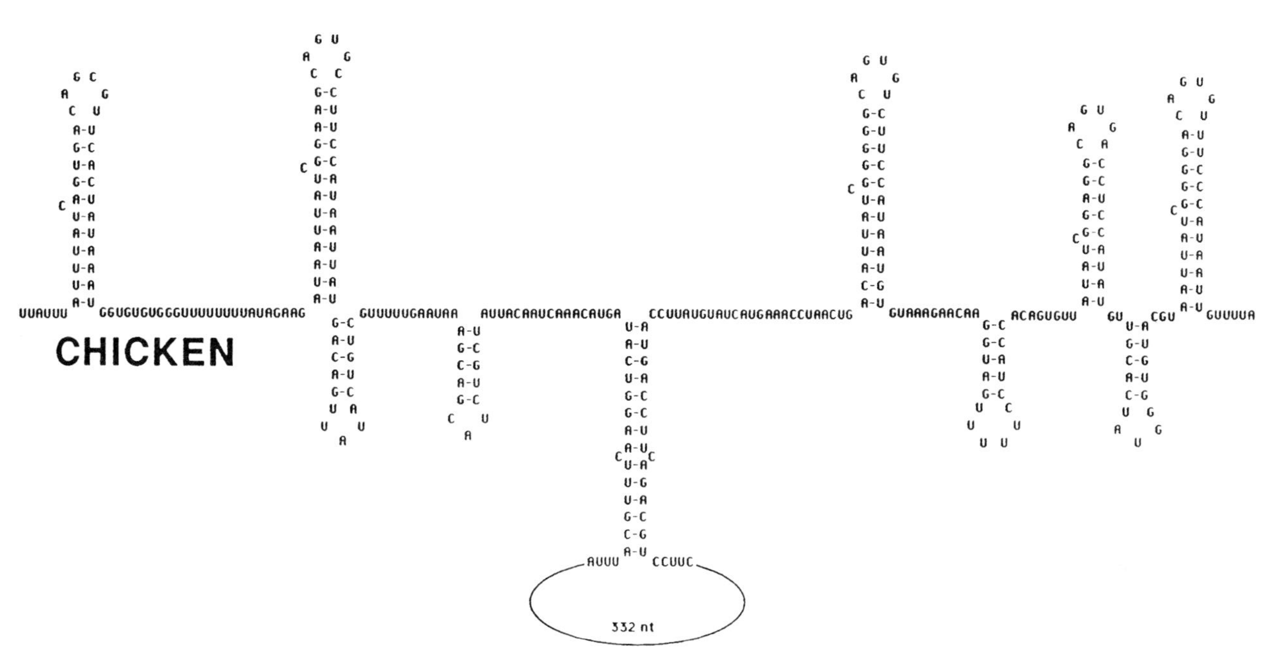

FIGURE 2 Potential secondary structures of portions of the 3′UTR of the human and chicken TfR mRNAs. That portion of the human TfR mRNA implicated in iron regulation (Casey *et al.*, 1988a) is depicted along with the corresponding region of the chicken TfR mRNA (Koeller *et al.*, 1989; Chan *et al.*, 1989). The large central loops (250 nucleotides in the human mRNA and 332 nucleotides in the chicken mRNA) where sequence similarity is <50% are not shown.

explain the ability of the ferritin IRE to impede translation despite its modest themodynamic stability compared with other 5′ UTR secondary structures that inhibit translation (Pelletier and Sonenberg, 1985; Kozak, 1986). Using rodent cell lysates, Leibold and Munro (1988) first reported the demonstration of a cellular factor that specifically interacted with an RNA transcript containing an IRE. This demonstration was accomplished by a modification of the gel electrophoretic mobility shift or band shift assay first utilized for demonstrating specific DNA–protein interactions. With similar methodology, Rouault *et al.* (1988) demonstrated that human cell lysates contained an IRE binding protein (IRE-BP). It was noted that there was an increased amount of binding activity present in the cytosol of cells that had been iron-starved when compared with cells exposed to normal levels of iron in culture. Since translation of ferritin mRNAs is inhibited under conditions of iron deprivation, this finding supported the notion that the IRE-BP represented a ferritin translational repressor (Caughman *et al.*, 1988). This protein's interaction with the ferritin IRE has since been shown to inhibit specifically the *in vitro* translation of ferritin mRNA (Walden *et al.*, 1988, 1989).

Human cell lysates contain both a higher affinity IRE-BP (calculated K_d of 10–30 p*M*) and a lower affinity IRE-BP (calculated K_d of 2–5 n*M*) (Haile *et al.*, 1989). Changes in the apparent number of IRE binding sites in response to iron availability were shown to involve alteration of the number of high affinity IRE binding sites. The ability to stimulate an increase in the number of high affinity binding sites by treatment of cells with an iron chelator did not require new protein synthesis (Hentze *et al.*, 1989). Thus, the high affinity form of IRE-BP could be recruited from a lower affinity pool of IRE-BP *via* a post-translational mechanism. One or more free sulfhydryl groups in the protein were required for binding to the RNA since IRE-binding activity could be abolished by the treatment of lysates with alkylating agents such as *N*-ethylmaleimide. Moreover, when the human cellular lysates were simply treated with either 2-mercaptoethanol or dithiothreitol, a large increase in the number of high affinity binding sites was observed. The high affinity IRE-binding activity could be abolished *in vitro* by treatment of the lysates with oxidizing agents such as diamide or copper orthophenanthroline. This inactivation of the high affinity binding site could be reversed by the subsequent addition of reducing agents. When the total number of IRE binding sites (high plus lower affinity) was measured, no significant change was observed upon reduction (Haile *et al.*, 1989). Thus, this change involved the specific conversion or recruitment to high affinity binding sites. IRE-BP activity has been detected in cell lysates from other species including birds, amphibians, fish, and insects (Rothenberger *et al.*, 1990), and interconversion of high affinity and low affinity pools of IRE-BP by redox reagents is a feature seen in all species examined to date (Barton *et al.*, 1990; Rothenberger *et al.*, 1990).

These observations led to the proposal of a "sulfhydryl switch" as a possible underlying mechanism for the regulation *in vivo* of the number of high affinity binding sites by iron. According to this model, the iron status of the cell alters the availability of one or more critical sulfhydryl groups in the IRE-BP by an oxidation–reduction reaction. Redox regulation has also been proposed for DNA binding proteins in the bacterial OxyR system (Storz *et al.*, 1990) and for the proteins encoded by the human protooncogenes c-*jun* and c-*fos* (Abate *et al.*, 1990).

The human IRE-BP has been isolated, and the cDNA for this protein has been cloned and successfully expressed in transfected mouse cells (Rouault *et al.*, 1989, 1990; Kaptain *et al.*, 1991). The deduced amino acid sequence of the human IRE-BP is nearly identifical (greater than 90% homology) to the sequence of the corresponding protein purified from rabbit liver by Walden *et al.* (1989). Interestingly, the human IRE-BP amino acid sequence is >30% homologous to that of mitochondrial aconitases from porcine heart and yeast, with >50% of the aligned amino acids being similar or identical (Rouault *et al.* 1991). Aconitase is an iron–sulfur (Fe–S) protein of the citric acid cycle that catalyzes the conversion of citrate to isocitrate. The crystal structure of aconitase is known (Robbins and Stout, 1989a, 1989b), and all of the active site residues of aconitase are conserved in the human IRE-BP. The similarity of the IRE-BP to aconitase suggests that the IRE-BP is also an Fe–S protein. Purified aconitase exists in both an inactive [3Fe–4S] state and an active [4Fe–4S] state. This characteristic is distinct from many other Fe–S proteins in which the form of the Fe–S center is relatively static (Beinert, 1990). Thus, aconitase provides a paradigm for a way in which the activity of a protein might be responsive to changes in iron availability, since when iron is abundant the [4Fe–4S] state would be likely to predominate. The IRE-BP has been shown directly to possess aconitase activity that is dependent upon the addition of iron (Kaptain *et al.*, 1991).

VI. The Rapid Turnover Determinant of the Transferrin Receptor mRNA

All available data are consistent with the coordinate regulation of ferritin mRNA translation and TfR mRNA stability being mediated by the interaction of an iron-regulated IRE-BP with the IREs of the two transcripts (see Fig. 3). This model envisions IRE-binding protein molecules interacting with the IRE RNA structures when iron is scarce. In the context of the 5′UTR of the ferritin mRNA, this interaction would attenuate translation, and in the context of the 3′UTR of the TfR mRNA, this interaction would protect the transcript from degradation. Such a model would account for the opposite directions of the changes in ferritin and TfR biosynthesis

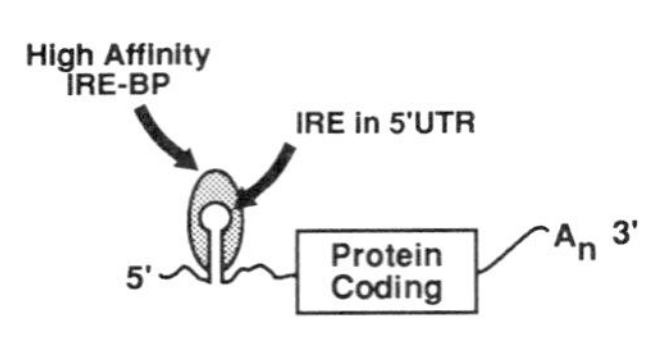

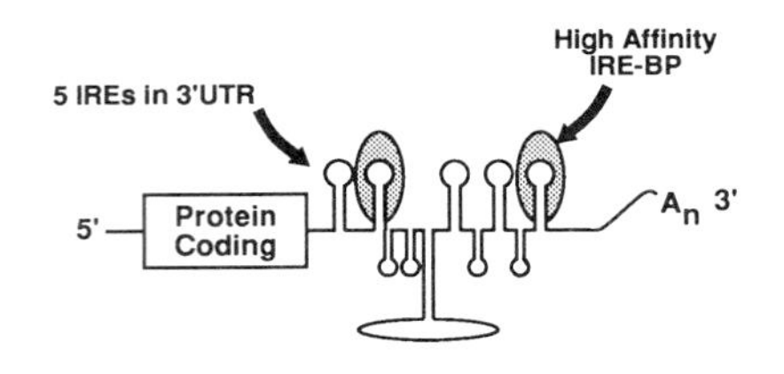

FIGURE 3 Coordinate regulation of ferritin mRNA translation and TfR mRNA stability. High-affinity interaction of the IRE-BP with the single IRE in the 5′UTR of the ferritin mRNA represses the translation of this mRNA. Analogous interaction of the IRE-BP with one or more of the IREs within the 3′UTR of the TfR mRNA represses the degradation of this transcript. Within cells, the IRE-BP is in its high-affinity state when iron is scarce.

in response to the common environmental stimulus of a change in iron availability. Although the rates of biosynthesis of ferritin and the TfR move in opposite directions in response to changes in iron, both RNAs are engaged by the IRE-BP under similar low iron conditions, and the effect of the interaction in both instances is inhibitory. The remainder of this chapter will focus on the regulated turnover of the TfR mRNA.

In contrast to the relatively simple regulatory structure (a single IRE) that encodes iron-dependent translational control, the regulatory region required for iron-dependent mRNA stability regulation appears to be much more complex. To understand the function of this region demands both structural and functional analyses of the 3′UTR regulatory element. The availability of TfR sequences from two relatively distant species (human and chicken) allows prediction of structure based on phylogenetic comparison. Only five differences exist within the 143 nucleotides that encode the five TfR IREs, and all of these differences are nondisruptive of the IRE structure motif (Casey *et al.*, 1988a, Koeller *et al.*, 1989; Chan *et al.*, 1989) (see Fig. 1). The comparison of human and chicken mRNA sequences also suggests that other structural motifs within the regulatory region may be critical. Several small non-IRE stem–loop structures and a long base paired structure lying between IREs B and C (see Fig. 2) are contained in the transcripts of both species. Some of these non-IRE sequence/structure motifs have been shown to be necessary for iron regulation (Casey *et al.*, 1989) (see Fig. 4). The sequence of a portion of the 3′UTR of the rat TfR mRNA has also been determined (Roberts and Griswold, 1990). It too contains IREs and, within the previously defined regulatory region, is very similar in sequence (and potential secondary structure) to the human and chicken mRNAs. Despite the striking similarity of IREs A and E to the corresponding elements of the

Construct Name	Regulatory Region Schematic	High Affinity IRE-BP Interaction	Regulatory "Phenotype"
3A-0	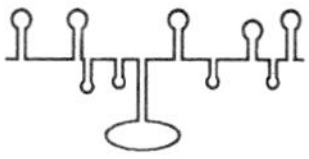	Yes	Regulated
3B-10/5A-60	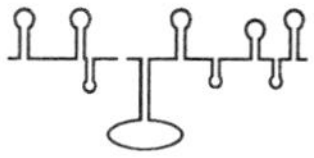	Yes	High & Unregulated
3A-44/5C-8	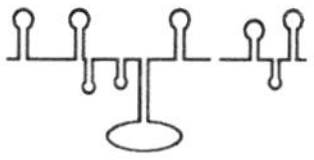	Yes	High & Unregulated
TRS-1	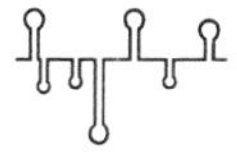	Yes	Regulated
TRS-3		No	High & Unregulated
TRS-4	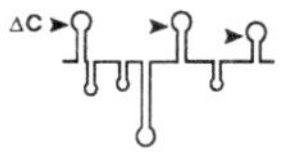 ΔC	No	Low & Unregulated

FIGURE 4 The regulation of TfR mRNA levels by iron requires IREs and a rapid turnover determinant within the 3′UTR. Constructs containing various portions of the 3′UTR of the TfR mRNA are depicted. Construct 3A-0 contains the portion of the 3′UTR of the human TfR mRNA shown in Fig. 2. In constructs 3B-10/5A-60 and 3A-44/5C-8 non-IRE elements have been deleted. In construct TRS-1, the region has been reduced in size from 678 nucleotides to 252 nucleotides by deletion of IREs A and E and deletion of the large central loop shown in Fig. 2. The three IREs of TRS-1 have been deleted in construct TRS-3. The 5′-most cytosine residue of the loop has been removed from each IRE of TRS-1 to yield construct TRS-4. High-affinity interaction with the IRE-BP was assessed by a gel-shift assay, and regulation was assessed by measuring mRNA levels after treatment with hemin or desferrioxamine (Casey *et al.*, 1989). Regulated constructs have a rapid turnover determinant whose activity is repressed by interaction between the IRE-BP and one or more of the IREs. The high and unregulated "phenotype" of construct 3B-10/5A-60, construct 3A-44/5C-8, and construct TRS-3 has been interpreted to indicate that the rapid turnover determinant is altered and is inactive irrespective of interaction with the IRE-BP. The low and unregulated "phenotype" of construct TRS-4 has been interpreted as indicative of an intrinsically active rapid turnover determinant that cannot be repressed since high-affinity interaction has been abolished by the removal of the three cytosine residues. The RNAs corresponding to TRS-1 and TRS-4 have been shown to adopt the depicted secondary structure (Horowitz and Harford, 1992) (see Fig. 6). Stem–loops depicted as pointing upward are IREs, whereas stem–loops pointing downward are non-IRE hairpins.

chicken mRNA and their ability to bind the IRE-BP (Koeller *et al.*, 1989), IREs A and E can be removed from the human TfR mRNA without apparent effect on iron regulation (Casey *et al.*, 1989). The 250 nucleotides of the human TfR mRNA depicted as the large central loop also can be deleted without apparent effect. The sequence in this 250-nucleotide region displays no significant similarity to that of the corresponding region of the chicken TfR mRNA. Elimination of this region along with IREs A and E reduces the size of the necessary regulatory element from 678 nucleotides to 252 nucleotides (see Fig. 4). The resultant 252-nucleotide element was termed TRS-1, and it mediated the full range of iron regulation of Tfr mRNA stability (Casey *et al.*, 1989). The iron-dependent regulation of the half-life of TRS-1 mRNA is shown in Fig. 5A. In cells treated with the iron source hemin, an mRNA half-life of 45–60 min is observed, whereas in cells treated with the iron chelator desferrioxamine the TRS-1 mRNA displays a half-life considerably greater than 3 hr (Koeller *et al.*, 1991).

Removal of the 100 nucleotides corresponding to the three IREs of TRS-1 produced an unregulated construct termed TRS-3 (Casey *et al.*, 1989) (see Fig. 4). The level of TRS-3 expression in transiently transfected cells was found to be relatively high (using a cotransfected plasmid to normalize) and completely unregulated by iron. In contrast, removal of three selected cytosine residues of TRS-1 (to generate a construct termed TRS-4) also eliminated regulation but gave rise to relatively low levels of expression in transiently transfected cells. The finding of a low level of TRS-4 mRNA in transfected cells was interpreted as indicating that this mRNA contained an intact rapid turnover determinant that could not be stabilized by interaction with the IRE-BP since removal of the three cytosines eliminates IRE-BP binding. The contrasting regulatory "phenotype" of TRS-3 , which lacks the IREs entirely, was interpreted as indicating that the nucleotides of the IREs overlap with those of the rapid turnover determinant. As shown in Fig. 5B, direct measurement of mRNA stability in stable transformants revealed that the TRS-3 mRNA is indeed intrinsically stable and that the TRS-4 mRNA is intrinsically unstable (Koeller *et al.*, 1991). The half-life of TRS-4 mRNA is indistinguishable from that of TRS-1 mRNA in hemin-treated cells. In contrast to TRS-1 mRNA, the stabilities of TRS-3 and TRS-4 mRNAs were not affected by manipulation of cellular iron availability. The regulatory differences among the various mutant TfR mRNAs can be readily explained by a model in which the regulatory region contains both a rapid turnover determinant and an element (or elements) that modulates the activity of the rapid turnover determinant (Casey *et al.*, 1989; Harford and Klausner, 1990). Although these components of the region may overlap physically, their functions can be individually ablated. In "high and unregulated" mRNAs, the rapid turnover determinant activity is missing. Some mutant TfR mRNAs that display this phenotype still bind the IRE-BP normally, whereas others do not (Casey *et al.*, 1989) (see Fig. 4). For this class

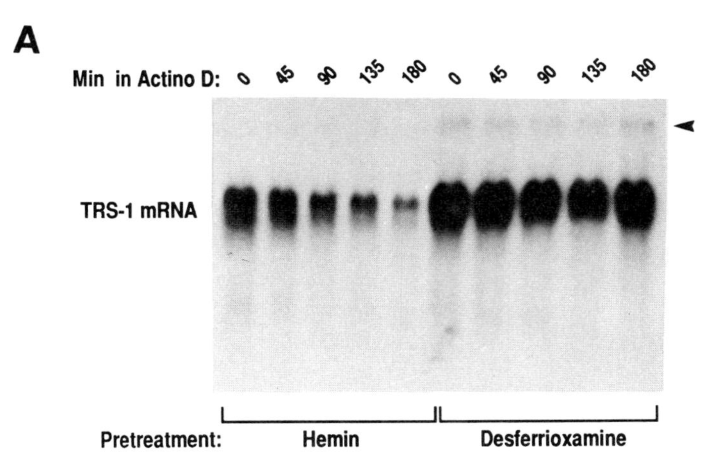

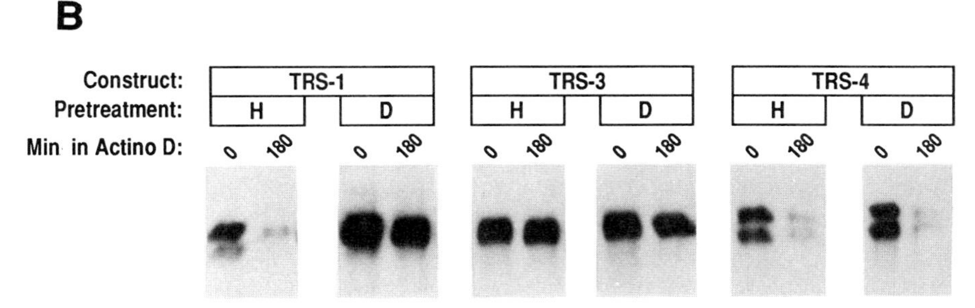

FIGURE 5 The degradation of TfR mRNA is iron regulated and alterations in sequence can result in two types of unregulated mRNAs. (A) Cytoplasmic mRNAs were isolated from mouse B6 fibroblasts stably expressing the human TfR by virtue of transfection with the plasmid TRS-1. Cells were treated for the indicated times with 4 μM actinomycin D following pretreatment for 16 hr with either 100 μM hemin or 100 μM desferrioxamine. Densitometry indicated that the half-life of the TRS-1 mRNA in hemin-treated cells was 45–60 min, whereas in desferrioxamine-treated cells the half-life appeared to be considerably greater than 3 hr. The arrowhead represents the migration position of the endogenous 4.9-kb mouse TfR mRNA. (B) The experimental design was as in Fig. 5A. Cells expressing TfR mRNAs encoded by plasmids TRS-1, TRS-3, and TRS-4 were compared. Only the level of TRS-1 mRNA was iron-regulated and displayed a half-life that was dependent on the iron status of the cell. Irrespective of iron status, TRS-3 mRNA was relatively stable (i.e., high and unregulated), and TRS-4 mRNA was relatively unstable (i.e., low and unregulated). [For more details, see Koeller *et al.* (1991).]

of altered mRNAs, the ability to bind the IRE-BP is irrelevant since the rapid turnover determinant is nonfunctional. For the "low and unregulated" TRS-4 mRNA, the rapid turnover determinant is functionally intact, but the transcript cannot be protected from degradation because high affinity binding of the IRE-BP is lost. This concept of a rapid turnover determinant that is functionally distinct from the IRE-BP binding site also explains why there is no iron regulation of the half-life of ferritin mRNA, which contains an IRE-BP binding site but evidently lacks a rapid turnover determinant.

VII. The Structure of the TfR mRNA Regulatory Region

In general, sequence-specific DNA binding proteins appear to recognize a linear stretch of double-stranded helical DNA. In contrast, the interaction of proteins with RNA is likely to depend on the secondary and tertiary structures of the RNA. Early deletion analysis of the regulatory region of the TfR mRNA (Casey *et al.*, 1988a) revealed that there existed nonoverlapping deletions and that each independently eliminated iron regulation. This observation was confirmed by finer deletion analysis (Casey *et al.*, 1989), which indicated that deletion of at least two well-separated, non-IRE sequence elements each independently abolished regulation, giving rise to a "high and unregulated" phenotype (see Fig. 4). Thus, both of these elements were necessary and neither was sufficient for rapid turnover when iron was present. One possibility is that the rapid turnover determinant of the TfR mRNA is a relatively complex structure in three dimensional space. Structural elements that are distant in the primary sequence of the mRNA may be brought into proximity by RNA folding. Together they may form the recognition determinant for the cellular machinery of degradation. Adjacent to or in the midst of this recognition determinant is positioned one or more of the IREs. Binding by the IRE-BP represses the activity of the recognition determinant, perhaps by sterically preventing it from interacting with other components of the cell's mRNA degrading machinery. It is also conceivable that the critical elements implicated by the deletion analyses are not themselves directly part of the recognition determinant but rather are distant structural elements whose deletion perturbs the overall geometry of this determinant. If so, they would be analogous to amino acids that are not part of the active site of an enzyme but whose removal abolishes enzymatic activity through an overall protein conformational change. The ability of single-stranded RNA to adopt complex secondary and tertiary structures makes indirect conformational effects a possibility that is not a major consideration in a DNA double helix.

In this chapter, all of the depictions of the regulatory regions of the TfR mRNA contain IRE stem–loop structures corresponding to those shown in Fig. 1. Owing to the palindromic nature of the IRE primary sequences and the presence of multiple IREs within the TfR mRNA, alternative secondary structures are possible. Indeed, an alternative secondary structure for a portion of the TfR mRNA has been proposed based on a computer algorithm for RNA folding and the results of certain deletions on iron regulation of TfR expression (Müllner and Kühn, 1988). In this other proposed structure, the IRE sequences do not fold as individual stem–loops but instead engage in alternative base pairing (specifically, the IRE B and IRE C sequences were proposed to base pair with each other). Despite the fact that this alternative structure contains relatively few of the specific base pairings of the structure depicted in Fig. 2, its calculated overall stability is quite similar (~60 kcal/mole for the TRS-1 RNA that is 252 nucleotides in length). The secondary structure of the portion of the TfR mRNA responsible for the regulation of the transcript's half-life has now been deduced by ribonuclease H cleavage directed by antisense oligodeoxyribonucleotides as well as by cleavage with other ribonucleases sensitive to RNA secondary structure (Horowitz and Harford, 1992) (see Fig. 6). These data indicate that a synthetic 252-nucleotide fragment comprising the regulatory region of TRS-1 RNA, and the comparable portion of the 2.7-kb cellular TRS-1 mRNA each contain all three of their IREs as individual stem–loops. Moreover, this secondary structure appears to be relatively static with little interconversion with other possible structures, involving longer range base pairing. The predominance of the structure containing three IRE stem–loops may result from the advantage of short-range base pairing (versus long-range base pairing) in determining RNA secondary structure (Abrahams *et al.*, 1990). TRS-4 RNA was shown to adopt a similar, if not identical, secondary structure to that of the TRS-1 RNA. Given that the IRE-BP can bind an RNA containing a single isolated IRE (Rouault *et al.*, 1988) and that this binding is dependent on the base pairing of the IRE stem (Leibold *et al.*, 1990), data indicating that as many as four molecules of the IRE-BP can associate simultaneously with a molecule of RNA containing five IREs (Müllner *et al.*, 1989) also favor an RNA structure in which the IREs exist as individual stem–loops.

FIGURE 6 Structure of TRS-1 RNA deduced by enzymatic cleavage. The 252-nt TRS-1 is shown in both panels. In the top panel, the bars juxtaposed to the structure indicate the sequences for which antisense oligodeoxyribonucleotides (antisense oligos) were prepared. Antisense oligos one at a time were allowed to anneal *in vitro* to TRS-1 RNA and cleavage by RNase H was assessed. Antisense oligos indicated by solid bars and plus signs resulted in RNase H cleavage, and it was inferred that the RNA complementary to the antisense oligo was single-stranded. Antisense oligos indicated by shaded bars and negative signs did not form RNase H cleavage sites, and the RNA complementary to these oligos was inferred to be engaged in intramolecular base pairing (i.e., was double-stranded). In the bottom panel are indicated the cleavage sites detected in the absense of antisense oligos using RNases that are sensitive to RNA secondary structure under what was judged to be primary cleavage conditions. Symbols for each RNase are given in the inset. [For more details, see Horowitz and Harford (1992).]

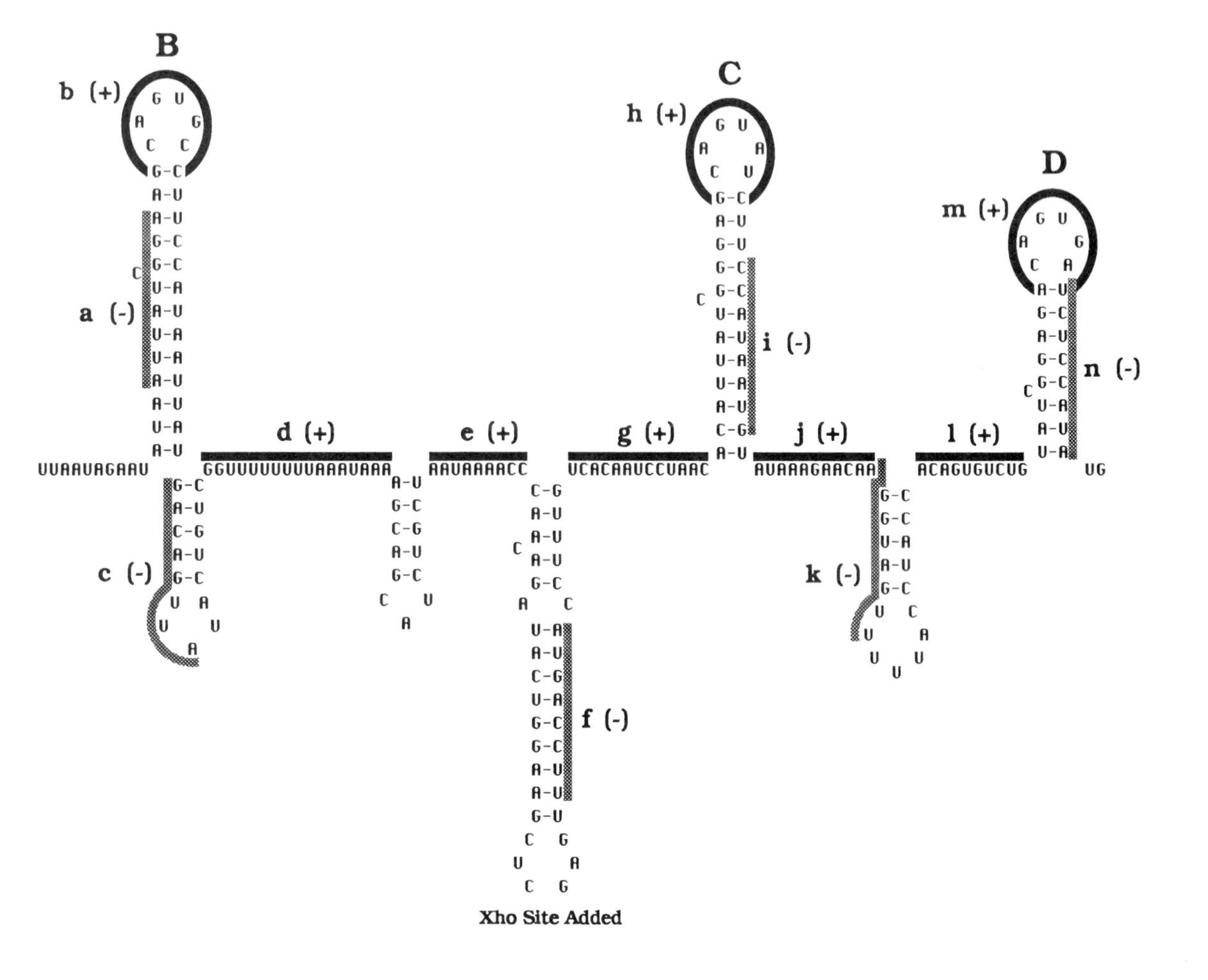

B
C
D
b (+)
a (-)
c (-)
d (+)
e (+)
f (-)
g (+)
h (+)
i (-)
j (+)
k (-)
l (+)
m (+)
n (-)
UUAAUAGAAU
GGUUUUUUUUAAAUAAA
AAUAAAACC
UCACAAUCCUAAC
AUAAAGAACAA
ACAGUGUCUG
Xho Site Added

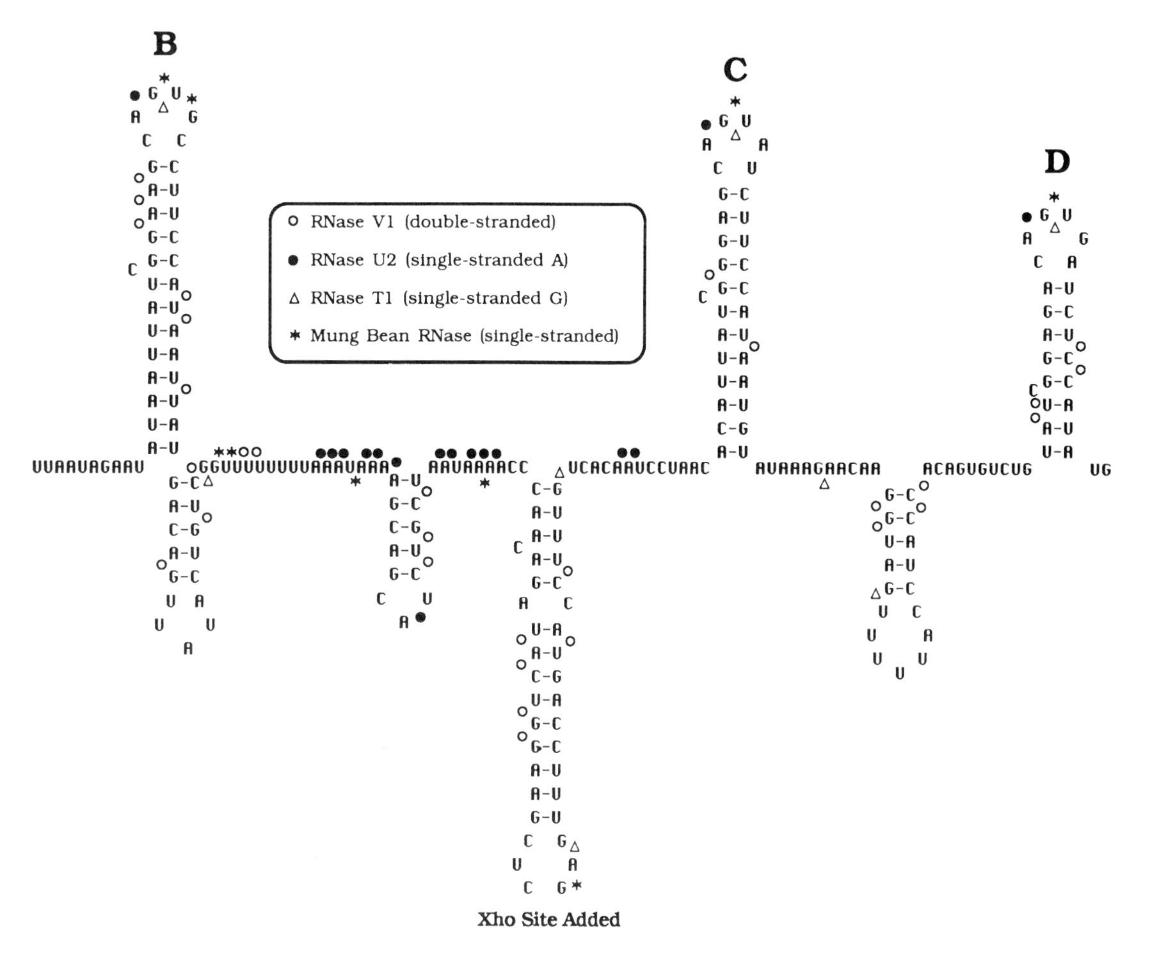

B
C
D
○ RNase V1 (double-stranded)
● RNase U2 (single-stranded A)
△ RNase T1 (single-stranded G)
* Mung Bean RNase (single-stranded)
UUAAUAGAAU
GGUUUUUUUUAAAUAAA
AAUAAAACC
UCACAAUCCUAAC
AUAAAGAACAA
ACAGUGUCUG
UG
Xho Site Added

Taken collectively, all of these data suggest that the secondary structure deduced for TRS-1 and TRS-4 represents that of the TfR rapid turnover determinant (Horowitz and Harford, 1992).

VIII. The Mechanism of TfR mRNA Decay

When the affinity of the IRE-BP for an IRE is decreased by iron, the TfR mRNA is destroyed. Virtually nothing is known concerning the nuclease(s) involved, but this statement can be made for most eukaryotic genes that are regulated at the level of mRNA stability. Detailed characterization of these nucleases remains an important objective. The simplest model for regulation of TfR mRNA expression would envision the IRE-BP as a steric protector of the transcript (see Fig. 3). Such a model would not mandate that the nuclease(s) involved in the degradation of the TfR mRNA be iron-regulated, nor would TfR mRNA have to be the only substrate of the nuclease(s).

It would seem likely that the IRE-BP prevents the nucleolytic attack that is rate-limiting with more rapid destruction of the transcript proceeding after this rate-limiting step. As is true for most RNA turnover, a large number of TfR mRNA breakdown products are not evident. One smaller RNA band has been seen on Northern blot analysis of samples from murine cells expressing the TRS-4 mRNA irrespective of iron treatment (Koeller *et al.*, 1991) (see Fig. 5B). An apparently identical RNA is observed in murine cells expressing the TRS-1 mRNA, but only upon hemin treatment of the cells. This smaller RNA species may represent a breakdown product since its presence correlates with rapid turnover of the TfR mRNA. We have noted a similar phenomenon involving the endogenous 4.9-kb TfR mRNA of a human cell line (Harford *et al.*, 1988). These smaller RNAs appear to represent truncated transcripts that are missing the poly(A) tail along with significant portions of the 3′UTR. The 3′ ends of these transcripts map within or near the IRE-containing region of each transcript. It is not clear whether these truncations reflect the primary site of cleavage by an endonuclease or a relatively stable pause in the processive action of a 3′ to 5′ exonuclease.

When transfectants expressing the intrinsically unstable TRS-4 mRNA were treated with cycloheximide (Chx), which inhibits translation elongation but leaves ribosomes on the mRNA, a marked increase in the level of TRS-4 mRNA was seen (Koeller *et al.*, 1991). It has also been reported that the human TfR mRNA expressed in mouse L cells is stabilized by Chx (Müllner and Kühn, 1988). A similar effect on the level of TRS-4 mRNA was observed when cells were treated with puromycin (Puro), which causes premature release of ribosomes and the nascent peptide chain. Thus, the TRS-4 mRNA level is increased by inhibition of protein synthesis irrespective of whether it is bound by ribosomes. As would be anticipated for a long-lived mRNA, no comparable effect of Chx on the cellular concentration of the intrinsically stable TRS-3 mRNA was seen. In cells expressing the iron-regulated TRS-1 mRNA, an effect of Chx on mRNA level was observed in cells that had been pretreated with the iron source hemin but not in cells treated with the iron chelator desferrioxamine. No effect on an unrelated mRNA (encoding actin) was seen under any regimen of iron treatment.

There exist a number of examples of rapidly degraded mRNAs whose levels are markedly increased in cells treated with inhibitors of protein synthesis. This phenomenon has been seen with several protooncogene transcripts, including c-*myc* (Linial *et al.*, 1985; Cole, 1986), c-*myb* (Thompson *et al.*, 1986), c-*fos* (Ramsdorf *et al.*, 1987), and c-*jun* mRNA (Ryseck *et al.*, 1988). These increases in mRNA levels, which are seen upon treatment of cells with a variety of translocation inhibitors that act by distinct mechanisms (i.e., inhibitors of either translation initiation or elongation), appear to result, at least in part, from substantial stabilization of the mRNAs.

At present, the mechanism by which protein synthesis inhibitors stabilize certain mRNAs remains obscure. It is thought that either a highly unstable protein is involved in the degradation of these mRNAs (a so-called trans-effect) or that translation of the mRNAs themselves is required for their decay (a cis-effect). These two possibilities are not necessarily mutually exclusive and each may be applicable in certain instances. One compelling example of a cis-effect of translation on mRNA stability is the autoregulation of β-tubulin mRNA degradation (Cleveland and Yen, 1989). Experimental evidence obtained in this system suggests that degradation of ribosome-bound β-tubulin mRNA is accelerated by cotranslational recognition of the nascent β-tubulin amino-terminal tetrapeptide in cells containing unassembled dimers of α- and β-tubulin. This recognition is presumed to activate an RNase that may be ribosome-associated. A ribosome-associated nuclease has also been invoked in models for regulated turnover of histone mRNAs (Ross *et al.*, 1986; Marzluff and Pandey, 1988; Cleveland and Yen, 1989). In the case of stabilization of c-*fos* mRNA by Chx, it has been suggested that the rapidity of the effect on mRNA stability is more consistent with a cis-effect (Wilson and Treisman, 1988).

The major difficulty in distinguishing between a cis- and a trans-effect of protein synthesis inhibitors is that inhibition of global protein synthesis will, by its nature, inhibit both the translation of the mRNA in question and the synthesis of the putative short-lived, trans-acting protein. The IRE/IRE-BP system has provided a means to distinguish between cis- and trans-effects of inhibition of protein synthesis. When an IRE exists in the 5′UTR of an mRNA (as in ferritin mRNAs) the translation of the mRNA is regulated by iron such that more translation occurs when iron is abundant and less when iron is scarce (Klausner and Harford, 1989; Theil, 1990; Harford and Klausner, 1990). When cells are treated with an iron source, the ferritin mRNA is found to be associated with polysomes, whereas ferritin mRNA exists in a smaller ribonucleoprotein complex in chelator-treated cells (Rogers and Munro, 1987). This effect is specific for mRNAs containing an IRE in the 5′UTR, and global protein synthesis is not affected by changes in iron availability. Thus, for an mRNA with an IRE in its 5′UTR, iron chelation can be used to attenuate translation in a way that is completely specific for IRE-containing mRNAs. If a short-lived mRNA had an IRE and was stabilized by global inhibition of protein synthesis, iron chelation should have a similar stabilizing effect only if translation of the mRNA itself was a requirement for mRNA degradation. If instead the effect of global inhibition of protein synthesis was due to the disappearance of a short-lived protein necessary for mRNA degradation, then changes in iron availability should be without effect since the synthesis of this putative short-lived protein should not be affected by adding or withdrawing iron.

Several constructs were designed to test whether iron could be used to regulate selectively the translation of a short-lived mRNA (Koeller *et al.*, 1991). Some constructs encoded mRNAs that contained a ferritin IRE within the 5′UTR and a rapid turnover determinant in the 3′UTR. All constructs utilized the TfR protein coding region (see Table 1). In one construct (termed 5′IRE-TRS-4), the 3′UTR was the TRS-4 element that had been shown to render the mRNA intrinsically unstable. In another construct (termed 5′IRE-TR-*fos*), this TRS-4 element in the 3′UTR was replaced with a 250-nucleotide segment of the human c-*fos* gene. This segment encodes the AU-rich region

TABLE 1

Construct	5′UTR	3′UTR	mRNA stability regulated by iron	mRNA translation regulated by iron
TRS-1	TfR	TRS-1	Yes	No
TRS-3	TfR	TRS-3	No (stable)	No
TRS-4	TfR	TRS-4	No (unstable)	No
TR-Δ3′UTR	TfR	None	No (stable)	No
TR-*fos*	TfR	c-*fos*	No (unstable)	No
5′IRE-TRS-4	TfR + IRE	TRS-4	No (unstable)	Yes
5′IRE-TR-*fos*	TfR + IRE	c-*fos*	No (unstable)	Yes

Note All constructs contained the promoter and coding region of the human transferrin receptor gene but differed from one another in their 5′UTR and/or their 3′UTR. All constructs were stably transfected into mouse fibroblasts. The mRNA stability was determined by the 3′UTR sequence, with "unstable" being operationally defined as displaying an mRNA half-life of less than 1 hr and "stable" being defined as displaying a half-life of greater than 3 hr. Only the construct TRS-1 produced an mRNA whose stability was iron-regulated. Translational regulation was determined by the sequence of the 5′UTR; only those constructs that produced an mRNA containing an IRE in the 5′UTR were regulated by iron at the level of translation. [For additional details, see Casey *et al.* (1989) and Koeller *et al.* (1991).]

that functions as one of the instability determinants in c-*fos* mRNA (Shyu *et al.*, 1989). When transferred to β-globin mRNA, the AU-rich region of c-*fos* mRNA confers a rapid degradation rate (Shaw and Kamen, 1986; Shyu *et al.*, 1989). The c-*fos* element also destabilized the chimeric TfR mRNA irrespective of the presence of an IRE within the 5′UTR (Koeller *et al.*, 1991).

When cells transfected with the 5′IRE-TR-*fos* construct were treated with Chx, a marked increase in the level of the encoded mRNA was seen. This effect was similar to that seen with the TRS-4 and TRS-1 constructs in iron-treated cells. A comparable Chx-induced increase in mRNA level was also seen with the construct 5′IRE-TRS-4. The increase in 5′IRE-TR-*fos* mRNA level after Chx treatment was shown to be the result of mRNA stabilization. Treatment of cells with an iron chelator reduced translation of 5′ IRE-TR-*fos* mRNA by a factor of ≥20 compared with the same mRNA in hemin-treated cells, but had only a negligible effect on the steady-state level of this mRNA (see Fig. 7) and on global protein synthesis. Direct measurement of the 5′IRE-TR-*fos* mRNA half-life revealed that the transcript was short-lived (half-life = 45–60 min) both in cells treated with hemin, where it is well translated, and in cells treated with deferrioxamine, where it is poorly translated (Koeller *et al.*, 1991) (see Fig. 8). The level of mRNA encoded by the construct 5′IRE-TRS-4, which contains the rapid turnover

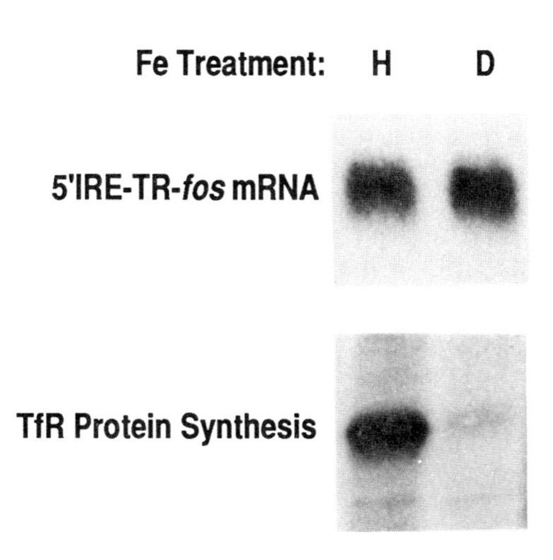

FIGURE 7 A ferritin IRE placed in the 5′UTR of a heterologous mRNA results in translational regulation. Cells expressing 5′IRE-TR-*fos* mRNA were treated with hemin (H) or desferrioxamine (D) as described in Fig. 5A. These treatments resulted in a marked difference (≥ 20-fold) in 5′IRE-TR-*fos* mRNA translation as judged by [^{35}S]methionine incorporation into the TfR encoded by this mRNA (lower panel), but had no effect on the level of the 5′IRE-*fos* mRNA as judged by Northern blot hybridization (upper panel). [For more details, see Koeller *et al.* (1991).]

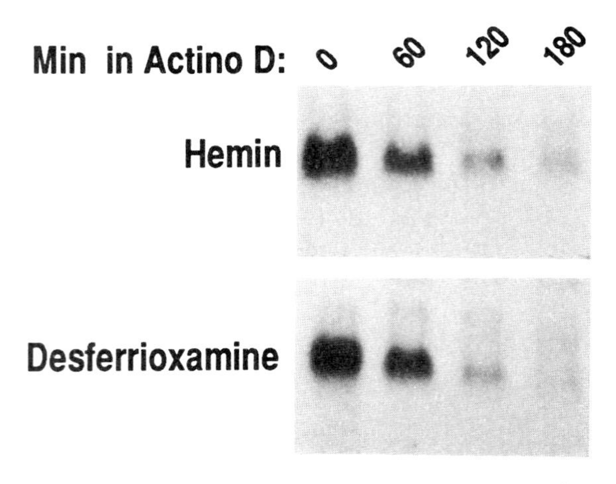

FIGURE 8 The half-life of 5′IRE-TR-*fos* mRNA is short irrespective of its translation. This transcript is short-lived due to the presence of the AU-rich region of the c-*fos* 3′UTR. Utilizing the experimental protocol described in Fig. 5, the half-life of the 5′IRE-TR-*fos* mRNA was determined to be 45–60 min irrespective of iron availability. Given the results shown in Fig. 7, these findings indicate that the rapid turnover of 5′IRE-TR-*fos* mRNA is not dependent upon its own translation. However, the 5′IRE-TR-*fos* mRNA is markedly stabilized by global inhibition of protein synthesis with cycloheximide or puromycin (Koeller *et al.*, 1991).

determinant of TfR rather than that of c-*fos* , also was not affected by changes in iron availability despite a 20-fold regulation of its translation.

The results of these experiments support the conclusion that the effect of global protein synthesis inhibitors on the stability of mRNAs containing the rapid turnover determinant of the TfR mRNA or the c-*fos* AU-rich element is due to the loss of a short-lived protein that participates in the decay of these transcripts rather than to a requirement for translation per se. Also supporting this conclusion is the fact that the stabilization of endogenous TfR mRNAs by protein synthesis inhibitors appears to be cell type specific, with certain cells displaying no observable stabilization over several hours of Chx treatment. However, all cell types examined modulate their TfR mRNA level in response to iron availability. Given that the level and/or half-life of a given protein may vary from cell type to cell type, the variable effect of Chx may reflect differences in the level or half-life of the short-lived protein involved in turnover of the TfR mRNA. It would seem more difficult to explain the cell type variation of Chx treatment by a cis-effect of the protein synthesis inhibitor, since one would then be forced to conclude that the fundamental mechanism by which cells modulate their TfR mRNA levels in response to iron is intimately tied to translation of the TfR mRNA in some cells, but independent of TfR mRNA translation in others.

There have been numerous suggestions that the stability of mRNAs can be influenced by the poly(A) tract at the 3′ end (Bernstein and Ross, 1989). In the case of c-*fos* mRNA, shortening or removal of the poly(A) precedes degradation of the remainder of the mRNA (Wilson and Treisman, 1988; Shyu *et al.*, 1991). Deletion of the AU-rich sequence from the 3′UTR of c-*fos* mRNA slows this process. Moreover, the removal of poly(A) from c-*fos* mRNA is slowed by inhibitors of translation. It has also been proposed that c-*myc* mRNA turnover proceeds by translation-dependent shortening of the poly(A) tail followed by a rapid decay of the remainder of the mRNA after the poly(A) tail is shortened beyond some critical length (Laird-Offringa *et al.*, 1990). These authors also concluded that poly(A) tail shortening was the translation-dependent step in c-*myc* mRNA degradation. Given these findings, it might be inferred that it is poly(A) tail shortening that requires the participation of the short-lived protein implicated by Koeller *et al.* (1991). It remains to be determined if the length of the poly(A) tail of the TfR mRNA is affected by global or selective inhibition of translation of these mRNAs or if iron treatment modulates poly(A) tail length of the TfR mRNA. The influence of poly(A) on mRNA stability is discussed in detail in Chapter 15.

IX. Summary and Perspectives

Cells utilize a common cis-acting RNA element (the IRE) and a common trans-acting protein (the IRE-BP) to modulate the expression of two genes involved in cellular iron homeostasis. The translation of the mRNA encoding ferritin is attenuated by iron deprivation as is the degradation of the mRNA encoding the TfR. To date, this is the only known example of this sort of overlap in post-transcriptional regulation of the expression of two genes. Although the levels of expression of the two genes are oppositely modulated by changes in iron availability, the IRE-BP serves as an iron-regulatable repressor in both instances. Clues as to how the IRE-BP might sense changes in cellular iron status have come from the observation that the IRE-BP is structurally similar to the mitochondrial enzyme aconitase (Rouault *et al.*, 1991) and possesses iron-dependent aconitase activity (Kaptain *et al.*, 1991). The latter finding strongly suggests that the IRE-BP, like aconitase, contains an iron–sulfur complex that can contain either three or four atoms of iron. When cellular iron is scarce, the IRE-BP is envisioned as being in the [3Fe–4S] state, where it is a more avid binder of IREs and therefore an effective repressor of ferritin mRNA translation and TfR mRNA degradation.

In the TfR mRNA, the IREs are adjacent to (and probably overlap with) a rapid turnover determinant. The simplest view of the action of the IRE-BP is that by binding the TfR IREs it sterically blocks access to this rapid turnover determinant. The rapid turnover determinant may be the site of an endonucleolytic cleavage, or it may in some way activate 3′ to 5′ exonucleolytic destruction of the TfR mRNA. For both transfected human TfR mRNAs expressed in mouse cells and the endogenous human TfR mRNA, a truncated transcript with properties expected of a degradation intermediate has been observed (Harford *et al.*, 1988; Koeller *et al.*, 1991). It is noteworthy that only one such intermediate is apparent and that the 3′ end of this putative intermediate appears to be relatively uniform. These features are equally consistent with a specific endonucleolytic cleavage or a well-defined pause in degradation by a 3′ to 5′ exonuclease. In either case, it would appear that shorter intermediates are extremely unstable.

It is possible that the structure of TfR mRNA that is recognized by the cellular degradative machinery is a complex entity in three-dimensional space. Deletion analyses are consistent with its comprising RNA sequence/structure elements that are well separated in the primary sequence of TfR mRNA. This finding raises the possibility that mRNA elements implicated by deletion analysis are not directly involved in interactions with trans-acting factors but instead play a role in bringing other elements into proper spatial juxtaposition. It is clear that there is much more to be learned about the TfR mRNA and its iron-regulated turnover. A critical goal in this endeavor is the establishment of a cell-free system that accurately reflects cellular turnover of this mRNA, i.e., one that is sensitive to the rather subtle mutations in the mRNA known to affect regulation *in vivo* (Casey *et al.*, 1989), inhibitable by interaction of the RNA with the high affinity form of the IRE-BP and dependent upon a short-lived protein (at least in mouse L cells). Such an *in vitro* turnover system would provide an avenue by which other components of the TfR mRNA degradative machinery might be identified and characterized.

Acknowledgments

I acknowledge my past and present colleagues in the Cell Biology and Metabolism Branch, whose work is cited here. I particularly want to acknowledge Rick Klausner, whose contributions, support, and most importantly friendship have been unwavering.

References

Abate, C., Patel, L., Rauscher, F. J., and Curran, T. (1990). Redox regulation of fos and jun DNA-binding activity in vitro. *Science* **249,** 1157–1161.

Abrahams, J. P., van den Berg, M., van Batenburg, E., and Pleij, C. (1990). Prediction of RNA secondary structure, including pseudoknotting, by computer simulation. *Nucleic Acids Res.* **18,** 3035–3044.

Aziz, N., and Munro, H. N. (1987). Iron regulates ferritin mRNA translation through a segment of its 5′ untranslated regulated region. *Proc. Natl. Acad. Sci. USA* **84,** 8478–8482.

Barton, H. A., Eisenstein, R. S., Bomford, A., and Munro, H. N. (1990). Determinants of the interaction between the iron-responsive element-binding protein and its binding site in rat L-ferritin mRNA. *J. Biol. Chem.* **265,** 7000–7008.

Beinert, H. (1990). Recent developments in the field of iron–sulfur proteins. *FASEB J.* **4,** 2483–2491.

Bernstein, P., and Ross, J. (1989). Poly(A), poly(A) binding protein and the regulation of mRNA stability. *Trends Biochem. Sci.* **14,** 373–377.

Casey, J. L., Hentze, M. W., Koeller, D. M., Caughman, S. W., Rouault, T. A., Klausner, R. D., and Harford, J. B. (1988a). Iron-responsive elements: Regulatory RNA sequences that control mRNA levels and translation. *Science* **240,** 924–928.

Casey, J. L., Di Jeso, B., Rao, K., Klausner, R. D., and Harford, J. B. (1988b). Two genetic loci participate in the regulation by iron of the gene for the human transferrin receptor. *Proc. Natl. Acad. Sci. USA* **85,** 1787–1791.

Casey, J. L., Koeller, D. M., Ramin, V. C., Klausner, R. D., and Harford, J. B. (1989). Iron regulation of transferrin receptor mRNA levels requires iron-responsive elements and a rapid turnover determinants in the 3′ untranslated region of the mRNA. *EMBO J.* **8,** 3693–3699.

Caughman, S. W., Hentze, M. W., Rouault, T. A., Harford, J. B., and Klausner, R. D. (1988). The iron-responsive element is the single elements responsible for iron-dependent translational regulation of ferritin biosynthesis: Evidence for function as the binding site for a translational repressor. *J. Biol. Chem.* **263,** 19048–19052.

Chan, L. N., Grammatikakis, N., Banks, J. M., and Gerhardt, E. M. (1989). Chicken transferrin receptor gene: Conservation 3′ noncoding sequences and expression in erythroid cells. *Nucleic Acids Res.* **17,** 3763–3771.

Chitamber, C. R., Massey, E. J., and Seligman, P. A. (1983). Regulation of transferrin receptor expression on human leukemic cells during proliferation and induction of differentiation. *J. Clin. Invest.* **72,** 1314–1325.

Cleveland, D. W., and Yen, T. J. (1989). Multiple determinants of eukaryotic mRNA stability. *New Biol.* **1,** 121–126.

Cole, M. D. (1986). The myc oncogene: Its role in transformation and differentiation. *Annu. Rev. Genet.* **20,** 361–384.

Crichton, R. R. (1991). Inorganic Biochemistry of Iron Metabolism. Ellis Horwood, New York.

Crichton, R. R., and Charloteaux-Wauters, M. (1987). Iron transport and storage. *Eur. J. Biochem.* **164,** 485–506.

Granick, S. (1946). Ferritin: Its properties and significance for iron metabolism. *Chem. Rev.* **38,** 379–395.

Gutteridge, J. M. C. (1989). Iron and oxygen: A biologically damaging mixture. *Acta Pediatr. Scand. Suppl.* **361,** 78–85.

Haile, D. J., Hentze, M. W., Rouault, T. A., Harford, J. B., and Klausner, R. D. (1989). Regulation of interaction of the iron-responsive element binding protein with iron-responsive RNA elements. *Mol. Cell. Biol.* **9,** 5055–5061.

Hanover, J. A., and Dickson, R. B. (1985). Transferrin: Receptor-mediated endocytosis and iron delivery. *In* Endocytosis (I. Pastan and M. C. Willingham, Eds.), pp.131–161. Plenum, New York.

Harford, J., Koeller, D., Casey, J., Hentze, M., Rouault, T., and Klausner, R. D. (1988). Iron-responsive elements within the 3′ untranslated region of the transferrin receptor mRNA. *J. Cell Biol.* **107**(3), 12a.

Harford, J. B., Casey, J. L., Koeller, D. M., and Klausner R. D. (1990). Structure, function and regulation of the transferrin receptor: Insights from molecular biology. *In* Intracellular Trafficking of Proteins (C. J. Steer and J. A. Hanover, Eds.), pp. 302–334. Cambridge Univ. Press, Cambridge.

Harford, J. B., and Klausner, R. D. (1990). Coordinate post-transcriptional regulation of ferritin and transferrin receptor expression: The role of regulated RNA-protein interaction. *Enzyme* **44,** 28–41.

Hentze, M. W., Keim, S., Papadopoulos, P., O'Brien, S., Modi, W., Drysdale, J., Leonard, W. J., Harford, J. B., and Klausner, R. D. (1986). Cloning, characterization, expression, and chromosomal localization of a human ferritin heavy-chain gene. *Proc. Natl. Acad. Sci. USA* **83,** 7226–7230.

Hentze, M. W., Caughman, S. W., Rouault, T. A., Barriocanal. J. G., Dancis, A., Harford, J. B., and Klausner, R. D. (1987a). Identification of the iron-responsive element for the translational regulation of human ferritin mRNA. *Science* **238,** 1570–1573.

Hentze, M. W., Rouault, T. A., Caughman, S. W., Dancis, A., Harford, J. B., and Klausner, R. D. (1987b). A cis-acting element is necessary and sufficient for translational regulation of human ferritin expression in response to iron. *Proc. Natl. Acad. Sci. USA* **84,** 6730–6734.

Hentze, M. W., Caughman, S. W., Casey, J. L., Koeller, D. M., Rouault, T. A., Harford, J. B., and Klausner, R. D. (1988). A model for the structure and functions of iron-responsive elements. *Gene* **72,** 201–208.

Hentze, M. W., Rouault, T. A., Harford, J. B., and Klausner, R. D. (1989). Oxidation-reduction and the molecular mechanism of a regulatory RNA-protein interaction. *Science* **244,** 357–359.

Horowitz, J. A., and Harford, J. B. (1992). The secondary structure of the regulatory region of the transferrin receptor mRNA deduced by enzymatic cleavage. *New Biol.* **4,** 330–338.

Huebers, H. A., and Finch, C. A. (1987). The physiology of transferrin and transferrin receptors. *Physiol. Rev.* **67,** 520–582.

Kaptain, S., Downey, W. E., Tang, C., Philpott, C., Haile, D., Orloff, D. G., Harford, J. B., Rouault, T. A., and Klausner, R. D. (1991). A regulated RNA binding protein also possesses aconitase activity. *Proc. Natl. Acad. Sci. USA* **88,** 10109–10113.

Klausner, R. D., van Renswoude, J., Kempf, C., Rao, K., Bateman, J. L., and Robbins, A. R. (1984). Failure to release iron from transferrin in a Chinese hamster ovary cell mutant pleiotropically defective in endocytosis. *J. Cell Biol.* **98,** 1098–1101.

Klausner, R. D., and Harford, J. B. (1989). cis–trans models for post-transcriptional gene regulation. *Science* **246,** 870–872.

Koeller, D. M., Casey, J. L., Hentze, M. W., Gerhardt, E. M., Chan, L. N., Klausner, R. D., and Harford, J. B. (1989). A cytosolic protein binds to structural elements within the iron regulatory region of the transferrin receptor mRNA. *Proc. Natl. Acad. Sci.USA* **86,** 3574–3578.

Koeller, D. M., Horowitz, J. A., Casey, J. L., Klausner, R. D., and Harford, J. B. (1991). Translation and the stability of mRNAs encoding the transferrin receptor and c-*fos*. *Proc. Natl. Acad. Sci. USA* **88,** 7778–7782.

Kozak, M. (1986). Influences of mRNA secondary structure on initiation by eukaryotic ribosomes. *Proc. Natl. Acad. Sci. USA* **83,** 2850–2854.

Kronke, M., Leonard, W. J., Depper, J. M., and Greene, W. C. (1985). Sequential expression of genes involved in human T lymphocyte growth and differentiation. *J. Exp. Med.* **161,** 1593–1598.

Kühn, L. C. (1989). The transferrin receptor: A key function in iron metabolism. *Schweiz. Med. Wochenschr.* **119,** 1319–1326.

Laird-Offringa, I. A., de Witt, C. L., Elfferich, P., and van der Eb, A. J. (1990). Poly(A) tail shortening is the translation-dependent step in c-*myc* mRNA degradation. *Mol. Cell. Biol.* **10,** 6132–6140.

Leibold, E. A., and Munro, H. N. (1987). Characterization and evolution of the expressed rat ferritin light subunit gene and its pseudogene family: Conservation of sequences within noncoding regions of ferritin genes. *J. Biol. Chem.* **262,** 7335–7341.

Leibold, E. A., and Munro, H. N. (1988). Cytoplasmic protein binds in vitro to a highly conserved sequence in the 5′ untranslated region of ferritin heavy- and light-subunit mRNAs. *Proc. Natl. Acad. Sci. USA* **85,** 2171–2175.

Leibold, E. A., Laudano, A., and Yu, Y. (1990). Structural requirements of iron-responsive elements for binding of the protein involved in both transferrin receptor and ferritin mRNA post-transcriptional regulation. *Nucleic Acids Res.* **18,** 1819–1824.

Linial, M. N., Gunderson, N., and Groudine, M. (1985). Enhanced transcription of c-*myc* in bursal lymphoma cells requires continuous protein synthesis. *Science* **230,** 1126–1132.

Mattia, E., Rao, K., Shapiro, D. S., Sussman, H. H., and Klausner, R. D. (1984). Biosynthetic regulation of the human transferrin receptor by desferrioxamine in K562 cells. *J. Biol. Chem.* **259,** 2689–2692.

Marzluff, W. F., and Pandey, N. B. (1988). Multiple regulatory steps control histone mRNA concentrations. *Trends Biochem. Sci.* **13,** 49–52.

McClelland, A., Kühn, L. C., and Ruddle, F. H. (1984). The human transferrin receptor gene: Genomic organization and complete primary structure of the receptor deduced from a cDNA sequence. *Cell* **39,** 267–274.

Müllner, E. W., and Kühn, L. C. (1988). A stem–loop in the 3′ untranslated region mediates iron-dependent regulation of transferrin receptor mRNA stability in the cytoplasm. *Cell* **53,** 815–825

Müllner, E. W., Neupert, B., and Kühn, L. C. (1989). A specific mRNA binding factor regulates the iron-dependent stability of cytoplasmic transferrin receptor mRNA. *Cell* **58,** 373–382

Munro, H. N., and Linder, M. C. (1978). Ferritin: Structure, biosynthesis, and role in iron metabolism. *Physiol. Rev.* **58,** 317–396.

Neilands, J. B. (1974). *In* Microbial Iron Metabolism (J. B. Neilands, Ed.), pp. 3–34. Academic Press, New York.

Owen, D., and Kühn, L. C. (1987). Noncoding 3′ sequences of the transferrin receptor gene are required for mRNA regulation by iron. *EMBO J.* **6,** 1287–1293.

Pelletier, J., and Sonenberg, N. (1985). Insertion mutagenesis to increase secondary structure within the 5′ noncoding region of a eukaryotic mRNA reduces translational efficiency. *Cell* **40,** 515–526.

Rahmsdorf, H. J., Schönthal, A., Angel, P., Litfin, M., Rüther, U., and Herrlich, P. (1987). Post-transcriptional regulation of c-*fos* mRNA expression. *Nucleic Acids Res.* **15,** 1643–1659.

Rao, K. K., Shapiro, D., Mattia, E., Bridges, K., and Klausner, R. (1985). Effects of alterations in cellular iron on biosynthesis of the transferrin receptor in K562 cells. *Mol. Cell. Biol.* **5,** 595–600.

Rao, K., Harford, J. B., Rouault, T., McClelland, A., Ruddle, F. H., and Klausner, R. D. (1986). Transcriptional regulation by iron of the gene for the transferrin receptor. *Mol. Cell. Biol.* **6,** 236–240.

Robbins, A. H., and Stout, C. D. (1989a). The structure of aconitase. *Proteins* **5,** 289–312.

Robbins, A. H., and Stout, C. D. (1989b). Structure of activated aconitase: Formation of the [4Fe–4S] clusterin the crystal. *Proc. Natl. Acad. Sci. USA* **86,** 3639–3643.

Roberts, K. P., and Griswold, M. D. (1990). Characterization of rat transferrin receptor cDNA: The regulation of transferrin receptor cDNA: The regulation of transferrin receptor mRNA in testes and in Seroli cells in culture. *Mol. Endocrinol.* **4,** 531–542.

Rogers, J., and Munro, H. N. (1987). Translation of ferritin light and heavy subunits mRNAs is regulated by intracellular chelatable iron levels in rat hepatoma cells. *Proc. Natl. Acad. Sci. USA* **84,** 2277–2281.

Ross, J., Kobs, G., Brewer, G., and Peltz, S. W. (1986). Properties of the exonuclease activity that degrades H4 histone mRNA. *J. Biol. Chem.* **262,** 9374–9381.

Rothenberger, S., Müllner, E. W., and Kühn, L. C. (1990). The mRNA-binding protein which controls ferritin and transferrin receptor expression is conserved during evolution. *Nucleic Acids Res.* **18,** 1175–1179.

Rouault, T. A., Hentze, M. W., Dancis, A., Caughman, W., Harford, J. B., and Klausner,

R. D. (1987). Influence of altered transcription on the translational control of human ferritin expression. *Proc. Natl. Acad. Sci. USA* **84,** 6335–6339.

Rouault, T. A., Hentze, M. W., Caughman, S. W., Harford, J. B., and Klausner, R. D. (1988). Binding of a cytosolic protein to the iron-responsive element of human ferritin messenger RNA. *Science* **241,** 1207–1210.

Rouault, T. A., Hentze, M. W., Haile, D. J., Haile, D. J., Harford, J. B., and Klausner, R. D. (1989). The iron-responsive element binding protein: A method for the affinity purification of a regulatory RNA-binding protein. *Proc. Natl. Acad. Sci. USA* **86,** 5768–5772.

Rouault, T. A., Tang, C. K., Kaptain, S., Burgess, W. H., Haile, D. J., Samaniego, F., McBride, W. O., Harford, J. B., and Klausner, R. D. (1990). Cloning of the cDNA encoding an RNA regulatory protein: The human iron responsive elements binding protein. *Proc. Natl. Acad. Sci. USA* **87,** 7958–7962.

Rouault, T. A., Stout, C. D., Kaptain, S., Harford, J. B., and Klausner, R. D. (1991). Structural relationship between an iron-regulated RNA-binding protein (IRE-BP) and aconitase: Functional implications. *Cell* **64,** 881–883.

Ryseck, R. P., Hirai, S. I., Yaniv, M., and Bravo, R. (1988). Transcriptional activation of c-*jun* during the G0/G1 transition in mouse fibroblasts. *Nature (London)* **334,** 535–537.

Shaw, G., and Kamen, R. (1986). A conserved AU sequence from the 3′ untranslated region of GM-CSF mRNA mediates selective mRNA degradation. *Cell* **46,** 659–667.

Shyu, A.-B., Greenberg, M. E., and Belasco, J. G. (1989). The c-*fos* transcript is targeted for rapid decay by two distinct mRNA degradation pathways. *Genes Dev.* **3,** 60–72.

Shyu, A.-B., Belasco, J. G., and Greenberg, M. E. (1991). Two distinct destabilizing elements in the c-*fos* message trigger deadenylation as a first step in rapid mRNA decay. *Genes Dev.* **5,** 221–231.

Storz, G., Tartaglia, L. A., and Ames, B. N. (1990). Transcriptional regulator of oxidative stress-inducible genes: Direct activation by oxidation. *Science* **248,** 189–194.

Stubbe, J. (1990). Ribonucleotide reductases. *Adv. Enzymol. Relat. Areas Mol. Biol.* **63,** 349–419.

Theil, E. C. (1987). Ferritin: Structure, gene regulation, and cellular function in animals, plants, and microorganisms. *Annu. Rev. Biochem.* **56,** 289–315.

Theil, E. C. (1990). Regulation of ferritin and transferrin receptor mRNAs. *J. Biol. Chem.* **265,** 4771–4774.

Thompson, C. B., Challoner, P. B., Neiman, P. E., and Groudine, M. (1986). Expression of the c-*myb* protooncogene during cellular proliferation. *Nature (London)* **319,** 374–380.

Trepel, J. B., Colamonici, O. R., Kelly, K., Schwab, G., Watt, R. A., Sausville, E. A., Jaffe, E. S., and Neckers, L. M. (1987). Transcriptional inactivation of c-*myc* and the transferrin receptor in dibutyryl cyclic AMP-treated HL60 cells. *Mol. Cell. Biol.* **7,** 2644–2648.

Walden, W. E., Daniels-McQueen, S., Brown, P. H., Gaffield, L., Russell, D. A., Bielser, D., Bailey, L. C., and Thach, R. E. (1988). Translational repression in eukaryotes: Partial purification and characterization of a repressor of ferritin mRNA translation. *Proc. Natl. Acad. Sci. USA* **85,** 9503–9507.

Walden, W. E., Patino, M. M., and Gaffield, L. (1989). Purification of a specific repressor of ferritin mRNA translation from rabbit liver. *J. Biol. Chem.* **264,** 13765–13769.

Ward, J. H., Kushner, J. P., and Kaplan, J. (1982). Regulation of HeLa cell transferrin receptors. *J. Biol. Chem.* **257,** 10317–10323.

Wilson, T., and Treisman, R. (1988). Removal of poly(A) and consequent degradation of c-*fos* mRNA facilitated by 3′ AU-rich sequences. *Nature (London)* **336,** 396–399.

Zahringer, J., Baliga, B. S., and Munro, H. N. (1976). Novel mechanism for translational control in regulation of ferritin synthesis by iron. *Proc. Natl. Acad. Sci. USA* **73,** 857–861.

12

Degradation of a Nonpolyadenylated Messenger: Histone mRNA Decay

WILLIAM F. MARZLUFF AND ROBERTA J. HANSON

I. Introduction

A. Early Studies of Regulation of Histone Synthesis

In mammals the synthesis of histone proteins is tightly coupled to DNA replication in the somatic cell cycle. Thus the synthesis of the two major components of the chromosomes is closely coordinated. The pioneering work of Drs. Gerald Mueller, Dieter Gallwitz, and Ted Borun and coworkers (Borun *et al.*, 1975; Butler and Mueller, 1973; Breindl and Gallwitz, 1974) in the late 1960s and early 1970s clearly defined many of the parameters affecting histone protein metabolism. These early results described many of the regulatory features of the coupling of DNA replication and histone synthesis. In particular, the rapid disappearance of histone mRNA when DNA synthesis is inhibited or when the S-phase ends was correctly attributed to an enhanced degradation of histone mRNA (Gallwitz, 1975). The requirement of continued protein synthesis for the accelerated degradation of histone mRNA was also described (Breindl and Gallwitz, 1974; Butler and Mueller, 1973). With the advent of gene cloning and the ability to study the molecular mechanisms involved in regulating histone mRNA levels, the early work has been corroborated and some aspects of the molecular basis of histone mRNA regulation have been elucidated. In this chapter, we review current models of the mechanism of histone mRNA degradation and the role histone mRNA half-life plays in the overall regulation of histone mRNA levels in animal cells.

B. Histone Gene and mRNA Structure

The replication-dependent histone mRNAs, which encode the majority of the histone proteins in dividing cells, are the only class of mRNAs that lack a poly(A) tail. These genes are also the only set of genes in higher animals that lack introns (Marzluff and Pandey, 1988). Histone genes are linked in large clusters extending over several hundred kilobases. In some organisms (e.g., frogs and newts) the histone genes are present in tandem repeats, with each repeat unit containing one copy of the gene for each of the five histone proteins. In other cases (e.g., mammals and birds), the genes for the five histone proteins have remained clustered, although they are arranged randomly (Marzluff, 1986; Maxson *et al.*, 1983). Histone mRNAs are among the smallest mRNAs in the cell, containing short (10–40 nt) 5′ untranslated regions and short (35–80 nt) 3′ untranslated regions. The 3′ end of histone mRNAs has been highly conserved in evolution (Birnstiel *et al.*, 1985; Marzluff, 1992) and is depicted in Fig. 1A. A sequence of about 20 nucleotides containing a potential stem–loop has been conserved. The histone mRNA ends in a short stretch (2–4 nt) of A and C residues and the stem–loop is preceded by a short A-rich region. This structure is represented in two dimensions as a stem–loop structure, with a six-base stem and a four-base loop, although the actual structure of the 3′ end has not been determined. The structure of both the stem and the loop have been conserved in evolution (Fig. 1A). The stem has two GC base pairs at the base, followed by three Py–Pu pairs and a UA base pair at the top of the stem. The loop sequence has an invariant U at position 3 of the loop, and the first base is U in all known histone mRNAs except those in *Caenorhabditis elegans*, where it is a C (Marzluff, 1992). The common 3′ end on all histone mRNAs provides a natural target for their coordinate regulation.

The 3′ end of histone mRNA is formed by an endonucleolytic cleavage, which is the only processing reaction involved in histone mRNA biosynthesis (Gick *et al.*, 1986). The 3′ processing signal consists of the stem–loop and a purine-rich region located 6–15 nt 3′ of the stem–loop (Fig. 1B). The pre-mRNA is cleaved between these two sequences, forming the mature histone mRNA and probably also releasing the RNA from the chromatin template. The processing reaction requires three components: the U7 snRNP, which recognizes the purine-rich sequence 3′ of the stem–loop (Strub and Birnstiel, 1986; Mowry and Steitz, 1987), a factor that recognizes the stem–loop (Vasserot *et al.*, 1989; Mowry *et al.*, 1989), and a third factor, which is heat-labile (Gick *et al.*, 1987). These three components have not yet been completely purified.

All of the histone mRNAs that end with the stem–loop structure are regulated coordinately with DNA synthesis and have been termed "replication-dependent" histone mRNAs. There is a second class of histone

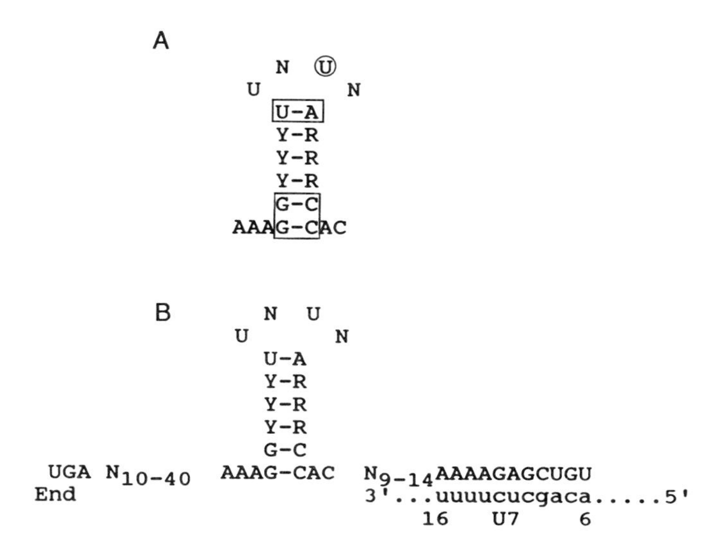

FIGURE 1 3′ end of histone mRNA. (A) The structure of the 3′ end of a typical mammalian histone mRNA is shown. The nucleotides that have been conserved in all metazoan histone mRNAs are circled and boxed. (B) The structure of the 3′ end of a typical mammalian histone pre-mRNA is shown. The base pairing between the U7 snRNP and the purine-rich region is indicated.

proteins, e.g., histone H3.3, often called "replacement" variants (Zweidler, 1984). These proteins are synthesized in nondividing cells or throughout the cell cycle in dividing cells (Wu and Bonner, 1981). These proteins are encoded by genes that contain introns and that are not closely related in nucleotide sequence to the replication-dependent histone genes. The mRNAs for the replacement variant histone H3.3 (Brush *et al.*, 1985; Wells and Kedes, 1985) and H2a.Z (Hatch and Bonner, 1990) are polyadenylated, and their levels are not regulated coordinately with DNA replication (Harris *et al.*, 1991; Sittman *et al.*, 1983a; Hatch and Bonner, 1990). There are also some genes encoding histone variants, e.g., histone H2a.X (Mannironi *et al.*, 1989), which can encode either a histone mRNA ending in the stem–loop or a longer polyadenylated histone mRNA, depending on which 3′ processing signal is used. In addition to the H2a.X gene, the genes for at least some histone H1 subtypes can also encode either an mRNA ending at the stem–loop or a longer mRNA with a poly(A) tail (Cheng *et al.*, 1989; Kirsh *et al.*, 1989). The different 3′ ends allow two mRNAs encoding identical proteins to be regulated independently. As we will discuss later, the levels of the polyadenylated histone mRNAs are not coupled to DNA synthesis, since they lack the stem–loop at the 3′ end.

II. Control of Histone mRNA Stability

A. Experimental Systems

Three general types of cell cultures have been utilized to study the regulation of histone mRNA levels. First, exponentially growing cell cultures have been used. Inhibiting DNA synthesis in these cultures using a drug that interferes either with DNA polymerase or deoxynucleotide metabolism results in a rapid disappearance of histone mRNA (Sittman *et al.*, 1983a; Graves and Marzluff, 1984). This provides a convenient, easily manipulated system that mimics in some respects the cessation of DNA replication at the end of the S-phase. The degree to which the signals generated by the inhibition of DNA synthesis in an S-phase cell are similar to those of the normal events that occur at the S–G2 boundary is not known.

Second, cells have been "synchronized" using a variety of techniques. Most common has been blockage at the G1–S boundary, followed by synchronous release of cells into the S-phase. This procedure is a convenient way to prepare large-scale cultures of cells highly enriched in S-phase cells (Heintz *et al.*, 1983; Morris *et al.*, 1991). The agents used to block cells are often the same antimetabolites used to inhibit DNA replication in exponential cultures. Since these drugs directly influence histone mRNA metabolism, the results obtained are not necessarily relevant to the events that occur in cycling cells at the G1–S phase boundary.

Studies of cells arrested in G_0, usually by serum deprivation, have also been made (DeLisle *et al.*, 1983; Stauber and Schümperli, 1988). Cell cycle mutants arrested in G1 under restricted conditions have also been used (Lüscher *et al.*, 1985; Gick *et al.*, 1987). While these cultures are often referred to as "synchronized" cells, many of the events that occur as a cell approaches the S-phase after growth arrest are clearly different from the events that occur in a continuously cycling cell. An advantage of this type of system is that the arrested cultures contain less than 1% S-phase cells, and the growth-arrested cells are clearly in a physiologically relevant state.

A limited number of studies have used naturally synchronous cells, selected either by mitotic shake-off or by elutriation (Alterman *et al.*, 1984; Harris *et al.*, 1991). These systems have been very useful for studying the events leading from the G1- to the S-phase, but due to the natural asynchrony that rapidly develops and the short length of the G2-phase, they have not been particularly useful for the study of the events that occur in G2, since enriched populations of G2 cells are difficult to obtain.

Recent studies strongly suggest that the regulation of histone mRNA half-life may be restricted to the period between the end of the S-phase and mitosis (Harris *et al.*, 1991). Direct measurement of the half-life of the small amount of histone mRNA present in G1-phase cells, using actinomycin D to inhibit transcription, suggested a similar half-life for histone mRNA in both G1- and S-phase cells. Similarly, the failure of protein synthesis inhibitors to greatly increase histone mRNA levels in G1-cells is consistent with the histone mRNA in G1-cells having a half-life of 45–60 min, in agreement with the half-life measured using actinomycin D (Harris *et al.*, 1991). Rapid degradation of histone mRNA is necessary to destroy the histone mRNA prior to mitosis. The great majority of studies of the mechanism of histone mRNA degradation have utilized inhibition of DNA synthesis in exponentially growing or S-phase-enriched cultures. The tacit assumption is that this experiment mimics the events that occur at the end of the S-phase.

B. Sequence Requirements for Histone mRNA Degradation

Histone mRNAs are naturally short-lived compared with many other mRNAs, with a half- life in exponentially growing or S-phase cells of 45 to 60 min. Treatment of cells with inhibitors of DNA replication results in a rapid 15- to 30-fold reduction of histone mRNA concentrations in 30 to 60 min (Gallwitz, 1975; Sittman *et al.*, 1983b). This can be accomplished only by a rapid specific degradation of histone mRNA, with a maximal half-life of 8–12 min. The response is rapid, occurring within 5 min of addition of the DNA synthesis inhibitor. The response is also reversible, with the normal half-life returning within 5–10 min after resumption of DNA synthesis (Graves and Marzluff, 1984).

The sequence requirements for histone mRNA degradation have been unambiguously determined. Using chimeric mRNAs containing histone and globin sequences, the 3′ end of the histone mRNA was shown to confer the response to inhibition of DNA synthesis. This accounts for the coordinate degradation of all histone mRNAs, since the 3′ end is the common structural feature unique to histone mRNAs. The stem–loop must be at the 3′ end of the histone mRNA to mediate rapid degradation; natural mRNAs (e.g., H2a.X) that contain the stem–loop sequence in the 3′ untranslated region but end in a downstream poly(A) sequence (Manniron *et al.*, 1989; Cheng *et al.*, 1989), or artificial genes with the stem–loop sequence present internally in the 3′ untranslated region (Graves *et al.*, 1987; Levine *et al.*, 1987) are not rapidly degraded when DNA synthesis is inhibited. A variety of experiments have ruled out the involvement of other sequences, including the 5′ untranslated region or coding region sequences, in the regulation of histone mRNA degradation (Graves *et al.*, 1987). The results obtained with a number of chimeric mRNAs are summarized in Fig. 2A.

The obvious interpretation of these experiments is that the stem–loop must be at the 3′ end of the mRNA for rapid degradation. However, one cannot rule out the possibility that the poly(A) sequence at the 3′ end of polyadenylated mRNAs containing an internal histone stem–loop protects the mRNA from degradation. One frustrating aspect of these studies is

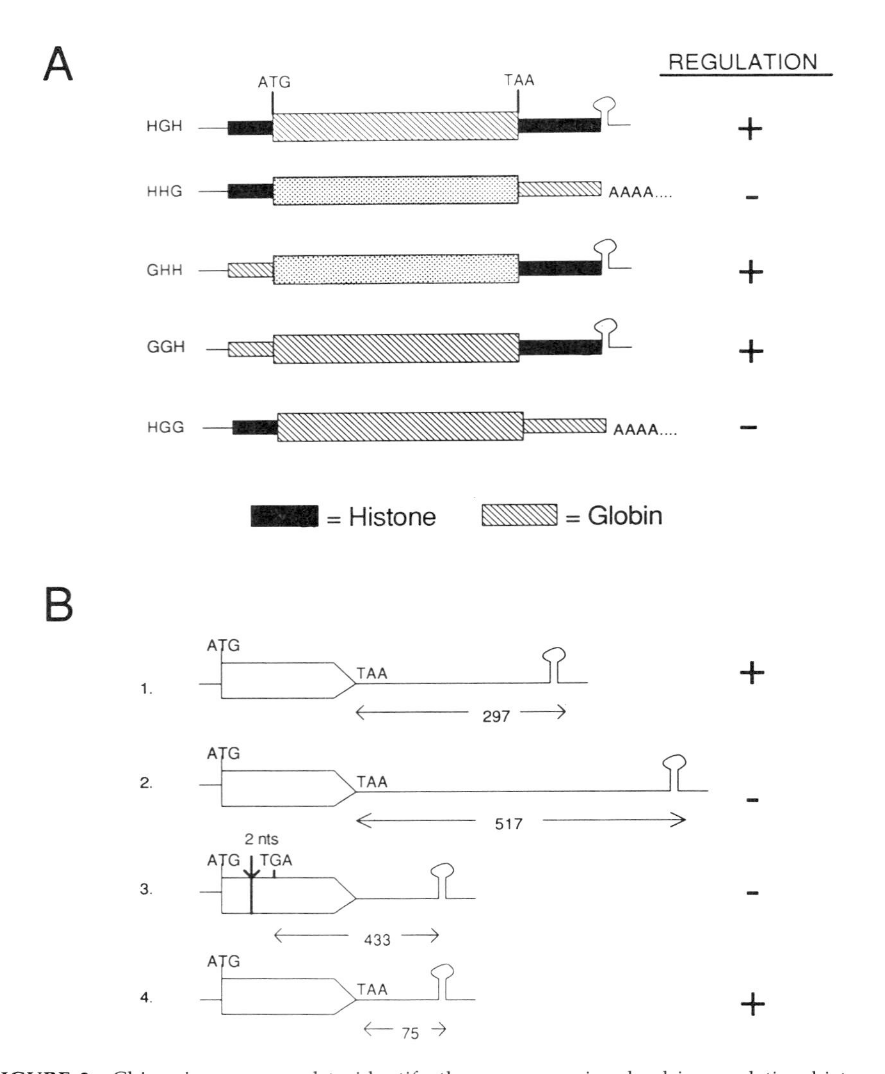

FIGURE 2 Chimeric genes used to identify the sequences involved in regulating histone mRNA degradation. (A) Histone–globin chimeric mRNAs. The promoter and 5′ untranslated region, the coding region, and the 3′ untranslated region of the mouse histone H2a and human a-globin genes were interchanged. The genes were introduced into either mouse L cells or CHO cells, and stable transformants were selected. Each gene encodes either a globin or a histone protein. The GGH gene has the globin promoter and 5′ untranslated region, the globin coding region; the mRNA was regulated properly; and a "– " indicates that the mRNA was stable when DNA synthesis was inhibited. See Levine *et al.* (1987) and Pandey and Marzluff (1987) for details. (B) Effect of the distance of the stem–loop from the termination codon. The structure of histone mRNAs containing long 3′ untranslated regions or termination codons within the coding region is shown. These genes all contain the H2a-614 histone 3′ end placed at different distances from the end of the open reading frame. The genes were introduced into mouse L cells, and stable transformants were selected. The distance from the termination codon to the stem–loop is given for each mRNA. The regulation of the mRNA is indicated as in A. [See Graves *et al.* (1987) for details.]

that it has not been possible to produce a stable mRNA that has a 3′ end other than poly(A) or a histone stem–loop.

C. Position of the 3′ End Relative to Translation Termination

There is a second important structural requirement for the alteration of histone mRNA half-life in response to inhibition of DNA synthesis. Drs. Reed Graves, Nunta Chodchoy, and Niranjan Pandey discovered that several synthetic mRNAs with histone 3′ ends were not rapidly degraded when DNA synthesis was inhibited (Graves *et al.*, 1987). These mRNAs all shared a common property: the stem–loop was located much farther from the translation stop codon than it is in any naturally occurring histone mRNA (Fig. 2B). Some of these mRNAs had stop codons early in the coding sequence, and others had long 3′ extensions on mRNAs that encoded normal histone proteins (Fig. 2B). Messenger RNAs deliberately constructed to differ from a normal histone mRNA in only one or two nucleotides, resulting in a frame shift that caused premature termination of translation, were degraded with the "normal" half-life of 45–60 min in the presence or absence of ongoing DNA synthesis. Premature translation termination at a site several hundred nucleotides before the stem–loop abolished the ability of the cell to rapidly degrade the histone mRNA. These experiments established the following requirements for proper regulation of histone mRNA degradation: (1) the stem–loop must be located at the 3′ end of the histone mRNA and (2) the stem–loop must be within 300 nt of the translation termination codon (Graves *et al.*, 1987) . The molecular mechanisms proposed for proper histone mRNA degradation must account for both of these requirements (Fig. 2B).

D. Coupling of Degradation and Translation

One of the early observations made was that histone mRNA was not degraded when cells were treated with inhibitors of protein synthesis prior to inhibition of DNA synthesis (Butler and Mueller, 1973). This result was confirmed when the effect of protein synthesis inhibitors on the levels of histone mRNA was studied using probes from cloned genes (Stimac *et al.*, 1984). Inhibitors of protein synthesis that affect initiation of translation (displacing the mRNA from polyribosomes) or inhibitors that affect elongation (freezing mRNA on polyribosomes) prevent the rapid degradation of histone mRNA (Stahl and Gallwitz, 1977; Graves and Marzluff, 1984). A common effect of inhibiting protein synthesis is to stabilize otherwise short-lived mRNAs such as oncogene mRNAs (Shyu *et al.*, 1989; Wisdom and Lee, 1990) and tubulin mRNAs (Yen *et al.*, 1988; Gay *et al.*, 1989; Gong and Brandhorst, 1988) (reviewed in Chapters 9 and 10). This makes it seem likely that there is a step or factor common to the degradation of all of these mRNAs that is affected by inhibition of protein synthesis.

The results obtained with inhibitors of protein synthesis can be (and have been) interpreted in several different ways. First, there could be a nuclease that is unstable and thus needs to be synthesized continuously. Alternatively, histone proteins could autoregulate the half-life of histone mRNA due to accumulation of an excess of histone proteins when DNA replication is stopped or when the cells exit the S-phase. Finally, accelerated degradation of histone mRNA may be directly coupled to the translation process, rather than being a result of inhibition of the synthesis of another protein.

Two lines of evidence suggest that the translation process itself is required for regulation of histone mRNA degradation. First, the requirement for a particular position of the termination codon with respect to the 3′ end of histone mRNA suggests that the ribosome must come within a certain distance of the stem–loop for proper regulation of the half-life of histone mRNA (Graves *et al.*, 1987). Second, the general effect of inhibition of protein synthesis on degradation of a large number of different mRNAs suggests a role for the translation process per se in mRNA degradation. This interpretation is supported by the experiments of Cleveland and co-workers (Yen *et al.*, 1988; Gay *et al.*, 1989) on tubulin mRNA degradation.

E. Pathway of Histone mRNA Degradation

Jeff Ross and co-workers have provided convincing evidence that the initial step in histone mRNA degradation is a cleavage near the 3′ end of the mRNA both *in vivo* (Ross *et al.*, 1986) and *in vitro* (Ross and Kobs, 1986). Within minutes after inhibition of DNA synthesis, histone mRNAs shorter at the 3′ end by about 5–15 nt were detected by nuclease protection assays (Ross *et al.*, 1986). The 5′ portion of the mRNA was still intact. The mRNAs were then completely degraded, but no other specific intermediates were detected *in vivo*. The initial intermediate that accumulates *in vivo* is consistent with an initial cleavage of the mRNA in the stem–loop sequence. This result is consistent with the genetic evidence described above that the stem–loop is the critical sequence directing histone mRNA degradation.

Histone mRNAs are degraded by a similar pathway *in vitro*. Incubation of polyribosomes containing histone mRNA from a human erythroleukemia cell line resulted in degradation of the histone mRNA by a ribosome-associated nuclease (Ross and Kobs, 1986). The same initial intermediate was detected *in vitro* as was seen *in vivo*. The exact pathway of degradation is not known and it is also not clear whether the initial cleavage is endonucleolytic or exonucleolytic. The intermediate terminates in the loop at the 3′ end of the mRNA, consistent with an initial endonucleolytic cleavage (Ross *et al.*, 1986). Purified histone mRNA is degraded 3′ to 5′ by a nuclease activity that can be extracted from polyribosomes (Peltz *et al.*, 1987; Ross *et al.*, 1987). Thus both the genetic and the biochemical evidence supports the conclusion that the initial step in histone mRNA degradation is removal of the 3′ end of histone mRNA.

Jeff Ross and co-workers have also measured the activity of an exonuclease that has been implicated in mRNA degradation. The activity of the exonuclease, measured *in vitro* after extraction from polyribosomes, is constant in G1- and S-phase cells and in cells treated with inhibitors of DNA or protein synthesis (Peltz *et al.*, 1989). Assuming that this exonuclease is important for mRNA degradation (and available evidence suggests that it is), then these experiments rule out the possibility that the levels of the nuclease are regulated and make it unlikely that the nuclease is very unstable and needs to be continually synthesized. Rather the accessibility of histone mRNA to the nuclease must be cell cycle regulated.

III. Biochemical Mechanisms Regulating Histone mRNA Degradation

A. A Protein Binds the 3′ End of Histone mRNA

Most nucleic acid sequences exert their function by interaction with specific proteins. The 3′ end shared by all histone mRNAs is an attractive potential binding site for a protein(s) that interacts specifically with histone mRNAs. Using a synthetic 30-nt RNA as a probe, proteins have been detected in both the nucleus and the cytoplasm of mouse myeloma cells, which bind specifically to the 3′ end of histone mRNA (Pandey *et al.*, 1991) (Fig. 3). The proteins were detected by their ability to alter the electrophoretic mobility of the synthetic RNA in nondenaturing polyacrylamide gels, an assay similar to that used to detect DNA-binding proteins. The complex formed with the nuclear proteins had a lower mobility than the complex formed with the polyribosomal proteins (Fig. 3A), suggesting that an additional component was present in the nuclear complex. Changing two nucleotides in the loop reduced binding to undetectable levels (Fig. 3A, lanes 3 and 4). The cytoplasmic binding activity was quantitatively associated with polyribosomes and was released only upon extraction with 0.5–0.8 *M* KCl. This suggests that the protein is a component of the histone mRNP and that, unlike other message specific RNA-binding proteins (e.g., the proteins that bind to ferritin mRNA (Rouault *et al.*, 1988) or to mRNAs containing AU-rich sequences (Malter, 1989)), there is no unbound protein present in the cytosol. Competition studies established that the complex was highly specific for the stem–loop structure, and changing the conserved U residues in the loop (Fig. 3A, lanes 3 and 4) or the sequence of the stem (Pandey *et al.*, 1991) reduced binding to undetectable levels. Changing the AU base

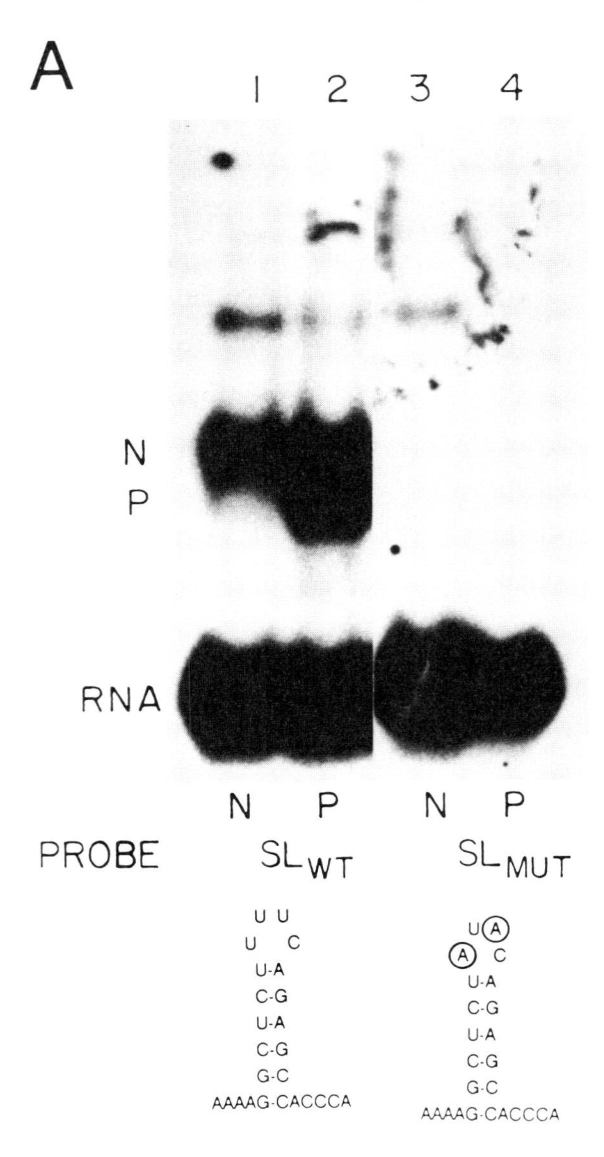

FIGURE 3 A protein binds specifically to the 3′ end of histone mRNA. (A) A 30-nt probe containing the stem–loop sequence (SL_{WT}, lanes 1 and 2) or a mutant stem–loop with two bases in the loop replaced (SL_{MUT}, lanes 3 and 4) was incubated with 40 μg of protein extracted from polyribosomes (P, lanes 2 and 4) or 40 μg of nuclear proteins (N, lanes 1 and 3). The complexes were resolved by electrophoresis on a nondenaturing gel and the specific nuclear (N) and polysomal (P) complexes detected (continues).

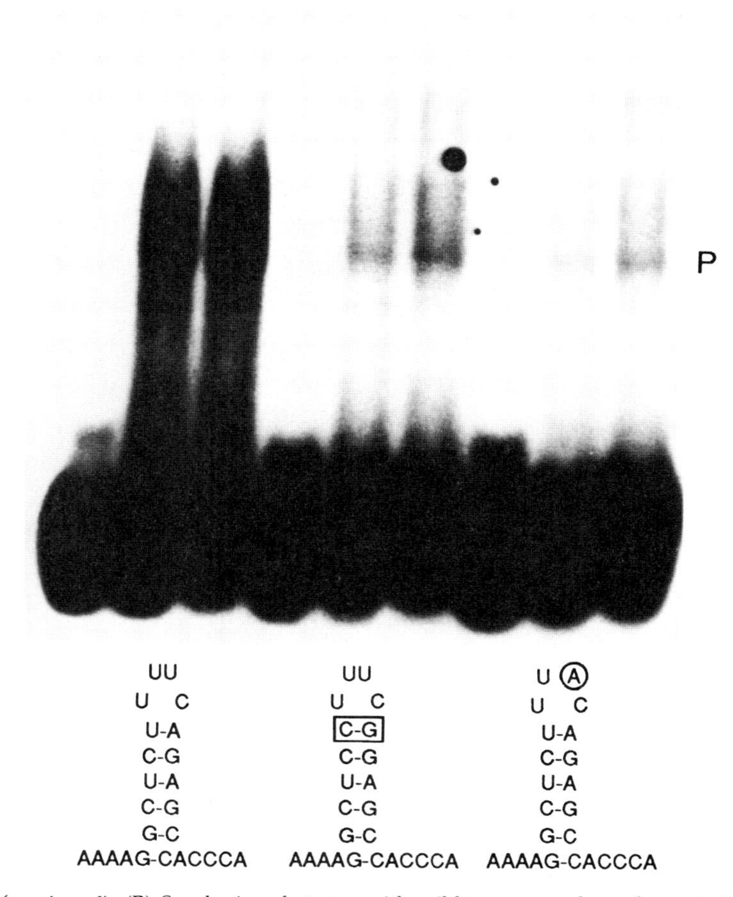

FIGURE 3 (continued). (B) Synthetic substrates with wild-type stem–loop (lanes 1–3) sequence, with the AU base pair at the top of the stem replaced with CG (lanes 4–6), and with the third base in the loop replaced with A (lanes 7–9) were incubated with buffer (lanes 1,4,7), 10 μgm (lanes 2,5,8), or 20 μg (lanes 3,6,9) of protein extracted from polyribosomes and the complexes resolved by electrophoresis. The different probes are shown under the figure. (continues).

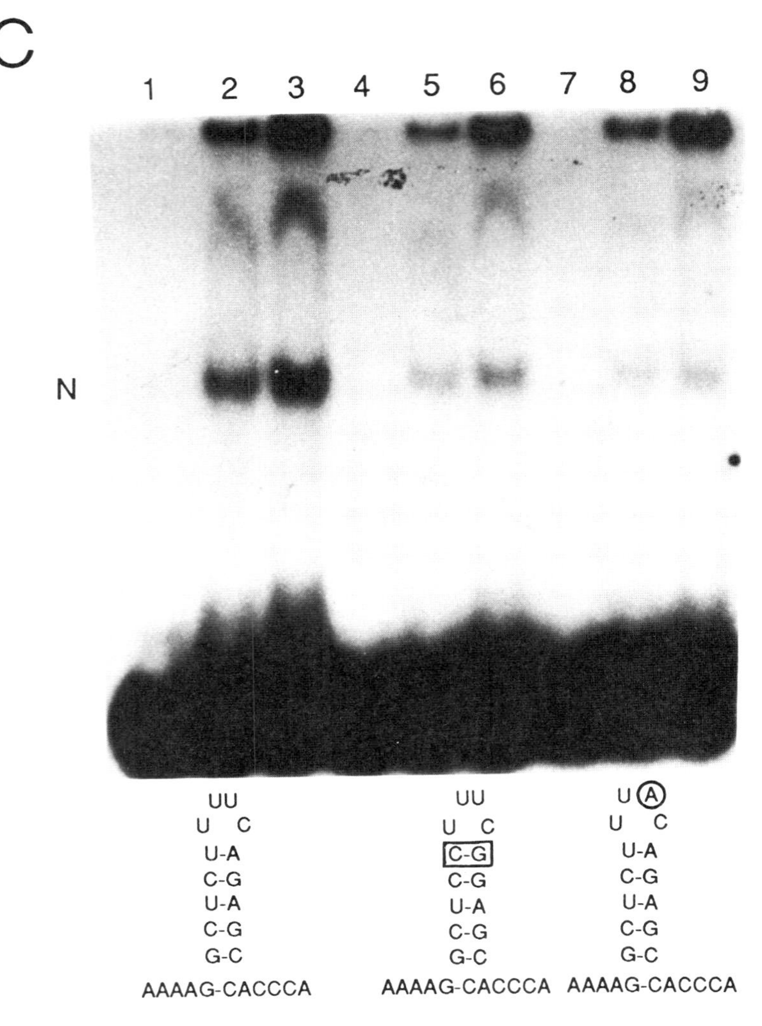

FIGURE 3 (continued). (C) The same substrates used in B were incubated in buffer (lanes 1,4,7) or with 20 μg (lanes 2,5,8) or 40 μgm of nuclear protein and the complexes resolved by gel electrophoresis. The different probes are shown under the figure. (Data taken from Pandey *et al.*, 1991).

pair at the top of the stem (Figs. 3B and 3C, lanes 4–6) or changing the third base in the loop to an A (Figs. 3B and 3C, lanes 7–9) reduced but did not abolish binding of both the nuclear and the polyribosomal proteins.

Cross-linking studies with UV light demonstrated that a protein of about 50 kDa, which we call the stem–loop binding protein (SLBP), could be cross-linked to the synthetic RNA probe, and this cross-linking was dependent on the stem–loop structure (Pandey *et al.*, 1991). The SLBP does not have detectable nuclease activity, but rather protects the synthetic RNA from attack by ribonuclease *in vitro* and may also protect the mRNA from nuclease attack *in vivo* (Pandey *et al.*, 1991). There is probably one SLBP bound to each histone mRNA.

Thus, in the cell, histone mRNA exists as an mRNP with at least one protein, the SLBP, bound to the 3′ end of all histone mRNAs. This mRNP is the physiological substrate for degradation of histone mRNA. Other mRNAs also exist as mRNPs, containing, for example, the poly(A)-binding protein (PABP) bound at the 3′ end. Removal of the poly(A) tail is the initial step in degradation of some mRNAs (Bernstein and Ross, 1989) (Chapter 7). Thus there are parallels between the initial steps in degradation of histone mRNAs and other mRNAs.

B. Role of the 50-kDa Stem–Loop Binding Protein in Histone mRNA Degradation

In the *in vitro* degradation experiments described above the substrate was polysomal histone mRNA, which probably is bound to the SLBP. While the role of the 50-kDa SLBP in the degradation process is completely unknown, there are two possible extreme models. First, the SLBP could protect the histone mRNA from degradation; if so, removal of the protein could be a necessary (and rate-determining) step in degradation of the mRNA (Fig. 4A). Alternatively, the protein–RNA complex could be the substrate for the nuclease that makes the initial cleavage. One then has to postulate that the structure of the SLBP–RNA complex is modified when DNA synthesis is inhibited, resulting in increased accessibility to a nuclease (Fig. 4B). Either of these models predicts an alteration of SLBP–RNA interaction prior to the initial cleavage of histone mRNA.

The nuclease responsible for the initial and/or the subsequent cleavage does not have to be specific for histone mRNA. Indeed the biochemical evidence suggests that the same nuclease is capable of degrading histone, c-*myc*, and globin mRNA (Ross and Kobs, 1986). The nuclease could be associated with the ribosome (even as a ribosomal "subunit," similar to the signal recognition particle) with the function of degrading any mRNA that has been "targeted" for degradation (see Fig. 4). This model is attractive since there is not a particular subset of ribosomes that translate only histone mRNAs; presumably any ribosome is capable of translating any mRNA. The probability that an mRNA is degraded in any particular round of translation is low, since even with a half-life of 8 min an mRNA could be translated an average of 50–100 times prior to degradation. Either there is a very low probability that the nuclease cleaves the mRNA during a particular translation cycle or, more likely, only a small (random) subset of ribosomes contains the nuclease.

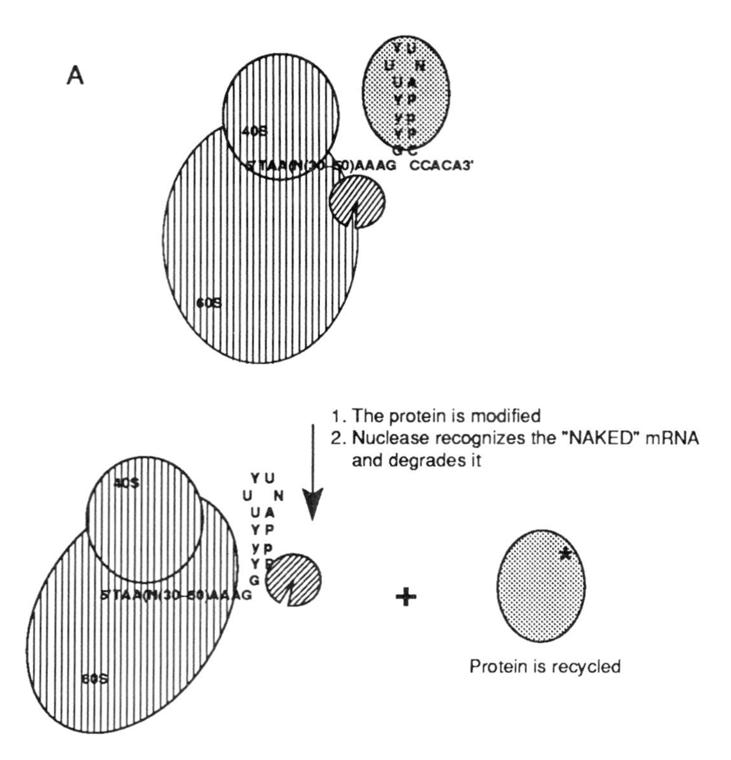

FIGURE 4 Models for histone mRNA degradation. (A) The stem–loop binding protein (SLBP) is depicted binding to the stem–loop at the 3′ end of the mRNA. When DNA synthesis is inhibited, the SLBP is removed from the mRNA, possibly after a covalent modification (*). The nuclease, which is associated with the ribosome, attacks the mRNA, degrading it 3′ to 5′. (B) The SLBP is bound to the 3′ end of the mRNA. When DNA synthesis is inhibited, the protein is modified, exposing the stem–loop to the ribosome-bound nuclease. Following 3′ to 5′ degradation of the mRNA, the protein is recycled (continues).

Figure 4 shows a possible model for the role of the 3′ end together with the 50-kDa protein in regulating histone mRNA degradation. Inhibition (or completion) of DNA synthesis leads to a modification of the 50-kDa protein. This could either remove the 50-kDa protein from the mRNA or simply alter the structure of the SLBP–RNA complex, resulting in enhanced accessibility of the stem–loop at the 3′ end of the histone mRNA to the nuclease associated with the ribosome. The distance effect is a result of the requirement that the ribosome bring the nuclease close to the 3′ end of the mRNA. Once the protein is removed from the mRNA (as a result of either modification or degradation of the mRNA), the protein is translocated back to the nucleus, where it presumably can be recycled (see below and Fig. 5).

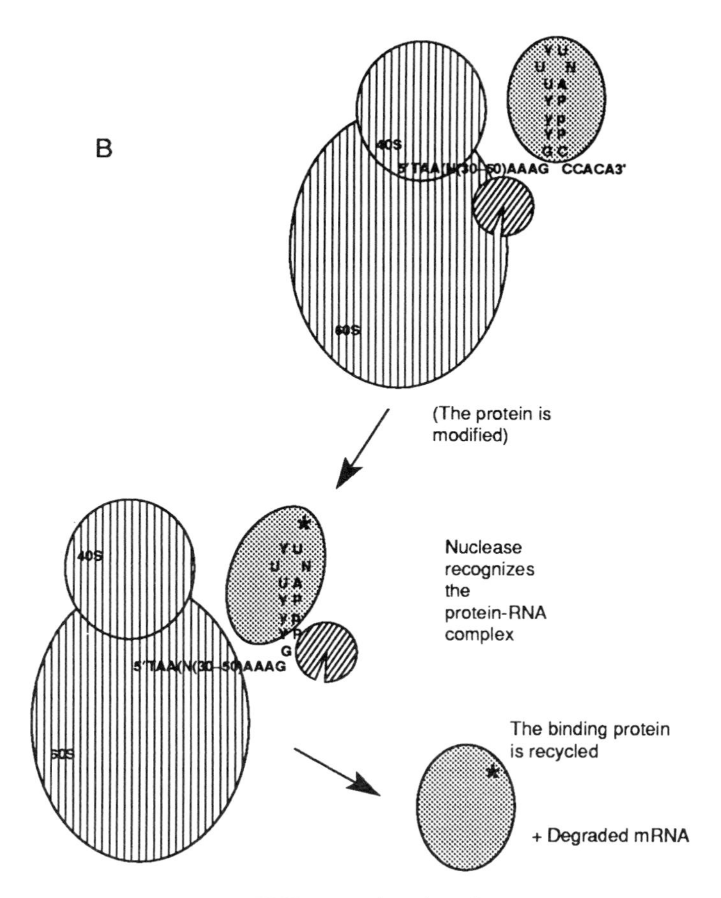

FIGURE 4 (continued).

C. Relationship between the Poly(A) Binding Protein and the Stem–Loop Binding Protein in mRNA Function

An obvious similarity between the PABP and the SLBP is the fact that both proteins bind to an unusual yet conserved feature of the RNA that is at the 3′ end of the untranslated region. In addition to the possible role of the poly(A) tail in mRNA degradation (Chapter 7), recent studies have demonstrated that the poly(A) tail has a number of important functions in translation (Sachs and Davis, 1989; Munroe and Jacobson, 1990; Jackson and Standart, 1990). During the maturation of *Xenopus* oocytes, cytoplasmic polyadenylation is an essential prerequisite for recruitment of mRNAs into

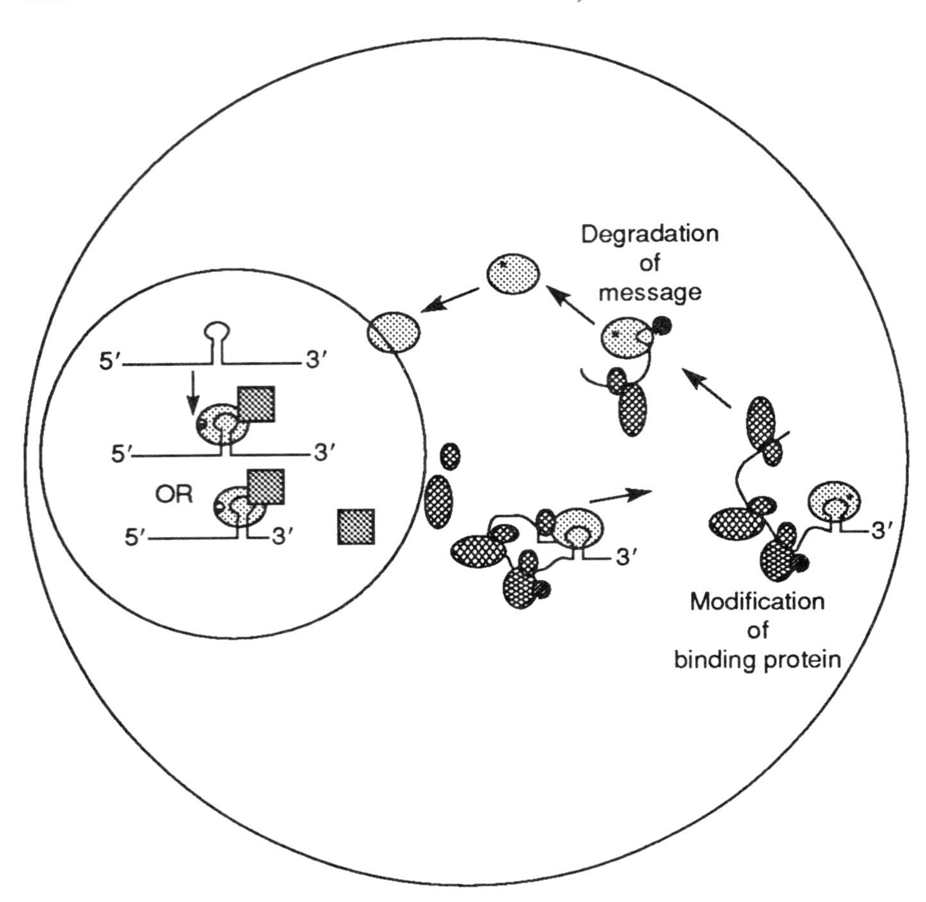

FIGURE 5 Model for functions of SLBP. The histone pre-mRNA is formed in the nucleus, where it associates with the SLBP. The SLBP could associate either before or after the histone pre-mRNA is processed. Since the SLBP has a higher affinity for the stem–loop when the stem–loop is located at the 3′ end than at an internal position in the mRNA (Pandey *et al.*, 1991), it may not associate with the stem–loop until after the mRNA is processed. The nuclear complex has an additional component. The mRNA exits the nucleus, where the SLBP assists in initiation of translation. After modification of the SLBP, the mRNA is degraded, and the SLBP enters the nucleus, where it can bind to another histone mRNA after reversal of the protein modification. ⁘ , ▩ , nuclear protein; *, modification; ● , nuclease.

polyribosomes (McGrew and Richter, 1990; Wickens, 1990). The initiation factor eIF-4F binds to the 5′ cap of mRNA and initiates the joining of initiation factors to the 40S ribosome, forming the ternary complex with the initiator tRNA (Rhoads, 1988). Recently Gallie demonstrated the PABP and the poly(A) tail interact not only with each other but also with the cap and cap binding protein (Gallie, 1991). PABP is necessary for cell viability, and cells lacking PABP have a defect in translation (Sachs and Davis, 1989). Thus the requirement of the poly(A) tail for translation is mediated by the poly(A) binding protein. This requirement for the poly(A) binding protein in translation is not seen in *in vitro* translation systems (Gallie, 1991). Any essential functions of the poly(A) tail in translation must be mediated by another sequence in the histone mRNA.

We have recently shown that the stem–loop in histone mRNA is also required for the efficient translation of histone mRNA *in vivo* (Sun *et al.*, 1992). We constructed genes that contain two 3′ end signals: a U1 snRNA 3′ end signal followed by a histone 3′ end signal. These genes produce two defined transcripts: one ending at the U1 snRNA 3′ end and the other, 100 nt longer, ending at the histone 3′ end. Both transcripts encode identical polypeptides. The transcripts ending at the histone 3′ end are all associated with polyribosomes, while the majority of the transcripts ending at the U1 snRNA 3′ end are not associated with polyribosomes (Sun *et al.*, 1992). Thus, it is likely that the SLBP fulfills the same functions that the PABP fulfills in translation.

A further parallel between the poly(A) tail and the 3′ end of histone mRNA is that the translation-dependent step in c-*myc* mRNA degradation is removal of the poly(A) tail (Laird-Offringa *et al.*, 1990). One prediction that arises from this observation is that the PABP and the SLBP will share some common domains, possibly including a site for binding of a nuclease that could remove the 3′ end of the mRNA, and a site for cap binding.

D. Relationship between the Nuclear Stem–Loop Binding Protein and the Cytoplasmic Stem–Loop Binding Protein

There is also a specific complex formed when nuclear proteins are incubated with the same synthetic 30-nt RNA stem–loop. Although this complex has a lower electrophoretic mobility than the polysomal complex, a 50-kDa nuclear protein of a molecular weight identical to that of the polysomal protein is cross-linked by UV light to the RNA (Pandey *et al.*, 1991). The nuclear and polyribosomal proteins are also similar as judged by partial protease digestion (our unpublished results with J. Sun). The data suggest that the nuclear complex consists of SLBP plus an additional factor. This larger complex associates with the histone mRNA (either before or after processing), but only the SLBP accompanies the histone mRNA to the cytoplasm, where it helps direct the histone mRNA onto polyribosomes (Fig. 5). The 50-kDa SLBP is thus an integral part of the histone mRNP throughout its lifetime. Indeed, removal of the 50-kDa SLBP from the mRNA may be the initial step in degradation of histone mRNA. The SLBP then returns to the nucleus, where it can bind with another histone transcript.

E. Autoregulation of Histone mRNA Stability by Histone Proteins?

A suggestion made early in the study of regulation of histone mRNA metabolism is that histone proteins may "autoregulate" the levels of histone mRNA (Butler and Mueller, 1973). The autoregulation model proposes that when DNA synthesis is inhibited, histone proteins accumulate (presumably in the cytoplasm) and the accumulation of histone proteins triggers the degradation of histone mRNA (Peltz and Ross, 1987; Sariban *et al.*, 1985). This interpretation was based originally on the stabilizing effect of inhibiting protein synthesis, which also supports the translation dependence model. We stress that these two models are not mutually exclusive; certainly the evidence strongly suggests that tubulin mRNA is autoregulated by a mechanism dependent on translation (Chapter 10) and the same could certainly be true for histone mRNA.

Mechanistically, one can distinguish between models in which histone proteins directly regulate histone mRNA degradation and models in which the histone proteins are intermediates in a signaling pathway but do not directly participate in the degradation process. However, these two possibilities may be very difficult to distinguish experimentally. Evidence supporting the involvement of histone proteins in histone mRNA degradation is twofold. First, the levels of histone mRNA are quantitatively closely coupled with the rate of DNA replication. Altering either the rate of DNA synthesis or the rate of histone protein synthesis slightly, using low doses of inhibitors of DNA or protein synthesis, causes the cell to adjust histone mRNA levels to balance the rates of DNA and histone synthesis (Bonner *et al.*, 1988; Sariban *et al.*, 1985). A convenient way to accomplish this is for the cell to monitor closely the levels of "free" histone protein. Second, Jeff Ross and colleagues (Peltz and Ross, 1987) have shown that addition of histone proteins to a cell-free mRNA degradation system results in a selective, rapid degradation of histone mRNA. These results are certainly consistent with an autoregulatory model of histone mRNA degradation, assuming that this cell-free system accurately reflects the events that occur in the cell.

Two considerations make it difficult to see how histone proteins could participate directly in regulation of histone mRNA degradation. First, it is not clear if histone proteins will accumulate in the cytoplasm when DNA replication is inhibited. Most studies suggest that histone proteins are targeted to the nucleus immediately after synthesis. Thus, it seems that if histone proteins are directly involved in regulation of mRNA degradation, then the import of histone proteins into the nucleus should also be disrupted by inhibition of DNA replication. Each of the histone mRNAs has the same 3′ end, and hence all are regulated by a common signal. The presence of a common signal suggests that if histone proteins do directly participate in regulation of histone mRNA levels, then the accelerated degradation should be initiated by an aggregate of several different histone proteins, e.g., the nucleosome core histone octamer. It is thus intriguing that Ross and co-workers used all four core histone proteins to obtain regulation of degradation *in vitro* (Peltz and Ross, 1987). In addition the histone proteins had to be incubated with a cytosol fraction before they could exert their effect (Peltz and Ross, 1987). This incubation could lead to octamer assembly, histone modification, or the generation of an intermediate signal. Thus it is very possible that accumulation of excess histone proteins is an essential part of the response leading to accelerated degradation of histone mRNA, but there could be an intermediate signal produced that then interacts directly with SLBP and/or the 3′ end of the histone mRNA.

F. Coupling of Histone mRNA Degradation and DNA Replication

One can readily imagine alternative signals for coupling DNA replication and histone mRNA levels. The inhibition or completion of DNA replication in the nucleus could generate a signal that is then communicated to the cytoplasm. Alternatively, the disruption of the supply of substrates for DNA replication, which are synthesized in the cytoplasm, could result in a signal that triggers histone mRNA degradation. A wide variety of inhibitors of DNA replication, including inhibitors of DNA polymerase (aphidicolin, which targets DNA polymerases α and δ) and inhibitors of deoxynucleotide metabolism (methotrexate, fluorodeoxyuridine, and hydroxyurea), have identical effects on histone mRNA metabolism (Graves and Marzluff, 1984). These results demonstrate that any reagent that rapidly inhibits DNA replication leads to the rapid degradation of histone mRNA. All of these reagents could also lead to an accumulation of histone protein, since cessation of DNA synthesis leads to cessation of chromatin assembly, and possibly an accumulation of excess histone proteins.

It is not possible to determine whether the signal regulating histone mRNA levels is generated by accumulation of excess histone proteins or whether the signal is generated by inhibiting DNA replication. The rapidity of the response strongly suggests a signaling system based on reversible covalent modification or an allosteric effect due to binding of a small molecule. Synthesis of macromolecules involved in the degradation process (e.g., a specific nuclease) is probably not required, and it is likely that none of the components of the degradation machinery are consumed during the process, since the regulation of histone mRNA half-life is readily reversible (Graves and Marzluff, 1984).

IV. Cell-Cycle Specific Signals and Histone mRNA Degradation

The accelerated degradation of histone mRNA is a cell-cycle regulated process. The signal resulting in histone mRNA degradation is generated at the S–G2 border. Strikingly, the half-life of the small amount of histone mRNA present in G1-phase cells is similar to the half-life of the mRNA in S-phase cells and is not abnormally low (Harris *et al.*, 1991; Morris *et al.*, 1991). In addition, there is evidence that the activation of histone mRNA biosynthesis, and hence histone mRNA accumulation, occurs prior to the entry of the cell into the S-phase (Harris *et al.*, 1991; DeLisle *et al.*, 1983). This result makes sense since it allows the cell to start to accumulate histone proteins (and more importantly the ability to rapidly synthesize histone proteins), which are required immediately upon initiation of DNA replication. It seems very likely that the capacity to rapidly degrade histone mRNA is present only after cells have entered the S-phase, and it follows that this capacity is lost (by destruction or inactivation of a key component) when cells go through mitosis (Harris *et al.*, 1991).

Many events in the cell cycle are regulated by interaction of various cyclins with cdc2 kinase and its analogs (Norbury and Nurse, 1989; Draetta, 1990). The timing of the degradation of histone mRNA does not correspond to that of any as yet identified cyclins or protein kinases. However, in the past 2 years, a large number of different cyclins involved in regulation of the G1- to S-phase transition have been identified, and one of these could be involved in the acquisition of the ability to rapidly degrade histone mRNA. A number of proteins (e.g., cyclins A and B) are rapidly degraded during mitosis, and it is possible that an essential component required for rapid destruction of histone mRNA is destroyed or inactivated at this time also. Only after resynthesis of this component during the next cell cycle would the cell then reacquire the ability to rapidly degrade histone mRNA.

V. Why Degrade Histone mRNA?

The extreme conservation of the 3′ end of histone mRNA in all animal species suggests that this structure has a critical function. The tight coordination of histone mRNA levels with DNA replication, efficiently accomplished by the unique 3′ end, serves to balance the production of histone proteins with DNA replication. Presumably, the variant histone mRNAs serve to replace damaged histones and possibly also to remodel chromatin during DNA repair. In contrast to the animal cell histone mRNAs, histone mRNAs in lower eukaryotes (Fahrner *et al.*, 1980; Bannon *et al.*, 1983) and in plants (Nakayama *et al.*, 1989) are polyadenylated. The mechanisms of cell-cycle regulation of histone mRNA in these organisms must be fundamentally different from the regulation in animal cells.

Histone mRNAs are destroyed and resynthesized in each mammalian cell cycle. One possible reason for the strict coordinate regulation of histone protein synthesis and DNA replication is that the accumulation of excess histone proteins is deleterious to the cell. In oocytes from many species, histone mRNAs and histone proteins accumulate in the absence of DNA synthesis to provide a supply of histone proteins for the remodeling of the sperm chromatin (Poccia *et al.*, 1984; Philpott *et al.*, 1991; Graves *et al.*, 1985; Kaye and Church, 1983) and for the initial cleavage cycles (Woodland and Adamson, 1977). In frog oocytes, which store massive amounts of histone proteins, the histone proteins are sequestered by binding to acidic proteins, nucleoplasmin, and N1 (Dilworth *et al.*, 1987), and it is possible that this specific storage mechanism prevents any deleterious effects of accumulation of excess histone proteins.

Much work remains to be done before we understand the molecular mechanism of histone mRNA degradation. In addition to learning about the chemistry of a fundamentally important reaction regulating gene expression, we also will need to learn about some of the signals that are produced as cells exit the S-phase and enter the G2-phase.

Acknowledgments

We thank Drs. Reed Graves, Niranjan Pandey, and Jianhua Sun for many helpful discussions and insights and for access to unpublished results. This work was supported by Grant GM29832 to W.F.M. and Postdoctoral Fellowship GM14368 to R.J.H. from the NIH.

References

Alterman, R. B., Ganguly, S., Schulze, D. H., Marzluff, W. F., and Skoultchi, A. I. (1984). Cell cycle regulation of mouse histone H3 mRNA metabolism. *Mol. Cell. Biol.* **4,** 123–132.

Bannon, G. A., Calzone, F. J., Bower, J. K., Allis, C. D., and Gorovsky, M. A. (1983). Multiple independently regulated, polyadenylated messages for histone H3 and H4 in Tetrahymena. *Nucleic Acids Res.* **11,** 3903.

Bernstein, P., and Ross, J. (1989). Poly(A), poly(A) binding protein and the regulation of mRNA stability. *Trends Biochem. Sci.* **14,** 373–377.

Birnstiel, M. L., Busslinger, M., and Strub, K. (1985). Transcription termination and 3′ processing: The end is in site. *Cell* **41,** 349–359.

Bonner, W. M., Wu, R. S., Panusz, H. T., and Muneses, C. (1988). Kinetics of accumulation and depletion of soluble newly synthesized histone in the reciprocal regulation of histone and DNA synthesis. *Biochemistry* **27,** 6542–6550.

Borun, T. W., Franco, G., Kozo, A., Zweidler, Z., and Baglioni, C. (1975). Further evidence of transcriptional and translational control of histone messenger RNA during the HeLa S3 cycle. *Cell* **4,** 59–67.

Breindl, M., and Gallwitz, D. (1974). On the translational control of histone synthesis. *Eur. J. Biochem.* **45,** 91–97.

Brush, D., Dodgson, J. B., Choi, O. R., Stevens, P. W., and Engel, J. D. (1985). Replacement variant histone genes contain intervening sequences. *Mol. Cell. Biol.* **5,** 1307–1317.

Butler, W. B., and Mueller, G. C. (1973). Control of histone synthesis in HeLa cells. *Biochim. Biophys. Acta* **294,** 481–496.

Cheng, G., Nandi, A., Clerk, S., and Skoultchi, A. I. (1989). Different 3′-end processing produces two independently regulated mRNAs from a single H1 histone gene. *Proc. Natl. Acad. Sci. USA* **86,** 7002–7006.

DeLisle, A. J., Graves, R. A., Marzluff, W. F., and Johnson, L. F. (1983). Regulation of histone mRNA production and stability in serum stimulated mouse fibroblasts. *Mol. Cell. Biol.* **3,** 1920–1929.

Dilworth, S. M., Black, S. J., and Laskey, R. A. (1987). Two complexes that contain histones are required for nucleosome assembly in vitro: Role of nucleoplasmin and N1 in Xenopus egg extracts. *Cell* **51,** 1009–1018.

Draetta, G. (1990). Cell cycle control in eukaryotes: Molecular mechanisms of cdc2 activation. *Trends Biochem. Sci.* **15,** 378–383.

Fahrner, K., Yarger, J., and Hereford, L. (1980). Yeast histone mRNA is polyadenylated. *Nucleic Acids Res.* **8,** 5725.

Gallie, D. R. (1991). The cap and poly(A) tail function synergistically to regulate mRNA translational efficiency. *Genes Dev.* **5,** 2108–2116.

Gallwitz, D. (1975). Kinetics of inactivation of histone mRNA in the cytoplasm after inhibition of DNA replication in synchronised HeLa cells. *Nature* **257,** 247–248.

Gay, D. A., Sisodia, S. S., and Cleveland, D. W. (1989). Autoregulatory control of β-tubulin mRNA stability is linked to translation elongation. *Proc. Natl. Acad. Sci. USA* **86,** 5763–5767.

Gick, O., Krämer, A., Keller, W., and Birnstiel, M. L. (1986). Generation of histone mRNA 3′ ends by endonucleolytic cleavage of the pre-mRNA in a snRNP-dependent in vitro reaction. *EMBO J.* **5,** 1319–1326.

Gick, O., Krämer, A., Vasserot, A., and Birnstiel, M. L. (1987). Heat-labile regulatory factor is required for 3′ processing of histone precursor mRNAs. *Proc. Natl. Acad. Sci. USA* **84,** 8937–8940.

Gong, Z., and Brandhorst, B. P. (1988). Stabilization of tubulin mRNA by inhibition of protein synthesis in sea urchin embryos. *Mol. Cell. Biol.* **8,** 3518–3525.

Graves, R. A., and Marzluff, W. F. (1984). Rapid, reversible alterations in histone gene transcription and histone mRNA levels in mouse myeloma cells. *Mol. Cell. Biol.* **4,** 351–357.

Graves, R. A., Marzluff, W. F., Giebelhaus, D. H., and Schultz, G. A. (1985). Quantitative and qualitative changes in histone gene expression during early mouse embryo development. *Proc. Natl. Acad. Sci. USA* **82,** 5685–5689.

Graves, R. A., Pandey, N. B., Chodchoy, N., and Marzluff, W. F. (1987). Translation is required for regulation of histone mRNA degradation. *Cell* **48,** 615–626.

Harris, M. E., Böhni, R., Schneiderman, M. H., Ramamurthy, L., Schümperli, D., and Marzluff, W. F. (1991). Regulation of histone mRNA in the unperturbed cell cycle: Evidence suggesting control at two posttranscriptional steps. *Mol. Cell. Biol.* **11,** 2416–2424.

Hatch, C. L., and Bonner, W. M. (1990). The human histone H2A.Z gene. Sequence and regulation. *J. Biol. Chem.* **265,** 15211–15218.

Heintz, N., Sive, H. L., and Roeder, R. G. (1983). Regulation of human histone gene expression: Kinetics of accumulation and changes in the rate of synthesis and in the half-lives of individual histone mRNAs during the HeLa cell cycle. *Mol. Cell. Biol.* **3,** 539–550.

Jackson, R. J., and Standart, N. (1990). Do the poly(A) tail and 3′ untranslated region control mRNA translation. *Cell* **62,** 15–24.

Kaye, P. L., and Church, R. B. (1983). Uncoordinate synthesis of histones and DNA by mouse eggs and preimplantation embryos. *J. Exp. Zool.* **226,** 231–237.

Kirsh, A. L., Groudine, M., and Challoner, P. B. (1989). Polyadenylation and U7 snRNP-mediated cleavage: Alternative modes of RNA 3′ processing in two avian histone H1 genes. *Genes Dev.* **3,** 2172–2179.

Laird-Offringa, I. A., De Wit, C. L., Elfferich, P., and Van der Eb, A. J. (1990). Poly(A) tail shortening is the translation-dependent step in c-*myc* mRNA degradation. *Mol. Cell. Biol.* **10,** 6132–6140.

Levine, B. J., Chodchoy, N., Marzluff, W. F., and Skoultchi, A. I. (1987). Coupling of replication type histone mRNA levels to DNA synthesis requires the stem–loop sequence at the 3′ end of the mRNA. *Proc. Natl. Acad. Sci. USA* **84,** 6189–6193.

Lüscher, B., Stauber, C., Schindler, R., and Schümperli, D. (1985). Faithful cell-cycle regulation of a recombinant mouse histone H4 gene is controlled by sequences in the 3′-terminal part of the gene. *Proc. Natl. Acad. Sci. USA* **82,** 4389–4393.

Malter, J. S. (1989). Identification of an AUUUA-specific messenger RNA binding protein. *Science* **246,** 664–666.

Manniron i, C., Bonner, W. M., and Hatch, C. L. (1989). H2A.X, a histone isoprotein with a conserved C-terminal sequence, is encoded by a novel mRNA with both DNA replication type and polyA 3′ processing signals. *Nucleic Acids Res.* **17,** 9113–9126.

Marzluff, W. F. (1986). Evolution of histone genes. *In* "DNA Systematic I: Evolution" (S. K. Dutta, Ed.), pp 139–169. CRC Press, Boca Raton, Florida.

Marzluff, W. F. (1992). Histone 3′ ends: Essential and regulatory functions. *Gene Exp.* **2,** 93–97.

Marzluff, W. F., and Pandey, N. B. (1988). Multiple levels of regulation of histone mRNA concentrations. *Trends Biochem. Sci.* **13,** 49–52.

Maxson, R., Mohun, T. J., Cohn, R., and Kedes, L. (1983). Expression and organization of histone genes. *Annu. Rev. Genet.* **17,** 239–277.

McGrew, L. L., and Richter, J. D. (1990). Translational control by cytoplasmic polyadenylation during *Xenopus* oocyte maturation: Characterization of *cis* and *trans* elements and regulation by cyclin/MPF. *EMBO J.* **9,** 3743–3751.

Morris, T. D., Weber, L. A., Hickey, E., Stein, G. S., and Stein, J. L. (1991). Changes in the stability of a human H3 histone mRNA during the HeLa cell cycle. *Mol. Cell. Biol.* **11,** 544–553.

Mowry, K. L., Oh, R., and Steitz, J. A. (1989). Each of the conserved sequence elements flanking the cleavage site of mammalian histone pre-mRNAs has a distinct role in the 3′-end processing reaction. *Mol. Cell. Biol.* **9,** 3105–3108.

Mowry, K. L., and Steitz, J. A. (1987). Identification of the human U7 snRNP as one of several factors involved in the 3′ end maturation of histone premessenger RNA's. *Science* **238,** 1682–1687.

Munroe, D., and Jacobson, A. (1990). mRNA poly(A) tail, a 3′ enhancer of translational initiation. *Mol. Cell Biol.* **10,** 3441–3455.

Nakayama, T., Ohtsubo, N., Mikami, K., Kawata, T., Tabata, T., Kanazawa, H., and Iwabuchi, M. (1989). *cis*-acting sequences that modulate transcription of wheat histone H3 gene and 3′ processing of H3 premature mRNA. *Plant Cell Physiol.* **30,** 825–832.

Norbury, C. J., and Nurse, P. (1989). Control of the higher eukaryote cell cycle by $p34^{cdc2}$ homologues. *Biochim. Biophys. Acta* **989,** 85–95.

Pandey, N. B., Sun, J.-H., and Marzluff, W. F. (1991). Different complexes are formed on the 3′ end of histone mRNA in nuclear and polysomal extracts. *Nucleic Acids Res.* **19,** 5653–5659.

Pandey, N. B., and Marzluff, W. F. (1987). The stem–loop structure at the 3′ end of histone mRNA is necessary and sufficient for regulation of histone mRNA stability. *Mol. Cell. Biol.* **7,** 4557– 4559.

Peltz, S. W., Brewer, G., Kobs, G., and Ross, J. (1987). Substrate specificity of the exonuclease activity that degrades H4 histone mRNA. *J. Biol. Chem.* **262,** 9382–9388.

Peltz, S. W., Brewer, G., Groppi, V., and Ross, J. (1989). Exonuclease activity that degrades histone mRNA is stable when DNA or protein synthesis is inhibited. *Mol. Biol. Med.* **6,** 227–238.

Peltz, S. W., and Ross, J. (1987). Autogenous regulation of histone mRNA decay by histone proteins in a cell-free system. *Mol. Cell. Biol.* **7,** 4345–4356.

Philpott, A., Leno, G. H., and Laskey, R. A. (1991). Sperm decondensation in Xenopus egg cytoplasm is mediated by nucleoplasmin. *Cell* **65,** 569–578.

Poccia, D., Greenough, T., Green, G. R., Nash, E., Erickson, J., and Gibbs, M. (1984). Remodeling of sperm chromatin following fertilization: Nucleosome repeat length and histone variant transitions in the absence of DNA synthesis. *Dev. Biol.* **104,** 274–286.

Rhoads, R. E. (1988). Cap recognition and the entry of mRNA into the protein synthesis initiation cycle. *Trends Biochem. Sci.* **13,** 52–56.

Ross, J., Peltz, S. W., Kobs, G., and Brewer, G. (1986). Histone mRNA degradation in vivo: The first detectable step occurs at or near the 3′ terminus. *Mol. Cell. Biol.* **6,** 4362–4371.

Ross, J., Kobs, G., Brewer, G., and Peltz, S. W. (1987). Properties of the exonuclease activity that degrades H4 histone mRNA. *J. Biol. Chem.* **262,** 9374–9381.

Ross, J., and Kobs, G. (1986). H4 histone messenger RNA decay in cell-free extracts initiates at or near the 3′ terminus and proceeds 3′ to 5′. *J. Mol. Biol.* **188,** 579–593.

Rouault, T. A., Hentze, M. W., Caughman, S. W., Harford, J. B., and Klausner, R. D. (1988). Binding of a cytosolic protein to the iron-responsive element of human ferritin messenger RNA. *Science* **241,** 1207–1210.

Sachs, A. B., and Davis, R. W. (1989). The poly(A) binding protein is required for poly(A) shortening and 60S ribosomal subunit-dependent translation initiation. *Cell* **58,** 857–867.

Sariban, E., Wu, R. S., Erickson, L. C., and Bonner, W. M. (1985). Interrelationships of protein and DNA syntheses during replication of mammalian cells. *Mol. Cell. Biol.* **5,** 1279–1286.

Shyu, A.-B., Greenberg, M. E., and Belasco, J. G. (1989). The c-*fos* transcript is targeted for rapid decay by two distinct mRNA degradation pathways. *Genes Dev.* **3,** 60–72.

Sittman, D. B., Graves, R. A., and Marzluff, W. F. (1983a). Structure of a cluster of mouse histone genes. *Nucleic Acids Res.* **11,** 6679–6697.

Sittman, D. B., Graves, R. A., and Marzluff, W. F. (1983b). Histone mRNA concentrations are regulated at the level of transcription and mRNA degradation. *Proc. Natl. Acad. Sci. USA* **80,** 1849–1853.

Stahl, H., and Gallwitz, D. (1977). Fate of histone messenger RNA in synchronized HeLa cells in the absence of initiation of protein synthesis. *Eur. J. Biochem.* **72,** 385–392.

Stauber, C., and Schümperli, D. (1988). 3′ processing of pre-mRNA plays a major role in proliferation-dependent regulation of histone gene expression. *Nucleic Acids Res.* **16,** 9399– 9413.

Stimac, E., Groppi, V. E., Jr., and Coffino, P. (1984). Inhibition of protein synthesis stabilizes histone mRNA. *Mol. Cell. Biol.* **4,** 2082–2087.

Strub, K., and Birnstiel, M. L. (1986). Genetic complementation in the Xenopus oocyte: Co-expression of sea urchin histone and U7 RNAs restores 3′ processing of H3 pre-mRNA in the oocyte. *EMBO J.* **5,** 1675–1682.

Sun, J., -H, Pilch, D. R., and Marzluff, W. F. (1992). The histone mRNA 3′ end is required for localization of histone mRNA to polyribosomes. *Nucleic Acids Res.* **20,** 6057–6066.

Vasserot, A. P., Schaufele, F. J., and Birnstiel, M. L. (1989). Conserved terminal hairpin sequences of histone mRNA precursors are not involved in duplex formation with the U7 RNA but act as a target site for a distinct processing factor. *Proc. Natl. Acad. Sci. USA* **86,** 4345–4349.

Wells, D., and Kedes, L. (1985). Structure of a human histone cDNA: Evidence that basally expressed histone genes have intervening sequences and encode polyadenylated mRNAs. *Proc. Natl. Acad. Sci. USA* **82,** 2834–2838.

Wickens, M. (1990). In the beginning is the end: Regulation of poly(A) addition and removal during early development. *Trends Biochem. Sci.* **15,** 320–324.

Wisdom, R., and Lee, W. (1990). Translation of c-*myc* mRNA is required for its post-transcriptional regulation during myogenesis. *J. Biol. Chem.* **265,** 19015–19021.

Woodland, H. R., and Adamson, E. D. (1977). The synthesis and storage of histones during oogenesis of *Xenopus laevis*. *Dev. Biol.* **57,** 118–135.

Wu, R. S., and Bonner, W. M. (1981). Separation of basal histone synthesis from S phase histone synthesis in dividing cells. *Cell* **27,** 321–330.

Yen, T. J., Machlin, P. S., and Cleveland, D. W. (1988). Autoregulated instability of β-tubulin mRNAs by recognition of the nascent amino terminus of a-tubulin. *Nature* **334,** 580–585.

Zweidler, A. (1984). Core histone variants of the mouse: Primary structure and expression. *In* "Histone Genes: Structure, Organization and Regulation" (G. Stein, W. Stein, and W. Marzluff, Eds.), pp. 373–395. Wiley, New York.

13

mRNA Turnover in *Saccharomyces cerevisiae*

STUART W. PELTZ AND ALLAN JACOBSON

I. Introduction

The yeast *Saccharomyces cerevisiae*, a simple yet typical eukaryote with 16 chromosomes and only 12 Mbp of single-copy DNA (Olson, 1991), has become an organism of choice for analyses of the regulation of gene expression. The ease with which biochemistry and genetics can be applied in this experimental system has led to great strides in our understanding of transcription, RNA splicing, translation, and protein processing (Baker and Schekman, 1989; Grunstein, 1990a, 1990b; Hicke and Schekman, 1990; Ruby and Abelson, 1991; Siegal and Walter, 1989; Young, 1991). Although work on yeast mRNA decay began more than two decades ago (Hutchison *et al.*, 1969; Kuo *et al.*, 1973), it has been only recently that this problem has attracted the efforts of a sufficiently large number of laboratories necessary for a concerted attack on the underlying regulatory mechanisms. The field has now reached a point where a simple and reliable procedure for determination of decay rates has been established, stable and unstable mRNAs have been identified, chimeric mRNAs have been utilized to delineate cis-acting instability determinants, purification and characterization of ribonucleases have begun, and mutants altered in stability regulation have been isolated and used to shed light on trans-acting factors. In this chapter we review all of this progress, finishing with a discussion of an important and, in retrospect, probably obvious conclusion that mRNA turnover and mRNA

Control of Messenger RNA Stability

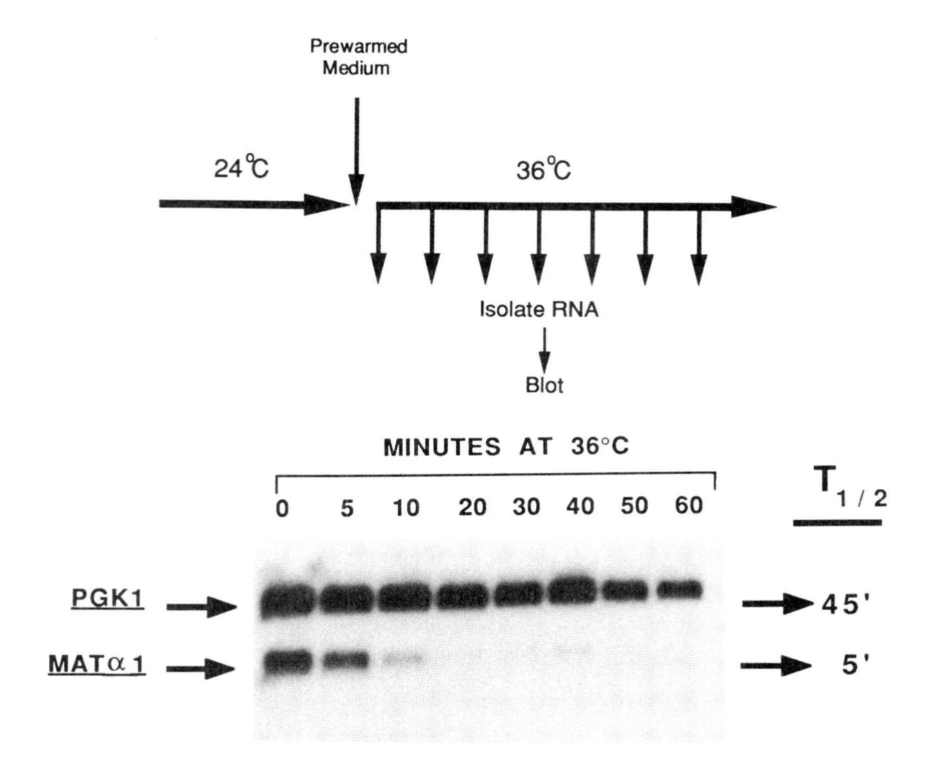

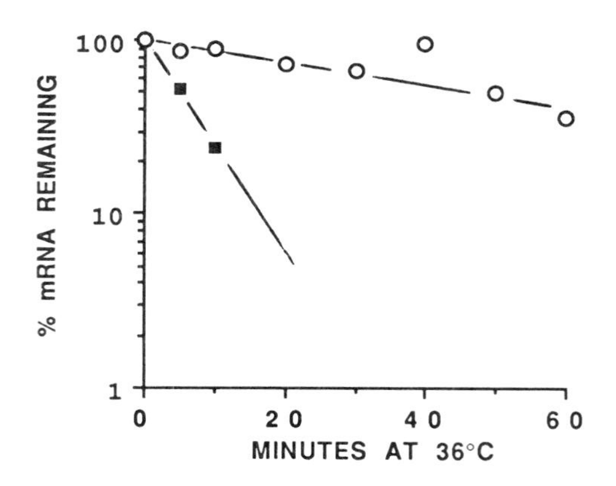

FIGURE 1 Use of a temperature-sensitive RNA polymerase II mutant to measure mRNA decay rates in yeast. (top) Summary of the procedure. In cells harboring the *rpb1-1* temperature-sensitive RNA polymerase II mutation, transcription is inactivated rapidly by a shift from 24 to 36°C. After transcription has been inhibited, decay rates of individual mRNAs are measured by RNA blotting (or nuclease protection) assays of RNA samples isolated at different times

translation are intimately linked. While this conclusion implies that the mechanisms of mRNA turnover may be significantly more complex than anticipated, we are confident that the powerful molecular biology of yeast will be capable of unraveling this complexity.

II. Measurement of mRNA Decay Rates in Yeast

A. Experimental Approaches to the Measurement of mRNA Half-Lives

mRNA decay rates have been measured in yeast by several different functional or chemical assays, including measurement of changes in mRNA labeling kinetics (pulse–chase and approach to steady-state labeling; Bach *et al.*, 1979; Hynes and Phillips, 1976; Kim and Warner, 1983; Losson *et al.*, 1983; Osley and Hereford, 1981; Zitomer *et al.*, 1979), quantitation of the disappearance of mRNA after the inhibition of transcription (Herrick *et al.*, 1990; Kuo *et al.*, 1973; Nonet *et al.*, 1987; Santiago *et al.*, 1986; Tonnesen and Friesen, 1973), or RNA processing (Chia and McLaughlin, 1979; Hutchison *et al.*, 1969; Koch and Friesen, 1979). Although *in vivo* labeling techniques can provide accurate measures of mRNA decay rates, they have several disadvantages, including a requirement for large quantities of radioactively labeled nucleic acid precursors, poor signal to noise ratios for mRNAs that are transcribed at low rates, a dependence on accurate measurements of nucleotide pool sizes, and a failure to provide information on mRNA integrity during the course of an experiment (see Herrick *et al.* (1990) and Parker *et al.* (1991) for further discussion). Recent emphasis has thus been on those procedures that measure mRNA decay rates subsequent to transcriptional inhibition. These include the use of (1) a temperature-sensitive (ts) RNA polymerase II mutant (Herrick *et al.*, 1990; Nonet *et al.*, 1987; Parker *et al.*, 1991), (2) the transcriptional inhibitors phenanthroline and thiolutin (Herrick *et al.*, 1990; Santiago *et al.*, 1986), and (3) the regulatable promoters of the yeast *GAL* genes (Parker *et al.*, 1991; Surosky and Esposito, 1992).

1. Inhibition of Transcription in a Temperature-Sensitive RNA Polymerase II Mutant

In strains harboring the temperature-sensitive *rpb1-1* allele of RNA polymerase II, a shift to 36°C leads to the rapid and selective cessation of mRNA synthesis and to a reduction in the steady-state levels of preexisting

---

after the temperature shift. (middle) Northern analysis of *PGK1* and *MATα1* mRNA levels at different times after the temperature shift. (bottom) Quantitation of the Northern blot. mRNA levels are normalized to the level at $t = 0$ and half-lives are obtained from the 50% intercept of the lines. Closed squares, *MATα1*; open circles, *PGK1*. From Parker and Jacobson (1990).

mRNAs (Nonet *et al.*, 1987). This observation led to the development of a routine procedure for the measurement of mRNA decay rates in which cells growing at 24°C were abruptly shifted to 36°C and changes in the relative amounts of individual mRNAs were quantitated by RNA blotting procedures (Herrick *et al.*, 1990; Nonet *et al.*, 1987; Parker *et al.*, 1991) (see Fig. 1, top). This procedure is applicable to mRNAs of any abundance class or transcription rate and is illustrated for the *MATα1* ($t_{1/2}$=5 min) and *PGK1* ($t_{1/2}$=45 min) mRNAs in Fig. 1 (middle and bottom). Changes in mRNA abundance after the temperature shift can be monitored by RNA dot blotting, RNase protection, Northern blotting, or even *in vitro* translation, but Northern blotting has an important advantage: it permits the simultaneous measurement of the decay rates for several mRNAs (provided that they differ sufficiently in molecular weight) *and* provides data on the integrity of each RNA sample. Potential disadvantages of this protocol are the possible loss of labile turnover factors in the absence of ongoing transcription and the possible complications of heat-shocking cells at 36°C. However, there

TABLE 1 mRNA Decay Rates in a ts RNA Polymerase II Mutant[a]

Unstable			Moderately stable			Stable		
mRNA	(kb)	$t_{1/2}$	mRNA	(kb)	$t_{1/2}$	mRNA	(kb)	$t_{1/2}$
HIS3	(0.83)	7.0	TCM1	(1.17)	11	ACT1	(1.25)	30
STE2	(1.70)	4.0	RP29	(0.62)	11	PGK1	(1.50)	45
DED1	(2.30)	4.0	cDNA74	(1.05)	17	CYH2	(0.60)	43
cDNA90	(1.10)	7.0	PAB1	(2.10)	11	RP51a	(0.58)	28
MFα1	(0.80)	5.0	MFA1	(0.45)	13	ADE3	(3.20)	50
MFA2	(0.45)	2.5	LEU2	(1.6)	14	PYK1		67
MATα1	(0.74)	5.0	GCN4	(1.5)	16	PDC1		43
FUS1	(1.80)	<3.0	CEP1	(1.7)	15	PFK2	(3.15)	29
URA3	(0.70)	3.0	HIS4	(2.3)	14	CRY1	(0.90)	25
URA4	(1.20)	4.0	CUP1	(0.50)	18	CRY2	(0.90)	25
HTB1	(0.61)	<4.0	URA5	(1.1)	13			
CDC4	(2.6)	5.0	cDNA39	(0.38)	19			
CLN3		5.0	PHO5	(0.62)	10			
STE3	(1.80)	3.5	TRP3		12			
PPR1	(2.90)	<3.0						
URA1		7.0						

[a] mRNA decay rates were measured by RNA blotting or RNase protection assays of RNA isolated at different times after a temperature shift in strains harboring the *rpb1-1* mutation. Decay rate phenotypes have been arbitrarily classified as unstable, moderately stable, and stable. The values in parentheses are mRNA sizes in kilobases (kb), determined by Northern blotting. [Data are from Heaton *et al.* (1992); Herrick (1989); Herrick and Jacobson (1992); Herrick *et al.* (1990); Leeds *et al.* (1991); Moore *et al.* (1991); Parker and Jacobson (1990); Peltz *et al.* (1992; 1993); and S. W. Peltz, A. H. Brown, J. L. Donahue, H. Feng, W. Sears, and A. Jacobson, unpublished exps.]

are no significant differences in mRNA decay rates measured in heat-shocked (to 36°C) and non-heat-shocked cells, and for most mRNAs, alternative procedures yield relative decay rates comparable to those obtained in temperature-shifted *rpb1-1* cells (Herrick *et al.*, 1990).

mRNA half-lives measured by this procedure range from approximately 1.0 to 60 min, (Table 1), a result consistent with the complex decay kinetics observed for the total mRNA population (Herrick *et al.*, 1990). mRNAs with a $t_{1/2}<7$ min have been arbitrarily defined as unstable; those with a $t_{1/2}=10–20$ min are defined as moderately stable; and those with a $t_{1/2}>20$ min are considered stable. Possible relationships between decay rate and function are suggested by the observations that (1) a significant number of the unstable mRNAs listed in Table 1 either encode regulated mRNAs (e.g., those involved in mating-type regulation) or low abundance mRNAs (e.g., those encoding pyrimidine biosynthetic enzymes) and (2) a significant number of the stable mRNAs encode abundant proteins (e.g., ribosomal proteins or glycolytic enzymes). It remains to be established whether such relationships are the exception or the rule.

2. *Inhibition of Transcription with Thiolutin*

At low concentrations (3–6 μg/ml), the antifungal agent thiolutin inhibits all three yeast RNA polymerases both *in vivo* and *in vitro* (Jimenez *et al.*, 1973; Tipper, 1973). In thiolutin-treated cells the most rapidly decaying component of the poly(A)$^+$ RNA population has a half-life of 4–5 min and the slowest decaying component has a half-life of approximately 60 min (Herrick *et al.*, 1990). Decay of the long-lived component is slightly slower than the comparable decay component in *rpb1-1* cells, and this difference is reflected in the decay rates of individual mRNAs. The *relative* decay rates of the majority of the mRNAs assayed in thiolutin-treated cells and in temperature-shifted *rpb1-1* cells are, however, similar (Herrick *et al.*, 1990). While the use of thiolutin may be a convenient substitute for the insertion of *rpb1-1* mutations into various strains, it is important to note that the drug does have pleiotropic effects that may complicate analysis of specific mRNAs. Most notable is the apparent induction of a stress response in which a small number of promoters are effectively activated, rather than inhibited, by this drug (Adams and Gross, 1991). Thiolutin is particularly useful for the analysis of mRNA decay rates in temperature-sensitive mutants that have defects in genes other than RNA polymerase II, e.g., mutants altered in genes that may affect mRNA stability (Minvielle-Sebastia *et al.*, 1991; Peltz *et al.*, 1992). Inactivation of the temperature-sensitive function in such mutants may not follow the same kinetics as inactivation of RNA polymerase II in *rpb1-1* strains, so construction of double mutants may not be a feasible approach to the analysis of mRNA decay rates. An example of the use of thiolutin to measure mRNA decay in a temperature-sensitive mutant is presented in Section V.A.

3. *Inhibition of Transcription with 1,10-Phenanthroline*

Another potent inhibitor of yeast RNA polymerases, 1,10-phenanthroline, has also been used to measure mRNA decay rates (Santiago *et al.*, 1986, 1987). Like many other transcriptional inhibitors (reviewed in Peltz *et al.*, 1991), this drug affects a number of cellular pathways, possibly because of its ability to chelate cellular zinc (Chang *et al.*, 1978; Chang and Yarbro, 1978; D'Aurora *et al.*, 1978; Krishnamurti *et al.*, 1980). Included among its pleiotropic effects are repression of the heat-shock-induced destabilization of the L25 and S10 transcripts (Herruer *et al.*, 1988) and apparent stabilization of the HSP82 transcript (Adams and Gross, 1991).

4. *Glucose Repression of* GAL *Promoters*

Glucose represses transcription of yeast *GAL* genes rapidly (Johnston, 1987). By making the appropriate promoter fusions (e.g., to the *GAL1* promoter), the decay rate of an individual (non-*GAL*) mRNA can also be measured by glucose inhibition of transcription (Parker *et al.*, 1991; Surosky and Esposito, 1992; S. W. Peltz, W. Sears, and A. Jacobson, unpublished observations). For a given transcript, half-lives measured by this procedure or by the ts RNA polymerase II procedure are comparable. Although the use of *GAL* promoters requires specific plasmid constructions, this method does eliminate concerns about possible secondary effects occurring during temperature-shift experiments. More over, it facilitates the analysis of specific mRNA decay rates in mutants with temperature-sensitive lesions in other (non-RNA polymerase) genes as well as the analysis of the decay rates of mRNAs encoding proteins whose continuous overexpression would inhibit cell growth.

B. Are mRNA Decay Rates Regulated?

Application of the aforementioned techniques yields relative mRNA half-life values that do not vary markedly under different growth conditions or with different genetic backgrounds. However, as has been observed for higher eukaryotes (Atwater *et al.*, 1990), the decay rates of some yeast mRNAs are regulated. Of these, the classic example is the regulation of histone mRNA turnover rates in response to increased histone gene dosage (Osley and Hereford, 1981) or to changes during the cell cycle (Hereford *et al.*, 1981; Lycan *et al.*, 1987; Xu *et al.*, 1990). Addition of an extra H2A-H2B gene pair causes a 2-fold increase in the turnover rate of H2B mRNA (Osley and Hereford, 1981). Post-transcriptional events, most likely changes in stability, cause 3- to 12-fold changes in histone transcript abundance throughout the cell cycle (Lycan *et al.*, 1987; Xu *et al.*, 1990). For the H2B mRNA, the latter changes are dependent on sequences present near the 3′

end of the transcript, including a short segment of the coding region (Xu *et al.*, 1990). Yeast histone mRNAs are polyadenylated (Fahrner *et al.*, 1980) and do not contain the 3′ stem–loop structures important for the regulation of histone mRNA stability in mammalian cells (Pandey and Marzluff, 1987) (see Chapter 12). Moreover, unlike mammalian histone mRNAs, changes in yeast histone mRNA abundance in response to inhibition of DNA synthesis appear to be primarily due to transcriptional regulation (Lycan *et al.*, 1987).

Decay of the mRNA for ribosomal protein L2 also appears to be autogenously regulated. Increasing the copy number of the L2A gene fails to cause a proportional increase in L2 mRNA levels and diminishes the relative representation of the L2B transcript. Vast overexpression of L2A (from a *GAL* promoter) drastically reduces the accumulation of the endogenous, highly homologous L2A and L2B mRNAs (Presutti *et al.*, 1991). These effects must be dependent directly on the L2 protein since an L2A gene containing a premature termination codon fails to alter L2 mRNA levels (Presutti *et al.*, 1991). Results of nuclease protection analyses suggest that overexpression of the L2A protein causes an endonucleolytic cleavage event, which generates a 5′-proximal L2A mRNA fragment that is significantly more stable than the 3′-proximal fragment (Presutti *et al.*, 1991).

The stability of other ribosomal protein transcripts has been reported to be transiently decreased by heat shock (Herruer *et al.*, 1988). Within 20 min after cells have been heat shocked the abundance of the L25 and S10 mRNAs decreases by 85%, a change consistent with the reduction in the respective mRNA half-lives from 9.0 to 2.5 min. This effect appears to be transient, since mRNA levels rapidly increase to control levels within 50 min after the temperature shift (Herruer *et al.*, 1988). It also may require ongoing transcription, since phenanthroline represses the heat-shock-induced destabilization of the L25 and S10 transcripts (Herruer *et al.*, 1988), and the half-lives of the RP29 and *TCM1* ribosomal protein mRNAs are not accelerated by heat shock in an *rpb1-1* mutant (Herrick *et al.*, 1990).

The decay rates of the meiosis-specific mRNAs encoded by the *SPO11* and *SPO13* genes decrease twofold when shifted from rich media to sporulation media lacking both glucose and a nitrogen source (Surosky and Esposito, 1992). Similarly, small changes in the decay rates of mRNAs encoding glycolytic enzymes occur when yeast cells are shifted from media containing glucose to media containing lactate (Moore *et al.*, 1991). These changes in mRNA decay rate may reflect specific regulatory mechanisms or may simply be attributable to changes in translation rates that occur when cells are shifted from one carbon source to another (see Section VI for a discussion of the role of translation in mRNA decay). Further support for the notion that mRNA decay rates will fluctuate as a function of translation (or growth) rates is derived from experiments that show that yeast mRNA decay rates increase as a direct function of growth temperature (Herrick *et al.*, 1990; Peltz *et al.*, 1992). An important lesson from these observations is that, in those instances in which mutants with apparent defects in mRNA turnover are isolated, it may be difficult to distinguish bona fide effects on turnover from the secondary consequences of growth rate alteration.

III. Are Nonspecific Determinants Important Effectors of the Differences between Stable and Unstable mRNAs?

A. Target Size: Effects of mRNA Length and Ribosome Loading

Possible determinants of mRNA stability can be subdivided into those that are specific and those that are nonspecific. Specific determinants are mRNA sequences or structures that either interact directly with components of the cellular turnover machinery or target mRNA to specific cellular sites in order to promote its decay. Nonspecific determinants are general mRNA features, such as overall size, which could contribute to mRNA decay rates by promoting or hindering random interactions with nonspecific nucleases. Table 1 lists the sizes of mRNAs for which decay rates have been measured using *rpb1-1* cells and the temperature shift procedure. These data show no correlation between mRNA size and stability. An inverse correlation between the size of a yeast mRNA and its stability has been reported in cells treated with phenanthroline (Santiago *et al.*, 1986, 1987), a result whose interpretation is complicated by the use of this transcriptional inhibitor. Further evidence for the lack of a correlation between mRNA size and stability has been obtained from experiments showing that large deletions do not increase the stability of *ACT1*, *PGK1*, or *STE3*/*ACT1* (chimeric) mRNAs (Heaton *et al.*, 1992). Of interest in the latter experiments are data that suggest that large deletions may, indeed, have the opposite effect; i.e., the decay rates of otherwise stable mRNAs are increased slightly by large deletions.

Target size (or availability) has also been considered as an explanation for decay rate differences among mRNAs that differ in their respective extents of ribosome loading. For example, a large number of experiments in several different systems, including yeast, have demonstrated a reduction in mRNA decay rates in the presence of cycloheximide or other inhibitors of translational elongation (Fort *et al.*, 1987; Herrick *et al.*, 1990; Kelly *et al.*, 1983; Stimac *et al.*, 1984; Wisdom and Lee, 1991). The significance of such observations is discussed in Section VI, but it should be noted here that it is unlikely that arrested ribosomes simply sterically "protect" mRNA from nucleases since (1) in normally growing cells, the ribosome packing density does not differ for stable and unstable mRNAs (Santiago *et al.*, 1987; Shapiro *et al.*, 1988) and (2) in those instances where ribosome density is reduced by premature translational termination, instability of the corresponding mRNA is not simply dependent on the existence of a ribosome-free zone (Peltz *et al.*, 1993). Experimental evidence for the latter conclusion is discussed in Section IV.

B. Role of the Poly(A) Tract

There is a long and contentious history regarding the possible role of the 3′ poly(A) tract as a determinant of mRNA stability. The possibility that the polyadenylation status of an mRNA influences its decay rate has been supported (Bernstein *et al.*, 1989; Brewer and Ross, 1988; Marbaix *et al.*, 1975; Mercer and Wake, 1985; Minvielle-Sebastia *et al.*, 1991; Nudel *et al.*, 1976; Wilson and Treisman, 1988) and refuted (Herrick *et al.*, 1990; Krowczynska *et al.*, 1985; Medford *et al.*, 1980; Palatnik *et al.*, 1980; Santiago *et al.*, 1987; Sehgal *et al.*, 1978; Shapiro *et al.*, 1988) by numerous studies in several organisms (see Peltz *et al.* (1991) and Sachs (1990) for reviews) (see also Chapters 9, 15). It is most likely that this discrepancy reflects the fact that the poly(A) tail is not, in itself, a determinant of mRNA stability. The conflicting results most likely represent *bona fide* differences in the way that different mRNAs *shorten* their poly(A) tails or respond to that shortening. In yeast, nuclear post-transcriptional processing adds a poly(A) tail of approximately 70–90 nt to a newly synthesized pre-mRNA (Groner *et al.*, 1974; Hynes and Philips, 1976; Minvielle-Sebastia *et al.*, 1991; Philips *et al.*, 1979; Saunders *et al.*, 1980; Sachs and Davis, 1989; Muhlrad and Parker, 1992; Lowell *et al.*, 1992; Sachs and Deardorff, 1992). Once the mRNA has entered the cytoplasm, the poly(A) tail is shortened. Experiments in temperature-shifted *rpb1-1* cells suggest that this shortening reaction is rapid, such that half of the poly(A)$^+$ mRNA is chased into the poly(A)$^-$ fraction within 7–10 min after a shift to 36°C (Herrick *et al.*, 1990; Minvielle-Sebastia *et al.*, 1991). Poly(A) shortening for most yeast mRNAs appears to either stop or reach a steady state (with competing elongation reactions) when poly(A) tails have been shortened to a length of 15–40 adenylate residues (Herrick *et al.*, 1990; Minvielle-Sebastia *et al.*, 1991; Philips *et al.*, 1979; Sachs, 1990; Muhlrad and Parker, 1992; Lowell *et al.*, 1992; Sachs and Deardorff, 1992). Such poly(A) shortening is not sufficient to trigger mRNA turnover because poly(A)-deficient *CYH2* or *PGK1* mRNAs, incapable of binding to oligo(dT)–cellulose, maintain their overall stability (Herrick *et al.*, 1990). However, the poly(A) tracts of some unstable mRNAs, exemplified by the *MFA2* mRNA (Muhlrad and Parker, 1992; Lowell *et al.*, 1992), can be shortened beyond the 15-nt limit. For these mRNAs, poly(A) shortening continues to the point of poly(A) removal, and the latter event appears to activate turnover of the remainder of the mRNA (Muhlrad and Parker, 1992; Lowell *et al.*, 1992). cis-acting sequences that control this deadenylation/turnover reaction appear to be localized to the respective 3′ UTRs (Muhlrad and Parker, 1992; Lowell et al., 1992). Two trans-acting factors essential for this reaction, poly(A)-binding protein (PAB) and poly(A) nuclease (PAN) have been purified and characterized extensively (Adam *et al.*, 1986; Sachs *et al.*, 1986, 1987; Sachs and Davis, 1989; Burd *et al.*, 1991; Sachs and Deardorff, 1992; Lowell *et al.*, 1992). Additional trans-acting factors that probably participate in this reaction are those encoded by the *RNA14* and *RNA15* genes (Minvielle-Sebastia *et al.*, 1991) (see Section V for additional discussion of trans-acting factors). The notion that poly(A) removal is a rate-limiting step for the turnover of a specific subset of yeast mRNAs is consistent with experimental results on the turnover of several unstable mammalian mRNAs (reviewed in Peltz *et al.*, 1991).

C. Nonrandom Distribution of Rare Codons

Bennetzen and Hall (1982) and Sharp *et al.* (1986) observed that highly expressed genes in yeast are biased toward the use of fewer than half of the 61 possible coding triplets. Such restricted codon usage correlates with the levels of the corresponding isoaccepting tRNAs (Aota *et al.*, 1988; Ikemura, 1982). Hoekema *et al.* (1987) replaced 164 high frequency codons of the yeast *PGK1* gene with rare codons and observed a concommitant 10-fold reduction in PGK protein synthesis and a 3-fold reduction in *PGK1* mRNA levels. Assuming comparable transcription rates of the two types of *PGK1* mRNA, this result would suggest a possible relationship between the percentage of rare codons present in a given mRNA and its rate of decay. Consistent with this prediction, unstable mRNAs contain a significantly higher percentage of rare codons than stable mRNAs (Herrick *et al.*, 1990). This correlation may reveal sequence constraints on the higher order structures of these mRNAs, an underlying mechanism for regulating mRNA decay rates, or the fact that some highly expressed genes have independently evolved high rates of transcription and translation and slow rates of mRNA decay. If there is a functional role for rare codons it may be related to sequence context. For example, clusters of rare codons, in a specific sequence context, may reduce translational elongation rates (Pedersen, 1984; Robinson *et al.*, 1984; Sorensen *et al.*, 1989) to the point that a "paused" ribosome could trigger a nucleolytic event or cause a ribosome frameshift (see Section VI). In this regard, it is interesting to note that structural determinants of mRNA instability in the *MATα1* and *HIS3* genes overlap with regions containing clusters of rare codons (Herrick, 1989; Parker and Jacobson, 1990) (see Section VI).

IV. cis-Acting Determinants of mRNA Instability

A. Experimental Strategy

cis-acting sequences that promote the rapid decay of unstable mRNAs have been identified in several experimental systems (see Peltz *et al.* (1991) for review). In an approach analogous to those used in other systems, such sequences have been identified in yeast by demonstrating that segments of unstable mRNAs will promote rapid mRNA decay when transferred to mRNAs that are normally stable (Jacobson *et al.*, 1990; Parker and Jacobson, 1990; Herrick and Jacobson, 1992; Heaton *et al.*, 1992). In general, chimeric genes encoding portions of a stable and an unstable mRNA are constructed (the reading frame across the hybrid junction is maintained) and introduced into cells (*rpb1-1* mutants) on single-copy centromere (*CEN*) vectors (selecting for a plasmid marker, e.g., *URA3*), and the decay rates of the resulting hybrid transcripts and their "parental" mRNAs are measured by RNA blotting or RNase protection assays using RNA isolated from temperature-shifted cells (see Fig. 2). Although both the *PGK1* and *ACT1* mRNAs have been used as stable reporter mRNAs, the latter is less preferable because overexpression of its gene product, or fragments thereof, may have adverse effects on cell growth (Herrick and Jacobson, 1992). The use of *CEN* vectors ensures that the chimeric gene is present in virtually all cells within a population and limits the number of gene copies per cell to an average of approximately one. The simplest expectation from this type of experiment is that instability elements will be modular; i.e., hybrid mRNAs that contain sequences sufficient to promote rapid turnover will themselves exhibit rapid decay kinetics. Hybrid mRNAs that are lacking in such sequences should be stable.

To apply this strategy it is necessary to rearrange and/or delete sequences from the stable *PGK1* and *ACT1* mRNAs. As a control, to demonstrate that fragments of stable mRNAs can maintain their stability even after their introduction into hybrid transcripts, decay rates of *ACT1/PGK1* chimeric mRNAs have been measured. These experiments show that chimeras consisting of segments of stable mRNAs are themselves stable (Heaton *et al.*, 1992), suggesting that stability is not a function of the overall structure of an mRNA or mRNP. Moreover, the stability of such chimeric mRNAs implies that the instability of hybrids formed between stable and unstable mRNAs is likely to be a consequence of cis-acting sequences derived from the unstable mRNA.

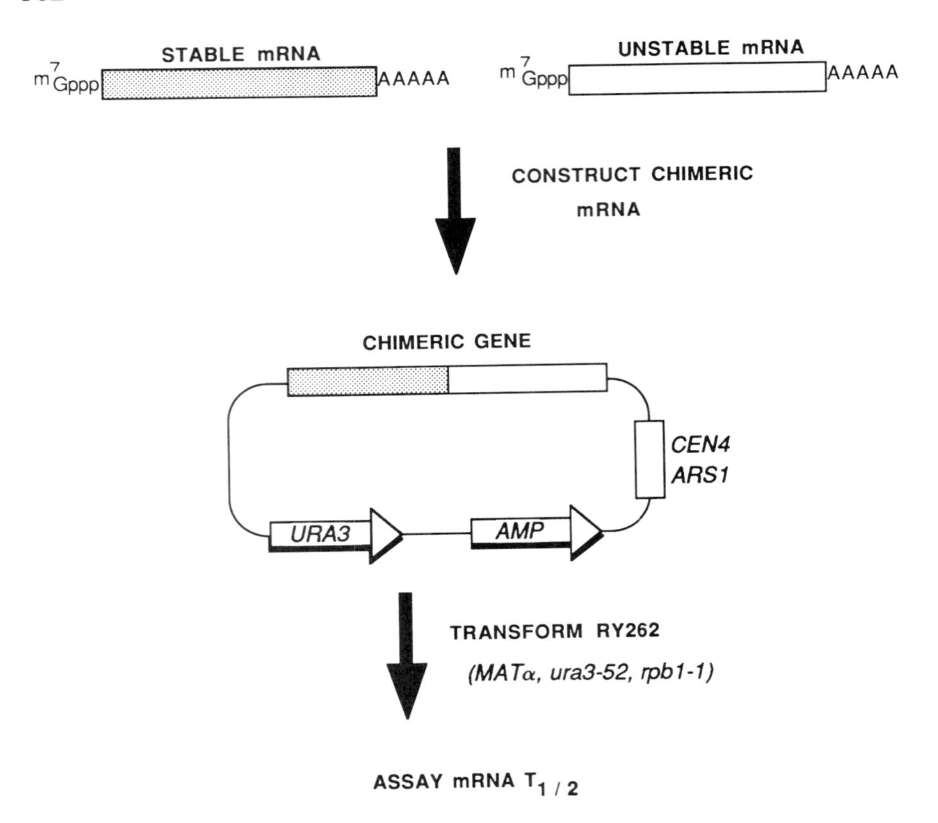

FIGURE 2 Experimental strategy for the identification of cis-acting determinants of mRNA stability or instability. Sequences from stable mRNAs are shown as closed boxes and sequences from unstable mRNAs are shown as open boxes.

B. cis-Acting Sequences Involved in the Rapid Decay of Inherently Unstable mRNAs

1. *The* MATα1 *mRNA Contains an Instability Element in Its Coding Region*

A chimeric mRNA, comprising 1205 5′-terminal nucleotides from the stable *PGK1* mRNA fused, in frame, to 580 3′-terminal nucleotides from the unstable *MATα1* mRNA, decays with a half-life (5 min) comparable to that of the unstable component of the chimera (Parker and Jacobson, 1990). A second gene fusion, in which the same sequences of *MATα1* were fused downstream of a fragment of the stable *ACT1* mRNA, produces a transcript that also decays rapidly ($t_{1/2}$ = 7.5 min). These results indicate that sequences sufficient to promote rapid mRNA degradation are contained within the 580 nucleotides of the *MATα1* mRNA present in the two hybrids.

Experiments in which the *PGK1* 3′-UTR was replaced with that from *MATα1* or in which the 3′ UTR of the *PGK1/MATα1* chimeric mRNA was replaced with that from *PGK1* demonstrated that the *MATα1* 3′ UTR is not sufficient or required for rapid decay (Parker and Jacobson, 1990). A series of in-frame deletions in which *MATα1* coding sequences were progressively removed from the *PGK1/MATα1* fusion gene showed that deletion of 52 nucleotides had little or no effect on the mRNA decay rate, but that deletion of 94 or 140 nucleotides caused a four- to fivefold increase in the half-life of the corresponding mRNA (Fig. 3) (Parker and Jacobson, 1990). mRNAs with

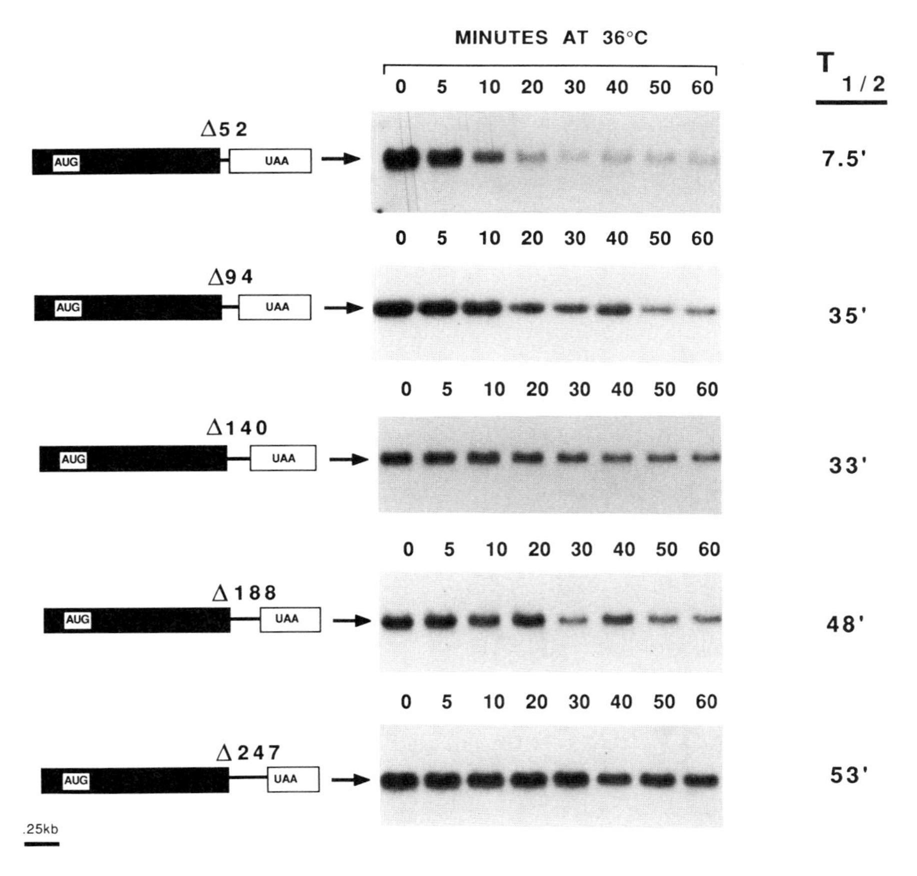

FIGURE 3 Deletion analysis of *MATα1* sequences contributing to rapid decay. Changes in mRNA decay rates resulting from deletion of *MATα1* sequences from a *PGK1/MATα1* fusion mRNA were quantitated in temperature-shift experiments with *rpbl-1* mutants as described in Fig. 1. In the schematic, closed boxes are *PGK1* sequences, open boxes are *MATα1* sequences, and deleted sequences are shown as single lines with the number of nucleotides deleted shown above each deletion. From Parker and Jacobson (1990).

larger deletions of 188 or 247 nucleotides decayed with kinetics similar to that of the *PGK1* mRNA ($t_{1/2}$ = 45 min). These results indicate that specific sequences within the 42 nucleotides between the end points of Δ52 and Δ94 are required for the rapid decay of the hybrid mRNAs; i.e., the *MATα1* mRNA contains an instability element in its coding region. Recent deletion experiments that map the 3′ boundary of this element show that the size of the complete element is approximately 150 nucleotides (R. Parker, personal communication).

2. *Instability Elements in Other Unstable mRNAs*

Experiments analogous to those described for *MATα1* have also been carried out with the unstable *HIS3*, *STE2*, *STE3*, and *CDC4* mRNAs (Herrick, 1989; Heaton *et al.*, 1992; Herrick and Jacobson, 1992; W. Sears, S. W. Peltz, and A. Jacobson, unpublished results). Sequences essential for rapid decay of chimeric mRNAs have been found in the coding regions of all four mRNAs. For example, large deletions from the coding regions of either *STE3/ACT1* or *HIS3/ACT1* chimeric genes resulted in three- to fourfold stabilization of the respective mRNAs (Heaton *et al.*, 1992; Herrick and Jacobson, 1992). For *HIS3* it was also shown that (1) a 411-nt segment important for the destabilization of a *HIS3/ACT1* chimeric mRNA is also required for the instability of bona fide *HIS3* mRNA and (2) insertion of the 411-nt segment into the entire *ACT1* gene had no destabilizing effect on the resultant hybrid mRNA (Herrick and Jacobson, 1992). The latter result, coupled with the absence of destabilizing activity in the *HIS3* 3′ UTR (Herrick and Jacobson, 1992), suggests that the instability determinants of *HIS3* mRNA are complex and involve not only the 411-nt coding region segment, but other parts of the coding region and/or 5′ UTR.

A further indication of the complexity of instability determinants was obtained in studies of the *STE3* mRNA. Whereas deletion of *STE3* sequences spanning codons 13–179 stabilizes a *STE3/ACT1* chimeric mRNA approximately threefold, deletion of the same sequences from bona fide *STE3* mRNA fails to stabilize that mRNA (Heaton *et al.*, 1992). This suggests that other regions of the *STE3* mRNA can stimulate mRNA decay independently of the coding region sequences. Experiments that support this conclusion and identify the *STE3* 3′ UTR as the location of the independent element showed that (1) replacement of the *STE3* 3′ UTR with similar sequences from the *ACT1* gene led to a twofold reduction in the mRNA decay rate and (2) substitution of the *STE3* 3′ UTR for the 3′ UTR of a stable mRNA transcribed from a *PGK1* gene with a large internal deletion (*PGKΔ182*) led to a five- to six-fold increase in the mRNA decay rate (Heaton *et al.*, 1992).

These are the first indications that yeast mRNAs, like those of mammalian cells (see Peltz *et al.* (1991) for review), can contain instability elements within 3′ UTRs. Experimental results suggesting the presence of 3′ UTR instability elements have also been obtained for the *STE2*, *HTB1*, and *MFA2* mRNAs (Herrick, 1989; Lowell *et al.*, 1992; Muhlrad and Parker, 1992; Xu *et al.*, 1990), whereas the 3′ UTRs of the *HIS3* and *MAT*α*1* mRNAs show no destabilizing activity (Herrick, 1989; Parker and Jacobson, 1990; Herrick and Jacobson, 1992). For the *MFA2* mRNA, analysis of poly(A) shortening both *in vivo* and *in vitro* indicates that the 3′ UTR controls the rate and extent of poly(A) shortening. mRNAs containing the *MFA2* 3′ UTR are apparently destabilized as a consequence of rapid and complete poly(A) removal (Lowell *et al.*, 1992; Muhlrad and Parker, 1992). For the transcript of the *HTB1* gene, sequences in the 3′ UTR and in the adjacent coding region are responsible, in part, for the periodic changes in histone mRNA abundance during the cell cycle (Xu *et al.*, 1990).

Whether the coding region instability elements also affect poly(A) removal or shortening remains to be addressed. However, the functions of the coding region and 3′ UTR elements must at least differ partially since (1) translation up to or through the *MAT*α*1* coding region element is required for its activity (Parker and Jacobson, 1990) (see below, Section VI), but such translation is not possible within the 3′ UTR and (2) the *MAT*α*1* coding region element is capable of destabilizing a chimeric mRNA containing almost 90% of the *PGK1* mRNA (Parker and Jacobson, 1990), whereas the *STE3* 3′ UTR element can only destabilize a *PGK1* transcript containing a large internal deletion (Heaton *et al.*, 1992).

It also remains to be determined whether all of the different coding region elements function via the same pathway. In this regard it is interesting to note that the instability elements of the *MAT*α*1* and *CDC4* mRNAs and a coding region segment of the unstable *CLN3* mRNA (which has yet to be tested for destabilizing activity) show a 10-nt homology (Fig. 4). The homologous sequence is not in the same reading frame in all three mRNAs, suggesting that, if important, its role is at the RNA level, not the polypeptide level. The possible role of this homologous sequence is discussed in Section VI.

MAT*α*1 **CAGTTtcGACAG**

CDC4 **CAGTTgaGACAG**

CLN3 **CAGTTccttGACAG**

FIGURE 4 Sequence homologies in the coding regions of unstable mRNAs. The 42-nt segment of the *MAT*α*1* mRNA shown to be essential for rapid mRNA decay (see Fig. 3) was compared with the sequences of other unstable mRNAs. Shown is a 10-nt homology (capital letters) between the *MAT*α*1* segment and portions of the coding regions of the unstable *CDC4* and *CLN3* mRNAs. In *CDC4* this region lies within sequences essential for rapid mRNA decay, whereas in *CLN3* the destabilizing activity of this region has yet to be tested (W. Sears, S. W. Peltz, and A. Jacobson, unpublished experiments).

C. cis-Acting Sequences Involved in Nonsense-Mediated mRNA Decay

1. *The Position of a Nonsense Codon Governs Its Effect on mRNA Decay*

In both prokaryotes and eukaryotes nonsense mutations in a gene can enhance the decay rate of the mRNA transcribed from that gene (Barker and Beamon, 1991; Baumann *et al.*, 1985; Daar and Maquat, 1988; Gaspar *et al.*, 1991; Gazalbo and Hohmann, 1990; Leeds *et al.*, 1991; Lim *et al.*, 1992; Losson and Lacroute, 1979; Maquat *et al.*, 1981; Morse and Yanofsky, 1969; Nilsson *et al.*, 1987; Pelsy and Lacroute, 1984). We use the term nonsense-mediated mRNA decay to describe this phenomenon. Early work on nonsense mutations in the yeast *URA3* gene (Losson and Lacroute, 1979) showed that (1) mRNA destabilization is linked to premature translational termination, because the nonsense-containing mRNA is stabilized in a strain containing an amber suppressor tRNA and (2) the extent of destabilization is position-dependent, because 5′-proximal nonsense mutations destabilized the *URA3* transcript to a greater degree than those that were 3′-proximal. To understand the mechanism of nonsense-mediated mRNA decay, amber mutations have been inserted at six different positions of the coding region of the abundant, stable mRNA encoded by the *PGK1* gene, and the decay rates of the resultant mutant *PGK1* mRNAs have been measured (Peltz *et al.*, 1993). As observed for the *URA3* mRNA, 5′-proximal nonsense mutations in *PGK1* mRNA accelerate the mRNA decay rate more than 3′-proximal mutations, although the relationship is nonlinear (Fig. 5 and Table 2) (Peltz *et al.*, 1993). Nonsense mutations that allow translation of only 55% or less of the *PGK1* coding sequence accelerate the *PGK1* mRNA decay rate approximately 12-fold. A nonsense mutation that allows translation of 67% of the protein coding region decreases the *PGK1* mRNA half-life only 4-fold, and nonsense mutations inserted in the last quarter of the *PGK1* transcript have no effect on mRNA decay (Fig. 5 and Table 2).

The discontinuous relationship between the position of a nonsense codon and its effect on mRNA decay rates suggests that sequences downstream of the nonsense codon may play a role in nonsense-mediated decay, either because they act as sites accessible to nuclease attack or because sequences in addition to a nonsense codon are required to trigger accelerated mRNA decay. A series of deletions that removed different amounts of the *PGK1* protein coding region downstream of an early nonsense mutation demonstrated that sequences 3′ of the nonsense codon are necessary to promote nonsense-mediated mRNA decay (Peltz *et al.*, 1993). Reinsertion of small regions of the deleted DNA into a *PGK1* nonsense allele lacking most of the protein coding region demonstrated that the 106-nt *Xba*I–*Hin*cII(1) fragment (see Fig. 5), when positioned downstream of the nonsense codon, can promote rapid mRNA decay (Peltz *et al.*, 1993). Although the 106-nt element is specific, it is not unique. Deletion of the 106-nt segment from an otherwise intact *PGK1* mRNA containing an early nonsense muta-

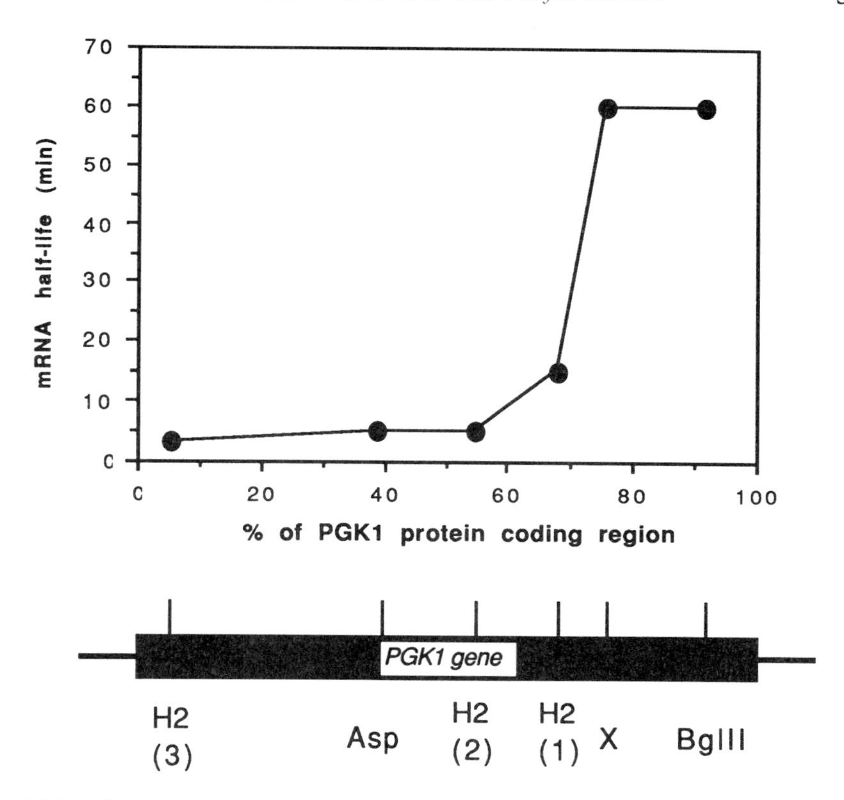

FIGURE 5 Nonsense mutations destabilize the *PGK1* mRNA. An oligonucleotide linker was used to insert amber mutations into six different positions of the *PGK1* mRNA corresponding to the position of six restriction enzyme digestion sites within the gene. To distinguish nonsense-containing transcripts from wild-type *PGK1* mRNA, the nonsense alleles were also tagged with a specific oligonucleotide in the 3′ UTR. mRNA half-lives were measured using the temperature shift procedure in *rpb1-1* mutants. Sites of amber mutations: H2, *Hinc*II; Asp, Asp718; X, *Xba*I; BglII, *Bgl*II. From Peltz *et al.* (1993).

tion did not stabilize the resultant transcript, indicating that there is a redundant 3′ cis-acting element(s) in the *PGK1* transcript that can promote nonsense-mediated mRNA decay (Peltz *et al.*, 1993).

Three observations suggest that the role of the downstream element in the nonsense-mediated mRNA decay pathway may be to promote translational reinitiation: (1) three AUG codons lie within the 106-nt sequence that serves as a functional downstream element; (2) insertion of a stem–loop structure, which inhibits both translation initiation and reinitiation, immediately downstream of a nonsense codon stabilizes an otherwise unstable *PGK1* transcript; and (3) inhibitors of amino acid biosynthesis that have been shown to reduce the capacity of cells to reinitiate translation at down-

TABLE 2 mRNA Decay Rates in Isogenic Strains That are either Wild-Type or Deleted for the *UPF1* Gene[a]

Transcript	Relative abundance UPF⁻/UPF⁺	$t_{1/2}$ (min)	
		UPF⁺	UPF⁻
his4-38	4.3	3.5	14
leu2-1	3.35	3.5	16
his4-713	1.0	14	14
his4-38-lacZ	5.0	3.5	20
HIS4-lacZ	1.0	12	13
HIS4	0.89	13	14
LEU2	1.0	14	14
MATα1	1.16	4.0	4.0
STE3	0.84	3.5	3.5
ACT1	1.14	37	34
PGK1	0.89	60	60
CUP1	1.0	18	18
eIF2	1.1	15	16
PAB1	1.2	13	11
TCM1	1.1	12.2	10.4
URA3	2.6	4.5	4.5
URA5	1.0	11	13
URA4	2.0	4.0	4.0
URA1	2.2	7.0	8.0
PPR1	2.4	1.0	3.0
PGK1	1.0	60	60
PGK(H2(3))UAG	20	3.0	60
PGK(Asp)UAG	12	5.0	60
PGK(H2(2))UAG	12	5.0	60
PGK(H2(1))UAG	4.0	15	60
PGK(Xba))UAG	1.0	60	60
PGK(BglII)UAG	1.0	60	60
GCN4	1.2	16.2	17.8

[a] mRNA decay rates were determined by Northern blotting assays of RNA isolated after a temperature-shift in strains harboring the *rpb1-1* mutation and either the wild-type *UPF1* gene or a deletion of the *UPF1* gene. Construction of *upf1Δ/rpb1-1* strains is described in Leeds *et al.* (1991). *Relative abundance* is the ratio of the steady-state mRNA levels in *upf1Δ* versus *UPF1*⁺ strains. The first four mRNAs are nonsense alleles of the respective genes. *his4-713* is a 3′-proximal nonsense allele of the *HIS4* gene. [Data are from Leeds *et al.* (1991); Peltz *et al.* (1993); and S. W. Peltz, A. H. Brown, J. Wood, A. Atkins, P. Leeds, M. Culbertson, and A. Jacobson., unpublished exps.]

stream start codons (as in the case of the *GCN4* transcript; Hinnebusch and Liebman, 1991) also stabilize mRNAs with nonsense mutations without affecting the decay of wild-type transcripts (S. W. Peltz, A. H. Brown, and A. Jacobson, unpublished results). If translational reinitiation is an important component in the nonsense-mediated mRNA decay pathway, then reducing the cell's capacity to reinitiate translation would be expected to affect that pathway.

2. *Availability of Functional Downstream Elements Cannot Account for the Position Effects*

When between 67 and 76.6%of the *PGK1* protein coding sequence is translated, the *PGK1* transcript becomes insensitive to nonsense-mediated mRNA decay (Fig. 5) (Peltz *et al.*, 1993). Since specific sequences are required downstream of a nonsense mutation to promote turnover, a possible explanation for the observation that 3′-proximal amber mutations do not promote nonsense-mediated mRNA decay is that they lack the necessary downstream element. This hypothesis predicts that the insertion of a functional downstream sequence element distal to a 3′-proximal nonsense mutation should promote rapid decay of its transcript. To test this hypothesis, sequences 3′ of the nonsense mutation inserted at 92% of the PGK1 protein coding region were replaced with the protein coding region and 3′ UTR from a PGK1 gene harboring downstream elements capable of promoting nonsense-mediated mRNA decay (Peltz *et al.*, 1993). The half-life of this transcript was unaffected by these insertions (Peltz *et al.*, 1993), indicating that 3′-proximal nonsense mutations are resistant to the nonsense-mediated mRNA decay pathway for reasons other than a lack of a specific downstream element. These results suggest the possibility that a factor required for the nonsense-mediated mRNA decay pathway is inactivated as the ribosome translocates across the *PGK1* transcript, either stochastically or as a consequence of translation through a specific region of the transcript.

D. cis-Acting Stabilizer Sequences

The 5′ cap notwithstanding (Piper *et al.*, 1987), there is only limited evidence for the existence of cis-acting stabilizer sequences in yeast mRNAs. Two different experiments summarized above are suggestive of possible stabilizer sequences within the coding region of the *PGK1* mRNA. In the first, Heaton *et al.* (1992) found that the *STE3* 3′ UTR could not destabilize an intact *PGK1* reporter mRNA, but could destabilize a *PGK1* mRNA with a large deletion of its coding region. In the second, Peltz *et al.* (1993) found that nonsense mutations in the *PGK1* mRNA were destabilizing only if they occurred within the first two-thirds of the transcript. As will be discussed below (see Sections V and VI), this transition may reflect the loss of a ribosome-associated factor required for nonsense-mediated mRNA decay.

If loss of such a factor is promoted by specific sequences, then those sequences can technically be considered to be stabilizers. In addition to these observations, Vreken *et al.* (1991) have shown that insertion of poly$(G)_{18}$, but not other ribopolymers, into the 3′ UTR of the *PGK1* mRNA causes a twofold stabilization of this mRNA. The significance of this observation is at present unclear. Since the poly(G) insertion, unlike insertions or deletions in the 3′ UTR of the stable *PYK1* mRNA (Purvis *et al.*, 1987), does not inhibit translation (Van den Heuvel *et al.*, 1990), it is unlikely that the observed stabilization is a secondary effect of a reduction in mRNA translatability (see Sections V and VI for a discussion of the relationships between translation and turnover).

V. Approaches to Identifying trans-Acting Factors Involved in mRNA Decay

A. Characterization of Conditional Lethal Mutants

One of the major advantages to using yeast as an experimental system is the opportunity to exploit its powerful genetics in order to dissect biochemical pathways. A strategy to identify trans-acting factors involved in the decay of inherently unstable mRNAs is to characterize conditional lethal mutants that affect the accumulation of such RNAs. A screen of a collection of temperature-sensitive mutants led to the identification of ts352, a mutant that accumulated moderately stable and unstable mRNAs after a shift from 23 to 37°C (Aebi *et al.*, 1990). Complementation of ts352 with a yeast genomic DNA library led to the identification of the gene responsible for the temperature-sensitive lesion (Aebi *et al.*, 1990). In the ts352 mutant, the gene coding for tRNA nucleotidyltransferase, the enzyme catalyzing 3′-terminal CCA addition to tRNAs, is defective. At the nonpermissive temperature ts352 accumulates tRNAs that are shorter than mature tRNAs, presumably because CCA termini are absent. The consequent reduction in the pool of functional tRNAs leads to a rapid decrease in the ability of ts352 cells to synthesize protein. Within 50 min of a shift to 37°C, ts352 cells have lost greater than 90% of their capacity to synthesize protein (as measured by ^{35}S-methionine incorporation in a brief labeling period). This block in protein synthesis is accompanied by a shift of mRNAs to heavier polysomes, a change identical to that observed in cycloheximide-treated cells (Peltz *et al.*, 1992), and diagnostic of a reduction in the rate of translational elongation (Rose and Lodish, 1976; Schneider *et al.*, 1984).

mRNA decay rates in ts352 were analyzed by inhibiting transcription with thiolutin either at 24°C or at different times after a shift to 36°C (Peltz *et al.*, 1992). At the permissive temperature (24°C), decay of the *CDC4*, *PAB1*, and *TCM1* mRNAs in ts352 occurred at the same rates as in wild-type cells. At 36°C, however, mRNA degradation in cells harboring the ts352 allele was slowed three- to fivefold when compared with cells containing the wild-type tRNA nucleotidyltransferase gene (Fig. 6). The extent of stabilization was directly related to the extent of protein synthesis inhibition. Stabilization was not limited to specific mRNAs, but affected all mRNAs examined, including those that were unstable and moderately stable. Comparable effects on mRNA decay rates were found in cycloheximide-treated cells.

Cycloheximide-induced increases in mRNA abundance and stability have been observed previously for a large number of mRNAs in different experimental systems (Fort *et al.*, 1987; Herrick *et al.*, 1990; Kelly *et al.*, 1983; Stimac *et al.*, 1984; Wisdom and Lee, 1991). The results from the ts352 studies suggest that such effects are not drug-related artifacts, but rather are indicative of a requirement for translational elongation in the decay of the mRNAs examined. These results do not differentiate between models that suggest that there is a labile factor that is degraded when protein synthesis is inhibited or that those that suggest mRNA decay elements must be translated in order to be active. Our own bias, supported by the experimental evidence discussed in Section VI, is that the act of translation is the important factor in this process.

The results with ts352 also suggest that screening procedures designed to identify mutations in mRNA decay pathways must be approached cautiously. Mutations that alter macromolecular synthesis may, indirectly, alter mRNA decay rates; i.e., it may often be difficult to determine whether a mutation that alters turnover rates affects a gene product that is directly or indirectly involved in the decay process.

B. Purification of Nucleases and Identification of Nuclease Mutants

Although several different ribonuclease activities have been identified in yeast cells (Dauber, 1973; Lee *et al.*, 1968; Mead and Oliver, 1983; Nakao *et al.*, 1968; Ohtaka *et al.*, 1963; Stevens, 1978, 1980, 1985, 1988; Udvardy *et al.*, 1972), a direct role for these nucleases in mRNA turnover has not been conclusively demonstrated. The best characterized yeast ribonuclease is a 5′→ 3′ exoribonuclease expressed from the *XRN1* gene (Stevens, 1978, 1980). This protein has been purified to near homogeneity and has a molecular weight of 160,000 kDa (Stevens and Maupin, 1987). The Xrn1p nuclease degrades RNAs with 5′ monophosphates, while capped RNAs are resistant to degradation (Stevens, 1978). RNAs harboring a 5′ triphosphate are approximately 10-fold poorer substrates for the Xrn1p than RNAs with a monophosphate at the 5′ end (Stevens, 1978; Stevens and Maupin, 1987).

The *XRN1* gene has been cloned and sequenced (Larimer and Stevens, 1990) and shown to be identical to several other independently isolated genes, i.e., *DST2*, *SEP1*, and *KEM1* (Dykstra *et al.*, 1990,1991; Kim *et al.*,

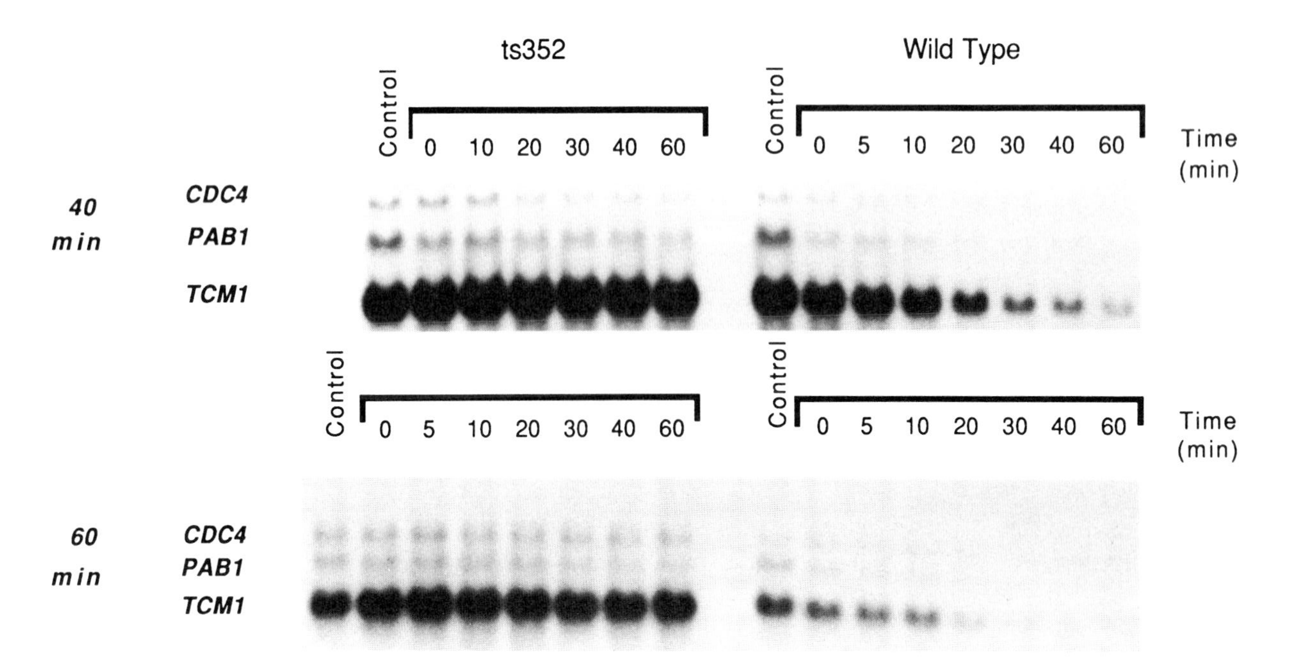

FIGURE 6 Analysis of mRNA decay in strains harboring the wild-type or ts352 allele of the tRNA nucleotidlytransferase gene. Cultures of wild-type or ts352 cells were shifted to 36°C, and mRNA decay rates were measured by addition of thiolutin at 40 or 60 min after the temperature shift. The relative levels of the *CDC4*, *PAB1*, and *TCM1* transcripts were measured by Northern blotting of RNAs isolated at different times after addition of thiolutin. Control lanes contain RNA isolated from cells prior to the temperature shift. From Peltz *et al.* (1992).

1990; Tishkoff *et al.*, 1990). The latter genes were thought to encode factors involved in nuclear fusion or in DNA strand exchange during genetic recombination (Dykstra *et al.*, 1990, 1991; Kim *et al.*, 1990; Tishkoff *et al.*, 1990) (see Chapter 17). Clearly, the *XRN1* gene product may be involved in many cellular pathways.

The *XRN1* gene is nonessential, but its deletion causes a slow growth phenotype (Larimer and Stevens, 1990) (see Chapter 17). Cells that are deleted for *XRN1* have altered ribosomal RNA processing and accumulate an internal transcribed spacer (ITS1) fragment of the pre-rRNA (Stevens *et al.*, 1991). Deletion of the *XRN1* gene from yeast cells appears to slow the decay of several mRNAs (A. Stevens, personal communication), but, as described above (see Sections II.B and V.A), it is difficult to determine whether such stabilization is due to a direct effect of the deletion of the *XRN1* gene on mRNA decay or whether it is a consequence of the slow growth phenotype or perhaps of altered protein synthesis. Since deletion of the *XRN1* gene affects rRNA biogenesis it will be important to determine whether translation is affected in an *xrn1*$^-$ strain (see Section VI for a discussion of the relationships between turnover and translation).

C. Factors That Interact with the Poly(A) Tract

As discussed in Section III.B, almost all yeast mRNAs are synthesized with a 70- to 90-nt 3′ poly(A) tail that is progressively shortened as the mRNA "ages" in the cytoplasm. For most mRNAs this poly(A) shortening reaction stops when a steady-state length of 15–40 adenylate residues is reached; however, for a limited number of mRNAs, shortening continues, leading to poly(A) removal and subsequent mRNA degradation (Herrick *et al.*, 1990; Hynes and Philips, 1976; Lowell *et al.*, 1992; Minvielle-Sebastia *et al.*, 1991; Muhlrad and Parker, 1992; Philips *et al.*, 1979; Sachs, 1990; Sachs and Davis, 1989; Saunders *et al.*, 1980). Recognizing the potential importance of the poly(A) shortening reaction for the turnover of specific mRNAs, Sachs and Deardorff (1992) exploited a novel property of PAB mutations in order to purify and characterize the PAN. Capitalizing on earlier work that had shown that *PAB1*$^-$ strains fail to shorten poly(A) tails (Sachs and Davis, 1989), Sachs and Deardorff (1992) were able to identify PAN activity by supplementing extracts from *PAB1*$^-$ cells with purified PAB. PAN was subsequently purified and shown to be a 135-kDa protein that requires the presence of PAB in order to degrade poly(A) *in vitro*. Sachs and Deardorff (1992) note appropriately that this result is of particular interest because it suggests that the degradative specificity of PAN (and, by analogy, other nonspecific nucleases) is conferred by the RNA-binding protein with which it interacts (in this case, PAB). Indirect evidence suggests that PAN is an exonuclease, but the possibility that PAN activates a nuclease activity in PAB cannot be excluded. The PAN/PAB complex normally shortens poly(A) tracts *in vitro* to a length of 15 to 25 nt. For the labile *MFA2* mRNA, however, a sequence element in the 3′ UTR allows complete deadenylation and decay of this transcript *in vitro* (Lowell *et al.*, 1992). These results are consistent with results from Muhlrad and Parker (1992), who have demonstrated that the 3′ UTR of the *MFA2* transcript is important in determining both the poly(A) shortening rate and the stability of the *MFA2* mRNA *in vivo*.

Two other genes that appear to affect poly(A) metabolism have been investigated by Minvielle-Sebastia *et al.* (1991). Mutants in *RNA14* and *RNA15* were isolated on the basis of their simultaneous cordycepin (3′-deoxyadenosine) sensitivity and temperature sensitivity (Bloch *et al.*, 1978). Compared with wild-type cells, mutants with lesions in either of these two genes show rapid reductions in poly(A) tail lengths and reduced stability of the *ACT1* mRNA at the nonpermissive temperature (Bloch *et al.*, 1979; Minvielle-Sebastia *et al.*, 1991). The rapid rate of poly(A) shortening observed at the nonpermissive temperature in strains harboring either the *rna14* or the *rna15* ts mutations is the same as the poly(A) shortening rate observed in cells where transcription has been inhibited either by addition of thiolutin or by mutation (*rpb1-1*) of RNA polymerase II. Since *rna14* and *rna15* mutants show no significant reductions in transcriptional activity, these results could be explained by (1) a defect in polyadenylation in the nucleus, (2) increased poly(A) shortening rates in the cytoplasm, or (3) loss of cytoplasmic poly(A) polymerase activity in the cytoplasm. Since the rapid poly(A) shortening phenotype is observed with both old and newly synthesized transcripts, it is unlikely that the *rna14* and *rna15* mutations simply affect nuclear polyadenylation. At present there is insufficient information to unequivocally differentiate between increased poly(A) shortening rates or decreased cytoplasmic poly(A) addition.

The *RNA14* and *RNA15* genes have been cloned and sequenced. The *RNA14* gene, located on chromosome II, encodes a 75-kDa protein and gives rise to one major (2.2 kb) and two minor (1.5 and 1.1 kb) transcripts whose respective functions are unknown. *RNA14* is not homologous to any other sequenced gene. The *RNA15* gene is on chromosome XVI, gives rise to a 1.2-kb transcript, and encodes a 33-kDa protein. The primary sequence of the *RNA15* gene is similar to those of known RNA- and DNA-binding proteins in that it contains an N-terminal segment with RNP1 and RNP2 consensus sequences (see Kenan *et al.*, (1991) for review) followed by glutamine- and asparagine-rich regions. Both *RNA14* and *RNA15* are single-copy essential genes. Overexpression of either *RNA14* or *RNA15* leads to suppression of both the *rna14* and the *rna15* mutations, indicating that the respective gene products are likely to interact (Minvielle-Sebastia *et al.*, 1991). However, the *RNA14* and *RNA15* gene products most likely do not interact with the proteins encoded by *RNA1* or *PAB1* since overexpression of *RNA14* or *RNA15* fails to suppress mutations in the former pair of genes (Minvielle-Sebastia *et al.*, 1991).

D. trans-Acting Factors Involved in the Nonsense-Mediated mRNA Decay Pathway

Nonsense suppressors in yeast are either tRNA mutants, capable of decoding a translation termination codon, or mutants with lesions in non-tRNA genes, which enhance the expression of nonsense-containing alleles by other mechanisms. The latter mutants include the allosuppressors, frameshift suppressors, and omnipotent suppressors (Surguchov, 1988; Hinnebusch and Liebman, 1991). At least one of these suppressors, *upf1*, acts by suppressing nonsense-mediated mRNA decay.

Mutants in the *UPF1* gene (and in *UPF2*, *-3*, and *-4*) were isolated on the basis of their ability to act as allosuppressors of the *his4-38* frameshift mutation (Culbertson *et al.*, 1980). The latter mutation is a single G insertion in the *HIS4* gene that generates a +1 frameshift and a UAA nonsense codon in the triplet adjacent to the insertion (Donahue *et al.*, 1981). At 30°C, but not 37°C, *his4-38* is suppressed by *SUF1-1*, which encodes a glycine tRNA capable of reading a four base codon (Mendenhall *et al.*, 1987). Mutations in *UPF1* allow cells that are *his4-38* and *SUF1-1* to grow at 37°C (Culbertson *et al.*, 1980). The basis of this suppression appears to be the loss of function of a trans-acting factor (Upf1p) essential for nonsense-mediated mRNA decay. Mutations in *UPF1* lead to the selective stabilization of mRNAs containing early nonsense mutations without affecting the decay rates of most other mRNAs (Table 2) (Leeds *et al.*, 1991). Thus, in a *UPF1* deletion mutant (*upf1Δ*), the *his4-38* mRNA is stabilized approximately 5-fold, half-lives of mRNAs from the *PGK1* early nonsense alleles are stabilized approximately 12-fold, and half-lives of the *PGK1* mRNAs with late nonsense codons or mRNAs from the wild-type *MATα1*, *STE3*, *LEU2*, *HIS4*, *PGK1*, *PAB1*, and *ACT1* genes are unchanged (Table 2). Regardless of position, all of the *PGK1* nonsense alleles have mRNA half-lives on the order of an hour in a *upf1Δ* strain (Peltz *et al.*, 1993), a result that indicates that the loss of *UPF1* function restores wild-type decay rates to mRNAs that would otherwise have been susceptible to the enhancement of decay rates promoted by early nonsense codons.

Suppression of nonsense-mediated mRNA decay in *upf1Δ* strains does not appear to result from enhanced read-through of the translation termination signal (Leeds *et al.*, 1991), nor does it appear to be specific for a particular nonsense codon. The ability of *upf1⁻* mutants to suppress *tyr7-1* (UAG), *leu2-1* (UAA), *leu2-2*, (UGA), *met8-1* (UAG), and *his4-166* (UGA) (Leeds *et al.*, 1992) indicates that they can act as omnipotent suppressors. Since many biosynthetic pathways do not require maximal levels of gene expression for cell survival (e.g., only 6% of the HIS4 gene product is required for growth on plates lacking histidine; Gaber and Culbertson, 1984), stabilization of nonsense-containing transcripts would allow synthesis of sufficient read-through protein to permit cells to grow on media lacking the relevant amino acids.

The *UPF1* gene has been cloned and sequenced and shown to be (1) nonessential for viability, (2) capable of encoding a 109-kDa protein with both zinc finger and nucleotide (GTP) binding site motifs, and (3) partially homologous to the yeast *SEN1* gene (Leeds *et al.*, 1992). The latter encodes a noncatalytic subunit of the tRNA splicing endonuclease complex (Winey and Culbertson, 1988), suggesting that Upf1p may also be part of a nuclease complex targeted to nonsense-containing mRNAs. It is unlikely, however, that its normal function is anticipatory, i.e., that it is solely involved in the degradation of mRNAs with premature nonsense codons. One role of the *UPF1* gene may be to regulate the decay rates of transcripts with upstream open reading frames, a conclusion that followed from an analysis of the steady-state levels and decay rates of mRNAs encoding gene products involved in the pyrimidine biosynthetic pathway (S. W. Peltz, A. H. Brown, J. Wood, A. Atkins, P. Leeds, M. Culbertson, and A. Jacobson, unpublished experiments). Transcription of several genes in this pathway (*URA1*, *URA3*, and *URA4*) is governed by the positive activator, *PPR1* (Losson *et al.*, 1983; Kammerer *et al.*, 1984; Liljelund *et al.*, 1984). The *PPR1* mRNA itself has a small (five codon) open reading frame upstream of the *PPR1* coding sequence; the termination codon for the upstream open reading frame overlaps with the ATG of the *PPR1* coding sequence (Losson *et al.*, 1983). Measurement of *PPR1* mRNA decay rates demonstrates that it decays threefold more slowly in a *upf1*$^-$ strain than in a *UPF1*$^+$ strain [$t_{1/2}=3$ min (*upf1*$^-$) vs $t_{1/2}=1$ min (*UPF1*$^+$); S. W. Peltz, A. H. Brown, J. Wood, A. Atkins, P. Leeds, M. Culbertson, and A. Jacobson, unpublished experiments], suggesting that upstream open reading frames, in addition to reducing the frequency of downstream translational initiation (Kozak, 1991; Hinnebusch and Liebman, 1991), may also destabilize specific transcripts. This destabilization mechanism, which would involve the *UPF1* gene product, would not necessarily pertain to all mRNAs with upstream open reading frames since there is no effect of *UPF1* status on the decay rate of the *GCN4* mRNA (Table 2; S. W. Peltz, A. H. Brown, J. Wood, A. Atkins, P. Leeds, M. Culbertson, and A. Jacobson, unpublished experiments).

The possibility also exists that mRNAs with splicing errors constitute another class of *UPF1* substrate; i.e., splicing fidelity would be maintained by the rapid turnover of RNAs in which processing errors have caused the insertion or deletion of one or more nucleotides. At present, there is no evidence to suggest an error-prone splicing process. However, a related phenomenon, premature transport of pre-mRNA from the nucleus, may occur (Yost and Lindquist, 1988; reviewed in Kozak, 1991). Translation of intron-containing RNAs would almost certainly lead to premature termination and, thus, could also be subject to *UPF1* regulation. Perhaps one general function of the *UPF1* pathway is to ensure that aberrant proteins do not accumulate in those instances in which regulated splicing (Baker, 1989) or incomplete RNA editing (Simpson and Shaw, 1989; Stuart, 1991) generates mRNAs lacking a functional open reading frame.

Regardless of its physiological role, identification of the *upf1* loss-of-function mutations provides a valuable tool to study the nonsense-mediated mRNA decay pathway. cis-acting elements that promote nonsense-mediated mRNA decay can now be defined as sequences that accelerate mRNA decay in wild-type cells, but that are inactivated in strains deleted for the *UPF1* gene.

As noted above, three other complementation groups of *UPF* mutants were identified in the same selection that identified *upf1* (Culbertson *et al.*, 1980). Since mutations in at least one of these genes, *UPF3*, also appear to reduce the rate of nonsense-mediated mRNA decay (Leeds *et al.*, 1992), it is likely that yeast genetics will prove invaluable in the dissection of this mRNA degradation pathway.

E. Do the Decay Pathways for Nonsense-Containing mRNAs and Inherently Unstable mRNAs Share Any trans-Acting Factors

The possibility exists that there are a limited number of mRNA decay pathways in the cell and that mRNAs can become substrates for these pathways by several different and independent routes. For example, it is conceivable that the sequences that promote rapid decay in mRNAs that are inherently unstable do so by promoting a shift in the translational reading frame, which, in turn, could ultimately make the mRNA in question a substrate for the nonsense-mediated decay pathway. Such confluence of decay pathways would be suggested if mutants, drugs, or cis-acting elements exist that have comparable effects on nonsense-mediated decay and the decay of inherently unstable mRNAs. At present, three results suggest that these two decay pathways are nonoverlapping: (1) the cis-acting element responsible for the decay of the inherently unstable *MATα1* mRNA still dictates rapid mRNA decay when inserted beyond the functional boundary for nonsense-mediated decay (Parker and Jacobson, 1990), (2) *upf1*$^-$ mutations affect the decay of mRNAs with early nonsense codons (or 5′ open reading frames), but do not affect the decay rates of mRNAs lacking such features (Leeds *et al.*, 1991), and (3) although cycloheximide treatment of cells will stabilize all types of mRNAs, removal of cycloheximide restores the rapid decay of inherently unstable mRNAs, but does not restore the rapid decay of mRNAs containing early nonsense codons (S. W. Peltz, A. H. Brown, and A. Jacobson, unpublished experiments). Although these results indicate that the decay of two types of unstable mRNAs occurs via different pathways, this conclusion applies only to the steps in these pathways that were analyzed in the aforementioned experiments. It is possible, for example, that these experiments discriminate among only the earliest steps in decay and that the two pathways merge after these initial steps have been completed.

VI. Why Are Translation and Turnover Intimately Linked?

Experiments described above indicate that inhibition of translational elongation can reduce mRNA decay rates and that premature translational termination can enhance mRNA decay rates. These observations suggest that the processes of mRNA translation and mRNA turnover must be intimately linked. Further evidence to support this conclusion includes experiments in yeast that show that (1) ribosome translocation up to or through the *MATα1* instability element is required for rapid decay of *PGK1/MATα1* and *ACT1/MATα1* chimeric mRNAs (Parker and Jacobson, 1990); (2) instability elements involved in the rapid decay of the *MATα1*, *HIS3*, *STE3*, *STE2*, and *CDC4* mRNAs have been localized to the coding regions of the respective mRNAs (Herrick, 1989; Parker and Jacobson, 1990; Herrick and Jacobson, 1992; Heaton *et al.*, 1992; W. Sears, S. W. Peltz, A. H. Brown, and A. Jacobson, unpublished experiments) (see Section IV); and (3) instability elements may be rich in rare codons (e.g., the 5′ boundary of the coding region instability element from the *MATα1* mRNA contains a stretch of six out of seven contiguous rare codons; Parker and Jacobson, 1990). In addition to these observations in yeast, experiments with mammalian cell-free systems indicate that a nuclease activity capable of degrading mammalian mRNAs *in vitro* is associated with polysomes (Peltz *et al.*, 1987; Ross and Kobs, 1986).

For the turnover of inherently unstable mRNAs, a requirement for both specific sequences (i.e., instability elements) *and* translation suggests several nonexclusive mechanisms that may be operative: (1) the initial recognition event may occur, not at the RNA level, but as with mammalian tubulin mRNAs (Cleveland, 1988), at the level of the nascent polypeptide as it emerges from the ribosome; (2) the ribosome itself may be critical to the degradation process, by delivering a nuclease (Graves *et al.*, 1987), activating a nuclease, or recognizing a specific sequence element; or (3) the passage of a ribosome through a specific region may alter the secondary structure of the mRNA or affect factors associated with the ribosome, in such a way as to expose mRNA sequences containing nuclease recognition sites that would normally not be available (Graves *et al.*, 1987). The concentration of rare codons in certain sequences required for rapid decay may induce translational pausing (Wolin and Walter, 1988). Alternatively, a translational pause may be induced by mRNA:rRNA base-pairing interactions. For example, possible complementarity between yeast 18S rRNA and 14 nucleotides near the beginning of the *MATα1* coding region instability element (including the region homologous to the *CDC4* and *CLN3* mRNAs; see Fig. 4) is shown in Fig. 7. This interaction is only hypothetical and must be weighed in light of models that suggest that this region of rRNA may be involved in intramolecular base-pairing (Dams *et al.*, 1988). However, the possibility of base-pairing between mRNA and rRNA merits attention because substantive evidence has emerged in recent years that rRNA may have a functional role in translation (Dahlberg, 1989; Noller, 1991). Regardless of the mechanism that induces pausing, a ribosome paused at a specific site may expose downstream nuclease recognition sites that could then be cleaved by either a soluble or a ribosome-bound or ribosome-activated

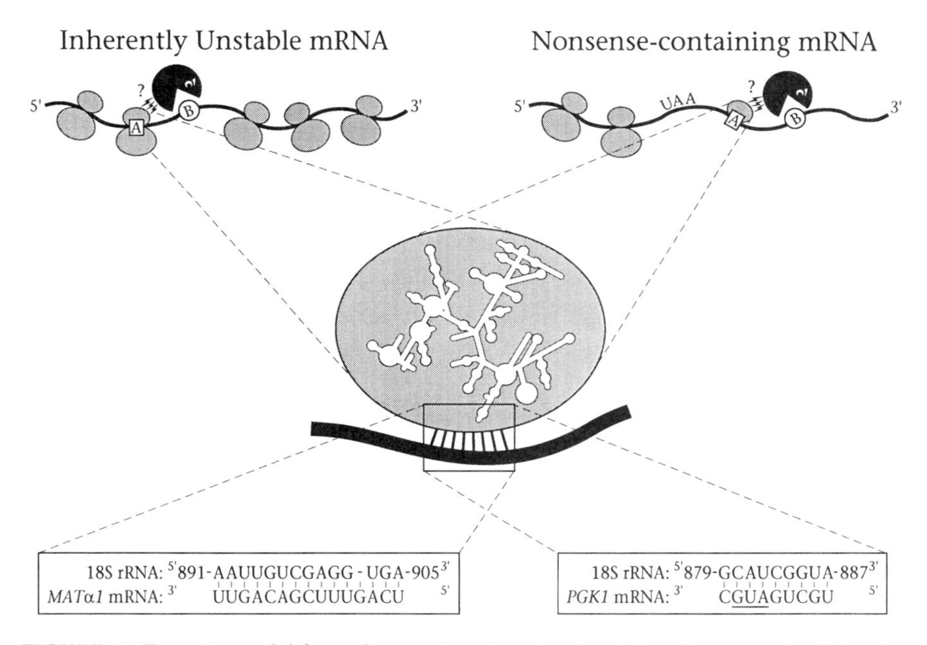

FIGURE 7 Two-site model for coding region cis-acting instability elements of mRNAs that are either inherently unstable or unstable by virtue of premature translational termination. The "A" site of either type of element is postulated to be a position of ribosome pausing, possibly as a consequence of mRNA:rRNA interactions. Examples of possible base-pairing are shown for a segment of yeast 18S rRNA and either the 5′-proximal region of the *MATα1* instability element (Parker and Jacobson, 1990) or a sequence flanking a downstream AUG codon required for destabilization of a *PGK1* mRNA that contains a premature translational termination codon (Peltz *et al.*, 1993). The "B" site indicates a position of endonucleolytic cleavage. Staggered lines between the 40S subunit and the putative nuclease are intended to suggest a possible interaction. Numbers flanking rRNA sequences refer to the nucleotide positions within the primary sequence (Rubtsov *et al.*, 1980; Mankin *et al.*, 1986). See text for details.

nuclease. We consider such a two-site model, in which the first site potentiates the cleavage mechanism and the second site is the actual position of the nucleolytic attack (see Fig. 7), to be consistent with the available data.

The role of translation in nonsense-mediated mRNA decay may be similar to that postulated in the decay of inherently unstable mRNAs, except that the number of ribosomes traversing the mRNA downstream of the nonsense codon is expected to be significantly less than that in a wild-type mRNA. The experiments summarized in Sections IV.C and V.D indicate that, in addition to a nonsense codon, a downstream sequence element is required to trigger rapid mRNA decay via this pathway, possibly by providing a site for translational reinitiation. While reinitiation may well be the event that triggers decay, it is tempting to speculate that an mRNA:rRNA interaction analogous to that proposed for the instability elements of inherently unstable mRNAs may also be involved (Fig. 7). The 105-nt downstream element of the *PGK1* mRNA (see Section IV.C) contains three AUGs. The first two are essential for nonsense-mediated mRNA decay and are bracketed by identical nucleotide sequences (Peltz *et al.*, 1993). The third AUG is not required for nonsense-mediated decay and is flanked by sequences that differ from the other two. Figure 7 shows that the sequences bracketing the first two AUGs are capable of base-pairing with 18S rRNA in almost the same region postulated to be involved in base-pairing with the *MAT*α*1* instability element. The nucleotides bracketing the third AUG will not accomodate this base-pairing scheme.

A nonsense codon and a downstream element are *sufficient* for nonsense-mediated mRNA decay provided that the ribosome has not traversed more than three-quarters of the *PGK1* coding region (see Section IV.C). Once beyond this position the ribosome is incapable of triggering rapid mRNA decay even if these sequences are placed downstream. Analysis of the decay rates of chimeric mRNAs suggests that, in addition to the aforementioned elements, the nonsense-mediated decay pathway is regulated by a sequence element that, when traversed by a ribosome, renders the mRNA insensitive to the pathway (Peltz *et al.*, 1993). We interpret these results to indicate that a translation or destabilization factor falls off the ribosome as a consequence of traversing the sequence element. A candidate for such a factor is the protein encoded by the *UPF1* gene (Upf1p; see Section V.D). As such, Upf1p could be (1) a translational initiation factor that is also required for a reinitiation event that may trigger mRNA degradation, (2) a nuclease, activated by a downstream reinitiation event, or (3) a factor that promotes an interaction with a specific nuclease or nuclease complex. By analogy to the "A" and "B" sites of inherently unstable mRNAs shown in Fig. 7, the A site in the nonsense-mediated pathway would be the site of translational reinitiation (or mRNA:rRNA base-pairing), and the B site would, again, be the site of cleavage. The two pathways would thus be quite similar, except that the nonsense-mediated pathway would be subject to regulation (activation by a nonsense codon and inhibition by a position effect), whereas decay of inherently unstable mRNAs would always be "on." Regulation of the nonsense-mediated pathway may be essential to prevent its activity during the course of normal translational termination. Evidence to suggest that this pathway is, indeed, inactive during normal translational termination is provided by the existence of relatively stable mRNAs with very short open reading frames, e.g., *CUP1* mRNA ($t_{1/2} = 18$ min; see Table 2). At present there is essentially no information available on yeast mRNA cleavage sites. Since substantial overexposure of RNA blots fails to reveal low abundance RNAs that would qualify as degradation intermediates (S.W. Peltz and A. Jacobson, unpublished experiments), it is possible that the sites of initial cleavage are heterogeneous or that the initial cleavage itself is rate limiting relative to the rate of degradation of the resulting mRNA fragments.

The models proposed here suggest an active role for the ribosome (or one of its subunits) in the mechanics of decay. Although such a role for the ribosome is plausible for those degradation events that are dependent on coding region sequences, it is more difficult to conceive of a ribosome-dependent activity for those events involving the 3′ UTR. The possibility that the role of the 3′ UTR in translational initiation (Jackson and Standart, 1990; Munroe and Jacobson, 1990; Sachs, 1990) may somehow be linked to the destabilization of specific mRNAs is intriguing, but, as of yet, unfounded.

Acknowledgments

This work was supported by a grant (GM27757) to A.J. from the National Institutes of Health and by a postdoctoral fellowship to S.W.P. from the American Cancer Society. We thank Christine Bonczek, Sharon Bowen, Agneta Brown, Michael Culbertson, Janet Donahue, He Feng, David Herrick, Steve Johnson, Peter Leeds, Roy Parker, Wendy Sears, Chris Trotta, and Ellen Welch for their contributions to the experiments described here and Roy Parker, Alan Sachs, Michael Culbertson, Audrey Stevens, Richard Surosky, Shelley Esposito, Al Brown, and Dick Raue for sending us preprints and reprints of their recent work.

References

Adam, S. A., Nakagawa, T., Swanson, M. S., Woodruff, T. K., and Dreyfuss, G. (1986). mRNA polyadenylate-binding protein: Gene isolation and sequencing and identification of a ribonucleoprotein consensus sequence. *Mol. Cell. Biol.* **6,** 2932–2943.

Adams, C. C., and Gross, D. S. (1991). The yeast heat shock response is induced by conversion of cells to spheroplasts and by potent transcriptional inhibitors. *J. Bacteriol.* **173,** 7429–7435.

Aebi, M., Kirschner, G., Chen, J.-Y., Vijayraghavan, U., Jacobson, A., Martin, N. C., and Abelson, J. (1990). Isolation of a temperature sensitive mutant with altered tRNA nucleotidyl transferase activities and cloning of the gene encoding tRNA nucleotidyl-transferase in the yeast *Saccharomyces cerevisiae*. *J. Biol. Chem.* **265,** 16216–16220.

Aota, S., Gojobori, T., Ishibashi, F., Maruyama, T., and Ikemura, T. (1988). Codon usage tabulated from the genBank genetic sequence data. *Nucleic Acids Res. Sequences Suppl.* **16,** r315–r402.

Atwater, J. A., Wisdom, R., and Verma, I. M. (1990). Regulated mRNA stability. *Annu. Rev. Genet.* **24,** 519–541.

Bach, M.-L., Lacroute, F., and Botstein, D. (1979). Evidence for transcriptional regulation of orotidine-5′-phosphate decarboxylase in yeast by hybridization of mRNA to the yeast structural gene cloned in *Escherichia coli*. *Proc. Natl. Acad. Sci. USA* **76,** 386–390.

Barker, G. F., and Beemon, K. (1991). Nonsense codons within the Rous sarcoma virus *gag* gene decrease the stability of unspliced viral RNA. *Mol. Cell. Biol.* **11,** 2760–2768.

Baker, B. (1989). Sex in flies: The splice of life. *Nature (London)* **340,** 521–524.

Baker, D., and Schekman, R. (1989). Reconstitution of protein using broken spheroplasts. *Methods Cell Biol.* **3,** 127–141.

Baumann, B., Potash, M. J., and Kohler, G. (1985). Consequences of frameshift mutations at the immunoglobulin heavy chain locus of the mouse. *EMBO J.* **4,** 351–359.

Bennetzen, J. L., and Hall, B. D. (1982). Codon selection in yeast. *J. Biol. Chem.* **257,** 3026–3031.

Bernstein, P., Peltz, S. W., and Ross, J. (1989). The poly(A)–poly(A)-binding protein complex is a major determinant of mRNA stability *in vitro*. *Mol. Cell. Biol.* **9,** 659–670.

Bloch, J. C., Perrin, F., and Lacroute, F. (1978). Yeast temperature sensitive mutants impaired in processing of poly(A)-containing RNAs. *Mol. Gen. Genet.* **165,** 123–127.

Brewer, G., and Ross, J. (1988). Poly(A) shortening and degradation of the 3′ A+U-rich sequences of human *c-myc* mRNA in a cell-free system. *Mol. Cell. Biol.* **8,** 1697–1708.

Burd, C. G., Matunis, E. L., and Dreyfuss, G. (1991). The multiple RNA-binding domains of the mRNA poly(A)-binding protein have different RNA-binding activities. *Mol. Cell. Biol.* **11,** 3419–3424.

Chang, C.-H., and Yarbro, J. W. (1978). Comparative effects of 1,10-phenanthroline on DNA and RNA synthesis in mouse spleen. *Life Sci.* **22,** 1007–1010.

Chang, C.-H., Yarbro, J. W., Mann, D. E., Jr., and Gautieri, R. F. (1978). Effects of 1,10-phenanthroline and a zinc complex of 1,10-phenanthroline on nucleic acid synthesis in mouse liver and spleen. *J. Pharmacol. Exp. Ther.* **205,** 27–32.

Chia, L-L., and McLaughlin, C. (1979). The half-life of mRNA in *Saccharomyces cerevisiae*. *Mol. Gen. Genet.* **170,** 129–135.

Cleveland, D. W. (1988). Autoregulated instability of tubulin mRNAs: A novel eukaryotic regulatory mechanism. *TIBS* **13,** 339–343.

Culbertson, M. R., Underbrink, K. M., and Fink, G. R. (1980). Frameshift suppression in *Saccharomyces cerevisiae*. II. Genetic properties of Group II suppressors. *Genetics* **95,** 833–853.

Daar, I. O., and Maquat, L. E. (1988). Premature translation termination mediates triosephosphate isomerase mRNA degradation. *Mol. Cell. Biol.* **8,** 802–813.

Dahlberg, A. (1989). The functional role of ribosomal RNA in protein synthesis. *Cell* **57,** 525–529.

Dams, E., Hendriks, L., Van de Peer, Y., Neefs, J.-M., Smits, G., Vandenbempt, I., and De Wachter, R. (1988). Compilation of small ribosomal subunit RNA sequences. *Nucleic Acids Res.* **16**(suppl.), r87–r173.

Dauber, D. (1973). Isolation and characterization of two nucleases from baker's yeast *Saccharomyces cerevisiae*. *Z. Lebensm.-Unters. Forsch.* **152,** 208–215.

D'Aurora, V., Stern, A. M., and Sigman, D. S. (1978). 1,10-Phenanthroline–cuprous ion complex, a potent inhibitor of DNA and RNA polymerase. *Biochem. Biophys. Res. Commun.* **80,** 1025–1032.

Donahue, T. F., Farabaugh, P. J., and Fink, G. R. (1981). Suppressible glycine and proline four base codons. *Science* **212,** 455–457.

Dykstra, C. C., Hamatake, R. K., and Sugino, A. (1990). DNA strand transfer protein β from yeast mitotic cells differs from strand transfer protein α from meiotic cells. *J. Biol. Chem.* **265,** 10968–10973.

Dykstra, C. C., Kitada, K., Clark, A. B., Hamatake, R. K., and Sugino, A. (1991). Cloning and characterization of DST2, the gene for DNA strand transfer protein β from *Saccharomyces cerevisiae. Mol. Cell. Biol.* **11,** 2583–2592.

Fahrner, K., Yarger, J., and Hereford, L. (1980). Yeast histone mRNA is polyadenylated. *Nucleic Acids Res.* **8,** 5725–5737.

Fort, P., Rech, J., Vie, A., Piechaczyk, M., Bonnieu, A., Jeanteur, P., and Blanchard, J. M. (1987). Regulation of *c-fos* gene expression in hamster fibroblasts: Initiation and elongation of transcription and mRNA degradation. *Nucleic Acids Res.* **15,** 5657–5667.

Gaber, R. F., and Culbertson, M. R. (1984). Codon recognition during frameshift suppression in *Saccharomyces cerevisiae. Mol. Cell. Biol.* **4,** 2052–2061.

Gaspar, M.-L., Meo, T., Bourgarel, P., Guenet, J.-L., and Tosi, M. (1991). A single base deletion in the *Tfm* androgen receptor gene creates a short-lived messenger RNA that directs internal translation initiation. *Proc. Natl. Acad. Sci. USA* **88,** 8606–8610.

Gozalbo, D., and Hohmann, S. (1990). Nonsense suppressors partially revert the decrease of the mRNA level of a nonsense mutant allele in yeast. *Curr. Genet.* **17,** 77–79.

Graves, R. A., Pandey, N. B., Chodchoy, N., and Marzluff, W. F. (1987). Translation is required for regulation of histone mRNA degradation. *Cell* **48,** 615–626.

Groner, B., Hynes, N., and Phillips, S. (1974). Length heterogeneity in the poly(adenylic acid) region of yeast messenger ribonucleic acid. *Biochemistry* **13,** 5378–5383.

Grunstein, M. (1990a). Nucleosomes: Regulators of transcription. *Trends Genet.* **6,** 395–400.

Grunstein, M. (1990b). Histone function in transcription. *Annu. Rev. Cell Biol.* **6,** 643–678.

Heaton, B., Decker, C., Muhlrad, D., Donahue, J., Jacobson, A., and Parker, R. (1992). Analysis of chimeric mRNAs identifies two regions within the *STE3* mRNA which promote rapid mRNA decay. *Nucleic Acids Res.* **20,** 5365–5373.

Hereford, L. M., Osley, M. A., Ludwig J. R., II, and McLaughlin, C. (1981). Cell-cycle regulation of yeast histone mRNA. *Cell* **24,** 367–375.

Herrick, D. (1989). Structural Determinants of mRNA Turnover in Yeast. Ph.D. thesis, University of Massachusetts Medical School, Worcester, Massachusetts.

Herrick, D., and Jacobson, A. (1992). A segment of the coding region is necessary but not sufficient for rapid decay of the *HIS3* mRNA in yeast. *Gene,* **114,** 35–41.

Herrick, D., Parker, R., and Jacobson, A. (1990). Identification and comparison of stable and unstable mRNAs in the yeast *Saccharomyces cerevisiae. Mol. Cell. Biol.* **10,** 2269–2284.

Herruer, M. H., Mager, W. H., Raue, H. A., Vreken, P., Wilms, E., and Planta, R. J. (1988). Mild temperature shock affects transcription of yeast ribosomal protein genes as well as the stability of their mRNAs. *Nucleic Acids Res.* **16,** 7917–7929.

Hicke, L., and Schekman, R. (1990). Molecular machinery required for protein transport from the endoplasmic reticulum to the Golgi complex. *Bioessays* **12,** 253–258.

Hinnebusch, A. G., and Liebman, S. W. (1991). Protein synthesis and translational control in *Saccharomyces cerevisiae. In* The Molecular and Cellular Biology of the Yeast *Saccharomyces,* Vol. I, Genome Dynamics, Protein Synthesis, and Energetics (J. R. Broach, J. R. Pringle, and E. W. Jones, Eds.), pp. 627–735. Cold Spring Harbor Laboratory Press, Cold Spring Harbor, New York.

Hoekema, A., Kastelein, R. A., Vasser, M., and deBoer, H. A. (1987). Codon replacement in the *PGK1* gene of *Saccharomyces cerevisiae:* Experimental approach to study the role of biased codon usage in gene expression. *Mol. Cell. Biol.* **7,** 2914–2924.

Hutchison, H. T., Hartwell, L. H., and McLaughlin, C. S. (1969). Temperature-sensitive yeast mutant defective in ribonucleic acid production. *J. Bacteriol.* **99,** 807–814.

Hynes, N. E., and Phillips, S. L. (1976). Turnover of polyadenylate-containing ribonucleic acid in *Saccharomyces cerevisiae. J. Bacteriol.* **125,** 595–600.

Ikemura, T. (1982). Correlation between the abundance of yeast transfer RNAs and the occurrence of the respective codons in protein genes. *J. Mol. Biol.* **158,** 573–597.

Jackson, R. J., and Standart, N. (1990). Do the poly(A) tail and 3′ untranslated region control mRNA translation. *Cell* **62,** 15–24.

Jacobson, A., Brown, A. H., Donahue, J. L., Herrick, D., Parker, R. and Peltz, S. W. (1990). Regulation of mRNA stability in yeast. *In* Post-Transcriptional Regulation of Gene Expression (J. E. G. McCarthy, and M. F. Tuite, Eds.), pp. 45–54. Springer-Verlag, Berlin.

Jimenez, A., Tipper, D. J., and Davies, J. (1973). Mode of action of thiolutin, an inhibitor of macromolecular synthesis in *Saccharomyces cerevisiae*. *Antimicrob. Agents Chemother.* **3,** 729–738.

Johnston, M. (1987). A model fungal gene regulatory mechanism: The GAL genes of *Saccharomyces cerevisiae*. *Microbiol. Rev.* **51,** 458–476.

Kammerer, B., Guyonvarch, A., and Hubert, J. C. (1984). Yeast regulatory gene *PPR1*. I. Nucleotide sequence, restriction map, and codon usage. *J. Mol. Biol.* **180,** 239–250.

Kelly, K., Cochran, B. H., Stiles, C. D., and Leder, P. (1983). Cell specific regulation of the *c-myc* gene by lymphocyte mitogens and platelet-derived growth factor. *Cell* **35,** 603–610.

Kenan, D. J., Query, C. O., and Keene, J. D. (1991). RNA recognition: Towards identifying determinants of specificity. *TIBS* **16,** 214–220.

Kim, C. H., and Warner, J. R. (1983). Messenger RNA for ribosomal proteins in yeast. *J. Mol. Biol.* **165,** 79–89.

Kim, J., Ljungdahl, P. O., and Fink, G. R. (1990). kem mutations affect nuclear fusion in *Saccharomyces cerevisiae*. *Genetics* **126,** 799–812.

Koch, H., and Friesen, J. D. (1979). Individual messenger RNA half lives in *Saccharomyces cerevisiae*. *Mol. Gen. Genet.* **170,** 129–135.

Kozak, M. (1991). Structural features in eukaryotic mRNAs that modulate the initiation of translation. *J. Biol. Chem.* **286,** 19867–19870.

Krishnamurti, C., Saryan, L. A., and Petering, D. H. (1980). Effects of ethylenediaminetetraacetic acid and 1,10-phenanthroline on cell proliferation and DNA synthesis of Erlich ascites cells. *Cancer Res.* **40,** 4092–4099.

Krowczynska, A., Yenofsky, R., and Brawerman, G. (1985). Regulation of messenger RNA stability in mouse erthroleukemia cells. *J. Mol. Biol.* **181,** 231–239.

Kuo, S. C., Cano, F. R., and Lampen, J. O. (1973). Lomofungin, an inhibitor of ribonucleic acid synthesis in yeast protoplasts: Its effect on enzyme formation. *Antimicrob. Agents Chemother.* **3,** 716–722.

Larimer, F. W., and Stevens, A. (1990). Disruption of the gene *XRN1*, coding for a 5′–3′ exoribonuclease, restricts yeast cell growth. *Gene* **95,** 85–90.

Lee, S. Y., Nakao, Y., and Bock, R. M. (1968). The nucleases of yeast. II. Purification, properties and specificity of an endonuclease of yeast. *Biochim. Biophys. Acta* **151,** 126–136.

Leeds, P., Peltz, S. W., Jacobson, A., and Culbertson, M. R. (1991). The product of the yeast *UPF1* gene is required for rapid turnover of mRNAs containing a premature translational termination codon. *Genes Dev.* **5,** 2303–2314.

Leeds, P., Wood, J. M., Lee, B.-S., and Culbertson, M. R. (1992). Gene products that promote mRNA turnover in *Saccharomyces cerevisiae*. *Mol. Cell. Biol.* **12,** 2165–2177.

Liljelund, P., Losson, R., Kammerer, B., and Lacroute, F. (1984). Yeast regulatory gene *PPR1*. II. Chromosomal localization, meiotic map, suppressibility, dominance/recessivity, and dosage effect. *J. Mol. Biol.* **180,** 251–265.

Lim, S.-K., Sigmund, C. D., Gross, K. W., and Maquat, L. E. (1992). Nonsense codons in human β-globin mRNA result in the production of mRNA degradation products. *Mol. Cell. Biol.* **12,** 1149–1161.

Losson, R., Fuchs, R. P. P., and Lacroute, F. (1983). *In vivo* transcription of a eucaryotic regulatory gene. *EMBO J.* **2,** 2179–2184.

Losson, R., and Lacroute, F. (1979). Interference of nonsense mutations with eukaryotic messenger RNA stability. *Proc. Natl. Acad. Sci. USA* **76,** 5134–5137.

Lowell, J. E., Rudner, D. Z., and Sachs, A. B. (1992). 3′-UTR-dependent deadenylation by the yeast poly(A) nuclease. *Genes Dev.* **6,** 2088–2099.

Lycan, D. E., Osley, M. A., and Hereford, L. M. (1987). Role of transcriptional and posttranscriptional regulation in expression in histone genes in *Saccharomyces cerevisiae. Mol. Cell. Biol.* **7,** 614–621.

Mankin, A. S., Skyrabin, K. G., and Rubtsov, P. M. (1986). Identification of ten additional nucleotides in the primary structure of yeast 18S rRNA. *Gene* **44,** 143.

Maquat, L. E., Kinniburgh, A. J., Rachmilewitz, E. A., and Ross, J. (1981). Unstable β-globin mRNA in mRNA-deficient $\beta°$ thalassemia. *Cell* **27,** 543–553.

Marbaix, G., Huez, G., Burny, A., Cleuter, Y., Hubert, E., Leclercq, M., Chantrenne, H., Soreq, H., Nudel, U., and Littauer, U. Z. (1975). Absence of polyadenylate segment in globin messenger RNA accelerates its degradation in *Xenopus* oocytes. *Proc. Natl. Acad. Sci. USA* **72,** 3065–3067.

Mead, D. J., and Oliver, S. G. (1983). Purification and properties of a double-stranded ribonuclease from the yeast *Saccharomyces cerevisiae. Eur. J. Biochem.* **137,** 501–507.

Medford, R. M., Wydro, R. M., Nguyen, H. T., and Nadal-Girard, B. (1980). Cytoplasmic processing of myosin heavy chain messenger RNA: Evidence provided by using a recombinant DNA plasmid. *Proc. Natl. Acad. Sci. USA* **77,** 5749–5753.

Mercer, J. F. B., and Wake, S. A. (1985). An analysis of the rate of metallothionein mRNA poly(A)-shortening using RNA blot hybridization. *Nucleic Acids Res.* **13,** 7929–7943.

Minvielle-Sebastia, L., Winsor, B., Bonneaud, N. and Lacroute, F. (1991). Mutations in the yeast *RNA14* and *RNA15* genes result in an abnormal mRNA decay rate. Sequence analysis reveals an RNA-binding domain in the *RNA15* protein. *Mol. Cell. Biol.* **11,** 3075–3087.

Moore, P. A., Sagliocco, F. A., Wood, R. M. C., and Brown, A. P. (1991). Yeast glycolytic mRNAs are differentially regulated. *Mol. Cell. Biol.* **11,** 5330–5337.

Morse, D. E., and Yanofsky, C. (1969). Polarity and the degradation of mRNA. *Nature* (*London*) **224,** 329–331.

Muhlrad, D., and Parker, R. (1992). Mutations affecting stability and deadenylation of the yeast *MFA2* transcript. *Genes Dev.* **6,** 2100–2111.

Munroe, D., and Jacobson, A. (1990). Tales of poly(A)—A review. *Gene* **91,** 151–158.

Nakao, Y., Lee, S. Y., Halvorson, H. O., and Bock, R. M. (1968). The nucleases of yeast. I. Properties and variability of ribonucleases. *Biochim. Biophys. Acta* **151,** 114–125.

Nilsson, G., Belasco, J. G., Cohen, S. N., and von Gabain, A. (1987). Effect of premature termination of translation on mRNA stability depends on the site of ribosome release. *Proc. Natl. Acad. Sci. USA* **84,** 4890–4894.

Noller, H. F. (1991). Ribosomal RNA and translation. *Annu. Rev. Biochem.* **60,** 191–227.

Nonet, M., Scafe, C., Sexton, J., and Young, R. (1987). Eucaryotic RNA polymerase conditional mutant that rapidly ceases mRNA synthesis. *Mol. Cell. Biol.* **7,** 1602–1611.

Nudel, U., Soreq, H., and Littauer, U. Z. (1976). Globin mRNA species containing poly(A) segments of different lengths. Their functional stability in *Xenopus* oocytes. *Eur. J. Biochem.* **64,** 115–121.

Ohtaka, Y., Uchida, K., and Salai, T. (1963). Purification and properties of ribonuclease from yeast. *J. Biochem.* **54,** 322–327.

Olson, M. V. (1991). Genome structure and organization in *Saccharomyces cerevisiae. In* The Molecular and Cellular Biology of the Yeast *Saccharomyces,* Vol. I, Genome Dynamics, Protein Synthesis, and Energetics (J. R. Broach, J. R. Pringle, and E. W. Jones, Eds.), pp. 1–39. Cold Spring Harbor Laboratory Press, Cold Spring Harbor, New York.

Osley, M. A., and Hereford, L. M. (1981). Yeast histone genes show dosage compensation. *Cell* **24,** 377–384.

Palatnik, C. M., Storti, R. V., Capone, A. K., and Jacobson, A. (1980). Messenger RNA stability in *Dictyostelium discoideum:* Does poly(A) have a regulatory role? *J. Mol. Biol.* **141,** 99–118.

Pandey, N. B., and Marzluff, W. F. (1987). The stem–loop structure at the 3′ end of histone mRNA is necessary and sufficient for regulation of histone mRNA stability. *Mol. Cell. Biol.* **7,** 4557–4559.

Parker, R., Herrick, D., Peltz, S. W., and Jacobson, A. (1991). Measurement of mRNA decay rates in *Saccharomyces cerevisiae. In* Methods in Enzymology: Molecular Biology of *Saccharomyces cerevisiae* (C. Guthrie, and G. Fink, Eds.), pp. 415–423. Academic Press, New York.

Parker, R., and Jacobson, A. (1990). Translation and a forty-two nucleotide segment within the coding region of the mRNA encoded by the *MAT*α*1* gene are involved in promoting rapid mRNA decay in yeast. *Proc. Natl. Acad. Sci. USA* **87,** 2780–2784.

Pedersen, S. (1984). *Escherichia coli* ribosomes translate *in vivo* with variable rate. *EMBO J.* **3,** 2895–2898.

Pelsy, F., and Lacroute, F. (1984). Effect of ochre nonsense mutations on yeast *URA1* stability. *Curr. Genet.* **8,** 277–282.

Peltz, S. W., Brewer, G., Kobs, G., and Ross, J. (1987). Substrate specificity of the exonuclease activity that degrades H4 histone mRNA. *J. Biol. Chem.* **262,** 9382–8388.

Peltz, S. W., Brewer, G., Bernstein, P., and Ross, J. (1991). Regulation of mRNA turnover in eukaryotic cells. *Crit. Rev. Euk. Gene Exp.* **1,** 99–126.

Peltz, S. W., Brown, A. H., and Jacobson, A. (1993). mRNA destabilization triggered by premature translational termination depends on three mRNA sequence elements and at least one trans-acting factor. Submitted for publication.

Peltz, S. W., Donahue, J. L., and Jacobson, A. (1992). A mutation in tRNA nucleotidyltransferase stabilizes mRNAs in *Saccharomyces cerevisiae. Mol. Cell. Biol.* **12,** 5778–5784.

Phillips, S. L., Tse, C., Serventi, I., and Hynes, N. (1979). Structure of polyadenylic acid in the ribonucleic acid of *Saccharomyces cerevisiae. J. Bacteriol.* **138,** 542–551.

Piper, P. W., Curran, B., Davies, W., Hirst, K., and Seward, K. (1987). *S. cerevisiae* mRNA populations of different intrinsic mRNA stability in unstressed and heat-shocked cells display almost constant m^7GpppA:m^7GpppG 5′-cap ratios. *FEBS Lett.* **220,** 177–180.

Presutti, C., Clafre, S.-A., and Bozzoni, I. (1991). The ribosomal protein L2 in *S. cerevisiae* controls the level of accumulation of its own mRNA. *EMBO J.* **10,** 2215–2221.

Purvis, I. J., Bettany, A. J. E., Loghlin, L., and Brown, A. J. P. (1987). The effects of alterations within the 3′ untranslated region of the pyruvate kinase messenger RNA upon its stability and translation in *Saccharomyces cerevisiae. Nucleic Acids Res.* **15,** 7951–7962.

Robinson, M., Lilley, R., Little, S., Emtage, J. S., Yarranton, G., Stephens, P., Millican, A., Eaton, M., and Humphreys, G. (1984). Codon usage can effect efficiency of translation of genes in *Escherichia coli. Nucleic Acids Res.* **12,** 6663–6671.

Rose, J. K., and Lodish, H. F. (1976). Translation in vitro of vesicular stomatitis virus mRNA lacking 5′-terminal 7-methylguanosine. *Nature (London)* **262,** 32–37.

Ross, J., and Kobs, G. (1986). H4 histone messenger RNA decay in cell-free extracts initiates at or near the 3′ terminus and proceeds 3′ to 5′. *J. Mol. Biol.* **188,** 579–593.

Rubtsov, P. M., Musakhanov, M. M., Zakharyev, V. M., Krayev, A. S., Skyrabin, K. G., and Bayev, A. A. (1980). The structure of the yeast ribosomal RNA genes. I. The complete nucleotide sequence of the 18S ribosomal RNA gene from *Saccharomyces cerevisiae. Nucleic Acids Res.* **8,** 5779–5794.

Ruby, S. W., and Abelson, J. (1991). Pre-mRNA splicing in yeast. *Trends Genet.* **7,** 79–85.

Sachs, A. (1990). The role of poly(A) in the translation and stability of mRNA. *Curr. Opinions Cell. Biol.* **2,** 1092–1098.

Sachs, A. B., Bond, M. W., and Kornberg, R. D. (1986). A single gene from yeast for both nuclear and cytoplasmic polyadenylate-binding proteins: Domain structure and expression. *Cell* **45,** 827–835.

Sachs, A. B., Davis, R. W., and Kornberg, R. D. (1987). A single domain of yeast poly(A)-binding protein is neccessary and sufficient for RNA binding and cell viability. *Mol. Cell. Biol.* **7,** 3268–3276.

Sachs, A. B., and Davis, R. W. (1989). The poly(A)-binding protein is required for poly(A) shortening and 60S ribosomal subunit dependent translation initiation. *Cell* **58,** 857–867.

Sachs, A. B., and Deardorff, J. A. (1992). Translation initiation requires the PAB-dependent poly(A) ribonuclease in yeast. *Cell* **70,** 961–973.

Santiago, T. C., Bettany, A. J. E., Purvis, I. J., and Brown, A. J. P. (1987). Messenger RNA stability in *Saccharomyces cerevisiae:* The influence of translation and poly(A) tail length. *Nucleic Acids Res.* **15,** 2417–2429.

Santiago, T. C., Purvis, I. J., Bettany, A. J. E., and Brown, A. J. P. (1986). The relationship between mRNA stability and length in *Saccharomyces cerevisiae. Nucleic Acids Res.* **14,** 8347–8360.

Saunders, C. A., Bostian, K. A., and Halvorson, H. O. (1980). Post-transcriptional modification of the poly(A) length of galactose-1-phosphate uridyl transferase mRNA in *Saccharomyces cerevisiae. Nucleic Acids Res.* **8,** 3841–3849

Schneider, R. J., Weinberger, C., and Shenk, T. (1984). Adenovirus VA1 RNA facilitates the initiation of translation in virus-infected cells. *Cell* **37,** 291–298.

Sehgal, P. B., Soreq, H., and Tamm, I. (1978). Does 3′-terminal poly(A) stabilize human fibroblast interferon mRNA in oocytes of *Xenopus laevis? Proc. Natl. Acad. Sci. USA* **75,** 5030–5033.

Shapiro, R. A., Herrick, D., Manrow, R. E., Blinder, D., and Jacobson, A. (1988). Determinants of mRNA stability in *Dictyostelium discoideum* amoebae: Differences in poly(A) tail length, ribosome loading, and mRNA size cannot account for the heterogeneity of mRNA decay rates. *Mol. Cell. Biol.* **8,** 1957–1969.

Sharp, P. M., Tuohy, T. M. F., and Mosurski, K. R. (1986). Codon usage in yeast: Cluster analysis clearly differentiates highly and lowly expressed genes. *Nucleic Acids Res.* **14,** 5125–5143.

Siegel, V., and Walter, P. (1989). Assembly of proteins in the endoplasmic reticulum. *Curr. Opinions Cell Biol.* **1,** 635–638.

Simpson, L., and Shaw, J. (1989). RNA editing and the mitochondrial cryptogenes of kinetoplastid protozoa. *Cell* **57,** 355–366.

Sorensen, M. A., Kurland, C. G., and Pedersen, S. (1989). Codon usage determines translation rate in *Escherichia coli. J. Mol. Biol.* **207,** 365–377.

Stevens, A. (1978). An exoribonuclease from *Saccharomyces cerevisiae:* Effect of modifications of 5′ end groups on the hydrolysis of substrates to 5′ nucleotides. *Biochem. Biophys. Res. Commun.* **81,** 656–661.

Stevens, A. (1980). Purification and characterization of a *Saccharomyces cerevisiae* exoribonuclease which yields 5′-monophosphates by a 5′–3′ mode of hydrolysis. *J. Biol. Chem.* **255,** 3080–3085.

Stevens, A. (1985). Pyrimidine-specific cleavage by an endoribonuclease of *Saccharomyces cerevisiae. J. Bacteriol.* **164,** 57–62.

Stevens, A., and Maupin, M. K. (1987). A 5′–3′ exoribonuclease of *Saccharomyces cerevisiae:* Size and novel substrate specificity. *Arch. Biochem. Biophys.* **252,** 339–347.

Stevens, A. (1988). mRNA decapping enzyme from *Saccharomyces cerevisiae:* Purification and unique specificity for long RNA chains. *Mol. Cell. Biol.* **8,** 2005–2010.

Stevens, A., Hsu, C. L., Isham, K. R., and Larimer, F. W. (1991). Fragments of the internal transcribed spacer 1 of pre-rRNA accumulate in *Saccharomyces cerevisiae* lacking 5′–3′ exoribonuclease 1. *J. Bacteriol.* **173,** 7024–7028.

Stimac, E., Groppi, V. E., Jr., and Coffino, P. (1984). Inhibition of protein synthesis stabilizes histone mRNA. *Mol. Cell. Biol.* **4,** 2082–2090.

Stuart, K. (1991). RNA editing in trypanosomatid mitochondria. *Annu. Rev. Microbiol.* **45,** 327–344.

Surguchov, A. P. (1988). Omnipotent nonsense suppressors: New clues to an old puzzle. *TIBS* **13,** 120–123.

Surosky, R. T., and Esposito, R. E. (1992). Early meiotic transcripts are highly unstable in *Saccharomyces cerevisiae. Mol. Cell. Biol.* **12,** 3948–3958.

Tipper, D. J. (1973). Inhibition of yeast ribonucleic acid polymerases by thiolutin. *J. Bacteriol.* **116,** 245–256.

Tishkoff, D., Johnson, A. W., and Kolodner, R. (1991). Molecular and genetic analysis of the gene encoding the *Saccharomyces cerevisiae* strand exchange protein Sep1. *Mol. Cell. Biol.* **11,** 2593–2608.

Tonnesen, T., and Friesen, J. D. (1973). Inhibitors of ribonucleic acid synthesis in *Saccharomyces cerevisiae:* Decay rate of messenger ribonucleic acid. *J. Bacteriol.* **115,** 889–896.

Udvardy, J., Farkas, G. L., Sora, S., and Marre, E. (1972). Purification and properties of a soluble ribonuclease from yeast cells in the stationary phase of growth. *Ital. J. Biochem.* **21,** 122–144.

Van den Heuvel, J. J., Planta, R. J., and Raue, H. A. (1990). Effect of leader primary structure on the translational efficiency of phosphoglycerate kinase mRNA in yeast. *Yeast* **6,** 473–482.

Vreken, P., van der Veen, R., de Regt, V. C. H. F., de Maat, A. L., Planta, R. J., and Raue, H. A. (1991). Turnover rate of yeast *PGK1* mRNA can be changed by specific alterations in its trailer structure. *Biochimie* **73,** 729–737.

Wilson, T., and Triesman, R. (1988). Removal of poly(A) and consequent degradation of *c-fos* mRNA facilitated by 3′ AU-rich sequences. *Nature (London)* **336,** 396–399.

Winey, M., and Culbertson, M. R. (1988). Mutations affecting the tRNA-splicing endonuclease activity of *Saccharomyces cerevisiae. Genetics* **118,** 607–617.

Wisdom, R., and Lee, W. (1991). The protein-coding region of c-*myc* mRNA contains a sequence that specifies rapid mRNA turnover and induction by protein synthesis inhibitors. *Genes Dev.* **5,** 232–243.

Wolin, S. L., and Walter, P. (1988). Ribosome pausing and stacking during translation of a eukaryotic mRNA. *EMBO J.* **7,** 3559–3569.

Xu, H., Johnson, L., and Grunstein, M. (1990). Coding and non-coding sequences at the 3′ end of yeast histone H2B mRNA confer cell-cycle regulation. *Mol. Cell. Biol.* **10,** 2687–2694.

Yost, H. J., and Lindquist, S. (1988). Translation of unspliced transcripts after heat shock. *Science* **242,** 1544–1547.

Young, R. A. (1991). RNA polymerase II. *Annu. Rev. Biochem.* **60,** 689–715.

14

Control of mRNA Degradation in Organelles

WILHELM GRUISSEM AND GADI SCHUSTER

I. Introduction

Mitochondria and chloroplasts are cellular organelles that contain genomes essential for their functions. While mitochondrial genomes vary in size from approximately 14 kB in animal cells to more than 2000 kB in the cells of some plants, the number and types of genes contained in these genomes are remarkably similar. In general, mitochondrial genomes encode ribosomal RNAs of mitochondrial ribosomes, a small, incomplete set of tRNAs, and 10–15 genes that encode a ribosomal protein, proteins important for oxidative phosphorylation, and proteins involved in RNA processing or intron transposition (for review, see Attardi and Schatz, 1988). In contrast, the genomes of chloroplasts, which are organelles found only in plants and algae, are circular molecules approximately 150 kB long that show less size variability. A chloroplast genome encodes approximately 120 different genes, considerably more than a mitochondrial genome (Ohyama *et al.*, 1986; Shinozaki *et al.*, 1986; Hiratsuka *et al.*, 1989). Chloroplast genes specify proteins for photosynthetic functions, several ribosomal proteins, and proteins involved in transcriptional and translational functions. Chloroplast genomes also encode the ribosomal RNAs of chloroplast-specific ribosomes and a complete set of tRNAs (for review, see Ozeki *et al.*, 1989; Sugiura, 1989). The content and arrangement of genes within transcription units are highly conserved among chloroplast genomes from highly divergent plant species (reviewed by Palmer, 1985). However, for both mitochondria and chloroplasts it is now clear that most organelle functions are

Control of Messenger RNA Stability

controlled by a large number of nuclear-encoded proteins that are imported into the organelles (Fig. 1).

The last few years have witnessed a rising interest in the mechanisms that control the spatial and temporal expression of mitochondrial and chloroplast genes in response to developmental and environmental cues. In plants, investigations have been spurred by the fact that chloroplast development in the light is accompanied by significant changes in the accumulation of different mRNAs. While initial efforts to analyze the control of plastid gene expression concentrated on the transcription of genes for photosynthetic proteins, it has been shown recently that considerable changes in mRNA stability are superimposed on the transcriptional regulation of chloroplast gene expression. Less is known about how mRNA synthesis and degradation are integrated in controlling mitochondrial gene expression. It is apparent, however, that the temporal and spatial control of organelle mRNA populations must be coordinated with the expression of nuclear genes for proteins that are integral parts of mitochondrial and chloroplast membrane complexes. This chapter discusses our present understanding of cis- and trans-acting factors that regulate the differential stability of organelle mRNAs (see Table 1 for genes discussed in this review).

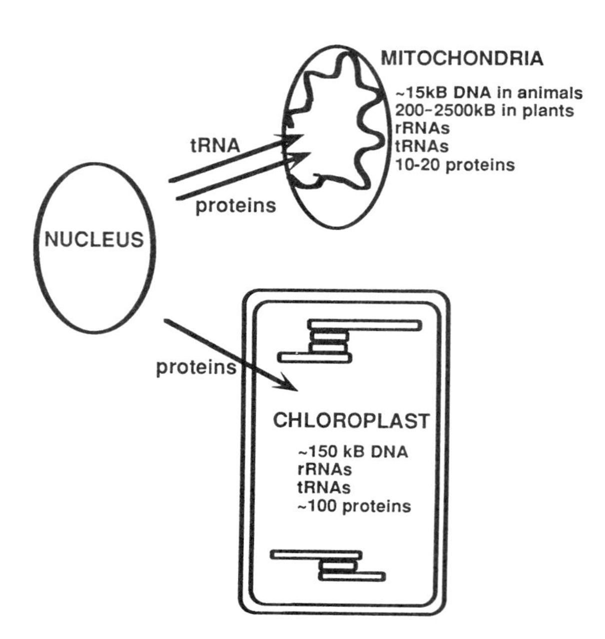

FIGURE 1 Genetic systems of chloroplast and mitochondria. Both organelle genomes encode a limited number of genes for genetic and enzymatic functions. Most, if not all, of the genes that regulate regulatory processes, however, are encoded by the nuclear genome. This includes regulatory proteins and enzymes involved in organelle RNA processing and decay. For details, see text.

II. Post-Transcriptional Control of mRNA Accumulation

A. Control of mRNA Accumulation during Chloroplast Development

1. Differential Accumulation of Plastid mRNAs after Illumination and during Leaf Development

During leaf development, the conversion of proplastids into photosynthetically active chloroplasts is one of the most significant morphogenetic changes of the organelle. The process is accompanied by the assembly of the major thylakoid protein complexes (photosystems I and II, ATP synthase, and cytochrome b_6/f complex), all of which consist of several protein subunits. The principal effector in the development of a functional chloroplast is light (reviewed by Tobin and Silverthorne, 1985), which acts through a complex signal transduction pathway that can be genetically disrupted (e.g., Chory *et al.*, 1989; Chory and Peto, 1990; Deng *et al.*, 1991). Light-induced modifications of populations of plastid transcripts for different photosynthetic proteins have been detected in a number of plant seedlings (for review, see Gruissem *et al.*, 1988; Mullet, 1988; Gruissem, 1989a). The RNAs of developing plastids roughly fall into three classes: RNAs that accumulate transiently or continuously during light-induced chloroplast development, RNAs whose levels remain approximately constant , and RNAs that decrease after illumination. Examples of this differential accumulation pattern of RNA populations are the mRNAs for the D1 protein of photosystem II (*psbA*) and the large subunit of ribulose-1,5-bisphosphate carboxylase (*rbcL*). There is a rapid and several-fold accumulation of *psbA* mRNA after illumination in several plant species, while *rbcL* mRNA levels, which are already high in dark-grown seedlings, do not change significantly in the light (e.g., Deng and Gruissem, 1987). Although differential RNA accumulation has now been reported for several plant species, no consistent pattern for specific chloroplast mRNAs has emerged from these studies, reflecting in part the adaptation of particular regulatory mechanisms for individual organelle mRNAs in different plants.

By comparing the degree of accumulation of several mRNAs during illumination with the relative transcription activity of their genes, it was shown that changes in mRNA stability are a significant factor in the control of chloroplast gene expression (Deng and Gruissem, 1987; Mullet and Klein, 1987). These run-on transcription studies demonstrated that in chloroplasts from evolutionary divergent plants such as spinach and barley, which are strikingly different in their patterns of leaf development, transcription activity and mRNA accumulation are uncoupled for several genes (for examples, see Table 2). The results further implied that change in mRNA stability is an important factor in establishing pools of specific mRNAs. However, in most cases it is not known exactly how changes in mRNA pools affect the translation of the corresponding mRNAs. Although the

TABLE 1 Mitochondrial and Chloroplast Genes and Proteins

Gene	Product
	Mitochondria
atp1	ATP synthase subunit 1
atp6	ATP synthase subunit 6
atp9 or *oli1*	ATP synthase subunit 9
cob	Cytochrome b
cox1, 2, 3	Subunits 1, 2, and 3 of cytochrome c oxidase
var1	Protein of the small ribosomal subunit (*S. Cerevisiae*)
rrn18	18 S ribosomal RNA
rrn26	26 S ribosomal RNA
	Chloroplast
atpBE	ATP synthase subunits β and ε
psbA	D1 protein of photosystem II
psbB	47-kDa chlorophyll binding protein of photosystem II
psbC	43-kDa chlorophyll binding protein of photosystem II
psbD	D2 protein of photosystem II
psaA/B	P700 photosystem I reaction center proteins
petD	Subunit IV of cytochrome b_6/f complex
rbcL	Large subunit of ribulose-1,5-bisphosphate carboxylase
rbs14	Ribosomal protein RPS14
rrn	Ribosomal RNA
	Nuclear
cab	Chlorophyll a/b binding protein of photosystem II
rbcS	Small subunit of ribulose-1,5-bisphosphate carboxylase

TABLE 2 Examples of Relative Transcription Rates and Accumulation of Plastid mRNAs during Light-Induced Chloroplast Development[a]

	Relative transcription rates (%)[c]		Relative mRNA accumulation (%)[d]	
Gene[b]	**Etioplast**[e]	**Chloroplast**[f]	**Etioplast**[e]	**Chloroplast**[f]
rbcL	100	100	91	100
rrn16	500	319	78	100
psaA	74	50	77	100
atpB	40	20	65	100
psbA	160	150	15	100

[a] The numbers in the table are based on results reported by Deng and Gruissem (1987).
[b] For designation of genes, see Table 1.
[c] Transcriptional activities are expressed as percentages of *rbcL* transcriptional activity, which was arbitrarily set to 100%.
[d] RNA accumulation in etioplasts is expressed as a percentage of accumulation in chloroplasts after 24 hr of illumination.
[e] Etioplasts were obtained from 3-day-old cotyledons of dark-grown spinach seedlings.
[f] Chloroplasts were obtained from cotyledons after illumination of 3-day-old dark-grown spinach seedlings for 24 hr.

precise mechanisms that control developmental and light-dependent changes in plastid mRNA turnover are unclear, there is no evidence that any of the known chloroplast genes encode proteins that are directly involved in the post-transcriptional control of mRNA stability. This raises the interesting possibility that most, if not all, of these control functions are performed by nuclear gene products that are imported into the organelle.

2. *Stability Changes of Specific Plastid mRNAs*

The developmental and light-dependent differences in chloroplast mRNA stability are best illustrated by the *psbA* and *rbcL* mRNAs. Both transcripts are monocistronic mRNAs transcribed from promoters that, in spinach chloroplasts, differ less than 2-fold in transcription initiation frequency based on *in vitro* studies using partially fractionated chloroplast RNA polymerase extracts (Gruissem and Zurawski, 1985) or run-on transcription assays in lysed plastids (Deng *et al.*, 1987). The relative transcription activity of the two genes remains essentially unchanged during cotyledon and leaf development, despite changes in the overall transcription activity of the organelle (Deng and Gruissem 1987). However, during cotyledon development in the dark, *rbcL* mRNA accumulates at 7- to 10-fold higher levels relative to *psbA* mRNA, despite the higher transcriptional activity of the *psbA* gene (Table 2). This strongly suggests that the *psbA* mRNA is rapidly degraded in the dark. In contrast, after illumination *psbA* mRNA accumulates within 24 hr to the level of the *rbcL* mRNA, which does not significantly change during this time. The rapid accumulation of *psbA* mRNA after illumination is consistent with a light-dependent stabilization of the transcript, because the relative transcriptional activities of the two genes do not change significantly during this time (for a more detailed discussion, see Gruissem *et al.*, 1988). Similarly, the *psbA* mRNA level increases between 3- and 4-fold over the level of the *rbcL* mRNA during leaf development, but again there is no correlative change in the transcriptional activity of the genes. Together, these differences represent an approximately 30- to 40-fold change in the level of *psbA* mRNA relative to that of the *rbcL* mRNA. This example demonstrates that mechanisms exist in chloroplasts, which can uncouple the accumulation of specific mRNAs from the transcriptional activity of their respective genes.

Changes in the stability of *rbcL* and *psbA* mRNA have also been demonstrated *in vivo* during leaf development (Klaff and Gruissem, 1991). Because pulse–chase experiments to measure mRNA decay are technically difficult in plants, inhibition of transcription with actinomycin D was used to determine changes in the half-life of *rbcL* and *psbA* mRNAs during leaf development. As discussed earlier, the combination of RNA blot analysis and run-on transcription has indicated that the stability of the *psbA* mRNA increases between threefold and fourfold relative to that of *rbcL* mRNA during the development of young leaves into mature leaves. As suggested by the similarity of the relative transcription rates for the two genes at both developmental stages (Deng and Gruissem, 1987), the direct *in vivo* measurements of mRNA half-life confirm that differences in *rbcL* and *psbA* mRNA stability can account for the differences in their accumulation during leaf development (Fig. 2). While the half-lives of *rbcL* and *psbA* mRNAs are similar in young leaves (4.5 and 4.7 hr, respectively), the half-life of the *psbA* mRNA increases to more than 10 hr in mature leaves; there is no corresponding increase in the half-life of the *rbcL* mRNA. These data on chloroplast mRNA half-lives are also interesting for another reason. Chloroplasts in plants are generally believed to have been derived from prokaryotic ancestors, and many features of the chloroplast genome support this view. In contrast to bacterial mRNAs, which have half-lives that generally range from a seconds to several minutes (e.g., Nilsson *et al.*, 1987), the half-lives

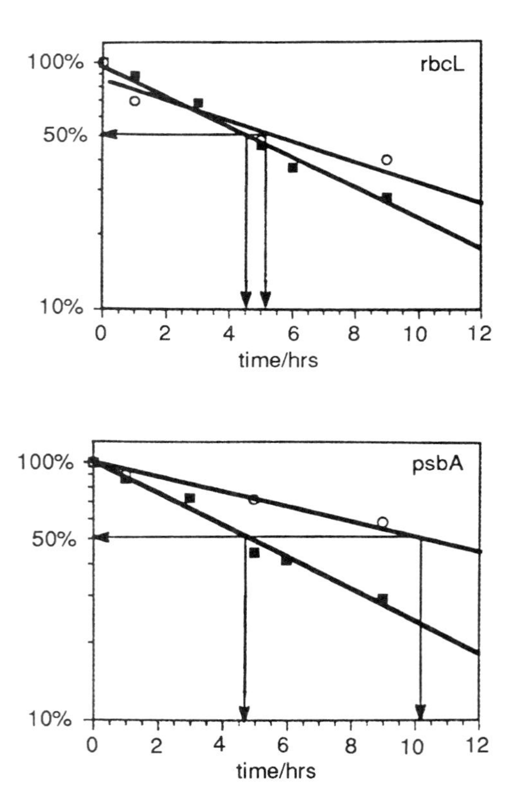

FIGURE 2 Changes in chloroplast mRNA stability during leaf development. The chloroplast *psbA* and *rbcL* mRNAs accumulate to different levels during leaf development, although there is no significant change in the relative transcriptional activity of their genes. The increased stability of the *psbA* mRNA relative to the *rbcL* mRNA can be detected in intact leaves following inhibition of transcription by actinomycin to monitor decay of both mRNAs. ■, young leaves; ○, mature leaves. [For details, see Deng and Gruissem (1987) and Klaff and Gruissem (1991).]

of the two chloroplast mRNAs are considerably longer and more similar to the half-lives of cytoplasmic mRNAs in higher eukaryotes. Thus, it is reasonable to assume that during the coevolution of chloroplasts with plant cells, mechanisms have developed that result in extended half-lives of plastid mRNAs.

B. RNA Half-Lives and Control of mRNA Accumulation in Mitochondria

Changes in stability and turnover of mitochondrial mRNAs are largely unexplored except for a few reports in which measurements of mitochondrial mRNA half-lives have been reported. In contrast, considerable information is available on the transcription of mitochondrial genomes, on RNA processing and intron splicing, and on specific nuclear proteins that affect the stability of individual mRNAs (reviewed by Attardi and Schatz, 1988; Grivell, 1989; Forsburg and Guarente, 1989) (also see below). Earlier studies in HeLa cells have shown that mitochondrial rRNAs and several different mRNAs are metabolically unstable, with half-lives ranging between 2.5 and 3.5 hr for rRNAs and between 25 and 90 min for mRNAs (Gelfand and Attardi 1981). These half-lives were determined by measuring the kinetics of labeling of individual RNA species with [5-^{3}H]uridine and the temporal changes in the specific activity of the precursor pool or by measuring the decay of labeled mitochodrial RNAs after blocking RNA synthesis with cordycepin (3′-deoxyadenosine). It appears that in HeLa cells all mitochondrial mRNAs processed from the precursor transcript of the H-strand (which encodes most of the mitochondrial genes compared with the L-strand) have similar half-lives that are considerably shorter than those of most cytoplasmic mRNAs. It has been suggested that the instability of mitochondrial RNAs is unlikely to reflect a requirement for rapid changes in gene expression, but rather may be a consequence of the high rate of mitochondrial DNA transcription (Attardi *et al.*,, 1982).

The short half-lives of mitochondrial mRNAs appear to be in conflict with results that demonstrate continued high rates of protein synthesis following inhibition of mitochondrial transcription. In enucleated cytoplasts of African green monkey kidney cells, chloramphenicol-sensitive [^{3}H]isoleucine incorporation into mitochondrial proteins continued at an essentially unchanged level for at least 20 hr following enucleation by centrifugation, and there was no detectable change in the profile of the synthesized proteins (England *et al.*, 1978). In mouse L (TK^{-}) cells, synthesis of seven discernable mitochondrial proteins continues without a significant change in rate for at least 48 hr following destruction of mitochondrial DNA containing 5-bromouracil by light, which reduces RNA synthesis by 85% (Lansman and Clayton, 1975). These results cannot be reconciled with the relatively low steady-state level of mitochondrial RNAs in the cells or with the fact that 30–40% of the mRNAs are found on ribosomes. Rather, it has been suggested that the stability of mitochondrial RNAs increases significantly after inhibition of transcription (Attardi *et al.*, 1982; Attardi and Schatz, 1988). This increase in stability would be consistent with a model that mitochondrial RNA turnover is coupled to RNA synthesis or RNA processing events and that mitochondrial RNA accumulation is partly controlled at the level of decay. Another implication of these results is that mitochondrial mRNA levels do not appear to be limiting for protein synthesis. Similar observations have been made in chloroplasts in which levels of DNA and mRNAs were reduced without a discernable effect on protein synthesis (Hosler *et al.*, 1989; Stern *et al.*, 1991).

Considerably less information is available on the half-lives of plant mitochondrial mRNAs or the mechanisms that control changes in their stability. Recent studies using transcription run-on assays in isolated mitochondria from maize tissue culture cells or etiolated seedlings have revealed differences in the transcription of several mitochondrial genes and in the accumulation of their corresponding mRNAs (Finnegan and Brown, 1990; Mulligan *et al.*, 1991). As in animal and yeast mitochondria, the plant rRNA genes are most highly transcribed, and rRNA constitutes the bulk of mitochondrial RNA (Table 3). Genes that encode proteins are transcribed at lower rates, with *atp1* showing the highest activity, followed by *atp6*, *atp9*, *cob*, and *cox3*. These mRNAs accumulate to significantly different levels that do not reflect their relative rates of synthesis, indicating that as in plant chloroplasts, mRNA accumulation can be uncoupled from RNA synthesis

TABLE 3 Examples of Relative Transcription Rates and Accumulation of Mitochondrial mRNAs in Plants[a]

Gene[b]	Relative transcription rate (%)[c]	Relative mRNA accumulation (%)[d]
rrn18	100	100
rrn26	65	83
atp9	16	61
atp1	46	18
cox3	7.5	14
cob	11	4.5
atp6	25.5	3

[a] The numbers in the table are approximate values based on the results reported by Mulligan *et al.* (1991).
[b] For designations of genes, see Table 1.
[c] The numbers represent the relative transcriptional activity as percentages of the *rrn18* transcriptional activity.
[d] The numbers represent the accumulation of RNA as percentages of *rrn18* RNA accumulation.

(Mulligan *et al.*, 1991). Together, these results demonstrate that mRNA stability and turnover are important in the regulation of mitochondrial RNA populations in animals and plants.

C. Regulation of Organelle mRNA Stability by Temporal and External Signals

Unlike mitochondria, chloroplasts can differentiate into nonphotosynthetic plastids that have different functions, such as starch-containing amyloplasts in root or chromoplasts in fruit, certain flower petals, and roots. Differentiation of nonphotosynthetic plastids from proplastids or chloroplasts is not a terminal process, as shown by the fact that specialized plastids maintain the capability to redifferentiate into chloroplasts (reviewed by Thomson and Whatley, 1980). Considering the differentiation capacity of plastids and the fact that the plastid genome encodes more than 50 genes for proteins involved in transcription and translation of plastid mRNAs, these plastids must maintain transcription and translation functions at all times to provide the components for basic genetic functions within the organelle (for a more detailed discussion, see Gruissem 1989a, 1989b). A similar argument, although not quite as compelling, could be made for mitochondria, which encode rRNA as a component of their translational machinery. Mitochondria in yeast, *Neurospora,* and maize also contain a gene that encodes a ribosomal protein, although this gene is not present in the mitochondrial genome of other organisms (see Attardi and Schatz, 1988). It is therefore of interest, especially in plants, to learn how the mRNAs for photosynthetic proteins are specifically targeted for degradation in nonphotosynthetic plastids in which the protein products are not needed.

1. Root Amyloplasts Accumulate mRNAs for Photosynthetic Proteins

In roots of spinach seedlings, mRNAs are detectable for *psbA, psaA, rbcL,* and *atpBE,* although their levels are considerably lower than those in cotyledons or leaves (Deng and Gruissem, 1988). In addition, the relative accumulation of individual mRNAs differs from that in the chloroplasts of leaves (Table 4). For example, *psbA* mRNA is highly abundant in spinach leaf chloroplasts and accumulates to approximately four times the level of *rbcL* mRNA, but in amyloplasts (and in etioplasts of dark-grown seedlings) *rbcL* mRNA levels exceed those for *psbA* mRNA. These differences in transcript accumulation are not reflected by the relative transcriptional activities of the genes, which, although lower, are similar in root amyloplasts when compared with leaf chloroplasts (Deng and Gruissem, 1987, 1988). The 5′ and 3′ end processing as well as splicing of introns in mRNAs for photosynthetic proteins is normal in amyloplasts (Deng and Gruissem, 1988; Barkan, 1989). These results suggest that the differential accumulation of a number of plastid mRNAs is regulated by modulating their stability. Furthermore,

TABLE 4 Examples of Plant mRNA Levels for Specific Photosynthetic Proteins in Different Plastid Types[a]

Plastid type	mRNA levels[b]					
	16S *rrn*	*psbA*	*psaA*	*atpBE*	*atpA*	*rbcL*
Etioplast (cotyledon)[c]	78	15	77	65	—	91
Etioplast (hypocotyl)[d]	5	2	4	7	—	2
Chloroplast (fruit)[e]	44	70	21	—	23	24
Chromoplast (fruit)[f]	16	20	ND	—	11	12
Amyloplast (root)[g]	1	0.1	0.5	1.7	—	0.3

[a] The data in this table were obtained from Deng and Gruissem (1987, 1988) and Piechulla *et al.* (1986).
[b] Numbers are expressed as percentages of mRNA levels in mature leaf chloroplasts.
[c] Etioplast mRNA levels were obtained from 3-day-old cotyledons of dark-grown spinach seedlings.
[d] Etioplast mRNA levels were obtained from hypocotyl segments of 3-day-old dark-grown spinach seedlings.
[e] Chloroplast mRNA levels were obtained from mature green tomato fruit.
[f] Chromoplast mRNA levels were obtained from red, ripe tomato fruit.
[g] Amyloplast mRNA levels were obtained from roots of hydroponically grown spinach plants.

they strongly imply that the plastid genome in plants is constitutively transcribed, thus emphasizing the importance of mRNA decay in post-transcriptional control of gene expression during plant development.

2. *Differential Decay Rates of mRNAs during Chromoplast Development*

The differentiation of chloroplasts into colored, carotenoid-containing chromoplasts is one of the most visible changes during ripening in many fruits. Concomitant with this process is the degradation of the chloroplast thylakoid membrane system and the disappearance of proteins for photosynthetic functions (e.g., Piechulla *et al.*, 1987). Despite these differentiation events, mRNAs for several photosynthetic proteins have been detected in chromoplasts of tomato fruit (Piechulla *et al.*, 1985, 1986) and pepper (Gounaris and Price, 1987). In tomato, the presence of these mRNAs correlates with the transcription activity of the corresponding genes (Gruissem *et al.*, 1988), suggesting that, as in root amyloplasts, the genes are constitutively transcribed in red fruit chromoplasts, although at a reduced level (Table 4). However, the individual mRNA levels are not correlated with the relative transcriptional activities of the genes, again suggesting that developmentally controlled changes in mRNA decay are responsible for the differential accumulation of the mRNAs. For example, *psbA* mRNA in ripe tomato fruit is still present at 20% of the leaf control levels (see Table 4). In comparison, the *psaA* mRNA, which encodes a core protein of photosystem

I, is undetectable in ripe fruit. The *rbcL* mRNA shows an intermediate pattern and in ripe tomato fruit is still present at approximately 13% of the leaf control level. In contrast, the *rbcL* and *psaA* mRNAs accumulate to approximately similar levels in fruit chloroplasts just before the onset of chromoplast differentiation (Gruissem, 1989c). The relative transcription rates of both genes are similar in both chloroplasts and chromoplasts, but the transcription activity is reduced in ripe tomato fruit. Although no precise mRNA half-live measurements have been obtained during fruit ripening, the uncoupling of mRNA accumulation from the transcriptional activities of the genes indicates that changes in plastid mRNA decay are controlled by the plant developmental program.

3. *Light Quality Affects the Stability of Specific Chloroplast mRNAs*

Photosynthetically active chloroplasts adapt to different environmental light qualities by modulation of thylakoid membrane complexes. For example, the chloroplasts from plants grown in photosystem II-sensitizing yellow light are enriched in photosystem I (PS I) complexes, while plants grown in red light preferentially absorbed by photosystem I are enriched in photosystem II (PS II) complexes (Melis, 1984; Glick *et al.*, 1986). These structural reorganizations of the components of the thylakoid membrane occur in response to the imbalance in light absorption between PS II and PS I and the resulting imbalance in electron flow between the two photosystems. The ultrastructural and biochemical changes caused by the two different light qualities are reflected in differences in the leves of the plastid mRNAs for PS II and PS I proteins (Glick *et al.*, 1986; Deng *et al.*, 1989). For example, in pea and spinach leaf chloroplasts grown under PS I-sensitizing red light, there is a 5- and 15-fold decrease, respectively, in the *psaA* mRNA, which encodes the PS I P700 chlorophyll *a*-apoprotein (Table 5). Other mRNAs show significantly less change in their accumulation in the different light conditions, suggesting that the reduced *psaA* mRNA level is specifically controlled and is not the result of general changes in chloroplast mRNA stability. Although there is a small decrease in the relative transcriptional activity of the *psaA* gene in chloroplasts from plants growing in red light, this decrease is not sufficient to explain the reduction in *psaA* mRNA accumulation (Deng *et al.*, 1989). It appears, therefore, that red light can activate a signal transduction pathway that results in the selective destabilization of the *psaA* mRNA, which is consistent with the reduced accumulation of the P700 chlorophyll *a*-apoprotein under these physiological conditions.

4. *Effect of Carotenoid Synthesis on Chloroplast mRNA Accumulation*

Besides being influenced by the developmental program and by light, chloroplast development and expression of genes for photosynthetic proteins can also be affected by alterations in biochemical pathways and photosynthetic intermediates (e.g., Harpster *et al.*, 1984; Thelander *et al.*, 1986;

TABLE 5 Relative Transcription Rates and mRNA Accumulation in Chloroplasts from Plants Growing under Different Light Qualities[a]

	Relative transcription rates (%)[c]		Relative mRNA accumulation (%)[d]	
Gene[b]	Yellow light[e]	Red light[f]	Yellow light[e]	Red light[f]
psaA	NC[g]	75	106	7
psbA	NC	NC	94	90
psbB	NC	NC	94	49
rbcL	NC	NC	109	44
atpB	NC	NC	112	56

[a] The numbers in the table are based on results reported by Deng *et al.* (1989).
[b] For designation of genes, see Table 1.
[c] Relative transcription rates are expressed as percentages of the transcriptional activities of the chloroplast genes from plants growing in white light.
[d] RNA accumulation is expressed as a percentage of RNA accumulation in chloroplasts from plants growing in white light.
[e] Yellow light preferentially excites photosystem II complexes.
[f] Red light preferentially excites photosystem I complexes.
[g] NC indicates that no change in relative transcription rates was detected.

Scolnick *et al.*, 1987; Sheen, 1990) (for review, see Gruissem, 1989a; Taylor, 1989). For example, carotenoids in higher plants protect chlorophyll from photooxidation under normal or high light conditions. The result of reduced carotenoid biosynthesis in plants is particularly dramatic and usually causes bleaching of chloroplasts due to photodestruction of chlorophyll, which is followed by plant death. Carotenoid-deficient mutants have altered chloroplast membrane structures and are deficient in chloroplast membrane proteins owing, in at least some cases, to a block in the synthesis of certain thylakoid membrane proteins or to their rapid turnover (Harpster *et al.*, 1984; Mayfield *et al.*, 1986). The carotenoid-deficient mutant phenotype can also be produced after application of specific herbicides, such as norflurazon, which block synthesis of carotenoids. The photooxidative destruction of chlorophyll, and possibly certain thylakoid membrane proteins, reduces the transcription activity of the nuclear *rbcS* and *cab* genes in both carotenoid-synthesis mutants and herbicide-treated plants, although the mechanism of this effect is not understood (Burgess and Taylor, 1988; Giuliano and Scolnik, 1988). The block in carotenoid biosynthesis does not completely inactivate the transcription of plastid genes even in fully bleached leaves, and consequently most of the mRNAs are still produced even though their levels are reduced relative to untreated leaves (Sagar and Briggs, 1990; Tonkyn *et al.*, 1992). Relative transcriptional activities and mRNA levels correlate well for most plastid genes in herbicide-treated leaves, although the stability of certain mRNAs (*rbcL, psbA,* and *psaA*) is altered, again indicating that the decay of plastid mRNAs is selectively and tightly controlled in response to different signals.

III. Roles of Nuclear Proteins in Organelle mRNA Stability

Research in several laboratories has demonstrated that the integration of the processes that lead to the formation of functional mitochondria and chloroplasts requires complex control mechanisms involving nuclear and organelle gene activity and interactions, RNA processing, protein synthesis, protein transmembrane transport, and assembly of protein complexes. As expected from the limited number of genes encoded by the mitochondrial and chloroplast genomes, regulatory processes in mitochondrial and chloroplast biogenesis are almost exclusively controlled by nuclear genes (see Table 6). In yeast, mutations in single nuclear genes can inactivate respiration, which results in a nuclear *petite* (*pet*) phenotype (reviewed by Tzagoloff and Myers, 1986; Tzagoloff and Dieckmann, 1990). Similarly, single nuclear mutations can affect chloroplast functions in algae and higher plants and are often lethal (reviewed by Somerville, 1986; Rochaix and Erickson, 1988). Higher plants can survive nuclear mutations affecting chloroplast function if the mutations are caused by unstable transposable elements. In such cases mosaic leaves are formed that support plant growth when sufficiently large sections have cells with functional chloroplasts. While a large number of nuclear mutations have been identified that affect many different processes in mitochondria and chloroplasts, including RNA processing and

TABLE 6 Nuclear Genes That Affect Organelle RNA Stability

Organelle	Nuclear gene	Organelle genes affected	Organism	References
Mitochondria	*CPB1*	*cob*	Yeast	McGraw and Tzagoloff, 1983; Dieckmann *et al.*, 1984; Dieckmann and Mittelmeir, 1987
	AEP2	*Oli1*	Yeast	Payne *et al.*, 1991
	cyt-4	Several (rRNA, mRNAs, and tRNA)	*N. Crassa*	Garriga *et al.*, 1984; Dovidson *et al.*, 1989; Turcq *et al.*, 1991
	SUV3	*var1, cob, cox1*	Yeast	Conrad-Webb *et al.*, 1991
Chloroplast	*NAC2*	*psbD*	*C. reinhardtii*	Kuchka *et al.*, 1989
	GE2.10	*psbB* and others	*C. reinhardtii*	Sieburth *et al.*, 1991
	6.2z5	*psbC*	*C. reinhardtii*	Sieburth *et al.*, 1991
	hcf-38	*atpB/E, psaA/B*	Maize	Barkan *et al.*, 1986

intron splicing (reviewed by Attardi and Schatz, 1988; Grivell, 1989; Chasan, 1991) or translation (reviewed by Fox, 1986; Rochaix and Erickson, 1988), there is a scarcity of nuclear mutations that specifically affect the stability of organelle mRNAs.

A. Nuclear Genes Affecting Mitochondrial mRNA Stability

At least five nuclear gene products are necessary and specific for the expression of the yeast mitochondrial cytochrome b gene (*cob*) (Dieckmann *et al.*, 1984; Dieckmann and Tzagollof, 1985; Rodell *et al.*, 1986). Of those, CBP1 and CBP2 are required for *cob* transcript maturation (Fig. 3). Mutations in *CBP2* block excision of the terminal intron from the *cob* mRNA precursor, while CBP1 is required for the stability of the mature *cob* transcript (McGraw and Tzagoloff, 1983). In *cbp1* mutant strains, the *cob* transcript is degraded after RNA processing releases $tRNA^{Glu}$ from the dicistronic precursor RNA. This demonstrates that the effect of CBP1 is specific (Dieckmann *et al.*, 1984). The analysis of transcripts produced by mitochondrial gene fusions, in which portions of the *cob* gene were replaced by analogous portions of the *oli1* gene (coding for ATP synthase subunit 9), has shown that CBP1 appears to stabilize the *cob* transcript via interactions with the 5′ untranslated leader sequence of the *cob* mRNA (Dieckmann *et al.*, 1984; Dieckmann

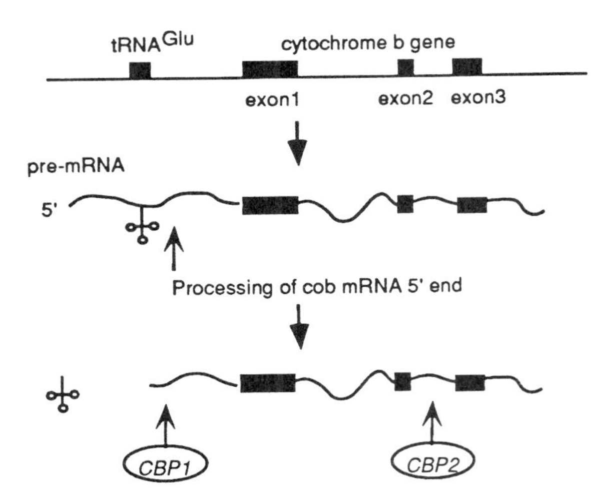

FIGURE 3 Processing and stability of the yeast mitochondrial cytochrome b (*cob*) mRNA. The *cob* mRNA is cotranscribed with the $tRNA^{Glu}$ gene into a polycistronic RNA from which the $tRNA^{Glu}$ is cleaved. The nuclear genes *CBP1* and *CBP2* are both involved in the *cob* mRNA maturation pathway. Mutations in *CBP2* block excision of the terminal intron in the cytochrome b coding sequence, while *CBP1* is required for the stability of the cytochrome b transcripts. [For details, see Mittelmeier and Dieckmann (1990).]

and Mittelmeier, 1987). The stabilization function of CBP1 does not require other sequences of the *cob* transcript, since transcripts from a *cob–oli1* fusion gene, in which the *cob* leader sequence was placed upstream of the *oli1* leader and coding region, are stable in wild-type strains but not *cbp1* mutant strains. Although the precise mechanisms by which CBP1 interacts with the *cob* mRNA 5′ leader and stabilizes the transcript are unknown, it has been speculated that CBP1 may negatively regulate a nuclease required for degradation of the *cob* transcript. Alternatively, CBP1 may regulate a specific processing event that makes the *cob* transcript resistant to nucleases (Mittelmeir and Dieckmann, 1990).

Recently, two temperature-sensitive nuclear *pet* mutations were found to affect the expression of the yeast mitochondrial *oli1* gene (Payne *et al.*, 1991). The nuclear genes *AEP1* and *AEP2* affect the translation (*AEP1*) and stability or processing (*AEP2*) of the *oli1* mRNA, and at the nonpermissive temperature both *aep1* and *aep2* mutants are found to be deficient in the synthesis of the ATP synthase subunit 9. In the *aep2*-ts mutant, a correlation is seen between the inhibition of subunit 9 synthesis and a rapid reduction in the level of mature *oli1* mRNA after a shift to the nonpermissive temperature, but the accumulation of other transcripts is not affected. Formation of the mature 0.9-kb *oli1* mRNA requires intergenic processing of a polycistronic transcript that includes sequences of $tRNA^{Ser}$ and *var1* mRNA located 3′ of the *oli1* coding region. Because all processing intermediates can be detected at normal levels, it appears that AEP2 specifically affects the stability of the fully processed *oli1* mRNA by mechanisms similar to those postulated for CBP1.

Two other nuclear genes have been identified in *Neurospora crassa* and yeast, whose products appear to have pleiotropic effects on the processing, splicing, and stability of mitochondrial mRNAs by mechanisms that are not well understood. The *cyt-4-1* mutation of *N. crassa* inhibits the 5′ and 3′ end processing of a number of mitochondrial RNAs, including 25S rRNA, tRNAs, and the *cob, coxI, coxII,* and ATPase 6 mRNAs, as well as the splicing of both group I introns in the *cob* mRNA (Dobinson *et al.*, 1989). As a result, several novel mitochondrial RNAs can be detected in *cyt-4* mutants, including an aberrant intron RNA, which could be normal degradation products that accumulate as a result of a defect in RNA turnover (Garriga and Lambowitz, 1984). Although the precise function of the *cyt-4* gene product is not known, it appears that it may directly or indirectly affect multiple components of the mitochondrial RNA processing/degradation system (Dobinson *et al.*, 1989). The *cyt-4*$^{+}$ gene was recently cloned and shown to encode a 120-kDa protein with significant similarity to the yeast SSD1/SRK1 protein and the *Saccharomyces pombe* DIS3 protein (Turcq *et al.*, 1992). It has been suggested that the *SSD1/SRK1* and *dis3*$^{+}$ genes encode novel phosphatase functions that regulate specific events during the cell cycle and chromosome segregation (Ohkura *et al.*, 1989; Sutton *et al.*, 1991;

Wilson *et al.*, 1991). If the *cyt-4* gene product has a similar function, then this provides the first evidence that protein phosphorylation may be important for the control of RNA metabolism in mitochondria. The yeast nuclear *suv3-1* mutation, which was initially isolated as a suppressor of a *VAR1* deletion mutant lacking a conserved 3′ end processing site, also causes the overaccumulation of several excised group I intron RNAs not detectable in $suv3^{+}$ strains. The suppressor also significantly lowers levels of the *cob* and *coxI* mRNAs (Conrad-Webb *et al.*, 1991). The function of the wild-type and suppressor gene products is not understood, but Conrad-Webb *et al.* (1991) speculate that the *SUV3* gene product could be a part of, or required for the assembly of, a complex responsible for transcription, processing, and translation of mitochondrial mRNAs. In the mutant *suv3-1* background, the specificity or efficiency of these processes could be altered.

B. Nuclear Genes Affecting Chloroplast mRNA Stability

Strains of the unicellular alga *Chlamydomonas reinhardtii* that carry single nuclear mutations affecting the stability of specific chloroplast mRNAs have been isolated. The nuclear *NAC2* locus appears to encode a protein that confers stability to the *psbD* mRNA, which encodes the D2 protein of PS II (Kuchka *et al.*, 1989). The defect in the *nac2-26* mutant causes accelerated decay of *psbD* mRNA but does not affect the stability of other chloroplast mRNAs for PS II proteins, although none of the these proteins accumulate. The *psbD* mRNA is cotranscribed with the second exon of the trans-spliced *psaA* mRNA, yet the *nac2-26* mutation does not affect the trans-splicing of the *psaA* mRNA, suggesting that trans-splicing and processing of the *psbD–psaA*-exon 2 RNA occurs faster than the degradation of the *psbD* mRNA (Fig. 4). The *psbD–psaA*-exon 2 RNA is degraded, however, in *nac2-26* double mutants in which trans-splicing is also inhibited. Based on this result, Kuchka *et al.* (1989) have suggested that the *NAC2* gene product may protect psbD mRNA from 5′ to 3′ nucleolytic degradation by specifically interacting with the *psbD* mRNA 5′ end, which contains several putative stem–loop structures.

Two other *Chlamydomonas* mutants have been isolated in which the accumulation of specific chloroplast mRNAs is affected (Sieburth *et al.*, 1991). Mutant *6.2z5* does not accumulate transcripts of the *psbC* gene, which encodes the 43-kDa chlorophyll-binding protein of PS II. Mutant *GE2.10* does not accumulate transcripts of the *psbB* gene, encoding the 47-kDa chlorophyll-binding protein of PS II, as well as several smaller transcripts from a region of the chloroplast genome proximal to *psbB*. Both *psbC* and *psbB* are transcribed normally in the mutants, and transcript accumulation from other genes is not affected. As in the case of *NAC2*, the function of the normal nuclear gene products is not known, although they all appear to be required for the specific stabilization of individual chloroplast transcripts.

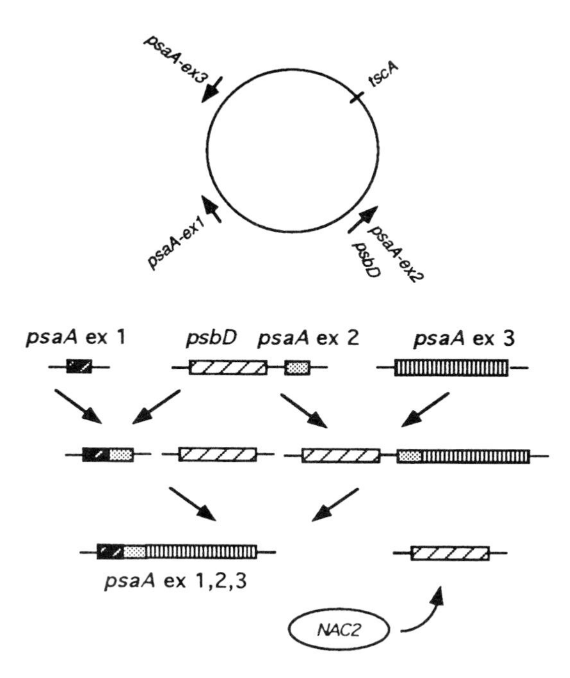

FIGURE 4 Processing and stability of the *Chlamydomonas* chloroplast *psbD* mRNA. The *psbD* mRNA is cotranscribed with exon 2 of the trans-spliced *psaA* mRNA. Several nuclear genes have been identified by mutations that interfere with correct trans-splicing of the *psaA* transcript, but that do not affect the normal accumulation of the *psbD* mRNA. The nuclear *NAC2* gene product is required for the stability of the *psbD* mRNA and apparently interacts with the 5′ end of the mRNA. Mutations in *NAC2* do not affect the trans-splicing process of the *psaA* mRNA, suggesting that processing of the *psaA* exon 2 from the *psbD–psaA* (exon 2) precursor is more rapid than the decay of the *psbD* mRNA in the *nac2* mutant. [For details, see Choquet *et al.* (1988) and Goldschmidt-Clermont *et al.* (1991).]

In contrast to *Chlamydomonas,* single nuclear mutations that specifically decrease the stability of individual chloroplast mRNAs have not been reported from higher plants, although several mutants have been isolated that affect other chloroplast functions, including the translation and/or stability of specific proteins required for photosynthetic functions (e.g., Miles, 1982; Barkan *et al.*, 1986; Gamble and Mullet, 1989) (reviewed by Somerville, 1986). In maize, several nuclear mutants have been isolated that are deficient in photosynthetic electron transport, but not in the accumulation of chlorophyll. These mutants, termed *hcf* because of their *h*igh *c*hlorophyll *f*luorescence, are all pleitropic for the reduction of a set of thylakoid membrane proteins (Barkan *et al.*, 1986). In most mutant seedlings, chloroplast RNAs are not altered in size or abundance, but in mutant *hcf-38, atpBE,* and *psaA/B* transcripts accumulate to lower levels (Barkan *et al.*, 1986).

IV. Role of Translation in Organelle mRNA Stability

Among several factors that determine the half-life of organelle mRNAs could be their association with ribosomes. The importance of polysome association for mRNA turnover has been reported for other genes, such as mammalian histone mRNAs (Graves *et al.*, 1987), the *MATα1* mRNA in yeast (Parker and Jacobson, 1990), the mammalian β-globin mRNA (Bandyopadhyay *et al.*, 1990), and the β-lactamase and the outer membrane protein (*ompA*) mRNA of *Escherichia coli* (Nilsson *et al.*, 1987). There is comparatively less information on how translation affects the half-life of organelle mRNAs, although it was generally assumed that ribosome association stabilizes mRNAs. The paucity of information is in part due to the fact that *in vitro* translation systems are not readily available, although protein synthesis in isolated organelles can be achieved with high fidelity. It is possible, however, to use organelle-specific translation inhibitors (e.g., lincomycin or chloramphenicol) *in vivo* to force mRNAs into polysome-bound and polysome-depleted states. In combination with transcription inhibitors such as actinomycin, it is then possible to estimate changes in mRNA half-lives as a function of polysome binding.

Using this approach with chloroplasts in intact spinach leaves, it was shown that in the presence of lincomycin, which inhibits an early step in polypeptide synthesis before polysome assembly, the decay of *psbA* and *rbcL* mRNAs is significantly reduced (Klaff and Gruissem, 1991). In contrast, chloramphenicol, which inhibits the peptidyl-transferase activity of ribosomes and thus stalls translation elongation, causes a more rapid decay of *psbA* and *rbcL* mRNAs (Fig.5). However, the decay rate in the presence of chloramphenicol was higher for the *psbA* mRNA in young leaves as compared with mature leaves, relative to *rbcL* mRNA. This finding suggests that selective chloroplast mRNA turnover could be a function of the developmental stage of the leaf or the chloroplast (Klaff and Gruissem, 1991). The results are consistent with the possibility that translation of both *psbA* and *rbcL* mRNAs may be critical for initiating and/or facilitating their turnover and that the two mRNAs decay by different mechanisms. Although this model may not be applicable to other organelle mRNAs, it suggests that mRNA decay in organelles, or at least in chloroplasts, differs from mRNA decay in bacteria, where mRNAs are generally more stable when associated with ribosomes (e.g., Schneider *et al.*, 1978) (see Chapter 6).

More direct evidence for a role of translation in chloroplast mRNA decay has been obtained from *Chlamydomonas* chloroplast mutants that cause premature termination of protein synthesis. In mutant *18-7G*, which contains a nonsense mutation close to the 5′ end of the *rbcL* gene, the *rbcL* mRNA accumulates to normal or increased levels (Spreitzer *et al.*, 1985). Similarly, in mutant *ac-u-g-2.3*, a single A/T base pair deletion causes early termination during translation of a PS I protein encoded by the *psaB* gene.

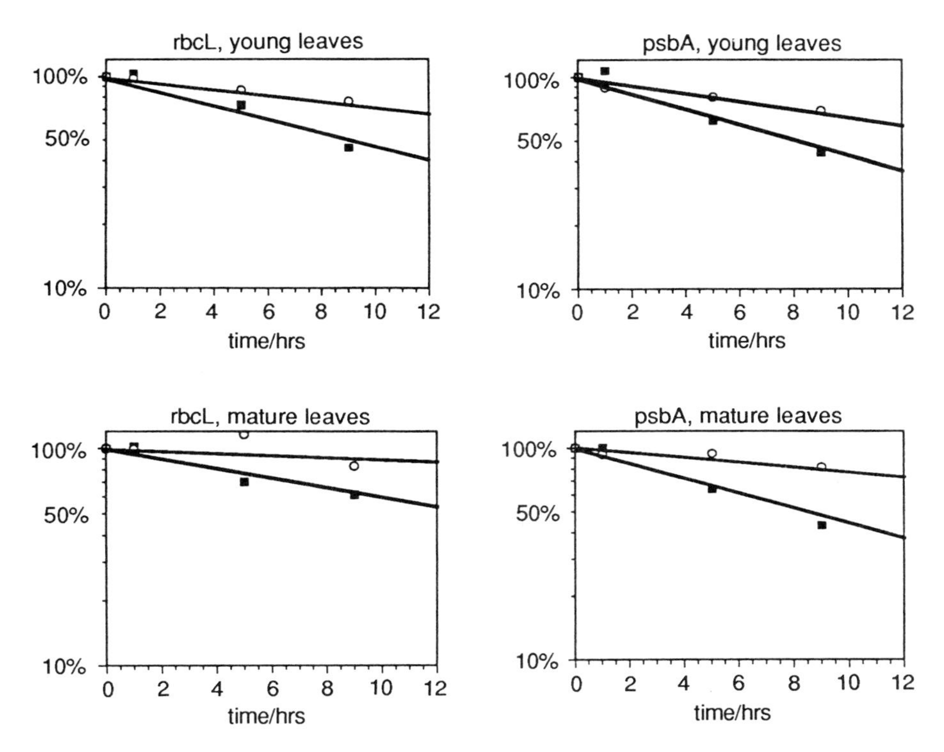

FIGURE 5 Effect of ribosomes on stability of chloroplast mRNAs. The decay rates of the *rbcL* and *psbA* mRNAs were determined in intact young and mature leaves following treatment with the translation inhibitors chloramphenicol and lincomycin. The data show that inhibition of translation with lincomycin has a general stabilizing effect on both *rbcL* and *psbA* mRNAs. Inhibition of translation with chloramphenicol accelerates the decay of both mRNAs, although the decay rate is different for the *psbA* mRNA in young and mature leaves. ■, chloramphenicol; ○, lincomycin. [For details, see Klaff and Gruissem (1991).]

In this mutant, the level of *psaB* mRNA increases approximately 2.5-fold relative to its level in wild-type cells. The increase in mRNA levels in *ac-u-g-2.3* is not due to enhanced transcription of the *psaB* gene, suggesting that premature translation termination increases the stability of the mRNA (A. Webber and S. Bingham, personal communication). These results are consistent with other observations in plant root amyloplasts and the *Chlamydomonas* chloroplast ribosome-deficient mutant *cr-1*, in which mRNAs for photosynthetic proteins are stable and accumulate, although they are specifically excluded from the polysomal RNA fraction (Deng and Gruissem, 1988; Liu *et al.*, 1989). In addition, a fraction of the *psbA* mRNA in higher plant chloroplasts accumulates in a polysome-free form, although the stability of this mRNA fraction relative to the polysome-bound mRNA is not known (Klein *et al.*, 1988; Klaff and Gruissem, 1991). Together, these results suggest that translation or polysome association of chloroplast mRNAs could be a critical determinant for their half-lives *in vivo*.

V. cis- and trans-Factors Affecting Organelle mRNA Stability

A. Function of 3′ Stem–Loop Structures in Organelle mRNAs

1. Chloroplast 3′ Stem–Loop Structures Do Not Function as Efficient Transcription Terminators

Most chloroplast mRNAs contain an inverted repeat (IR) sequence in their 3′ untranslated region that could fold into a stable stem–loop structure (Fig. 6). In prokaryotic cells, 3′ stem–loop structures are known to participate in rho-dependent and rho-independent transcription termination (reviewed by Platt, 1986). In addition to their function in termination, it is also possible that IR sequences located at the 3′ end of mRNA molecules function as RNA processing sites or protective structures against degradation, as has been suggested for such sequences in photosynthetic and nonphotosynthetic bacteria and animal cells (e.g., Birchmeier *et al.*, 1984; Belasco *et al.*, 1985; Mott *et al.*, 1985; Wong and Chang, 1986; Newbury *et al.*, 1987; Chen and Orozco, 1988). Chloroplast IR sequences for individual mRNAs are highly conserved between evolutionary divergent plants (Zurawski and Clegg, 1987), although there are considerable differences in the IR sequences from different chloroplast mRNAs (Fig. 6). The role of chloroplast IR sequences in transcription termination has been examined directly in a chloroplast transcription extract (Stern and Gruissem, 1987). In this study, several different chloroplast IR sequences were cloned between a strong promoter and a reporter tRNA gene to assay their transcription termination efficiency. It was found that transcription of such constructs resulted in efficient read-through of the IR sequences, demonstrating that these IR sequences do not have a general function as transcription terminators. Chloroplast RNA polymerase can, however, terminate efficiently at known bacterial terminator regions, indicating that the enzyme can terminate at IR sequences (Stern and Gruissem, 1987; Chen and Orozco, 1988). In addition, chloroplast RNA polymerase can efficiently terminate after transcribing certain tRNA genes, even though these tRNAs are not flanked by 3′ IR sequences (Stern and Gruissem, 1987; Tonkyn and Gruissem, 1992).

2. 3′ Stem–Loop Structures Function in mRNA 3′ End Processing

Chloroplast mRNA 3′ ends are distinct and map just downstream of the IR sequences. In the absence of efficient transcription termination, chloroplast mRNA 3′ ends therefore must be created by processing. By

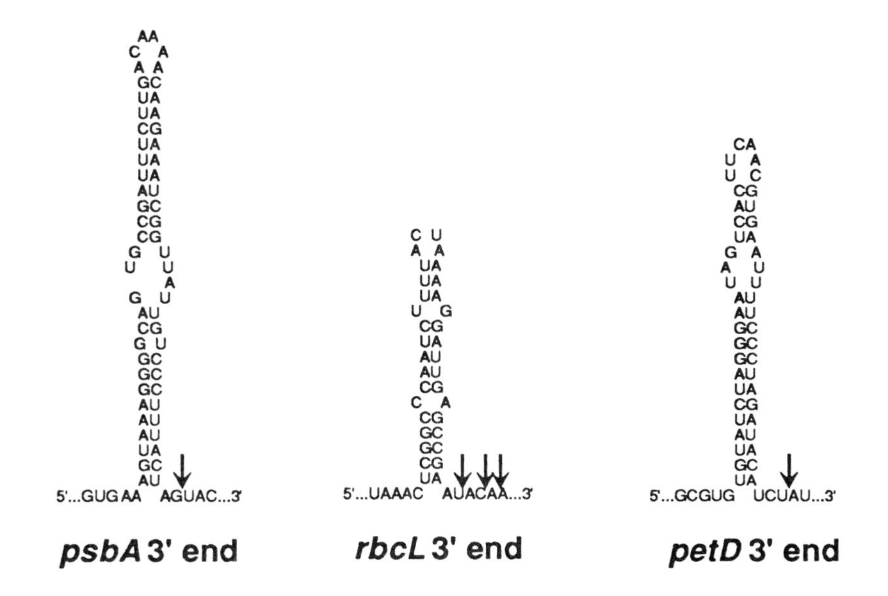

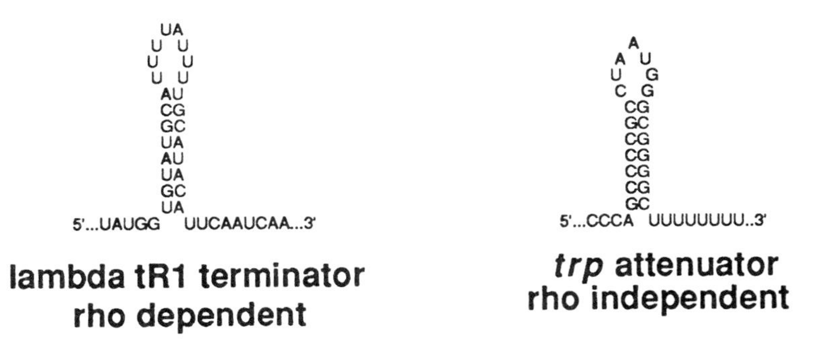

FIGURE 6 Stem–loop structures for chloroplast and bacterial mRNAs. Most chloroplast mRNAs contain inverted repeat sequences in their 3′ untranslated region that are highly conserved between different plants and can fold into stem–loop structures. In contrast to the bacterial stem–loop structures that function as transcription terminator sequences in *E.coli* [for details, see Rosenberg *et al.* (1978) and Yanofsky (1981)], the chloroplast sequences do not support efficient transcription termination, but are required for the stability of the mature mRNA. The 3′ ends of the chloroplast mRNAs indicated by arrows can be established by specific and precise processing events that appear to be part of the mRNA maturation pathway. [For details, see Stern and Gruissem (1987).]

impeding 3′-exonucleolytic digestions, the 3′ stem–loop structures could function both as processing signals and as stabilizing structures that increase the half-life of the mature mRNA. When synthetic 3′ RNA precursors encompassing the IR and distal sequences of *rbcL, psbA, petD,* or *rps14* are added to a chloroplast protein extract, correct 3′ ends are generated by rapid 3′ to 5′ exonucleolytic processing. These ends are indistinguishable from the 3′ ends of the mature mRNAs found *in vivo* (Stern and Gruissem, 1987). These results indicate that nuclease(s) exist in chloroplasts that, alone or in combination with other proteins and cis-acting RNA sequences, have a precise mRNA 3′ end processing function. Recently, two specific ribonuclease activities have been identified in spinach chloroplasts that affect the processing and stability of the *petD* mRNA 3′ end. One activity is an exonuclease that catalyzes the conversion *in vitro* of the mRNA 3′ end precursor to the mature RNA ending with the inverted repeat. In this respect, the chloroplast enzyme differs from the *E.coli* exoribonucleases polynucleotide phosphorylase (E.C. 2.7.7.8) and ribonuclease II, which have similar functions in bacterial mRNA processing (see Chapter 2), but do not correctly recognize and process the *petD* 3′ precursor RNA (Stern and Gruissem, 1989). In addition, an endonuclease activity (EndoC2) cleaves the *petD* RNA both at the termination codon and at the mature mRNA 3′ end *in vitro* (Fig. 7), resulting in the removal of the 3′ IR and rapid degradation

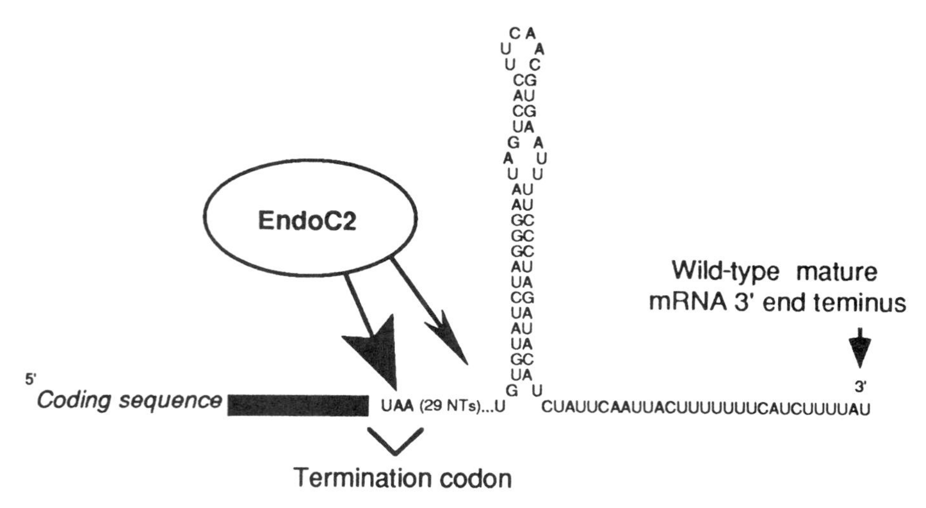

FIGURE 7 Turnover of chloroplast mRNAs may be initiated by specific nucleolytic cleavage reactions that remove the 3′ stem–loop structure. An endonuclease activity has been identified in the chloroplast that cleaves the *petD* mRNA at two specific sites 5′ of the inverted repeat. Both cleavage events separate the 3′ stem–loop structure from the *petD* mRNA coding region *in vitro*, but the relevance of this specific nucleolytic reaction for the stability of the *petD* mRNA *in vivo* is currently not known. [For details, see Hsu-Ching and Stern (1991b).]

of the upstream RNA (Hsu-Ching and Stern, 1991b). This endonuclease may therefore have direct relevance to the decay of the *petD* mRNA *in vivo*.

The processing of mRNA 3′ ends in chloroplasts seems to differ from the formation of mRNA 3′ ends in yeast mitochondria, where the 3′ end of many mRNAs coincide with the dodecamer motif AAUAAUAUUCUU. It appears that this sequence is recognized as a site for a specific endonucleolytic cleavage that establishes a stable mRNA 3′ end and not as a site for transcription termination by the mitochondrial RNA polymerase (reviewed by Grivell, 1989). Transcripts with sequences downstream of the dodecamer motif have not been detected. Little information is available on the 3′ termini of plant mitochondrial mRNAs, but it has been suggested that certain partially conserved primary and secondary structural elements at the 3′ end may function as transcription termination of processing signals (Schuster *et al.*, 1985).

3. 3′ Stem–Loop Structures Are Required for mRNA Stability

cis-regulatory sequences and/or structures have been identified that are important for the processing, stability, and/or degradation of some mRNAs (reviewed by Brawerman, 1989), although the precise molecular mechanisms by which these cis-regulatory RNA sequences control stability are not well understood. There is now good evidence from *in vitro* and *in vivo* studies that chloroplast mRNA 3′ IR sequences are critical for the stability of mRNAs. Earlier experiments have shown that 3′ IR sequences from different chloroplast mRNAs can stabilize upstream RNA sequences *in vitro* (Stern and Gruissem, 1987; Stern *et al.*, 1989) (reviewed in Gruissem *et al.*, 1988). RNAs that do not contain a 3′ IR sequence are rapidly degraded in a chloroplast protein extract, indicating that the stem–loop structure can function as a barrier against 3′ exonucleolytic decay. Progressive deletions into the 3′ IR RNA sequences reduce the stability of the upstream RNA segments, as shown for the *petD* mRNA, indicating that the potential to form a stem–loop structure is a minimum requirement for mRNA stability *in vitro* (Stern *et al.*, 1989). Certain single point mutations and nucleotide transversions that reduce the theoretical ΔG of stem–loop formation also affect the processing and/or stability of the 3′ IR *in vitro* (Fig. 8) (Stern and Gruissem, 1989; Stern *et al.*, 1989; Hsu-Ching and Stern, 1991). Most strikingly, mutations in the *psbA* 3′ IR loop that convert the wild-type sequence ACAAAAA to CUUCGGU result in a significant increase in the stability of the processed mRNA 3′ end, but do not change the rate of precursor processing (Adams and Stern, 1990). The sequence CUUCGG is a common motif in the IRs of prokaryotic organisms and confers increased thermodynamic stability to the stem–loop structure (Tuerk *et al.*, 1988).

The direct demonstration of the importance of a 3′ IR for mRNA stability *in vivo* has essentially confirmed the results obtained from the *in vitro* experiments (Stern *et al.*, 1991). In this case it was shown that partial or

petD 3' IR wild type and mutant sequences

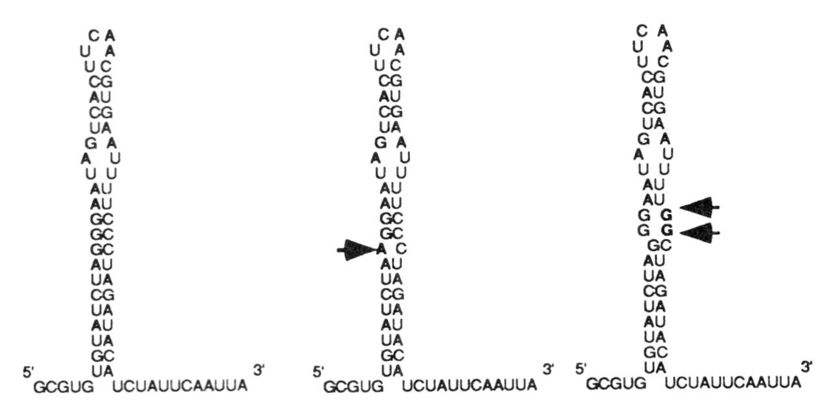

Name	wild type	*petD*3	*petD*2
ΔG°	-28.1	-19.8	-16.8
Stability	98	89	51

psbA 3' IR wild type and mutant sequences

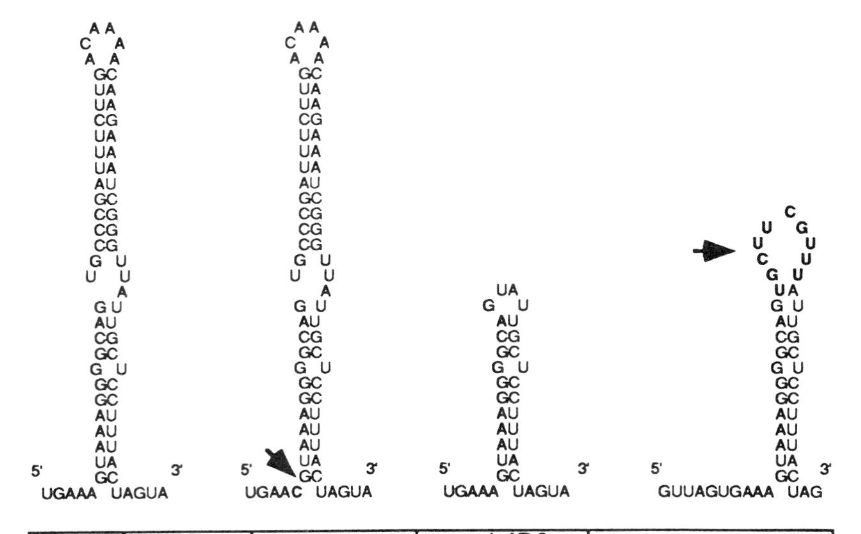

Name	wild type	*psbA*1C	*psbA*D2	*psbA*D1
ΔG°	-35.6	-33.1	-16.3	-19.4
Stability	73	58	57	29

complete deletion of the 3′ IR sequence in the *atpB* gene in transformed *Chlamydomonas* chloroplasts leads to a 60 to 80% reduction in the accumulation of the *atpB* mRNA, without affecting *atpB* transcription rates. This result is consistent with the model that 3′ IR sequences in chloroplast mRNAs are critical for mRNA accumulation and stability. However, these results do not provide information on the possible function of these cis-acting RNA sequences in plastid mRNA decay.

The 3′ untranslated region of several plant mitochondrial mRNAs can potentially fold into thermodynamically stable secondary structures. For example, the 3′ ends of the rice mitochondrial *atp9, cob-1,* and *orf25/cox3* transcripts all map to inverted repeats that can form tandem stem–loop structures (E. Kaleikau and V. Walbot, personal communication). The function of the secondary structures in mitochondrial transcription termination, mRNA 3′ end processing, and/or mRNA stability is presently not well established. However, it appears that, as is the case for chloroplast mRNA, 3′ stem–loop structures in mitochondrial mRNAs may also be critical for their stability. A comparison of the rice mitochondrial *cob-1* and *cob-2* mRNAs has shown that they are identical in the 5′ upstream region and through most of the coding region, but differ in their 3′ end. Only *cob-1* has a sequence at its mapped 3′ end that could fold into a tandem stem–loop structure. The proposed function of this structure in mRNA stability is consistent with the finding that only transcripts from *cob-1* accumulate, although both *cob-1* and *cob-2* are transcribed (E. Kaleikau, C. Andre and V. Walbot, personal communication). Thus, it appears that, as in chloroplasts, 3′ stem–loop structures in plant mitochondrial mRNAs can function as barriers against rapid 3′ exonucleolytic decay.

B. Chloroplast 3′ Stem–Loop Structures Interact with Different Proteins

The combined *in vitro* and *in vivo* approaches have clearly established a role for chloroplast 3′ IR sequences in conferring mRNA stability, but the mechanisms that regulate differential mRNA decay are still largely unknown. It is reasonable to argue that the effectiveness of the 3′ IR as a stabilizing element may be modulated during plant development or in response to environmental signals. Such modulation could be achieved through participation of additional trans-acting factors such as nucleases or other RNA-binding proteins. This would be consistent with the observation

---

FIGURE 8 Mutations in chloroplast mRNA 3′ inverted repeat sequences affect the stability of the resulting stem–loop structures. This figure shows examples of mutations in the *petD* and *psbA* mRNA 3′ inverted repeat sequences and their affect on the accumulation of the mutant RNAs. Although some mutations result in significant changes of the predicted ΔG^{0} of the RNAs, these changes are not strictly related to the accumulation of the mutant RNAs. [For details, see Stern *et al.* (1989) and Adams and Stern (1990).]

that the accumulation of individual plastid mRNAs *in vivo* and the relative stabilities of the plastid mRNAs *in vitro* do not necessarily reflect the theoretical free energy of the 3′ IRs *in vitro* (Stern and Gruissem, 1987). In addition, specific mutations in the 3′ IR sequences can also affect the stability of the RNAs *in vitro,* even though these mutant IR sequences are predicted by computer modeling to form stable stem–loop structures. For example, a mutation in the *petD* 3′ IR that changes the sequence UUCCCUA in the stem to UUGGCUA (*petD2*) and lowers the calculated ΔG of the stem–loop structure from -28.1 kcal/mole for the wild-type sequence to -16.8 kcal/mole for the mutant results in a 50% decrease in the accumulation of the RNA processed from the *petD* RNA precursor. In comparison, wild-type *rbcL,* whose 3′ IR sequence can fold into a stem–loop structure with a predicted ΔG of -16 kcal/mole, is stable under conditions under which the stability of the *petD2* stem–loop structure is reduced. Other *petD* 3′ IR sequence mutations that also lower the predicted ΔG of the resulting stem–loop structures have little or no effect on the stability of the RNA processed from a precursor molecule (Stern and Gruissem, 1987; Stern *et al.*, 1989). These results suggest that specific mutations in the 3′ IR sequences can either enhance the accessibility of a nuclease to the mRNA 3′ end or interfere with the binding of specific proteins required for precise 3′ end processing and/or stabilization of the stem–loop structures.

The first demonstration that the chloroplast 3′ stem–loop structures interact with proteins was accomplished by RNA mobility shift and UV cross-linking assays. In the presence of a nonspecific competitor, RNA–protein complexes form with the 3′ ends of *petD, rbcL,* and *psbA* mRNA in a chloroplast extract that supports mRNA 3′ end processing (Stern *et al.*, 1989) The formation of the complexes is specific, as indicated by the fact that competition with homologous and heterologous RNAs results in increased mobility or no mobility change, respectively, of the RNA–protein complexes. These experiments have also shown that both nonspecific and mRNA-specific RNA-binding proteins participate in mRNA 3′ end complex formation, since heterologous 3′ IR RNAs do not completely abolish protein binding. The proteins responsible for the mobility shifts of the *petD, rbcL,* and *psbA* 3′ IR RNAs have been identified directly by cross-linking chloroplast proteins to the radioactively labeled 3′ IR RNAs with UV light (Fig. 9). The results from these experiments have confirmed that the differences observed in gel mobility shift competitions can be correlated with the interaction of different sets of proteins with the 3′ IR RNAs. All three 3′ IR RNAs interact with a protein (or group of proteins) of 28 kDa, which must therefore be considered a protein with general affinity for different chloroplast mRNA 3′ ends (Schuster and Gruissem, 1991) (The estimated M_r of the RNA-binding proteins is based on their migration in SDS–polyacrylamide gels after UV cross-linking and RNase digestion.) A protein of 24 kDa also binds to all three 3′ IR RNAs, although the labeling intensity of this protein is

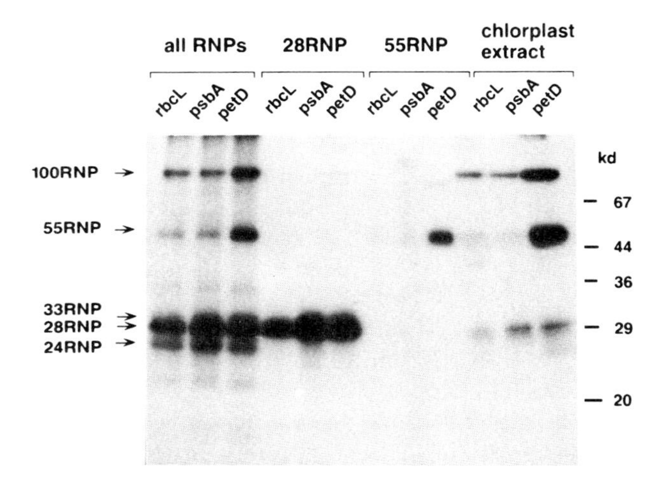

FIGURE 9 Spinach chloroplast RNA binding proteins. Radiolabeled RNAs corresponding to the 3′ end (3′ IR) of the chloroplast *petD, psbA,* and *rbcL* mRNAs were UV-crosslinked to proteins in a chloroplast RNA processing extract (chloroplast extract), to the purified 28RNP and 55RNP RNA-binding proteins, and to a protein fraction enriched for RNPs (all RNPs). Binding conditions were similar to those that facilitate effective mRNA 3′ end processing, except that polyU was included to inhibit the processing reaction. Following UV crosslinking, the RNAs were digested with RNase, and the indirectly labeled proteins were analyzed by SDS–polyacrylamide gel electrophoresis. The 55RNP was isolated by heparin–agarose chromatography. The fraction enriched for all five RNPs (24RNP, 28RNP, 33RNP, 55RNP, and 100RNP) was isolated using chromatography on a ssDNA–cellulose column and elution with 0.2 and 1.0 *M* KCl, respectively. Note that the 28RNP and 24RNP bind all three different RNAs, whereas the 55RNP binds preferentially to the *petD* RNA.

different when cross-linked to different RNAs (S. Abrahamson and W. Gruissem, unpublished results). In contrast, two other proteins of 55 and 100 kDa appear to bind specifically to the *petD* 3′ IR RNA, since their labeling intensities are significantly lower when cross-linked to the *rbcL* and *psbA* RNAs (Stern *et al.*, 1989). It should be noted, however, that the indirect labeling of proteins by UV cross-linking to radiolabeled RNA depends on the number of contact sites and the number of labeled uridine residues that react with the protein and are protected from RNase digestion; therefore, the labeling intensities do not necessarily reflect the binding affinities of the proteins directly.

The binding sites of the proteins interacting with the *petD* 3′ IR RNA have been mapped in some detail (Hsu-Ching and Stern, 1991). It appears that the 55-kDa protein that binds specifically to the *petD* mRNA 3′ end (designated a 57-kDa protein by Hsu-Ching and Stern, 1991) interacts with a duplicated sequence motif, AUUYNAUU, that is located 17–30 nucleotides 5′ and 2–7 nucleotides 3′ of the *petD* stem–loop structure and that is highly conserved in higher plants (see Fig. 7). This sequence motif is similar to the AUUUA sequence motif found in many AU-rich elements that facilitate the decay of several mammalian protooncogene and lymphokine mRNAs (reviewed by Brawerman, 1987, 1989) (see Chapter 9). The 55-kDa protein may interfere with the activity of a specific endonuclease that cleaves the *petD* 3′ IR RNA in spinach chloroplast protein extracts at a site close to or overlapping the 55-kDa protein binding site located upstream of the *petD* stem–loop structure (Stern *et al.*, 1990). Based on these results, Hsu-Ching and Stern (1991) suggest that the 55-kDa protein may be critical to the stability of the *petD* mRNA 3′ end. A 33-kDa protein (previously identified as a 29-kDa protein; Stern *et al.*, 1989) appears to recognize the double-stranded stem of the 3′ IR RNA and may interact with the 55-kDa protein. In the absence of the 55-kDa protein, the 33-kDa protein also fails to bind (Hsu-Ching and Stern, 1991). Thus, in combination with the 55-kDa protein, the 33-kDa protein may stabilize the secondary structure of the stem–loop. In contrast, the 24-, 28- (designated 28 and 32 kDa by Hsu-Ching and Stern, 1991), and 100-kDa proteins do not appear to require specific sequences to bind RNA, although their precise binding sites on the 3′ IR RNAs have not been established.

C. A Class of Nuclear-Encoded Chloroplast RNA-Binding Proteins Has Conserved RNA-Binding Domains and Novel Acidic Domains

Based on its binding affinity for the different *petD, rbcL,* and *psbA* 3′ IR RNAs, the spinach chloroplast 28-kDa (28RNP) protein was purified by a combination of FPLC and RNA affinity chromatography (Schuster and Gruissem, 1991). The protein contains two amino acid sequence domains that share extensive homology with the conserved RNA-binding domains found in a number of other RNA-binding proteins, such as U1A snRNP or *Drosophila sex-lethal* (reviewed by Bandziulis *et al.*, 1989; Mattaj 1989; Kenan *et al.*, 1991). Using single-stranded DNA chromatography, several other RNA-binding proteins with sequences similar to those of conserved RNA binding domains have been isolated from tobacco and spinach chloroplasts. The molecular masses of the tobacco chloroplast proteins range from 28 to 33 kDa, and their RNA-binding domains share extensive amino acid sequence homologies, suggesting that they are encoded by a family of recently diverged genes (Li and Sugiura, 1990; Ye *et al.*, 1991). A new 24-kDa protein from spinach chloroplasts also contains two sequence motifs related to the conserved RNA-binding domain and, like the 28RNP, interacts with the 3′ ends of different chloroplast mRNAs (S. Abrahmason and W. Gruissem, unpublished results). A cDNA encoding a 28-kDa RNA-binding protein was recently isolated from *Arabidopsis* (R. Hell and W. Gruissem, unpublished results). The amino acid sequences in the two RNA-binding domains of the 28RNPs from *Arabidopsis*, spinach, and tobacco are highly conserved, suggesting that the proteins have similar functions in chloroplast RNA metabolism in evolutionary divergent plants (Fig. 10). Although the precise function of the RNA-binding domains is not known, X-ray crystollagraphic and mutational analyses of the U1A snRNP suggest that the RNP1 and RNP2 consensus sequences within the RNA-binding domain form contact sites to the U1 RNA (Nagai *et al.*, 1990; Scherly *et al.*, 1990; Hoffman *et al.*, 1990), which is part of the intron splicing complex (reviewed by Ruby and Abelson, 1991). It is possible that the RNP1 and RNP2 consensus sequences within the RNA-binding domain represent general, sequence-independent RNA contact sites, which would be consistent with the spinach chloroplast 28RNP and 24RNP properties of binding to different 3′ IR RNAs. In contrast to the cytoplasmic and nuclear RNA-binding proteins with conserved RNA-binding domains, the chloroplast proteins have an acidic N-terminal domain that may function in protein–protein interactions. For example, the 46-amino-acidic N-terminal domain of the spinach 28RNP is 35% glycine and glutamic acid, and contains no basic amino acids (Schuster and Gruissem, 1991).

Immunodepletion of the spinach 28RNP from a chloroplast extract capable of mRNA 3′ end processing has provided insights into the possible function of the protein. In the chloroplast protein extract, mRNA 3′ end precursors containing the 3′ IR are rapidly processed to RNAs with the mature 3′ end. In the absence of the 28RNP, the processing reaction is inhibited, and the decay of the 3′ end precursor RNA is accelerated (Schuster and Gruissem, 1991). These results suggest that the 28RNP has a direct role in mRNA 3′ end processing and stability. It is conceivable that the 28RNP interacts with a specific nuclease via the acidic N-terminal domain to direct correct mRNA 3′ end processing, which results in a stable mRNA ending with a 3′ stem–loop structure. In the absence of the 28RNP–nuclease complex, the mRNA 3′ end precursor is degraded nonspecifically by nucleases that fail to recognize the structural and/or sequence elements in the RNA that normally direct 3′ end processing. This model is consistent with the observation that isolation of the 28RNP from the chloroplast extract by immunoaffinity chromatography with 28RNP-specific IgG, but not with the preimmune serum IgG control, results in the copurification of a nuclease activity that in combination with the 28RNP has mRNA 3′ end processing activity. Various control experiments have excluded the possibility that the 28RNP itself has nuclease activity (Schuster and Gruissem, 1991).

The function of the other chloroplast RNA-binding proteins is not known, but based on their conserved RNA-binding domains, they are likely to participate in specific RNA processing events. For example, several chloroplast mRNA precursors are interrupted by introns that are efficiently removed. On the basis of their similarity to specific hnRNP proteins involved in pre-mRNA splicing, Li and colleagues (Li and Sugiura, 1990; Ye *et al.*, 1991) have proposed that the chloroplast RNA-binding proteins may

PROTEIN		RNP-CS2		RNP-CS1	
Spinach					
24K domain 1	lK	**IFVGNL**	PfnVDsaeLaqLFqaaGtVEmiRiymiiDwkeS	**RGFGFVTM**	SsveEveaAaqqfnnyeLdGRtirVtedsh
2	nR	**VFVGNL**	sWkVDddaLktLFsetGdVVeAkVIYDrDTGRS	**RGFGFVTY**	nsanEvntAiesldqvdLnGRsiRVtaaea
28K domain 1	aK	**LFVGNL**	PYdVDsekLaqiFdaaGvVEiAEVIYNtETDRS	**RGFGFVTM**	dtveEaekAvellnqydIwmeqltVNktrr
2	yR	**VYVGNL**	PWdVDtsrLeqLFsehGkVVsARVVsDRETGRS	**RGFGFVTM**	SsesEvndAiaaldgqtLdGRavrVNvaee
Arabidopsis					
28K domain 1	aK	**LFGVNL**	PYdVDsqaLamLFeqaGtVEisEVIYNRDTDqS	**RGFGFVTM**	StveEaekAvekfnsfeVnGRrltVVNraa
2	fR	**IYVGNL**	PWdVDsgrLerLFsehGkVVdARVVsDRETGRS	**RGFGFVQM**	SnenEvnvAiaaldgonLeGRAIKVNvarn
Tobacco					
28K domain 1	aK	**LFVGNL**	PYdIDseqLaqLFqqaGvVEiAEVIYNRETDRS	**RGFGFVTM**	StveEadkAvelysqydLnGRlltVNkaap
2	yR	**IYVGNI**	PWdIDdrrLeqvFsehGkVVsARVVfDRESGRS	**RGFGFVTM**	SseaEmseAianldqqtLdGRtirVNaaee
29A domain 1	lK	**IFVGNL**	PfsaDsaaLaeLFeraGnVEmvEVIYDklTGRS	**RGFGFVTM**	SskeEveaAcqqfnqyeLdGRalrVNsqpp
2	nR	**VYVGNL**	aWqVDqdaLetLFseqGkVVdAKVVYDRDSGRS	**RGFGFVTY**	SsaeEvnnAiesldqvdLnGRairVspeae
29B domain 1	lK	**LFVGNL**	PfsVDsaaLaqLFeraGnVEmvEVIYDklTGRS	**RGFGFVTM**	StkeEveaAeqqfnqyeIdGRairVNaqpa
2	nR	**VYVGNL**	sWqVDdlaLkeLFseqGnVVdAKVVYDRDSGRS	**RGFGFVTY**	SsskEvndAidslnqvdLdGRsirVsaaee
31K domain 1	aK	**LFVGNL**	PYdVDseqLarLFeqaGvVEiAEVIYNRDTDqS	**RGFGFVTM**	StveEaekAvemynrydVnGRlltVNkaar
2	yR	**IYVGNI**	PWgIDdarLegLFsehGkVVsARVVYDRETGRS	**RGFGFVTM**	aseaEmsdAianldgqsLdGRtirVNvaed
33K domain 1	gR	**LYVGNL**	PfsmtssqLseiFaeaGtVanyEIVYDRvTDRS	**RGFAFVTM**	gsveEakeAirlfdgsqVgGRtvkVNfpev
2	hK	**LYVANL**	sWaltsqgLrdaFadqpgfmsAkVIYDRssGRS	**RGFGFITF**	SsaeamnsAldtmneveLeGRplrlNvagq
RNP CONSENSUS	K	**LFVGNL**	PY VD L LF G VE AEVIYDRETDRS	**RGFGFVTM**	S E A L GR VN
	R	**Y**	W I V R V N DSG		V

FIGURE 10 RNA binding domains of the chloroplast RNP-CS-type proteins. The most highly conserved segments of these domains are denoted RNP-CS2 and RNP-CS1 (Bandziulis *et al.*, 1989) and are printed in uppercase type, as are other conserved amino acid residues located throughout the domain. Protein sequences are listed for the RNA binding domains of the spinach 24RNP (S. Abrahamson and W. Gruissem, unpublished results), spinach 28RNP (Schuster and Gruissem, 1991), Arabidopsis 28RNP (R. Hell and W. Gruissem, unpublished results), and the tobacco 28-, 31-, 33-, and 29-kDa (29A and 29B) proteins (Li and Sugiura, 1990; Ye *et al.*, 1991).

participate in the formation of complexes that, like hnRNP particles and spliceosomes, are required for the removal of chloroplast mRNA introns. It is possible, however, that like the 28RNP, certain of the proteins may also be involved in the 3′ end processing of different classes of mRNAs or the precise intergenic processing of polycistronic precursors to monocistronic mRNAs. To what extent the proteins also affect the decay of chloroplast mRNAs is not clear, but it is quite likely that incorrect processing and splicing events due to altered or missing RNA-binding proteins could result in accelerated mRNA decay.

VI. Conclusions

The results from recent studies have demonstrated that chloroplasts and mitochondria have developed, in concert with nuclear genomes, precise pathways to establish and control the levels of functional organelle mRNAs. It appears that organelles place considerable emphasis on post-transcriptional mechanisms that regulate mRNA processing and stability as well as temporal changes in mRNA decay. While most of the molecular details are still elusive, it is apparent that most, if not all, of the post-transcriptional events are under the control of the nucleus, because all known organelle proteins that are part of the RNA processing and degradative machinery are encoded by nuclear genes. It is striking, however, that only in *Chlamydomonas* and yeast have single nuclear genes been identified that specifically affect the accumulation of a single mRNA in chloroplasts and mitochondria, respectively. It is conceivable that similar nuclear mutations are lethal and therefore difficult to isolate in multicellular organisms or that mutant screens have not been saturated to recover these mutations. Alternatively, as the result of complex developmental programs and environmental factors, organelle mRNA processing and stability may be regulated by different mechanisms in multicellular plants and animals. The pleiotropic nature of most nuclear mutations that affect chloroplast mRNA functions is consistent with this view. Despite the possible differences in mechanisms, it is clear that the regulation of differential mRNA stability/decay is an important control step in chloroplast and mitochondrial gene expression.

Acknowledgments

The authors thank Drs. Robert Hayes, Rüdiger Hell, and Susan Abrahamson for their critical comments, suggestions, and help in preparing the manuscript, and several colleagues for communicating their unpublished results. Research in the authors' laboratory was supported by the National Science Foundation, the Department of Energy, and the UC Berkeley

NSF Plant Development Center. G.S. was supported by a fellowship from the Weizmann Foundation.

References

Adams, C. C., and Stern, D. B. (1990). Control of mRNA stability in chloroplasts by 3′ inverted repeats: Effects of stem and loop mutations on degradation of *psbA* mRNA in vitro. *Nucleic Acids Res.* **18,** 6003–6010.

Attardi, G., and Schatz, G. (1988). Biogenesis of mitochondria. *Annu. Rev. Cell Biol.* **4,** 289–333.

Attardi, G., Cantatore, P., Crews, S., Gelfand, R., Merkel, C., Montoya, J., and Ojala, D. (1982). A comprehensive view of mitochondrial gene expression in human cells. *In* Mitochondrial Genes (P. Slonimski, G. Borst, and G. Attardi, Eds.), pp. 51–71. Cold Spring Harbor Laboratories, Cold Spring Harbor, New York.

Bandyopadhyay, R., Coutts, M., Krowczynska, A., and Brawerman, G. (1990). Nuclease activity associated with mammalian mRNA in its native state: Possible basis for selectivity in mRNA decay. *Mol. Cell. Biol.* **10,** 2060–2069.

Bandziulis, R. J., Swanson, M. S., and Dreyfuss, G. (1989). RNA-binding proteins as developmental regulators. *Genes Dev.* **3,** 431–437.

Barkan, A. (1989). Tissue-dependent plastid RNA splicing in maize: Transcripts from four plastid genes are predominantly unspliced in leaf meristems and roots. *Plant Cell* **1,** 437–445.

Barkan, A., Miles, D., and Taylor, W. C. (1986). Chloroplast gene expression in nuclear, photosynthetic mutants of maize. *EMBO J.* **5,** 1421–1427.

Belasco, J. G., Beatty, J. T., Adams, C. W., von Gabain, A., and Cohen, S. N. (1985). Differential expression of photosynthetic genes in *R. capsulata* results from segmental differences in stability within the polycistronic *rxcA* transcript. *Cell* **40,** 171–181.

Birchmeier, C., Schumperli, D., Sconzo, G., and Birnstiel, M. L. (1984). 3′ editing of mRNAs: Sequence requirements and involvement of a 60-nucleotide RNA in maturation of histone mRNA precursors. *Proc. Natl. Acad. Sci. USA* **81,** 1057–1061.

Brawerman, G. (1987). Determinants of messenger RNA stability. *Cell* **48,** 5–6.

Brawerman, G. (1989). mRNA decay: Finding the right targets. *Cell* **57,** 9–10.

Burgess, D. G., and Taylor, W. C. (1988). The chloroplast affects the transcription of a nuclear gene family. *Mol. Gen. Genet.* **214,** 89–96.

Chasan, R. (1991). Splices and edits—RNA processing in plants. *Plant Cell* **3,** 1045–1047.

Chen, L.-J., and Orozco, E. M. J. (1988). Recognition of prokaryotic transcription terminators by spinach chloroplast RNA polymerase. *Nucleic Acids Res.* **16,** 8411–8431.

Choquet, Y., Goldschmidt-Clermont, M., Girard-Bascou, J., Kück, U., Benoun, P., and Rochaix, J. (1988). Mutant phenotypes support a trans-splicing mechanism for the expression of the tripartite *psaA* gene in the *C. reinhardtii* chloroplast. *Cell* **52,** 903–913.

Chory, J., and Peto, C. A. (1990). Mutations in the *DET1* gene affect cell-type-specific expression of light-regulated genes and chloroplast development in *Arabidopsis*. *Proc. Natl. Acad. Sci. USA* **87,** 8776–8780.

Chory, J., Peto, C., Feinbaum, R., Pratt, L., and Ausubel, F. (1989). *Arabidopsis thaliana* mutant that develops as a light-grown plant in the absence of light. *Cell* **58,** 991–999.

Conrad-Webb, H., Perlman, P. H., Zhu, H., and Butow, R. (1991). The nuclear *SUV3-1* mutation affects a variety of post-transcriptional processes in yeast mitochondria. *Nucleic Acids Res.* **18,** 1369–1376.

Deng, X. W., and Gruissem, W. (1987). Control of plastid gene expression during development: The limited role of transcriptional regulation. *Cell* **49,** 379–387.

Deng, X.-W., and Gruissem, W. (1988). Constitutive transcription and regulation of gene expression in non-photosynthetic plastids of higher plants. *EMBO J.* **7,** 3301–3308.

Deng, X.-W., Stern, D. B., Tonkyn, J. C., and Gruissem, W. G. (1987). Plastid run-on transcription. Application to determine the transcriptional regulation of spinach plastid genes. *J. Biol. Chem.* **262,** 9641–9648.

Deng, X.-W., Tonkyn, J. C., Peter, G. F., Thornber, J. P., and Gruissem, W. (1989). Post-transcriptional control of plastid mRNA accumulation during adaptation of chloroplasts to different light quality environments. *Plant Cell* **1,** 645–654.

Deng, X.-W., Caspar, T., and Quail, P. H. (1991). *cop1*: A regulatory locus involved in light-controlled development and gene expression in *Arabidopsis*. *Genes Dev.* **5,** 1172–1182.

Dieckmann, C. L., and Tzagollof, A. (1985). Assembly of the mitochondrial membraine system. *CBP6*, a yeast nuclear gene necessary for synthesis of cytochrome *b*. *J. Biol. Chem.* **260,** 1513–1520.

Dieckmann, C. L., and Mittelmeier, T. M. (1987). *Curr. Genet.* **12,** 391–397.

Dieckmann, C., Homison, G., and Tzagoloff, A. (1984). Assembly of mitochondrial membrane system: Nucleotide sequence of the yeast nuclear gene (*CBP1*) involved in 5′ end processing of cytochrome *b* pre-mRNA. *J. Biol. Chem.* **259,** 4732–4738.

Dobinson, K. F., Henderson, M., Kelley, R. L., Collins, R. A., and Lambowitz, A. M. (1989). Mutations in nuclear gene *cyt-4* of *Neurospora crassa* result in pleiotropic defects in processing and splicing of mitochondrial RNAs. *Genetics* **123,** 97–108.

England, J. D., Costantino, P., and Attardi, G. (1978). Mitochondrial RNA and protein synthesis in enucleated African green monkey cells. *J. Mol. Biol.* **119,** 455–462.

Finnegan, P. M., and Brown, G. G. (1990). Trancriptional and post-transcriptional regulation of RNA levels in maize mitochondria. *Plant Cell* **2,** 71–83.

Forsburg, S. L., and Guarente, L. (1989). Communication between mitochondria and the nucleus in regulation of cytochrome genes in the yeast *Saccharomyces cerevisiae*. *Annu. Rev. Cell Biol.* **5,** 153–180.

Fox, T. D. (1986). Nuclear gene products required for translation of specific mitochondrially coded mRNAs in yeast. *Trends Genet.* **2,** 97–100.

Gamble, P. E., and Mullet, J. E. (1989). Translation and stability of proteins encoded by the plastid *psbA* and *psbB* genes are regulated by a nuclear gene during light-induced chloroplast development in barley. *J. Biol. Chem.* **264,** 7236–7243.

Garriga, G., and Lambowitz, A. M. (1984). RNA splicing in neurospora mitochondria: Self splicing of a mitochondrial intron in vitro. *Cell* **39,** 631–641.

Gelfand, R., and Attardi, G. (1981). Synthesis and turnover of mitochondrial RNA in HeLa cells: The mature ribosomal and messenger RNA species are metabolically unstable. *Mol. Cell Biol.* **1,** 497–511.

Giuliano, G., and Scolnik, P. A. (1988). Transcription of two photosynthesis-associated nuclear gene families correlates with the presence of chloroplasts in leaves of the variegated tomato *ghost* mutant. *Plant Physiol.* **86,** 7–9.

Glick, R. E., McCauley, S. W., Gruissem, W., and Melis, A. (1986). Light quality regulates expression of chloroplast genes and assembly of photosynthetic membrane complexes. *Proc. Natl. Acad. Sci. USA* **83,** 4287–4291.

Goldschmidt-Clermont, M., Choquet, Y., Girard-Bascou, J., Michel, F., Schirmer-Rahire, M., and Rochaix, J. (1991). A small chloroplast RNA may be required for trans-splicing in Chlamydomonas reinhardtii. *Cell* **65,**135–143

Gounaris, I., and Price, C. A. (1987). Plastid transcripts in chloroplasts and chromoplasts of *Capsicum annuum*. *Curr. Genet.* **12,** 219–224.

Graves, R. A., Pandy, N. B., and Chodchoy, N. (1987). Translation is required for regulation of histone mRNA degradation. *Cell* **48,** 615–626.

Grivell, L. A. (1989). Nucleo-mitochondrial interactions in yeast mitochondrial biogenesis. *Eur. J. Biochem.* **182,** 477–493.

Gruissem, W. (1989a). Chloroplast gene expression: How plants turn their plastids on. *Cell* **56,** 161–170.

Gruissem, W. (1989b). Chloroplast RNA: Transcription and processing. *In* The Biochemistry of Plants, pp. 151–191. Academic Press, San Diego.

Gruissem, W. (1989c). Regulation of gene expression in non-photosynthetic plastids of higher plants. *In* Physiology, Biochemistry, and Genetics of Nongreen Plastids (C. D. Boyer, J. C. Shannon, and R. C. Hardison, Eds.), pp. 227–240. American Society of Plant Physiologists, Maryland.

Gruissem, W., and Zurawski, G. (1985). Analysis of promoter regions for the spinach chloroplast *rbc*L, *atp*B and *psb*A genes. *EMBO J.* **4,** 3375–3383.

Gruissem, W., Barkan, A., Deng, X.-W., and Stern, D. (1988). Transcriptional and post-transcriptional control of plastid mRNA levels in higher plants. *Trends Genet.* **4,** 258–263.

Harpster, M. H., Mayfield, S. P., and Taylor, W. C. (1984). Effects of pigment-deficient mutants on the accumulation of photosynthetic proteins in maize. *Plant Mol. Biol.* **3,** 59–71.

Hiratsuka, J., Shimada, H., Whittier, R., Ishibaxhi, T., Sakamoto, M., Mori, M., Kondo, C., Honji, Y., Sun, C., Meng, B., Li, Y., Kanno, A., Nishizawa, Y., Hirai, A., Shinozaki, K., and Sugiura, M. (1989). The complete sequence of the rice *(Oryza sativa)* choroplast genome: Intermolecular recombination between distinct tRNA genes accounts for a major plastid DNA inversion during the evolution of the cereals. *Mol. Gen. Genet.* **217,** 185–194.

Hosler, J. P., Wurtz, E. A., Harris, E. H., Gillham, N. W., and Boynton, J. E. (1989). Relationship between gene dosage and gene expression in the chloroplast of *Chlamydomonas rienhardtii. Plant Physiol.* **91,** 648–655.

Hsu-Ching, C., and Stern, D. B. (1991a). Specific binding of chloroplast proteins in vitro to the 3′ untranslated region of spinach chloroplast *petD* mRNA. *Mol. Cell. Biol.* **11,** 4380–4388.

Hsu-Ching, C., and Stern, D. B. (1991b). Specific ribonuclease activities in spinach chloroplasts promote mRNA maturation and degradation. *J. Biol. Chem.* **266,** 24205–24211.

Kenan, D. J., Query, C. C., and Keene, J. D. (1991). RNA recognition: Towards identifying determinants of specificity. *Trends Gen.* **16,** 214–220.

Klaff, P., and Gruissem, W. (1991). Changes in chloroplast mRNA stability during leaf development. *Plant Cell* **3,** 517–529.

Klein, R. R., Mason, H. S., and Mullet, J. E. (1988). Light-regulated translation of chloroplast proteins. I. Transcripts of *psaA-psaB, psbA* and *rbcL* are associated with polysomes in dark grown and illuminated barley seedlings. *J. Cell Biol.* **106,** 289–3301.

Kuchka, M. R., Goldschmidt-Clermont, M., van Dillewijn, J., and Rochaix, J.-D. (1989). Mutation at the Chlamydomonas nuclear NAC2 locus specifically affects stability of the chloroplast *psbD* transcript encoding polypeptide D2 of PSII. *Cell* **58,** 869–876.

Lansman, R. A., and Clayton, D. A. (1975). Mitochondrial protein synthesis in mouse L-cells: Effect of selective nicking of mitochondrial DNA. *J. Mol. Biol.* **99,** 777–793.

Li, Y., and Sugiura, M. (1990). Three distinct ribonucleoproteins from tobacco chloroplasts: Each contains a unique amino terminal acidic domain and two ribonucleoprotein consensus motifs. *EMBO J.* **9,** 3059–3066.

Liu, X., Hosler, J. P., Boynton, J. E., and Gillham, N. W. (1989). mRNA for two ribosomal proteins are preferentially translated in the chloroplast of *Chlamydomonas reinhardtii* under conditions of reduced protein synthesis. *Plant Mol. Biol.* **12,** 385–394.

Mattaj, I. W. (1989). A binding consensus: RNA-protein interactions in splicing, snRNPs, and sex. *Cell* **57,** 1–3.

Mayfield, S. P., Nelson, T., Taylor, W. C., and Malkin, R. (1986). Carotenoid synthesis and pleiotropic effects in carotenoid-deficient seedlings of maize. *Planta* **169,** 23–32.

McGraw, P., and Tzagoloff, A. (1983). Assembly of the mitochondrial membrane system: Characterization of a yeast nuclear gene involved in the processing of cytochrome b pre-mRNA. *J. Biol. Chem.* **258,** 9459–9468.

Melis, A. (1984). Light regulation of photosynthetic membrane structure, organization and function. *J. Cell Biochem.* **24,** 271–285.

Miles, C. D. (1982). The use of mutations to probe photosynthesis in higher plants. *In* Methods in Chloroplast Molecular Biology (R. Hallick, M. Edelman, and N. H. Chua, Eds.), pp. 75–109. Elsevier, New York

Mittelmeir, T. M., and Dieckmann, C. L. (1990). CBP1 function is required for stability of a hybrid *cob-oli1* trascript in yeast mitochondria. *Curr. Genet.* **18,** 421–428.

Mott, J. E., Galloway, J. L., and Platt, T. (1985). Maturation of *Escherichia coli* tryptophan operon mRNA: Evidence for 3′ exonucleolytic processing after rho-dependent termination. *EMBO J.* **4,** 1887–1891.

Mullet, J. E. (1988). Chloroplast development and gene expression. *Annu. Rev. Plant Physiol. Plant Mol. Biol.* **39,** 475–502.

Mullet, J. E., and Klein, R. R. (1987). Transcription and RNA stability are important determinants of higher plant chloroplast RNA levels. *EMBO J.* **6,** 1571–1579.

Mulligan, R. M., Leon, P., and Walbot, V. (1991). Transcriptional and posttrascriptional regulation of maize mitochondrial gene expression. *Mol. Cell. Biol.* **11,** 533–543.

Newbury, S. F., Smith, N. H., Robinson, E. C., Hiles, I. D., and Higgins, C. F. (1987). Stabilization of translationally active mRNA by prokaryotic REP sequences. *Cell* **48,** 297–310.

Nilsson, G., Belasco, J. G., Cohen, S. N., and von Gabain, A. (1987). Effect of premature termination of translation on mRNA stability depends on ribosome release. *Proc. Natl. Acad. Sci. USA* **84,** 4890–4894.

Ohkura, H., Kinoshita, N., Miyatani, S., Toda, T., and Yanagida, M. (1989). The fission yeast *dis2+* gene required for chromosome disjoining encodes one of two putative type 1 protein phosphatases. *Cell* **57,** 997–1007.

Ohyama, K., Fukuzawa, H., Kohchi, T., Shirai, H., Sano, T., Sano, S., Umesono, K., Shiki, Y., Takeuchi, M., Chang, Z., Aota, S., Inokuchi, H., and Ozeki, H. (1986). Chloroplast gene organization deduced from complete sequence of liverwort *Marchantia polymorpha* chloroplast DNA. *Nature (London)* **322,** 572–574.

Ozeki, H., Umesono, K., and Inokuchi, H. (1989). The chloroplast genome of plants: A unique origin. *Genome* **31,** 169–174.

Palmer, J. D. (1985). Comparative organization of chloroplast genomes. *Annu. Rev. Genet.* **19,** 325–354.

Parker, R., and Jacobson, A. (1990). Translation and a 42-nucleotide segment within the coding region of the mRNA encoded by the MATα1 gene are involved in promoting rapid mRNA decay in yeast. *Proc. Natl. Acad. Sci. USA* **87,** 2780–2784.

Payne, M. J., Schweizer, E., and Lukins, H. B. (1991). Properties of two nuclear *pet* mutants affecting expression of mitochondrial *oli1* gene in *Saccharomyces cerevisiae. Curr. Genet.* **19,** 343–351.

Piechulla, B., Imlay, K. R. C., and Gruissem, W. (1985). Plastid gene expression during fruit ripening in tomato. *Plant Mol. Biol.* **5,** 373–384.

Piechulla, B., Pichersky, E., Cashmore, A. R., and Gruissem, W. (1986). Expression of nuclear and plastid genes for photosynthesis-specific proteins during tomato fruit development and ripening. *Plant Mol. Biol.* **7,** 367–376.

Piechulla, B., Glick, R. E., Bahl, H., Melis, A., and Gruissem, W. (1987). Changes in photosynthetic capacity and photosynthetic protein pattern during tomato fruit ripening. *Plant Physiol.* **84,** 911–917.

Platt, T. (1986). Transcription termination and the regulation of gene expression. *Annu. Rev. Biochem.* **55,** 339–372.

Rochaix, J.-D., and Erickson, J. (1988). Function and assembly of photosystem II: Genetic and molecular analysis. *Trends Biochem. Sci.* **13,** 56–59.

Rodell, G., Michaelis, U., Forsbach, V., Kreike, J., and Kaudewitz, F. (1986). Molecular cloning of the yeast nuclear genes CBS1 and CBS2. *Curr. Genet.* **11,** 47–53.

Rosenberg, M., Court, D., Shimatate, H., Brady, C., and Wulff, D. L. (1978). The relationship between function and DNA sequence in an intercistronic regulatory region of phage λ. *Nature (London)* **272,** 414.

Ruby, S. W., and Abelson, J. (1991). Pre-mRNA splicing in yeast. *Trends Gent.* **7,** 79–85.

Sagar, A. D., and Briggs, W. R. (1990). Effects of high light stress on carotenoid-deficient chloroplasts in *Pisum sativum*. *Plant Physiol.* **94,** 1663–1670.

Schneider, E., Blundell, M., and Kennell, D. (1978). Translation and mRNA decay. *Mol. Gen. Genet.* **160,** 121–129.

Schuster, G., and Gruissem, W. (1991). Chloroplast mRNA 3′ end processing requires a nuclear encoded RNA-binding protein. *EMBO J.* **10,** 1493–1502.

Schuster, W., Hiesel, R., Isaac, P., Leaver, C. J., and Brennecke, A. (1985). Transcript termini of messenger RNAs in higher plants mitochondria. *Nucleic Acid Res.* **14,** 5943–5954.

Scolnick, P. A., Hinton, P., Greenblatt, I. M., Giuliano, G., Delanoy, M. R., Spector, D. L., and Pollock, D. (1987). Somatic instability of carotenoid biosynthesis in the tomato *ghost* mutant and its effect on plastid development. *Planta* **171,** 11–18.

Sheen, J. (1990). Metabolic repression of transcription in higher plants. *Plant Cell* **2,** 1027–1038.

Shinozaki, K., Ohme, M., Tanaka, M., Wakasugi, T., Hayashida, N., Matsubayashi, T., Zaita, N., Chumwongse, J., Obokata, J., Yamaguchi-Shinozaki, K., Ohto, C., Torazawa, K., Meng, B. Y., Sugita, M., Deno, H., Kamogashira, T., Yamada, K., Kusuda, J., Takaiwa, F., Kato, A., Tohdoh, N., Shimada, H., and Sugiura, M. (1986). The complete nucleotide sequence of the tobacco chloroplast genome: Its gene organization and expression. *EMBO J.* **5,** 2043–2049.

Sieburth, L. E., Berry-Lowe, S., and Schmidt, G. W. (1991). Chloroplast RNA stability in *Chlamydomonas*: Rapid degradation of *psbB* and *psbC* transcripts in two nuclear mutants. *Plant Cell* **3,** 175–189.

Somerville, C. R. (1986). Analysis of photosynthesis with mutants of higher plants and algae. *Annu. Rev. Plant Physiol.* **37,** 467–507.

Spreitzer, R. J., Goldschmidt-Clermont, M., Rahire, M., and Rochaix, J. D. (1985). Nonsense mutations in the chlamydomonas chloroplast gene that codes for the large subunit of ribulosebisphosphate carboxylase/oxygenase. *Proc. Natl. Acad. Sci. USA* **82,** 5460–5464.

Stern, D. B., and Gruissem, W. (1987). Control of plastid gene expression: 3′ inverted repeats act as mRNA processing and stabilizing elements, but do not terminate transcription. *Cell* **51,** 1145–1157.

Stern, D. B., and Gruissem, W. (1989). Chloroplast mRNA 3′ end maturation is biochemically distinct from prokaryotic mRNA processing. *Plant Mol. Biol.* **13,** 615–625.

Stern, D. B., Jones, H., and Gruissem, W. (1989). Function of plastid mRNA 3′ inverted repeats: RNA stabilization and gene-specific protein binding. *J. Biol. Chem.* **264,** 18742–18750.

Stern, D. B., Hsu-Ching, C., Adams, C. C., and Kindle, K. L. (1990). Post-transcriptional control of gene expression in chloroplasts. *In* Post-Transcriptional Control of Gene Expression (J. E. G. McCarthy, and M. F. Tuite, Eds.), Vol. 49, pp. 73–82. Springer-Verlag, New York.

Stern, D. B., Radwanski, E. R., and Kindle, K. L. (1991). A 3′ stem–loop structure of the *Chlamydomonas* chloroplast *atp*B gene regulates mRNA accumulation in vivo. *Plant Cell* **3,** 285–297.

Sugiura, M. (1989). The chloroplast chromosomes in land plants. *Annu. Rev. Cell Biol.* **5,** 51–70.

Sutton, A., Immanuel, D., and Arndt, K. T. (1991). The SIT4 protein phosphatase functions in late G_1 for progression into S phase. *Mol. Cell. Biol.* **11,** 2133–2148.

Taylor, W. C. (1989). Regulatory interactions between nuclear and plastid genomes. *Annu. Rev. Plant Physiol. Plant Mol. Biol.* **40,** 211–233.

Thelander, M., Narita, J. O., and Gruissem, W. (1986). Plastid differentiation and pigment biosynthesis during tomato fruit ripening. *Curr. Top. Plant Biochem. Physiol.* **5,** 128–141.

Thomson, W. W., and Whatley, J. M. (1980). Development of nongreen plastids. *Annu. Rev. Plant Physiol.* **31,** 375–394.

Tobin, E. M., and Silverthorne, J. (1985). Light regulation of gene expression in higher plants. *Annu. Rev. Plant Physiol.* **36,** 569–593.

Tonkyn, J. C., and Gruissem, W. (1993). Transcription and mRNA accumulation of a partially duplicated spinach chloroplast ribosomal protein operon. *Mol. Gen. Genet.*, submitted for publication.

Tonkyn, J. C., Deng, X.-W., and Gruissem, W. (1992). Regulation of plastid gene expression during photooxidative stress. *Plant Physiol.* **99,** 1406–1415.

Tuerk, C., Gauss, P., Thermes, C., Groebe, D. R., Gayle, M., Guild, N., Stormo, G., d'Aubenton-Carafa, Y., Uhlenbeck, O. C., and Tinoco, I. Jr. (1988). CUUCGG hairpins: Extraordinarily stable RNA secondary structures associated with various biochemical processes. *Proc. Natl. Acad. Sci. USA* **85,** 1364–1368.

Turcq, B., Dobinson, K. F., Serizawa, N., and Lambowitz, A. M. (1992). A protein required for RNA processing and splicing in Neurospora mitochondria is related to gene products involved in cell cycles protein phosphatase functions. *Proc. Natl. Acad. Sci. USA* **89,** 1676–1680.

Tzagoloff, A., and Dieckmann, C. (1990). PET genes of *Sacchaomyces cerevisiae*. *Microbiol. Rev.* **54,** 211–225.

Tzagoloff, A., and Myers, A. M. (1986). Genetics of mitochondrial biogenesis. *Annu. Rev. Biochem.* **55,** 249–288.

Wilson, R. B., Brenner, A. A., White, T. B., Engler, M. J., Gaughran, J. P., and Tatchell, K. (1991). The *Saccharomyces cerevisiae* SRK1 gene, a suppressor of *bcy1* and *ins1*, may be involved in protein phospatase function. *Mol. Cell. Biol.* **11,** 3369–3373.

Wong, H. C., and Chang, S. (1986). Identification of a positive retroregulator that stabilizes mRNAs in bacteria. *Proc. Natl. Acad. Sci. USA* **83,** 3233–3237.

Ye, L., Li, Y., Fukami-Kobayashi, K., Go, M., Konishi, T., Watanabe, A., and Sugiura, M. (1991). Diversity of ribonucleoprotein family in tobacco chloroplasts: Two new chloroplast ribonucleoproteins and a phylogenetic tree of ten chloroplast RNA-binding domains. *Nucleic Acids Res.* **19,** 6485–6490.

Zurawski, G., and Clegg, M. T. (1987). Evolution of higher-plant chloroplast DNA-encoded genes: Implications for structure–function and phylogenetic studies. *Annu. Rev. Plant Physiol.* **38,** 391–418.

Yanofsky, C. (1981). Attenuation in the control of expression of bacterial operons. *Nature (London)* **289,** 751–758.

15

Control of Poly(A) Length

ELLEN J. BAKER

I. Background: Poly(A) and the Poly(A)-Binding Protein

A. Introduction

Almost all articles concerning poly(A) begin with a sentence reminding the reader that after all the years since poly(A) was first described, its functions still remain unclear. While still true, there is reason for optimism that the field is on the verge of having to change that opening statement. New approaches and an expanding community of investigators have resulted in a sharply accelerated pace of progress in this area. There is increasingly persuasive evidence that poly(A), in association with its poly(A)-binding protein, has an important role in the initiation of protein synthesis. There have been several recent review articles addressing the question of poly(A) function with particular reference to its role in translation (Sachs, 1990; Jackson and Standart, 1990; Munroe and Jacobson, 1990), and most of this material will not be covered again here. On the other hand, evidence supporting a role for poly(A) in mRNA stability remains less convincing, and the question of a function for poly(A) in the control of mRNA stability will be reevaluated. A separate, but related, question regarding poly(A) function is the physiological significance of poly(A) length heterogeneity, and the complex metabolism that gives rise to it. The major aim of this article is to review information on the factors that affect the length of both nuclear and cytoplasmic poly(A) and the evidence for messenger-RNA-specific control of poly(A) length.

Control of Messenger RNA Stability

B. Poly(A) Occurrence

Poly(A) tails are a feature of mRNAs in all eukaryotic organisms, and short 3′ terminal poly(A) tracts have been reported to occur on mRNAs in prokaryotes and organelles as well (Brawerman, 1981; Littauer and Soreq, 1982). In eukaryotes, a tract of poly(A) is added on to the 3′ terminus of most newly transcribed messenger RNA molecules in the nucleus. The length of the newly synthesized poly(A) tract is organism-specific, ranging from a maximum length of about 90 residues in *S. cerevisiae* (Groner and Phillips, 1975) to about 300 in vertebrates (Brawerman, 1981). The only known nonviral eukaryotic mRNAs to be produced in an unadenylated form are a subset of histone mRNAs. Many early studies revealed that translatable RNAs occurred in the cytoplasm in both poly(A)-containing and poly(A)-deficient forms, defined operationally by the ability to bind to affinity matrices, most often oligo(dT)–cellulose or poly(U)–Sepharose. Analysis of translation products of poly(A)-containing and poly(A)-deficient mRNAs indicated that, with the exception of histone mRNAs, all abundant poly(A)-deficient mRNAs have polyadenylated counterparts (reviewed in Brawerman, 1981). Although some saturation hybridization studies have indicated that poly(A)-deficient mRNA populations may contain rare sequences not present in the poly(A)-containing population, no specific examples have yet been found (see Fung *et al.* (1991) for discussion). It is assumed that all poly(A)-deficient mRNAs are derived from polyadenylated versions by cytoplasmic poly(A) shortening.

The minimal poly(A) length retained quantitatively on poly(U)–Sepharose and oligo(dT)–cellulose ranges from about 10 to 30 residues depending upon chromatography conditions (Littauer and Soreq, 1982). Thus, poly(A)-deficient fractions described in the literature contain RNA molecules with both short poly(A) and no poly(A), as well as partially degraded mRNA fragments. In this Chapter, the term *poly(A)-deficient* will be used to describe the heterogeneous fractions obtained by affinity chromatography, which will be assumed to include no molecules with poly(A) greater than 30 residues; the term *oligo(A)* will be used for tracts of less than 20 adenylates; and the term *unadenylated* will be used when there are no detectable terminal adenylates. The term *deadenylated* will be reserved for cases in which a poly(A) removal process is implied.

C. The Poly(A)–Poly(A)-Binding Protein Complex

In the cytoplasm of probably all eukaryotic cells poly(A) is complexed with a specific protein, now referred to simply as the poly(A)-binding protein (PABP). The single PABP gene in yeast has been shown to be essential for cell viability (Sachs *et al.*, 1987). PABPs have been identified in diverse eukaryotic organisms, although sometimes poly(A) is complexed with more than one major protein (e.g., Manrow and Jacobson, 1986; Swiderski and Richter, 1988; Drawbridge *et al.*, 1990). The interesting possibility that multicellular organisms may have multiple, cell type-specific PABPs has been raised by the finding that the Arabidopsis genome contains at least 20 nonidentical PABP-like genes, one of which clearly exhibits organ-specific expression (D. Belostotsky and R. Meager, personal communication). There is also at least one poly(A)-binding protein in the nucleus. In *Saccharomyces cerevisiae*, a nuclear PABP and the cytoplasmic PABP are products of the same gene, the nuclear version being a proteolytic product of the larger cytoplasmic protein; in human cells, the major detectable nuclear PABP appears to be a different gene product (Setyono and Greenberg, 1981). The minimal binding site for the yeast PABP has been determined to be 12 adenylate residues, and multiple cytoplasmic PABPs can bind to a single poly(A) tract with a packing density of about one PABP per 25–27 residues (Baer and Kornberg, 1983; Sachs *et al.*, 1987). This "beads-on-a-string" structure seems to be the actual form of the poly(A)–PABP complex *in vivo* (Baer and Kornberg, 1980).

Genes or cDNAs encoding PABPs from several different organisms have been sequenced (Adam *et al.*, 1986; Sachs *et al.*, 1986; Grange *et al.*, 1987; Zelus *et al.*, 1989; Lafrère *et al.*, 1990; Burd *et al.*, 1991; Wang *et al.*, 1992), revealing highly related proteins that are members of a family of RNA-binding proteins characterized by a conserved RNA-binding domain (RNP-CS-type proteins; Bandziulis *et al.*, 1989). All PABPs contain four tandem RNA-binding domains in the N-terminal two-thirds of the protein. A single binding domain of the yeast protein has been shown to be sufficient for poly(A)-binding *in vitro;* moreover, a PABP fragment that contained only part of one RNA-binding domain was able to restore cell viability to a PABP-depleted yeast strain (Sachs *et al.*, 1987). The functional significance of four domains remains a question. Analysis of the binding characteristics of the four domains of the yeast protein shows that they are not equivalent (Nietfeld *et al.*, 1990; Burd *et al.*, 1991). One of these studies showed that the region of the protein that exhibited high-affinity poly(A) binding *in vitro* was not the region shown by Sachs *et al.* (1987) to restore viability to PABP-depleted cells (Burd *et al.*, 1991). This finding raises the interesting possibility that the essential function of this protein is not mediated by its poly(A)-binding activity. A variety of evidence indicates that the PABP does not remain statically bound to poly(A), but is capable of continual exchange (Greenberg, 1981; Sachs *et al.*, 1987). The redundancy in RNA-binding domains may confer on PABP the ability to move to different poly(A) tracts (Sachs *et al.*, 1987) and/or to simultaneously bind other RNA sequences.

There is accumulating evidence that the PABP (or the poly(A)–PABP complex) participates in the initiation of protein synthesis, probably acting at the level of 60S ribosomal subunit addition (reviewed in Jackson and Standart, 1990; Sachs, 1990; Munroe and Jacobson, 1990; Sachs and Deardorff, 1992). How it functions in this capacity and whether it must be complexed with poly(A) in order to function are not known. Among the more persuasive lines of evidence for this role is the finding that PABP-depleted yeast cells show severe deficiencies in translation initiation, and several suppressor mutations that allow yeast cells to grow in the absence of PABP do so by altering the 60S ribosomal subunit (Sachs and Davis, 1989). This function of the PABP has to be considered when contemplating other possible roles for the PABP in poly(A) metabolism (see Section III.D). It seems likely that cytoplasmic poly(A) functions mainly as a "carrier" of the PABP and that the RNP complex itself is the biologically functional unit. However, it is possible that poly(A) has functions independent of the PABP, and vice versa.

II. Poly(A) Addition in the Nucleus

A. Biochemistry of the Nuclear Polyadenylation Process

The remarkable progress in elucidating the biochemistry of the nuclear polyadenylation process has been recently reviewed (Manley, 1988; Jacob *et al.*, 1990; Wickens, 1990a; Wahle, 1992). Only a brief introduction is possible here, with a focus on those data that pertain directly to the question of how the length of the added poly(A) may be controlled.

1. Poly(A) Signals

Polyadenylation of polymerase II transcripts is obligatorily linked to an endonucleolytic cleavage of the primary transcript, which creates the substrate for polyadenylation. Current knowledge of the nucleotide sequences that signal cleavage and polyadenylation has been recently reviewed (Proudfoot, 1991). These sequences are well defined only in the case of mammalian cells. Two sequence elements have been shown to constitute the minimum requirements: a nearly ubiquitous AAUAAA hexamer that normally lies within 10–30 nucleotides from the 3′ end of the transcribed portion of all polyadenylated mammalian mRNAs and a less conserved, GU-rich region downstream of the site of cleavage and polyadenylation, which is required for the cleavage reaction. The AAUAAA sequence often appears at the 3′ terminus of mRNAs from a variety of other vertebrate and nonvertebrate organisms. Plant mRNAs also usually have an AAUAAA or a close variant of this sequence near their 3′ ends (Hunt *et al.*, 1987). *S. cerevisiae* mRNAs do not contain this sequence near their termini. Multiple sequence elements seem to be involved in directing polyadenylation in *S. cerevisiae*, but so far, extensive mutagenesis of these elements has failed to define strict sequence requirements (Irniger *et al.*, 1991; Proudfoot, 1991).

Given the widespread occurrence of AAUAAA-like sequences among diverse organisms, it seemed relatively safe to assume that it functions as a polyadenylation signal where it occurs. However, the surprising finding has been made that, although *S. pombe* mRNAs contain AAUAAA sequences, or close variants, near their mapped 3′ termini, these sequences apparently are not the functional polyadenylation signals used by this organism (Humphrey *et al.*, 1991). Furthermore, *S. cerevisiae* and *S. pombe*, very distantly related organisms, have interchangeable poly(A) signals, suggesting the occurrence of a "lower eukaryotic" mode of polyadenylation exemplified by these organisms (Humphrey *et al.*, 1991). This finding also raises the possibility that the prevalent AAUAAA sequence may have additional functions.

2. *Properties of the Nuclear Polyadenylation Reactions*

The coupled processes of primary transcript cleavage and polyadenylation are reproducible *in vitro* using whole cell extracts or nuclear extracts from HeLa cells and yeast (Manley, 1988; Butler and Platt, 1988; Butler *et al.*, 1990). The ability to uncouple cleavage and polyadenylation reactions in HeLa extracts, such that precleaved substrates are recognized and polyadenylated properly, has permitted experiments that address the nature of the polyadenylation reaction itself. The following properties have been demonstrated:

The AAUAAA signal is required for polyadenylation, as well as for cleavage. A single nucleotide (nt) mutation in the AAUAAA sequence (AAGAAA) can completely prevent the polyadenylation of precleaved substrate molecules. Sheets *et al.* (1990) tested all possible (18) single nt variations from this signal in the HeLa extract and found that each change resulted in impaired polyadenylation efficiency, most often severe. The mildest mutation, AUUAAA, is also the most common natural variant.

The AAUAAA sequence is not only necessary, but also essentially sufficient for polyadenylation *in vitro*. Wigley *et al.* (1990) have defined the minimal substrate for polyadenylation *in vitro:* it consists of a mere 11 nucleotides. While the only specific sequence requirement is AAUAAA, a minimum number of nucleotides, optimally eight or more, must separate this signal from the substrate terminus.

Poly(A) elongation is biphasic. Only the first phase of polyadenylation is dependent on the AAUAAA hexamer. The second phase is independent of this sequence, but dependent on a minimal 10-residue oligo(A) primer polymerized during the first phase (Sheets and Wickens, 1989).

3. *Components of the Polyadenylation Machinery*

Fractionation of HeLa extracts has identified three major components of the polyadenylation machinery.

The poly(A) polymerase (PAP). The enzyme that actually catalyzes the polymerization of poly(A) has turned out to be the same enzyme whose activity was first described 30 years ago and has been purified from many sources to various extents in the intervening years (Jacob and Rose, 1983; Edmonds, 1990). PAP has been shown to participate *in vitro* in both the AAUAAA-dependent polyadenylation reaction (Christofori and Keller, 1989; Ryner *et al.*, 1989; Bardwell *et al.*, 1990; Wahle, 1991a) and the poly(A) elongation from oligoadenylated substrates (Wahle, 1991b). Purified mammalian PAPs show no preference for AAUAAA-containing RNA substrates and will add adenylates to any RNA primer (Edmonds, 1990). Both the AAUAAA-mediated and oligo(A)-mediated specificities are now known to be conferred by distinct "specificity factors" (see later).

The cleavage and polyadenylation factor (CPF). CPF (also called SF, specificity factor, or PF2, polyadenylation factor 2) confers AAUAAA-sequence specificity to the first phase of polyadenylation (Christofori and Keller, 1988; Gilmartin and Nevins, 1989; Takagaki *et al.*, 1989), stimulating the polymerizing activity of PAP. The factor has recently been purified to near homogeneity from both calf thymus and HeLa cells (Bienroth *et al.*, 1991), and consists of a large, almost 500-kDa, complex composed of four different polypeptides. CPF recognizes and binds directly to the AAUAAA sequence (Keller *et al.*, 1991).

Poly(A)-binding protein II (PABII). PABII is a newly discovered factor that endows the PAP with specificity for oligoadenylated substrates in the second phase of polyadenylation (Wahle, 1991b). PABII was purified based on its ability to stimulate synthesis of long poly(A) tails (ca. 200 adenylates) in reactions containing PAP and a highly purified preparation of CPF that together stimulate the polymerization of only short poly(A) tails. Poly(A) alone can serve as a substrate for the PABII-stimulated elongation reaction. Sequenced peptides derived from the 49-kDa protein reveal two motifs found in the RNA-binding domains of RNP-CS-type proteins (Bandziulis *et al.*, 1989), but otherwise there are no similarities to other known protein sequences, including that of the human cytoplasmic PABP. While the second phase of polyadenylation is independent of the CPF and its binding site, polyadenylation is most efficient when an intact AAUAAA and CPF are also present (Wahle, 1991b). Interestingly, biphasic polyadenylation is a property of the highly purified vaccinia virus PAP catalytic subunit itself, (which operates in the cytoplasm, not nucleus). The transition between phases has been shown to involve a switch from a rapid, primer-dependent, processive activity which adds adenylates to a final length of about 35, to a slow nonprocessive extension (Gershon and Moss, 1992).

B. Control of Nuclear Poly(A) Length

Termination of polyadenylation does not occur at a specific length, but occurs over a limited range of lengths, both *in vivo* and in polyadenylation reactions *in vitro*. It is not yet known what factor(s) control this range. PAP itself shows no intrinsic limitation in the length of poly(A) it will polymerize under nonspecific conditions; purified bovine PAP has been shown to add up to 1000 adenylate residues or more (e.g., Bardwell *et al.*, 1990), and the yeast PAP was shown to add over 300 (Lingner *et al.*, 1991a). It remains possible that post-translational modifications of PAP may affect the extent of polyadenylation *in vivo*. For example, phosphorylated and unphosphorylated forms of PAP display markedly different polymerizing abilities *in vitro* (Rose and Jacob, 1980).

Length control is reproducible in extracts, at least under some conditions. The unfractionated HeLa nuclear extract has been shown to reproduce a mechanism that limits the length of poly(A) to a maximum of about 200 residues (Sheets and Wickens, 1989). In this extract, precleaved substrates were shown to receive poly(A) tails in the range of 150–200 residues. Substrates that carried presynthesized poly(A) tracts of about 30, 45, 60,120, and 175 residues each received poly(A) to a maximum final length of about 200 residues, while RNAs carrying over 200 adenylates did not receive additional poly(A). Extracts from yeast have been shown to add about 60–80 adenylates to cleaved substrates, reproducing the approximate length distribution of newly synthesized poly(A) observed in yeast cells (Butler and Platt, 1989; Butler *et al.*, 1990). Because the cleavage and polyadenylation reactions have not been uncoupled in yeast extracts, it is not yet known whether polyadenylation is biphasic in this organism. The relatively short poly(A) tract typical of lower eukaryotes could be due to the absence of the second phase of polyadenylation.

Another nuclear poly(A) elongation process has been described in mammalian systems: a slow extension following the rapid polymerization event. The experiments described earlier, which demonstrated an upper limit of about 200 adenylates added by HeLa extracts (Sheets and Wickens, 1989), examined poly(A) length after a relatively short incubation (30 min). Using a longer incubation in a combined cleavage–polyadenylation assay, Moore and Sharp (1985) noted the occurrence of two kinetically distinguishable stages in the addition of poly(A): a rapid (15–30 min) addition of a poly(A) tract averaging 130 nt in their assay, followed by a slower addition of 200 to 600 more adenylates over the next 3–4 hr. Wahle (1991b), using highly purified PAP, CPF, and PABII, observed a similarly rapid addition of about 200 adenylates, followed by a slow extension of poly(A). Preliminary studies (reported in Wahle, 1991b) indicate that the abrupt deceleration in polymerization rate at a length of about 200 adenylates is dependent on CPF and its AAUAAA binding site as well as PABII; extension to about 400 adenylates proceeded steadily in the absence of either CPF or AAUAAA. This is an interesting finding, in that the extension of an oligoadenylated substrate itself is not dependent on CPF and AAUAAA (Sheets and Wickens, 1989; Wahle, 1991b). It was proposed that the transition to a slow polymerization rate may correspond to the physical release of CPF from its binding site (Wahle, 1991b). A slow nuclear poly(A) extension has also been shown to occur *in vivo* and appears distinct from rapid *de novo* addition due to its insensitivity to cordycepin (3′-deoxyadenosine; Brawerman and Diez, 1975). Figure 1 illustrates the stages of nuclear poly(A) addition in mammalian cells, as defined in cell-free extracts.

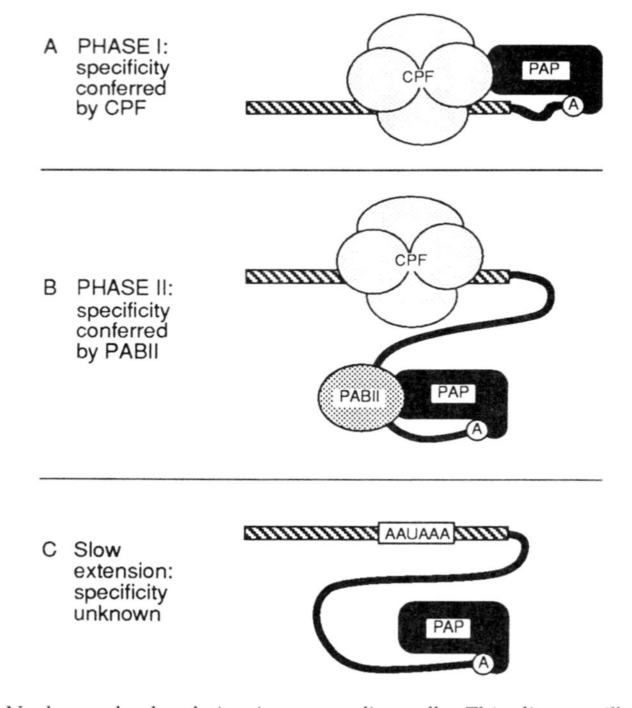

FIGURE 1 Nuclear polyadenylation in mammalian cells. This diagram illustrates only a subset of required factors: those shown to be involved in the polyadenylation of precleaved RNA substrates *in vitro*. *In vivo*, additional components are involved in the coupled cleavage reaction. The crosshatched box represents the 3′ portion of an RNA transcript; the black line represents poly(A). (A) Phase I—PAP (poly(A) polymerase) activity and specificity are dependent on the CPF (cleavage and polyadenylation factor) complex and its binding site, the AAUAAA hexamer. (B) Phase 2—PAP activity and specificity are dependent on PABII (poly(A)-binding protein II) and an oligo(A) primer, and are independent of the CPF and AAUAAA hexamer. (C) Slow elongation—the specificity of this extension process is unknown. It could be due to nonspecific polymerization by PAP. In A and B, PAP is shown in direct contact with each specificity factor, but this association has not been demonstrated experimentally.

C. Is the Extent of Poly(A) Addition Regulated *in Vivo?*

Although the distribution of poly(A) lengths observed on pulse-labeled RNA and nuclear RNA in a variety of cells is very homogeneous compared with poly(A) lengths in the cytoplasm, it actually encompasses a fairly broad range (e.g., Sheiness *et al.*, 1975). Variation in the length distribution of newly acquired poly(A) may be a property of particular mRNAs, but has not been widely investigated. There are a number of cases in somatic cells in which stimulus-induced alterations in poly(A) lengths on specific mRNAs have been reported (Muschel *et al.*, 1986; Paek and Axel, 1987; Robinson *et al.*, 1988; Carrazana *et al.*, 1988; Zingg *et al.*, 1988; Zingg and Lefebvre, 1989; Carter and Murphy, 1989; Krane *et al.*, 1991; Wu and Miller, 1991). Tissue-specific poly(A) lengths on particular mRNAs also occur (Ivell and Richter, 1984; Ivell *et al.*, 1986; Pastori *et al.*, 1992). In most cases, it is not known what aspect of mRNA metabolism is responsible for the changes. Altered nuclear polyadenylation may be responsible in at least two cases, based upon the finding that nuclear transcripts also show the altered poly(A) length. One case is the control of the cyclic alterations in the length of the poly(A) tail on vasopressin mRNA in the rat suprachiasmatic nucleus, which is the location of the endogenous circadian pacemaker in mammals (Robinson *et al.*, 1988; Carter and Murphy, 1989). Two distinct poly(A) lengths are observed for cytoplasmic vasopressin mRNAs, a single mRNA with a 240-nt tail during the light period, and two mRNAs with 240- and 30-nt tails during the dark period. Both mRNA species are also detectable in nuclear RNA (Carter and Murphy, 1992). Murphy *et al.* (1992) have similarly found that a 100- to 150-nt difference in poly(A) length on cytoplasmic rat growth hormone mRNA in the presence and absence of thyroid hormone is also apparent in mature nuclear RNAs. The main concern in interpretation of such data stems from the difficulty of preparing nuclei uncontaminated by cytoplasmic material, although suitable precautions were taken in these studies. Reports of exceptionally long poly(A), such as the 400-nt poly(A) tail carried by vasopressin mRNA under conditions of osmotic stress (Zingg *et al.*, 1988; Carrazana *et al.*, 1988), may be best explained by extended nuclear polyadenylation (Murphy and Carter, 1990), although cytoplasmic extension (Section IV) remains a possibility. While the occurrence of particularly short poly(A) is explicable by known cytoplasmic shortening activity, it remains unclear whether all poly(A)-deficient mRNA is actually derived from precursors with long poly(A) tails. Serum albumin mRNA in *Xenopus* liver, which has been shown to have a very short oligo(A) tract in the cytoplasm (Schoenberg *et al.*, 1989), may be synthesized with this short tail; longer poly(A) is not observed in nuclear preparations under conditions in which the unspliced primary transcript is readily detectable (D. Schoenberg, personal communication).

Two factors that might affect the final length of newly polymerized poly(A) are (1) sequence variations in the polyadenylation signal and its context, which might affect the time of transition from rapid polymerization to slow elongation, and (2) the length of time required for complete processing of the mRNA. An RNA that remains for a longer time in the nucleus may remain a substrate for the slow poly(A) extension process. Two treatments have been shown to result in a general and dramatic increase in nuclear poly(A) length: exposure to high levels of actinomycin D (Diez and Brawerman, 1974; Brawerman and Diez, 1975) and amino acid starvation (Brawerman, 1973). The latter effect has not been explained, but the effect of actinomycin D may be due to the extended residence of transcripts in the nucleus under these conditions (Brawerman and Diez, 1975). The possibility of a nuclear deadenylation activity cannot be eliminated, although *in vivo* manifestations of such an activity have never been described. Deadenylation activities have, however, been isolated from nuclear fractions (Müller *et al.*, 1976; Åström *et al.*, 1991), and highly purified PAP itself has been shown to exhibit poly(A) hydrolase activity in the absence of ATP (Abraham and Jacob, 1978).

III. Cytoplasmic Poly(A) Metabolism

A. Description of the Poly(A) Shortening Process

Upon entry into the cytoplasm, the poly(A) sequence on newly synthesized mRNA is subject to a degradative process that causes its progressive shortening. Poly(A)-shortening appears to have two kinetic components, at least in mammalian cells: (1) size reduction proceeds most rapidly during the first hours after entry into the cytoplasm, during which time the distribution of lengths remains quite narrow; and (2) beyond this period, the rate of shortening slows, but shorter and more heterogeneous poly(A) lengths are generated (Brawerman, 1981). It is not known whether one or multiple enzyme activities are involved in these processes. A two-component shortening process has not been described for the shorter poly(A) tails of lower eukaryotes. However, the yeast poly(A) nuclease (PAN) has been shown to digest poly(A) tails in two kinetically distinct phases *in vitro* (Lowell *et al.*, 1992; Section III.E).

The rate of poly(A) shortening has been quantified for only a few specific mRNAs. Typical mammalian two-component rates of shortening have been measured for metallothionein mRNA in rat liver (Mercer and Wake, 1985) and a transfected β-globin gene mRNA in mouse fibroblasts (Shyu *et al.*, 1991). The initial rates of shortening were about 20 A/hr for the metallothionein mRNA, and closer to 50 A/hr for the globin mRNA, each followed by decelerating rates of loss at least sixfold slower than the initial rates. The rates of poly(A) shortening on two tubulin mRNAs in *Chlamydomonas* were measured to be about 30–40 A/hr (Baker *et al.*, 1989), in the same range as the initial rate in mammalian cells. Poly(A) shortening is faster in yeast cells (Herrick *et al.*, 1990). The kinetics of poly(A) shortening on total yeast mRNA following a block to transcription indicate an average rate of shortening greater than 100 residues per hour (Minvielle-Sebastia *et al.*, 1991), but the rate of deadenylation may vary considerably for different yeast mRNAs (C. Decker and R. Parker, personal communication). For example, deadenylation of the MFA2 mRNA is particularly rapid at 11–14 A/min (660–840 A/hr) (Muhlrad and Parker, 1992). mRNAs encoding the protooncogenes c-*fos* and c-*myc* have been shown to undergo particularly rapid poly(A) loss in mammalian cells (Laird-Offringa *et al.*, 1989; Wilson and Treisman, 1988; Swartwout and Kinniburgh, 1989; Shyu *et al.*, 1991). Complete removal of long poly(A) tails from c-*fos* mRNA can take place within 30 min or less, representing a rate of loss exceeding 400 A/hr, if the residues are being removed sequentially. However, the rapidity with which the c-*fos* poly(A) tail becomes heterogeneous in length is suggestive of random endonucleolytic cleavage (Wilson and Treisman, 1988).

The rapid rate of poly(A) loss on c-*myc* and c-*fos* mRNAs in mammalian cells and the MFA2 mRNA in yeast cells shows that different mRNAs can be subject to different rates of poly(A) loss within the same cellular environment. Additional evidence that this is true was presented in a study that showed that newly synthesized β-actin mRNA and tubulin mRNA were chased into poly(A)-deficient fractions significantly faster than two other mRNA species in mouse cells (Krowczynska *et al.*, 1985). However, there is now evidence that β-actin mRNA categorized as poly(A)-deficient may actually possess a substantial poly(A) tail (see below). The most dramatic and quantitative example of differences in the rate of poly(A) loss on different mRNA species is a comparison of poly(A) loss on rabbit β-globin and human c-*fos* mRNAs expressed via the c-*fos* serum-inducible promotor in mouse fibroblasts, shown in Fig. 2A (Shyu *et al.*, 1991). Both newly synthesized mRNAs had poly(A) tails of about 200–230 residues. Over the course of 6 hr, the β-globin poly(A) tail shortened to about 65 residues and then remained stable thereafter ("BBB" in this figure), while the c-*fos* mRNA tail was completely removed in as short a time as 10 min.

The poly(A)-shortening process seems not to lead to completely deadenylated mRNA in most cases; rather mRNAs with short poly(A) tails accumulate as the major (stable) end products. The predominant steady-state poly(A) length is a remarkably similar 40–65 nt in many organisms (Brawerman, 1981). However, analysis of the distribution of individual mRNAs in different poly(A) length classes shows that the mean steady-state poly(A) length does vary for different mRNA species (Morrison *et al.*, 1979; Palatnik *et al.*, 1979; Shapiro *et al.*, 1988). In extensive studies of factors that might affect the final steady-state length of poly(A) in *Dictyostelium*

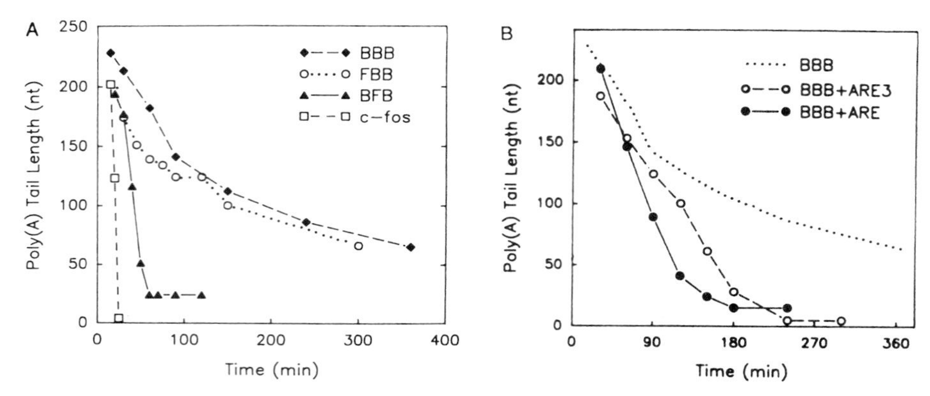

FIGURE 2 Comparison of the rate and extent of poly(A) shortening for human β-globin mRNA, mouse c-*fos* mRNA, and β-globin/c-*fos* chimeric mRNAs. mRNAs were expressed in NIH3T3 cells via the c-*fos* serum-inducible promotor, which drives a brief burst of transcription in response to serum stimulation. Poly(A) lengths were calculated from Northern blots. The convention used for naming chimeric mRNAs is a sequence of three letters identifying the origin of the 5′ UTR, coding region, and 3′ UTR sequences in each construct: B, β-globin; F, c-*fos*. (A) The coding region of c-*fos* mRNA contains sequences directing rapid deadenylation. The complete β-globin mRNA (◆) and a chimera composed of c-*fos* 5′ sequences, globin coding, and 3′ sequences (○) both exhibit slow deadenylation. The complete c-*fos* mRNA (□) and a chimera composed of globin 5′ and 3′ sequences and c-*fos* coding sequences (▲) both exhibit rapid deadenylation. (B) Insertion of the c-*fos* 3′ AU-rich element into β-globin mRNA increases its rate and extent of deadenylation. A β-globin mRNA in which the c-*fos* AU-rich element (ARE) has been inserted into the 3′ UTR region (●) shows significantly faster and more complete deadenylation than the unaltered β-globin mRNA (···). Introduction of three point mutations into the inserted c-*fos* AU-rich element (○) does not alter the ability of the element to confer accelerated deadenylation on β-globin mRNA, although it does strongly inhibit the destabilization of the mRNA (see text). [Reprinted with permission from Shyu *et al.* (1991).]

(Palatnik *et al.*, 1980; Shapiro *et al.*, 1988), it was determined that, in general, the final steady-state length correlates inversely with mRNA stability; i.e., unstable mRNAs tended to have longer poly(A) tails due to their relatively brief exposure to the cytoplasmic shortening activity, and vice versa. The extent of deadenylation is also subject to regulation; in *Dictyostelium*, poly(A) tails on most vegetative cell mRNAs are simultaneously shortened by 25–30 residues within minutes after induction to differentiate (Palatnik *et al.*, 1984). Specific mRNAs that accumulate to a large extent in a poly(A)-deficient form have been identified (Sonenshein *et al.*, 1976; Medford *et al.*, 1980; Swartwout *et al.*, 1987; Schoenberg *et al.*, 1989; Herrick *et al.*, 1990). In most of these cases, the criterion for poly(A) deficiency was inability to bind to oligo(dT)–cellulose. The accumulation of a large proportion of actin mRNA unable to bind to oligo(dT)–cellulose has been documented in a variety of cells (e.g., Sonenshein *et al.*, 1976; Palatnik *et al.*, 1984; Krowczynska *et al.*, 1985; Geoghegan and McCoy, 1986; Peppel and Baglioni, 1991). Bandyopadhyay and Brawerman (1992) have now shown that "poly(A)-deficient" β-actin mRNA isolated from mouse sarcoma ascites cells actually carries a sizable poly(A) tract of 60–90 adenylate residues, but that only a portion of this tract is available for interaction with oligo(dT). RNAse H digestion of poly(A)–oligo(dT) hybrids was shown to remove only the terminal 20–40 residues of the poly(A) tract. Melting of secondary structure permitted access to the proximal region of the poly(A) tract and resulted in complete digestion. The β-actin mRNA 3′ untranslated region (UTR) contains two extensive U-rich regions; thus, it is likely that the portion of the poly(A) tract unavailable for oligo(dT)-binding is engaged in base-pairing interactions with one or both of these regions. It will be of great interest to learn whether a similar 3′ terminal structure occurs *in vivo*.

A number of studies in which the poly(A) tail lengths on specific mRNAs were measured have revealed a virtual absence of poly(A) tails below a certain minimal length, which is usually estimated to be about 10–30 adenylates (Mansbridge *et al.*, 1974; Kelly and Cox, 1982; Krowczynska and Brawerman, 1986; Kleene *et al.*, 1984; Mercer and Wake, 1985; Carter *et al.*, 1989; Muhlrad and Parker, 1992). Upon close inspection, the minimal length cutoff may be quite sharp, 18 adenylates in the case of rabbit β-globin mRNA (Krowczynska and Brawerman, 1986), 10–12 adenylates in the case of the a-mating factor (MFA2) mRNA of yeast (Muhlrad and Parker, 1992). The only specific mRNAs known to shorten to the unadenylated state during their normal metabolism are c-*fos* mRNA (Wilson and Treisman, 1988; Shyu *et al.*, 1991), c-*myc* mRNA (Laird-Offringa *et al.*, 1989), and the *Chlamydomonas* α- and β-tubulin mRNAs (Baker, unpublished data). It remains possible that mRNAs which exhibit a minimal poly(A) length may be subject to complete poly(A) removal, but be rapidly degraded in the unadenylated state. For this to be true, it would have to be assumed that poly(A) shortening to the stable, minimal length and complete removal are kinetically distinguishable steps, an assumption for which there is new evidence (Lowell *et al.*, 1992; Section III.E).

It is not known what stabilizes the length of the relatively short poly(A) tails typical of steady-state mRNA. Short poly(A) tails may be preferentially protected from, or not a substrate for, the usual shortening activity. The recent characterization of a poly(A)-nuclease from yeast indicates that the enzyme itself may be intrinsically limited in the extent of deadenylation that it can efficiently catalyze (Sachs and Deardorff, 1992; Lowell *et al.*, 1992; Section III.E). Similarly, a deadenylation activity from *Xenopus* oocytes displays the property of requiring a minimal poly(A) tract as a substrate (Varnum and Wormington, 1990). Stabilization of short poly(A) tails could also result from an equilibrium between poly(A)-shortening and poly(A)-

elongation activities. A cytoplasmic polyadenylation activity that results in addition of an average of six to eight terminal adenylates on steady-state mRNAs has been shown to be active in mammalian cells (Diez and Brawerman, 1973; Sawicki *et al.*, 1977). No net cytoplasmic poly(A) elongation, such as that which occurs in gametes and embryos (Section IV), has been attributed to this activity. However, it remains a possibility that some reported poly(A) elongations (Section II.C) occur via cytoplasmic polyadenylation.

B. Sequence Information for the Rate and Extent of Poly(A) Loss

Information that directs rapid and complete deadenylation has been identified in the sequence of the c-*fos* mRNA. In the first study to identify such sequences, Wilson and Treisman (1988) found that removal of an AU-rich region from the 3′ UTR of a human c-*fos* mRNA slowed its rate of poly(A) shortening and possibly changed the mode of this process. Normally, poly(A) tails on newly synthesized c-*fos* mRNA are subject to a nucleolytic process that rapidly generates a very heterogeneous length distribution, ranging from zero to near-full length; poly(A) tails on c-*fos* mRNAs missing the AU-rich region shorten gradually, the length distribution remaining fairly homogeneous with time. Shyu *et al.* (1991) analyzed the poly(A) metabolism of chimeric mRNAs composed of different regions of the β-globin and c-*fos* mRNAs, and found that two different regions of the c-*fos* mRNA could independently accelerate poly(A) shortening when introduced into β-globin mRNA: the AU-rich region from the 3′ UTR identified by Wilson and Treisman and also a 320-nt segment from the c-*fos* coding region. Figure 2A shows that a chimeric mRNA containing the c-*fos* coding region, flanked by β-globin 5′ and 3′ UTR sequences ("BFB") is deadenylated nearly as rapidly and completely as the complete c-*fos* mRNA. Figure 2B shows that the insertion of the c-*fos* AU-rich element alone into the 3′ UTR of a β-globin mRNA ("BBB + ARE") dramatically increases its rate and extent of deadenylation, although not to the same degree as the complete c-*fos* mRNA. Significantly, these same 3′ UTR and coding region sequence elements have also been found to function as mRNA destabilizing elements (see Section V.A). Although similar AU-rich sequences have been identified in the 3′ UTRs of a number of unstable mRNAs, they do not all necessarily function to facilitate rapid deadenylation. For example, an AU-rich sequence in the 3′ UTR of β-interferon mRNA does not promote rapid poly(A) loss, although it does promote rapid mRNA degradation (Whittemore and Maniatis, 1990).

Muhlrad and Parker (1992), using a genetic approach, have identified a region in the 3′ UTR of the yeast MFA2 mRNA involved in stimulating its unusually rapid deadenylation and degradation. Extensive mutagenesis of this region did not result in identification of a critical sequence element, but rather, yielded data that indicate a functional requirement for unstructured RNA in this region. *In vitro* studies suggest that the processes of poly(A) shortening and terminal deadenylation in yeast may be controlled by different sequences. This conclusion stems, in part, from the observation that certain alterations of MFA2 3′ UTR sequence allow rapid poly(A) shortening, but inhibit terminal deadenylation (Lowell *et al.*, 1992). A tentative conclusion from these studies is that an AU-rich or A-rich region may be required roughly 60 nucleotides upstream of the poly(A) tail to allow complete terminal deadenylation, because placement of certain other sequences in this position can inhibit the process.

A sequence element that may act to promote deadenylation has been identified in the 3′ UTRs of at least two mRNAs involved in pattern formation in *Drosophila* embryos (Wharton and Struhl, 1991). The 3′ UTRs of the *bicoid* (*bcd*) and *hunchback* (*hb*) mRNAs contain one and two copies, respectively, of a sequence that has been defined functionally as the *nos* response element (NRE; GUUGUNNNNNAUUGUA), because the product of the *nanos* (*nos*) gene can block the translation of mRNAs that contain this element. Comparison of the polyadenylation state of the *bcd* mRNA in the presence and absence of *nos* revealed that this mRNA became extensively deadenylated in the presence of *nos*, but was predominantly polyadenylated in its absence. It is not known whether deadenylation is the direct result of *nos* activity or a consequence of translational inhibition.

Two models have been proposed to explain how sequence information in the 3′ UTR of an mRNA might act to accelerate the rate of poly(A) degradation. Ross and colleagues base their model on the principle that the PABP protects poly(A) from degradation and that the AU-rich 3′ UTR regions of unstable mRNAs promote dissociation of the poly(A)–PABP complex, perhaps by serving as alternate PABP binding sites (Brewer and Ross, 1988; Bernstein *et al.*, 1989). Wilson and Treisman (1988) proposed that the poly(A) tail forms base pairs with AU-rich regions in the 3′ UTR and that a specific nuclease cleaves at mismatched regions, thus degrading the poly(A) by endonucleolytic cleavages. More complicated models may be envisioned, involving additional protein factors that either promote or inhibit deadenylation by interaction with the 3′ UTR elements. The *nos* gene product may represent such a factor. These models do not address how sequences in the coding region might promote deadenylation. It has been noted that the coding region sequence implicated in the deadenylation and destabilization of c-*fos* mRNA encodes the protein dimerization motif of the c-*fos* protein. Similarly, a coding region destabilization element in the c-*myc* mRNA encodes that protein's dimerization motif and must be translated to be functional (Wisdom and Lee, 1991). These observations suggest a model in which the deadenylation process may be stimulated by interaction of a regulatory protein factor with the nascent polypeptide itself (Laird-Offringa, 1992).

It will be interesting to see how other mRNAs that exhibit unusual poly(A) structure or metabolism embody that information. The fact that vertebrate actin mRNAs have extremely conserved 3′ UTRs (Yaffe *et al.*, 1985) implies the functional importance of this region, which may include such information. Likewise, mRNAs encoding the *Xenopus* serum proteins albumin and γ-fibrinogen have a region of striking similarity in the sequences surrounding the AAUAAA polyadenylation signal (Pastori *et al.*, 1991). These otherwise unrelated mRNAs share the properties of having very short oligo(A) tails and being coordinately destabilized by estrogen. The discovery of sequence information that determines the characteristics of poly(A) removal for some mRNAs does not imply that all mRNAs carry comparable information. It seems more likely that there may exist a "default" poly(A) shortening process, similar to that described in *Xenopus* oocytes (see Section IV.B.2), which degrades poly(A) in a relatively uniform manner in the absence of sequence information that specifies otherwise.

C. Translation and Poly(A) Shortening

Poly(A) shortening appears to proceed independently of active translation for most mRNAs. The effect of inhibitors of protein synthesis on the poly(A)-shortening process has been looked at many times. The majority of studies showed no difference in poly(A)-shortening rates for total poly(A) or poly(A) on specific mRNAs as a result of translation inhibition, indicating that there is not a general coupling between the two processes in somatic cells (Brawerman, 1973; Sehgal *et al.*, 1978; Merkel *et al.*, 1976; Adams and Jeffrey, 1978; Mercer and Wake, 1985; Berger *et al.*, 1985; Carter *et al.*, 1989). Comparison of poly(A)-shortening in polysomal versus nontranslating mRNP fractions in sea urchin embryos and developing spermatids has indicated a possible requirement of a polysomal location for poly(A)-shortening in some instances (Dworkin *et al.*, 1977; Kleene, 1989) (Section IV.D).

Three cases have been reported in which normal poly(A)-shortening is dramatically affected by inhibition of translation. The usual rapid deadenylation of c-*fos* mRNA (Wilson and Treisman, 1988) and c-*myc* mRNA (Laird-Offringa *et al.*, 1990) is inhibited in the presence of cycloheximide. The effect is clearly quite specific for these mRNAs, since many studies have failed to detect a general effect of protein synthesis inhibitors on poly(A)-shortening in mammalian cells. Thus active translation seems to be required for these two examples of rapid poly(A) removal. In contrast, the usual slow processive loss of poly(A) on tubulin mRNAs in *Chlamydomonas* is accelerated in cycloheximide, such that a large proportion of tubulin mRNA is rapidly and completely deadenylated (Baker *et al.*, 1989). In this case, active translation appears to protect the tubulin mRNAs from rapid poly(A) removal. Figure 3 shows the marked effect of translational inhibition by cyclohexi-

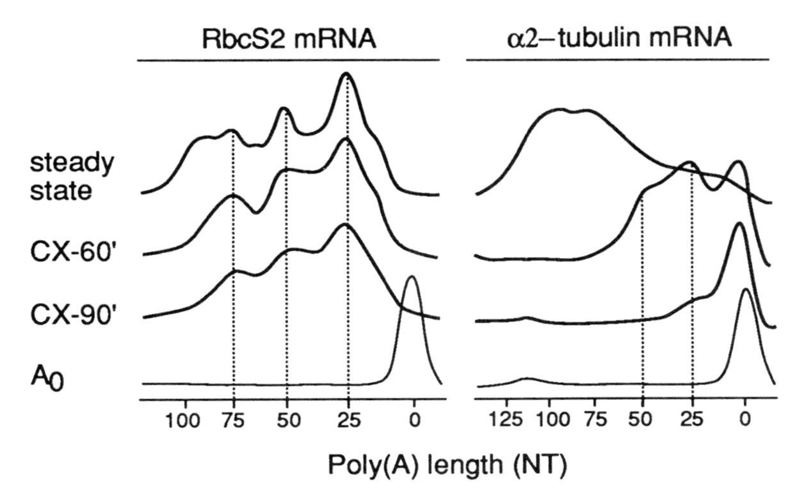

FIGURE 3 Poly(A) length distributions at steady state and following cycloheximide addition for two mRNAs in *Chlamydomonas*. Densitometer tracings of Northern blots of oligonucleotide-RNAse H-generated 3′ fragments of the RbcS2 mRNA (ribulose bisphosphate carboxylase, small subunit) and the α2-tubulin mRNA. CX, cycloheximide; A_0, length of completely deadenylated fragment. Fragment lengths were calculated using RNA molecular weight markers, and poly(A) lengths were determined by subtracting the known length of the completely deadenylated fragment, A_0, from the fragment lengths. RbcS2 mRNA shows a distinct accumulation of poly(A) lengths in multiples of 25 residues. No mRNA with poly(A) shorter than 12–15 residues is ever observed. The pattern of discrete lengths and the minimal poly(A) length is not significantly altered by inhibition of protein synthesis. α2-mRNA shows a very different poly(A) distribution at steady state, with no evidence of discrete poly(A) lengths. Inhibition of protein synthesis causes rapid loss of the poly(A) tail leaving most mRNAs unadenylated or with short oligo(A) tails (rapid deadenylation is not characteristic of this mRNA under normal conditions (Baker *et al.*, 1989)). Discrete poly(A) lengths of 50 and 25 residues are intermediates in the deadenylation process.

mide on the steady-state poly(A) length distribution of an α-tubulin mRNA in *Chlamydomonas*. That the effect is mRNA-specific is demonstrated by the essentially unchanging poly(A) length distribution exhibited by the RbcS2 (ribulose bisphosphate carboxylase small subunit) mRNA under the same conditions.

An interesting finding, given the translational dependence of rapid c-*myc* poly(A) shortening *in vivo*, is the observation that the poly(A) tail on polysomal c-*myc* mRNA is much more vulnerable to degradation than is the poly(A) tract on a γ-globin mRNA in a nontranslating cell-free extract (Brewer and Ross, 1988, 1989). Although it has not been demonstrated that the activity that digests the c-*myc* poly(A) tail *in vitro* is the nuclease that actually operates *in vivo*, these results do reveal a conspicuous difference between the c-*myc* and γ-globin poly(A) tails, perhaps reflecting a structural dissimilarity that underlies their different metabolisms *in vivo*.

D. Role of the Poly(A)-Binding Protein in Poly(A) Metabolism

There are two quite different functions in poly(A) metabolism that have been attributed to the PABP. The first, which is the more intuitively obvious one, is that the presence of PABP on poly(A) protects this sequence from rapid degradation. The second, seemingly antithetical role, is that the PABP is required for poly(A) shortening.

The main studies indicating a role of the PABP in poly(A) stability have demonstrated that polysomal poly(A) exists in a state that is highly resistent to digestion by a variety of endonucleases and exonucleases (e.g., Bergmann and Brawerman, 1977; Müller *et al.*, 1978). Resistance of poly(A) to 3′ exonucleases was shown to depend on other regions of the mRNP complex, and it could be abolished by a cytoplasmic factor, which acted neither as a nuclease nor as a protease (Bergmann and Brawerman, 1977). These findings suggested that the protection phenomenon being studied was not simply a nonspecific shielding effect, but that specific RNA–protein interactions were involved in controlling accessibility of the poly(A) tail. In a new *in vitro* approach using more defined components, Bernstein *et al.* (1989) have studied the protective effect of PABP against a ribosome-associated nuclease activity from human erythroleukemia cells, which was shown to degrade unadenylated, but not polyadenylated transcripts (Peltz *et al.*, 1987). It was found that endogenous PABP in the ribosomal salt wash (RSW) fraction containing this activity was a necessary component in endowing relative stability on polyadenylated molecules. Depletion of PABP from the RSW caused the normally stable polyadenylated β-globin RNA to be degraded, and protection could be restored by readdition of either yeast or human PABP, but not by adding a nonspecific RNA-binding protein, *E. coli* SSB. A striking feature of the degradation of "unprotected" globin mRNA was that the main intermediates of degradation were mRNAs with progressively decreasing poly(A) tail lengths with an accumulation of unadenylated, full-length molecules. These observations suggest that a prominent nuclease activity in the RSW is a poly(A)-specific exonuclease that can be deterred by PABP. It seems unlikely that poly(A) shortening could have been catalyzed by a nonspecific exonuclease in this assay; the presence of a block to continuous digestion at the poly(A)–mRNA junction would have to be invoked to explain the accumulation of unadenylated molecules. An enzyme activity that specifically digests 3′ poly(A) tracts, stopping at the poly(A)–mRNA junction, has recently been described (Åström *et al.*, 1991) (Section III.E).

That the PABP confers protection against degradation of poly(A) in cells is suggested by studies that show that poly(A) tracts accumulate *in vivo* in a series of discrete lengths that correspond, at least roughly, to multimers of 25–30. This fragment length is similar to the known packing density of PABP on polysomal poly(A) *in vivo* (Baer and Kornberg, 1980, 1983). The periodicity of both rabbit and mouse globin mRNA poly(A) tails has been well documented (e.g., Mansbridge *et al.*, 1974; Alquist and Kaesberg, 1979; Kelly and Cox, 1982; Krowczynska and Brawerman, 1986), but reports of other specific mRNA poly(A) periodicities are scarce (Laird-Offringa *et al.*, 1989; Baker *et al.*, 1989). Total yeast mRNA was reported to display a distinct periodicity in length classes (Groner and Phillips, 1975), but that phenomenon has not been apparent in more recent electrophoretic analyses (e.g., Sachs and Davis, 1989; Minvielle-Sebastia *et al.*, 1991). The poly(A) length distribution of RbcS2 mRNA, shown in Fig. 3, reveals a striking accumulation of molecules bearing poly(A) tails of 25, 50, and 75 residues. Periodicity in poly(A) lengths due to bound PABP could arise by endonucleolytic clipping between bound proteins or by the action of a 3′ exonuclease that could rapidly hydrolyze terminal poly(A) stretches left bare by release of PABP. The conferral of partial protection at discrete distances along the poly(A) tract indicates that the protective protein would have to be phased from an origin near the mRNA–poly(A) junction.

A prediction of a role for the PABP in protecting poly(A) from digestion is that cells depleted of PABP should have shorter poly(A) tails. A study that examined the effect of PABP depletion in yeast cells, by placing the wild-type PABP gene under control of a repressible promotor, demonstrated just the opposite effect: a dramatic increase in the average length of cytoplasmic poly(A) tails, such that they resembled the length distribution of newly synthesized tails (Sachs and Davis, 1989). Several controls supported the conclusion that PABP depletion must be inhibiting the poly(A) shortening process itself, and these authors concluded that the PABP is required for poly(A) shortening in *S. cerevisiae*. This was, of course, a surprising result, but it is not necessarily inconsistent with a role in poly(A) protection (see later). This study also raised a disconcerting question about the functional significance of poly(A)-shortening, since a number of suppressor mutations that permitted cells to grow in the absence of PABP did so without restoring poly(A)-shortening.

E. Enzymology of Cytoplasmic Deadenylation

The value of yeast genetics as a tool with which to identify gene products involved in RNA metabolism has been amply demonstrated recently. New studies in yeast have identified what is undoubtedly an authentic poly(A)-shortening enzyme, as well as three other "cofactors" that probably function in the process of poly(A)-shortening.

A poly(A) nuclease has been purified from extracts of a PABP-deficient yeast strain, based on the ability of PABP to restore poly(A)-shortening. The nuclease possesses the unprecedented property of digesting only poly(A) associated with the PABP; thus it recognizes only a ribonucleoprotein as a substrate. Complete reactions can be assembled with synthetic transcripts, PABP, and the purified enzyme (named PAN for PABP-dependent poly(A) nuclease). PAN is likely to be the authentic poly(A) degrading enzyme

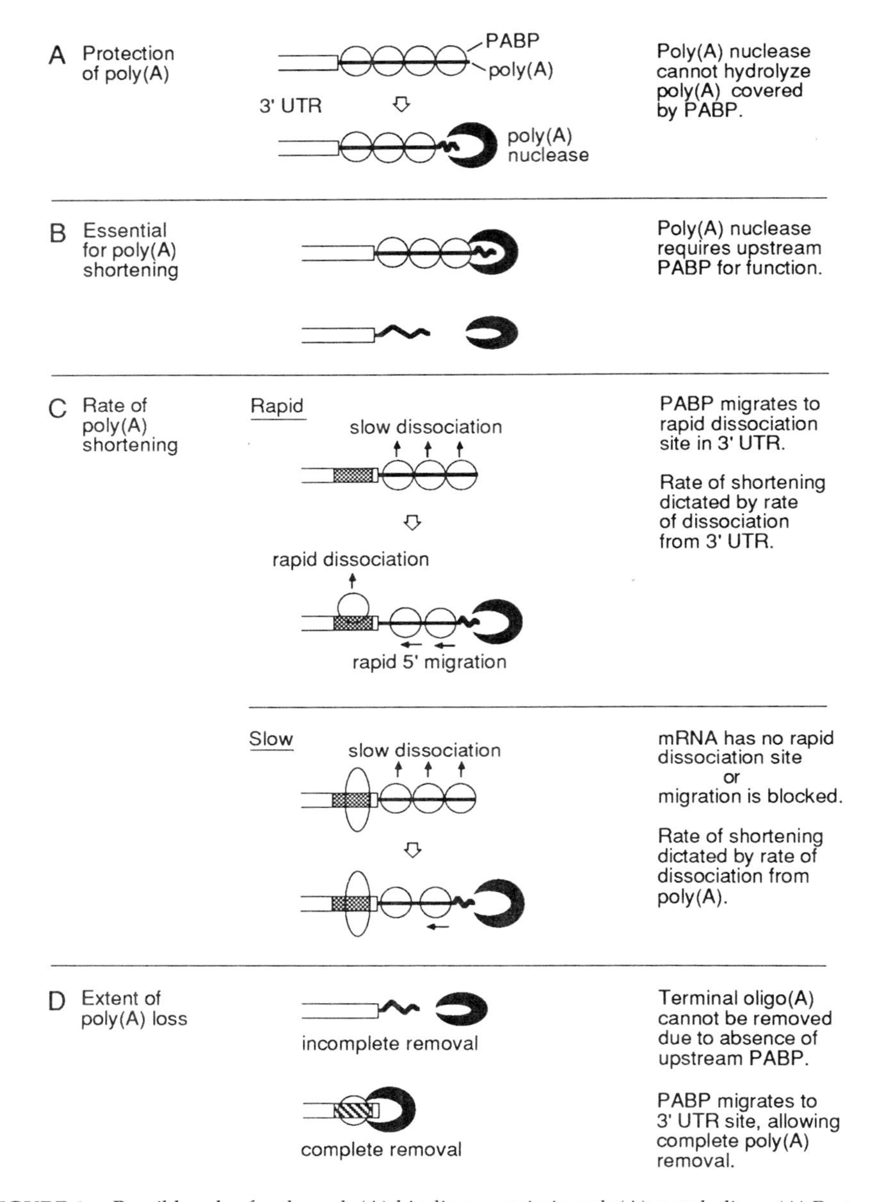

FIGURE 4 Possible roles for the poly(A)-binding protein in poly(A) metabolism. (A) Protection of poly(A) against nuclease digestion has long been considered a likely function for the PABP. Evidence for a protective effect is reviewed in Section IIID. (B) Evidence that the PABP may be an essential component in the poly(A)-shortening process can be reconciled with a protective role by assuming that the poly(A) nuclease can digest only sequences downstream of a PABP. (C) One model for how PABP could be involved in controlling the rate of poly(A) loss for different mRNAs is illustrated. This model proposes that sequences in the 3′ UTR facilitate migration and rapid dissociation of PABP molecules, resulting in a rate of PABP loss faster than would occur by direct dissociation of PABP from poly(A). (D), Migration of a terminal PABP into the 3′ UTR could also explain the ability of some mRNAs to experience (continues)

which operates *in vivo,* because conditional mutations in PAN cause an increased accumulation of full length poly(A) tails (Sachs and Deardorff, 1992). Poly(A) alone is a substrate for the purified enzyme, but sequences upstream from the poly(A) tract influence the rate and mode of PAN activity. PAN digests poly(A) in a manner that suggests it acts as a distributive exonuclease on most substrates. However, the enzyme appears to act as a *processive* exonuclease on a substrate containing 3′ UTR sequences of the MFA2 mRNA (Lowell *et al.*, 1992). The basis for the enhanced processivity is not yet known, but could involve an increased affinity of the requisite PABP for the MFA2 substrate. Two pieces of evidence indicate that PAN may not be strictly a poly(A) nuclease. (1) RNAs that terminate in 100 adenylates followed by 32 nucleotides of random RNA were efficiently deadenylated. (2) PAN was able to digest an RNA substrate that contained an AU-rich region from the MFA2 3′ UTR (without a terminal poly(A) tract). Because PABP is capable of binding to this 3′ UTR region (shown by a filter-binding assay), this finding suggests the interesting possibility that PAN may be able to digest any 3′ terminus with which PABP can associate. Thus, an mRNA which contains a PABP binding site in the 3′ UTR might be subject to both deadenylation and degradation of 3′ UTR sequences by PAN (Lowell *et al.*, 1992).

Significantly, PAN was shown to digest poly(A) in two kinetically distinct stages on several substrates composed of different yeast mRNA 3′ UTR sequences (Lowell *et al.*, 1992). Poly(A) shortening to a short tail (10–25 residues, similar to the minimal length of poly(A) observed *in vivo*) was generally much more rapid than subsequent terminal deadenylation. *In vitro,* PAN is able to slowly digest short poly(A) tracts in a PABP-*independent* fashion, but whether or not this reaction occurs *in vivo* is not certain. The requirement for the PABP in the poly(A)-shortening process in yeast and the usual inability of PAN to efficiently remove the poly(A) segment most proximal to the poly(A)–RNA junction together suggest that PAN requires a PABP 5′ to its site of action. An mRNA that contains a PABP binding site in its 3′ UTR may remain a substrate for efficient PABP-dependent poly(A)removal. Simultaneous roles for PABP in poly(A) protection and poly(A) degradation are easily envisioned. PABP could protect poly(A) with which it is associated, but stimulate the degradation of downstream sequences that become exposed by loss or migration of a terminal PABP. Figure 4 illustrates a number of possible roles for the PABP in poly(A) metabolism.

(continued)
complete poly(A) removal, while other mRNAs retain oligo(A) tails. The idea that loss of poly(A)-bound PABP may occur via migration into 3′ UTR regions was first proposed in a model by Bernstein and Ross (1989). Models proposing the requirement of an upstream PABP for poly(A) nuclease activity and its possible consequences for the extent for poly(A) loss were suggested by Sachs and Deardorff (1992); Lowell *et al.* (1992).

In addition to the PABP itself, two other proteins have been identified that may be involved in control of the poly(A)-shortening process in yeast (Minvielle-Sebastia *et al.*, 1991). Their genes were isolated by complementation of thermosensitive mutants that exhibit a dramatic shortening of poly(A) tails at the restrictive temperature. The RNA14 and RNA15 genes are both single copy and are essential for cell viability. The RNA14 gene product shows no significant homology to any known protein; but the RNA 15 product has a characteristic RNA-binding domain at its N terminus. Although it has been demonstrated that the phenotype observed is not due to a substantial block in transcription and that the mutant cells appear to have normal poly(A) polymerase activity (Block *et al.*, 1978), it has not yet been shown with certainty that the defect in these mutants is in the cytoplasmic control of poly(A) length.

Very little is known about the nucleases and associated factors that operate in other cells to effect poly(A) removal, nor even how many different activities may be operative. Although nucleases that show poly(A) specificity or preference have been isolated from mammalian cells (e.g., Müller *et al.*, 1976; Schröder *et al.*, 1980), there has been no clear assignment of *in vivo* functions. An activity has recently been identified in HeLa cells that demonstrates true poly(A) specificity on a mixed polymer (Åström *et al.*, 1991, 1992). This activity was shown to degrade progressively only the poly(A) tracts of substrates composed of 5′-capped nonpoly(A) sequences followed by pure poly(A) tails of 30 residues. It exhibits properties that distinguish it from all previously described poly(A) nucleases (discussed in Åström *et al.*, 1992). There seem to be no specific sequence requirements in the RNA body for deadenylation, a property consistent with a default poly(A)-shortening activity described in *Xenopus* oocytes (Section IV.B.2). A question concerning the *in vivo* function of this activity is raised by the fact that the activity was maximal in a chromatographic fraction of a nuclear, not cytoplasmic, extract. Since the nuclear fraction could well have contained particulate material from the cytoplasm that cosedimented during fractionation, it is important that the actual cellular location be determined.

IV. Polyadenylation and Deadenylation in Gametes and Early Embryos

A. Occurrence and Significance

Cytoplasmic modification of poly(A) length is an extremely widespread phenomenon in both vertebrate and invertebrate developing gametes and early embryos. mRNA-specific polyadenylation and deadenylation occur during meiotic maturation of oocytes and fertilization of eggs (Rosenthal and Ruderman, 1987; Rosenthal and Wilt, 1987; Wickens, 1990b), during spermatid differentiation (Iatrou and Dixon, 1977; Kleene *et al.*, 1984; Schäfer *et al.*, 1990), and possibly in the sperm-to-oocyte switch in *Caenorhabditis elegans* hermaphrodites (Ahringer and Kimble, 1991). In these cases, the poly(A) length changes are unambiguously cytoplasmic, since transcription is effectively shut down in most instances. Many studies, some presented below, strongly support the hypothesis that changes in polyadenylation are involved in control of mRNA transfer between nontranslating and translating compartments in developing systems. Overall, cytoplasmic polyadenylation correlates with translational activation of dormant messages, and deadenylation with translational inactivation, although there are exceptions. There is good evidence that cytoplasmic polyadenylation is a prerequisite for, not a consequence of, translational recruitment in mouse and frog oocytes and in *Drosophila* spermatids (Fox *et al.*, 1989; McGrew *et al.*, 1989, Vassalli *et al.*, 1989; Schäfer *et al.*, 1990). mRNA deadenylation seems not to be used as a mechanism to rapidly destabilize specific mRNAs, at least in *Xenopus* oocytes, since deadenylated mRNAs can be maintained stably in nonpolysomal form (e.g., Dworkin and Dworkin-Rastl, 1985; Hyman and Wormington, 1988; Paris and Phillippe, 1990). These developmental systems have provided the most convincing existing evidence that differences in poly(A) length can be highly significant to mRNA function. They also provide the opportunity to investigate the biochemistry of cytoplasmic poly(A) modification and the basis for its selectivity. It remains unknown whether the activities that operate in gametes and embryos occur in other cells.

Frog and mouse oocytes have proved to be ideal experimental systems in which to address the mechanism and function of cytoplasmic polyadenylation and deadenylation, and by far the greatest progress has been made studying meiotic maturation in *Xenopus* oocytes. Oocytes contain a store of mRNAs that are selectively translated during oogenesis and early development. Some mRNAs are not translated in immature oocytes and are recruited into polysomes concommitant with maturation-induced polyadenylation (Dworkin and Dworkin-Rastl, 1985; Paynton *et al.*, 1988; Huarte *et al.*, 1987) while other mRNAs are translated in immature oocytes and removed from polysomes concommitant with maturation-induced deadenylation (Bachvarova *et al.*, 1985; Hyman and Wormington, 1988). The experimentally powerful features of oocytes are that they can be induced to mature *in vitro* and that synthetic RNA molecules injected into oocytes function in a sequence-specific manner as substrates for polyadenylation or deadenylation. Figure 5A shows a time course of polyadenylation of a synthetic RNA substrate in maturing *Xenopus* oocytes. This synthetic substrate is a mere 110-nt transcript composed, in part, of sequences from a natural substrate RNA. Most experiments discussed below involved the use of chimeric or truncated versions of natural substrates. For simplicity, the actual RNA construct will not be specified, except where relevant.

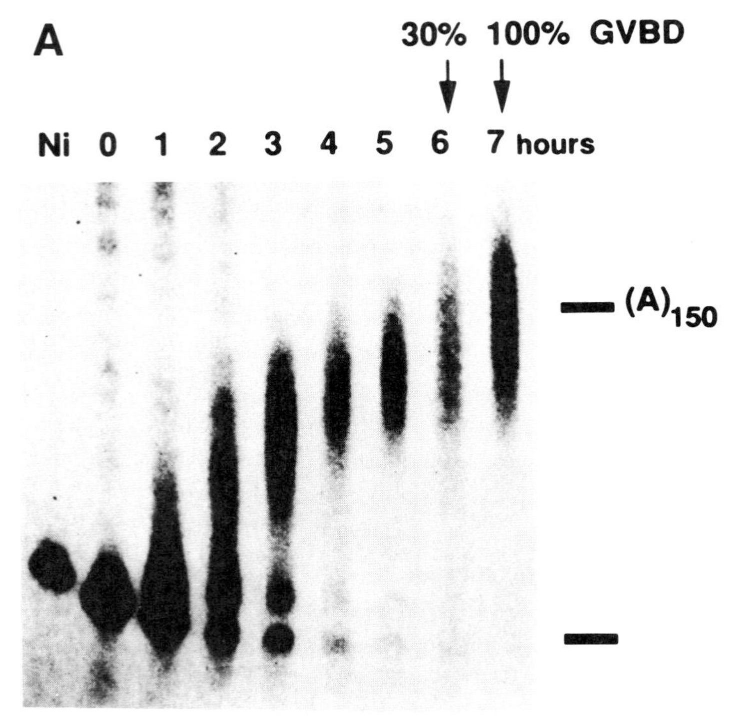

FIGURE 5 Activation of polyadenylation of an RNA substrate transcribed *in vitro* in *Xenopus* oocytes and oocyte extracts. The competent substrate RNA, sGb/B4, is a 110-nucleotide transcript composed of 5′ noncoding sequence from the *Xenopus* β-globin mRNA and the CPE-AAUAAA region from the B4 mRNA, a natural substrate for maturation-induced polyadenylation. (A) Time course of polyadenylation during maturation of oocytes. Labeled transcript was injected into oocytes that were then induced to mature by incubation in progesterone. The increase in size of the injected RNAs as a function of time, due to polyadenylation, is demonstrated by electrophoresis of RNA samples. Ni, noninjected transcript. The percentage of oocytes that had undergone germinal vesicle breakdown (GVBD) is indicated, showing that the onset of polyadenylation precedes this event. (B) Role of cyclin and $p34^{cdc2}$ kinase in the activation of polyadenylation. Extracts prepared from immature oocytes, incapable of polyadenylation, were supplemented with lysates prepared from baculovirus-infected insect cells containing baculovirus wild-type proteins only (Bac-WT, control), baculovirus-expressed clam cyclin B (Bac-cyclin B), or both clam cyclin B and human $p34^{cdc2}$ (Bac-cyclin B/cdc2). After an incubation period, labeled sGb/B4 substrate was added. Both of the latter two lysates activated polyadenylation (lanes 2 and 3). Depletion of extract $p34^{cdc2}$ by the affinity matrix, p13, prevented the activation of polyadenylation (lane 5), while a mock depletion by protein A–Sepharose did not (lane 4). Activation of polyadenylation in depleted extracts could be restored by addition of baculovirus-expressed cdc2 (lane 6). The lower panel shows the ability of the extracts to phosphorylate histone H1, a substrate for $p34^{cdc2}$, and provides a measure of relative $p34^{cdc2}$ kinase activity in each extract (continues). [Reprinted with permission from Paris and Richter (1990) (A), and Paris *et al.* (1991) (B).]

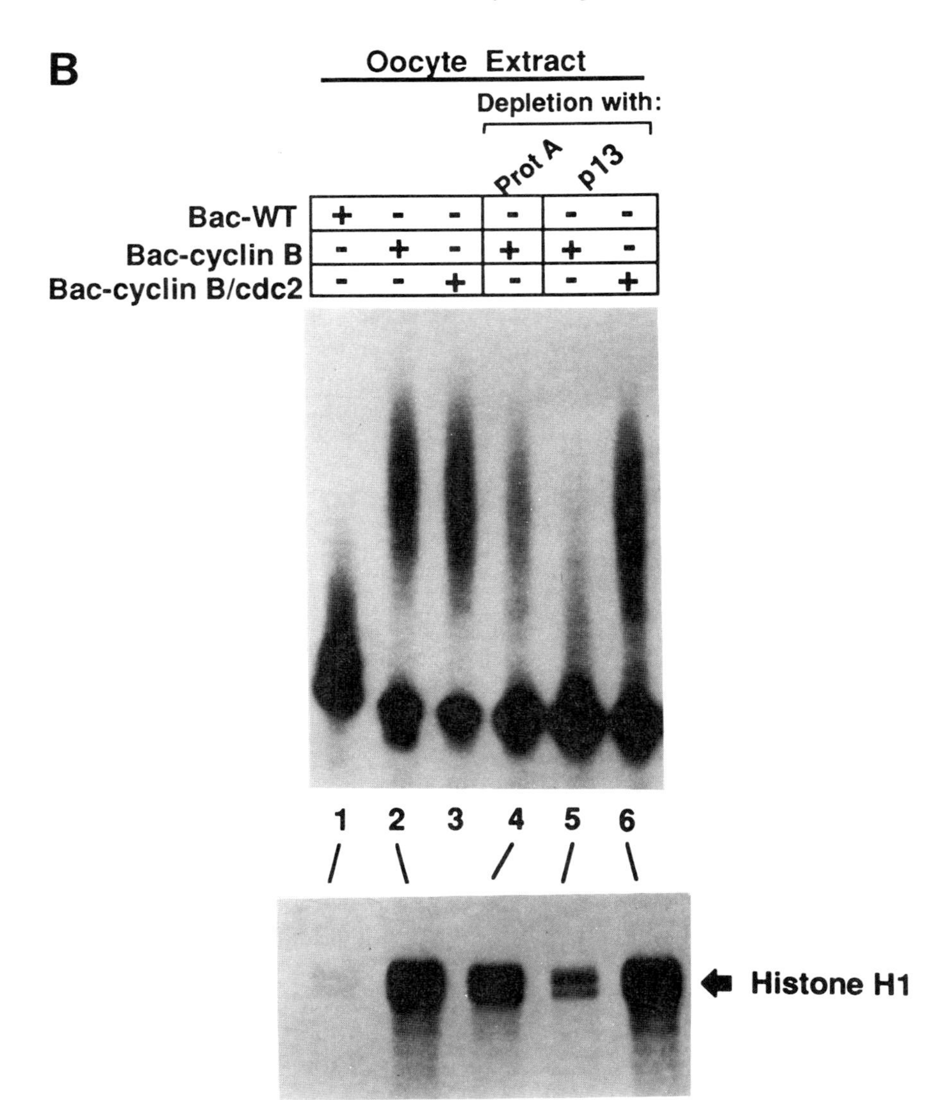

FIGURE 5 (continued).

B. Sequence Information for Cytoplasmic Adenylation and Deadenylation

1. *Cytoplasmic Polyadenylation Signals*

Sequences responsible for cytoplasmic polyadenylation in both frog and mouse oocytes have been localized to the terminal portions of the 3′ UTR. In the frog, two discrete sequence elements have been identified. One element is the hexanucleotide, AAUAAA, which is also required for nuclear polyadenylation (Fox *et al.*, 1989; McGrew *et al.*, 1989). A point mutation in the hexanucleotide sequence (AAGAAA) that abolishes nuclear polyadenylation also abolishes cytoplasmic polyadenylation (Fox *et al.*, 1989; McGrew and Richter, 1990). The hexamer sequence is not sufficient for polyadenylation, as would be expected since essentially all mRNA termini contain this sequence. The second important sequence element, which has been termed the CPE (cytoplasmic polyadenylation element; McGrew and Richter, 1990), is generally positioned 4–13 nt upstream of the AAUAAA sequence and varies among the different substrate mRNAs. All described or suspected CPEs have in common an extreme U-richness; e.g., two CPEs have been shown to be UUUUUUAUAAAG (McGrew and Richter, 1990) and UUUUUAAU (Paris and Richter, 1990). Extensive mutation of CPE elements has been shown to reduce, but not abolish polyadenylation (Fox and Wickens, 1990; McGrew and Richter, 1990), indicating that CPE sequence requirements are quite flexible. These two sequence elements (AAUAAA and the CPE) are not only required, but virtually sufficient for signaling polyadenylation. A 65-nt portion of adenovirus L3 mRNA, which by chance includes both a known CPE and an AAUAAA sequence in reverse order, was shown to be a competent substrate for polyadenylation (Fox *et al.*, 1989). Maternal mRNAs that undergo specific polyadenylation at fertilization or during early developmental stages in the frog embryo have also been identified (Paris *et al.*, 1988; Paris and Philippe, 1990). Sequence elements required for embryonic polyadenylation and translational recruitment have been identified for one of these mRNAs called C12: they are the hexanucleotide AAUAAA and a dodecauridine element. The latter sequence was also shown to be necessary for repression of polyadenylation during oocyte maturation (Simon and Richter, 1992).

In the mouse oocyte, the AAUAAA hexanucleotide is also a necessary but insufficient sequence element for polyadenylation and recruitment (Vassalli *et al.*, 1989; Huarte *et al.*, 1992). The mouse equivalent of the CPE has recently been identified and is also located in the 3′ UTR (Vassalli *et al.*, 1989; Huarte *et al.*, 1992; Sallés *et al.*, 1992). The CPE of the tissue-type plasminogen activator (tPA) mRNA is contained in an extremely AU-rich region of the 3′ UTR that includes two copies of an AUUUUAAU motif and two other copies of AUUUUA. One copy of the AUUUUAAU motif, in combination with the AAUAAA hexamer, is a sufficient signal for polyadenylation (Sallés *et al.*, 1992). Another substrate for maturation-induced polyadenylation (HPRT mRNA; Paynton *et al.*, 1988) also contains an AUUUUAAU sequence, in addition to a UUUUAAAU sequence similar to a *Xenopus* CPE. The signals for polyadenylation are quite conserved, since the HPRT mRNA is recognized as a substrate by *Xenopus* oocytes (Paris and Richter, 1990).

On the other hand, a sequence required in *Drosophila* for polyadenylation and polysomal recruitment of a spermatid-specific mRNA, called Mst87F, is unrelated to the oocyte signals; it is a 12-nt element in the 5′ UTR (Schäfer *et al.*, 1990). Significantly, the sequencing of six other genes that form a family with Mst87F reveals an almost perfect conservation of this 5′ element. A 5′ UTR signal for translational activation is not common to all developing sperm, since the element for translational activation of at least one mouse spermatid-specific mRNA has been localized to the 3′ UTR (Braun *et al.*, 1989). However, this recruitment process does not involve polyadenylation (see Section IV.D).

2. *Sequence Information Directing Deadenylation*

There is now evidence for two different kinds of deadenylation activity in oocytes: a maturation-induced activity that requires no specific sequence information and a sequence-specific activity in immature oocytes. Reports from two laboratories have concluded that the maturation-induced deadenylation activity is a default reaction in the *Xenopus* oocyte (Fox and Wickens, 1990; Varnum and Wormington, 1990). That is, in the absence of information to the contrary, an RNA is deadenylated during maturation. These studies showed that injected, polyadenylated RNAs could serve as substrates for the maturation-induced deadenylation activity and that both natural and unnatural substrates were subject to the deadenylation activity, including transcripts of random prokaryotic vector sequences (Fox and Wickens, 1990) and a synthetic C–G homopolymer (Varnum and Wormington, 1990). Escape from default deadenylation requires sequence information, and it turns out to be the same information that is required to direct cytoplasmic polyadenylation. Mutations of the AAUAAA hexamer or the CPE that abolish or impair polyadenylation result in deadenylation of polyadenylated versions of the same RNAs. Fox and Wickens (1990) have proposed two models for how the sequence information directing polyadenylation might prevent deadenylation: (1) compensation, in which poly(A)-shortening occurs to all mRNAs, but a more efficient poly(A)-polymerizing activity results in net elongation on CPE-bearing molecules; and (2) protection, in which poly(A) added cytoplasmically is resistant to deadenylation. Protection could be conferred by interaction with factors that bind to CPE-containing mRNAs or possibly by covalent modification of the poly(A) tract itself.

A second deadenylation activity, which is sequence-specific, has recently been described (Huarte *et al.*, 1992). This activity is operative in growing and primary (immature) mouse oocytes, and promotes extensive deadenylation of targeted mRNAs during a period in which poly(A) tails on other mRNAs are only modestly shortened. It is likely that the substrates for this activity are mRNAs destined for translational dormancy in the oocyte. These authors show that tPA mRNA is synthesized in growing oocytes with a long, 300- to 400-nt poly(A) tail that is then shortened to a 30- to 40-adenylate tail by the sequence-specific deadenylase activity. Thus the tPA mRNA is subject to both sequence-specific deadenylation in the immature oocyte and sequence-specific polyadenylation in the maturing oocyte. Experiments with mutant transcripts revealed that the sequence elements controlling these two events are overlapping, but not identical. The only critical difference may be that polyadenylation requires the AAUAAA hexamer, while deadenylation does not. In light of these findings, the motif defined as a CPE may more appropriately be called an ACE (adenylation control element; Huarte *et al.*, 1992). Moreover, the mouse tPA mRNA is resistant to maturation-induced default deadenylation, as is true for polyadenylation substrates in *Xenopus* oocytes. Huarte *et al.* (1992) observed that presynthesized poly(A) tails on transcripts containing a mutated hexamer sequence (AAGAAA) were shortened to the usual 30–40 residues in the immature oocyte, but then completely deadenylated in the mature oocyte. This indicates that escape from default deadenylation in the mouse oocyte, as in *Xenopus*, requires polyadenylation in the oocyte. Figure 6 illustrates the three different enzymatic activities that alter the polyadenylation state of maternal mRNAs in oocytes.

C. Control of Poly(A) Length

It has been suggested that the natural variation of CPE sequences on maternal mRNAs might lead to a spectrum of poly(A) lengths (Fox and Wickens, 1990). A survey of experimental results from many papers reveals a wide variation in the length of poly(A) added to different natural or derived substrate RNAs, ranging from only about 50 (Paris and Richter, 1990) to 500 or more (Huarte *et al.*, 1987; McGrew *et al.*, 1989). The length of poly(A) polymerized does seem to be under the control of 3′-terminal sequences of maternal mRNAs. In a controlled study, polyadenylation was compared for three similar chimeric mRNAs, composed of 5′ β-globin sequences and CPE/AAUAAA regions of three different natural substrates (Paris and Richter, 1990). It was found that the proportion of adenylated molecules and the length of poly(A) that was added differed for each construct. There are also many observations that indicate that sequences other than the CPE per se have an effect on the length of poly(A) polymerized. For example, a complete copy of a natural substrate, called G10, received about 100 adenylates after injection, whereas a chimeric mRNA

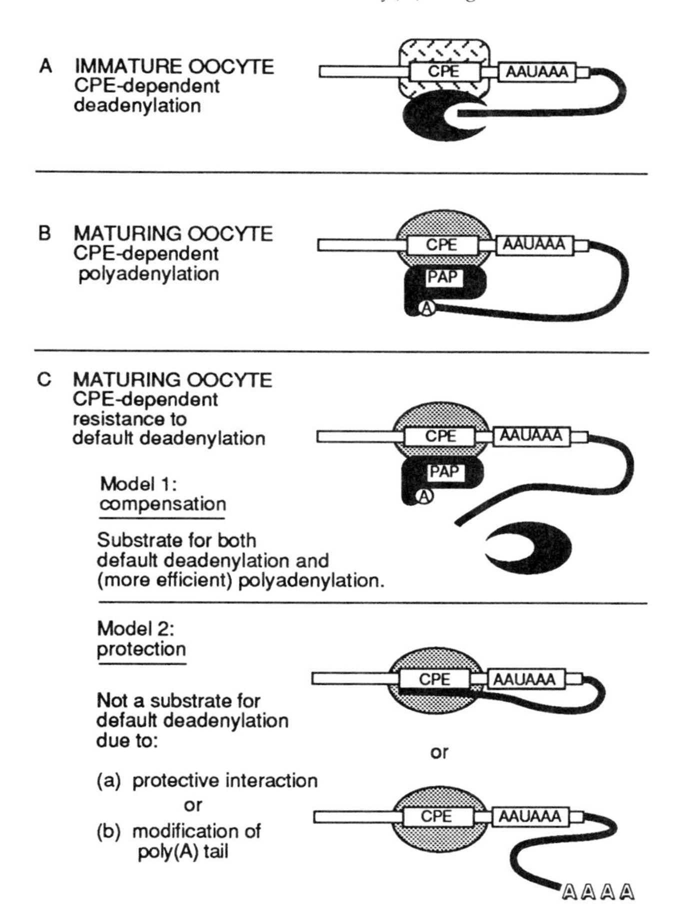

FIGURE 6 Enzymatic activities that control poly(A) length on CPE (cytoplasmic polyadenylation element)-bearing mRNAs in oocytes. (A) CPE-dependent deadenylation catalyzes the extensive deadenylation of specific mRNAs in growing and immature mouse oocytes. (B) mRNAs with a CPE are subject to poly(A) extension in oocytes undergoing meiotic maturation. (C) Most mRNAs are subject to a default deadenylation process in the mature oocyte, but mRNAs bearing a CPE are resistant. Two models for how CPE-bearing molecules escape from default deadenylation are illustrated (Fox and Wickens, 1990). Model 1, the RNA is a substrate for both default deadenylation activity and a compensating polyadenylation activity. Model 2, cytoplasmically acquired poly(A) tracts are protected from deadenylation by association with other regions of the mRNA or a protein, or by chemical modification of the poly(A) tract. There is no evidence that the poly(A) nuclease or poly(A) polymerase is in direct contact with a CPE-binding protein as diagrammed here.

consisting of β-globin sequences and the G10 CPE/AAUAAA region received about 200 adenylates (McGrew *et al.*, 1989). Similarly, the efficiency of polyadenylation for substrates containing one or more copies of the mouse tPA CPE was lower than that for transcripts containing the intact wild-type sequence (Sallés *et al.*, 1992). There is evidence for an intrinsic length limitation to polyadenylation for some substrates, but not others. An mRNA called B4, which normally receives a 150-nt poly(A) tail, is not further polyadenylated by oocytes when it carries a presynthesized poly(A) tail of that length (Paris and Richter, 1990) while a different mRNA, G10, receives additional poly(A) even if it already has a long poly(A) tail (McGrew *et al.*, 1989). This distinction may reflect fundamental differences in the way polysomal recruitment occurs for these RNAs (see later).

The extent of poly(A) removal by the default deadenylation activity has not been measured at high resolution in all studies, but it appears that most of the poly(A) is removed by this activity in both *Xenopus* and mouse oocytes (Huarte *et al.*, 1992; Varnum *et al.*, 1992). Although deadenylation is a default reaction, the efficiency of deadenylation can be affected by sequences in the substrate. Complete elimination of polyadenylation (by mutation of the AAUAAA) was shown to result in very effective deadenylation, while partial inhibition of polyadenylation (by a CPE mutation) resulted in less effective deadenylation (Fox and Wickens, 1990). It was also noted that, while extensive mutations in the 3′ UTR of a natural substrate, the s22 ribosomal protein mRNA, did not prevent default deadenylation, at least one mutant was clearly less efficiently deadenylated (Varnum and Wormington, 1990). It is unlikely that variation in the extent of deadenylation reflects an equilibrium between competing polyadenylating and deadenylating activities, at least in *Xenopus* oocytes, as the two activities operate in different time frames, with polyadenylation preceding deadenylation (Fox and Wickens, 1990; Varnum *et al.*, 1992).

D. Role of Poly(A) in Translational Recruitment

While experiments indicate that the extent of poly(A) addition and removal is sequence-specific and may be quite variable for different mRNAs, the question remains, do these variations in poly(A) length have functional significance? The answer may be, in some cases, yes and, in other cases, no. Two different mechanisms by which cytoplasmic polyadenylation may effect recruitment have been described. For some translationally repressed mRNAs, it is clear that it is the absence of a sufficiently long poly(A) tail per se that limits their translational competence, since addition of presynthesized poly(A) tails to injected transcripts allows their translation in the immature oocyte (Vassalli *et al.*, 1989; Paris and Richter, 1990; Huarte *et al.*, 1992). However, the inability of *Xenopus* G10 mRNA to be translated in immature oocytes is apparently not due simply to the lack of a long poly(A) tail. Injected G10 mRNAs carrying presynthesized poly(A) tails of 100–325 adenylates remained untranslatable in immature oocytes, but were recruited after additional extension during maturation. This is not because the G10 mRNA requires an extremely long poly(A) tail, since addition of even shorter poly(A) tracts renders it recruitable in a maturing oocyte (McGrew *et al.*, 1989). The actual process of polyadenylation may be required for recruitment of this mRNA. An mRNA that is polyadenylated during early embryogenesis, the C12 mRNA, also appears to require active polyadenylation for its recruitment (Simon and Richter, 1992). A similar requirement for active polyadenylation may apply in the activation of the spermatid-specific Mst87F mRNA in *Drosophila*, since this mRNA already carries a long, 140-nt poly(A) tail in the untranslated state (Schäfer *et al.*, 1990). A rather discrete range of poly(A) lengths may be critical for the translational activation of some mRNAs. The *Xenopus* B4 mRNA poly(A) tail is normally extended to about 150 nt during maturation. When B4-derived RNAs containing poly(A) tracts ranging from 150 to 400 adenylates were injected into immature oocytes, only those RNAs carrying tails of about 150 nt were found to be polysomal (Paris and Richter, 1990), suggesting that a poly(A) tail can be too long for effective recruitment. Consistent with this possibility is the apparent intrinsic limit to the length of poly(A) added to this particular mRNA *in vivo* (Section IV.C).

That the relatively few mRNAs analyzed show such striking differences in their polyadenylation and recruitment properties undoubtedly underlies a highly message-specific process in oocytes, in which both the timing and the efficiency of translational activation and inactivation are regulated. mRNA-specific regulation of the polyadenylation process is also indicated by the finding that the CPEs of two different *Xenopus* mRNAs, B4 mRNA (Paris and Richter, 1990) and G10 mRNA (McGrew and Richter, 1990), bind to different protein factors. Progress in understanding the mechanism(s) that triggers cytoplasmic polyadenylation is highlighted by recent studies that demonstrate a role for $p34^{cdc2}$ kinase in this process. McGrew and Richter (1990) first demonstrated that injection of cyclin mRNA into immature oocytes could induce polyadenylation of substrate RNAs. Cyclin is known to induce meiotic maturation by activating $p34^{cdc2}$ kinase. Paris *et al.* (1991) have now shown that cyclin induces polyadenylation via the same route in oocyte extracts. Figure 5B shows that polyadenylation of a B4 RNA substrate is activated in immature oocyte extracts by addition of baculovirus-expressed cyclin B and that polyadenylation can be prevented by depletion of endogenous $p34^{cdc2}$ kinase from the extract. Interestingly, the B4 RNA CPE-binding protein was found to be a substrate for a $p34^{cdc2}$-like kinase activity, suggesting that polyadenylation of B4 mRNA may be triggered by phosphorylation of this protein.

Less is known about the relationship between deadenylation and translational inactivation. For example, it is not yet known whether removal from polysomes corresponds to loss of poly(A) below a critical length. In immature mouse oocytes, tPA mRNA is synthesized with a long, 300- to 400-adenylate tail, and the sequence-specific deadenylation activity shortens it to 30–40 adenylates, which is apparently insufficient for translation (Huarte *et al.*, 1992). Although in most cases deadenylation correlates with release from polysomes, the process of spermiogenesis in the mouse is a striking, and potentially illuminating, exception to the rule. The poly(A) on mRNAs for a number of sperm-specific proteins are observed to become shorter, not longer, concommitant with translational activation (Kleene *et al.*, 1984; Heidaran and Kistler, 1987; Kleene, 1989). In the stored state, each of the mRNAs carries a remarkably homogeneous poly(A) tract, ranging from 120 to 160 residues for the different mRNAs, while the polysomal forms of these mRNAs possess heterogeneous 30- to 160-nucleotide poly(A) tracts. It is likely that poly(A)-shortening is a consequence of translational activation in this case; a shortened poly(A) tail is not a prerequisite for translation, since the polysomal population of activated mRNAs includes the complete range of poly(A) lengths, including full length. One interpretation is that poly(A) on stored mRNAs may remain in a completely protected state until the mRNA is released for translation. This may represent an alternative mechanism for keeping mRNAs in a translationally inactive state.

E. Enzymology of Polyadenylation and Deadenylation in Oocytes

1. Comparison of Nuclear and Cytoplasmic Polyadenylation

Because the enzymology of cytoplasmic polyadenylation is just beginning to be investigated, it is not yet known to what extent the nuclear and cytoplasmic processes share components. Egg extracts that accurately polyadenylate exogenously added substrate RNAs will greatly facilitate this investigation (Paris and Richter, 1990). It is known though that cytoplasmic polyadenylation in *Xenopus* oocytes is not carried out by nuclear machinery released during maturation at germinal vesicle breakdown (GVBD), because polyadenylation activity appears well before GVBD, as demonstrated in Fig. 5A (Fox *et al.*, 1989; Paris and Richter, 1990), and enucleation of immature oocytes does not prevent cytoplasmic polyadenylation at maturation (Fox *et al.*, 1989). The fact that both processes are dependent on the AAUAAA hexamer suggests related, if not identical, machinery for poly(A) polymerization. So far, however, no AAUAAA-binding protein has been detectable by cross-linking under conditions under which a CPE-binding protein is readily detectable (McGrew and Richter, 1990). There are also some obvious differences in the two processes. Cytoplasmic polyadenylation is clearly selective, while nuclear polyadenylation is not. Cytoplasmic polyadenylation in *Xenopus* oocytes apparently does not consist of the two stages that occur in the nucleus, in which the second stage is independent of sequence information in the mRNA. An RNA with mutated hexanucleotide

(AAGAAA) and a "primer" poly(A) tract was found not to be further extended, indicating continued dependence on the AAUAAA element (Paris and Richter, 1990). Continued dependence on the CPE can be inferred, since most mRNAs in the oocyte have at least short poly(A) tails, but are not extended. The kinetics of polyadenylation are also extremely different in the cytoplasm and nucleus. Cytoplasmic polyadenylation of both endogenous and injected mRNAs in *Xenopus* oocytes (Paris and Richter, 1990; McGrew and Richter, 1990) and mouse oocytes (Vassalli *et al.*, 1989) is a very slow and apparently processive process, requiring hours to attain maximal poly(A) length (e.g., see Fig. 5A), while poly(A) addition in the nucleus (Merkel *et al.*, 1976), even to injected substrates (Fox *et al.*, 1989), is very rapid. It seems most likely that some component(s) of the polyadenylation machinery is rate limiting in the cytoplasm. Competition for rate-limiting shared factors required for polyadenylation may impart some of the specificity observed in the timing and extent of polyadenylation (Paris and Richter, 1990).

2. *Comparison of Deadenylation in Oocytes and Somatic/Vegetative Cells*

The default deadenylation described for *Xenopus* oocytes and the deadenylation described in nongametic cells have many similarities. Both processes are slow. The deadenylation process in both mouse and frog oocytes requires several hours (Paynton *et al.*, 1988; Varnum and Wormington, 1990). Generally, neither form of deadenylation requires translation of the substrate RNA. A clearly nonpolysomal β-globin RNA fragment was an effective substrate for default deadenylation in the oocyte (Varnum and Wormington, 1990), and endogenous nonpolysomal ribosomal protein mRNAs are subject to deadenylation (Hyman and Wormington, 1988). Default deadenylation in the oocyte may be more complete than deadenylation in somatic cells, although there is not yet enough information to draw a general conclusion. Interestingly, there appears to be a minimum length requirement for maturation-induced deadenylation of injected transcripts: a poly(A) tail of 20 nt was not removed, while a poly(A) tail of 42 nt was (Varnum and Wormington, 1990). This lower limit may represent a constraint similar to that which prevents complete deadenylation of some mRNAs in other cells.

The sequence-specific deadenylation activity in immature mouse oocytes also catalyzes a slow loss of poly(A) and, like the somatic cell activity, leaves a minimal 30- to 40-nucleotide poly(A) tract, at least on tPA-derived substrates (Huarte *et al.*, 1992).

A surprising new finding is that the maturation-induced deadenylation activity in *Xenopus* oocytes requires a nuclear factor and that the onset of deadenylation is triggered by release of the factor at the time of germinal vesicle breakdown (Varnum *et al.*, 1992). Enucleation of immature oocytes was found to prevent the appearance of deadenylation activity in mature oocytes. Apparently the requisite factors for deadenylation are already present in the immature oocyte, since mixing nuclear and cytoplasmic extracts from immature oocytes, which individually have no detectable deadenylation activity, reconstitutes the activity *in vitro*. A requirement for a nuclear factor in deadenylation, which had been considered to be a purely cytoplasmic function, was unexpected. The relationship of this finding to the deadenylation process in somatic cells is unknown.

V. The Role of Poly(A) in mRNA Stability

A. Evidence That Poly(A) Has a Protective Function

A distinction can be made between two possible ways in which poly(A) may be considered to stabilize mRNAs: (1) *specific* stabilization, in which poly(A) removal is a step in the normal degradation pathway and, as such, is a prerequisite for subsequent steps; or (2) *nonspecific* stabilization, in which poly(A) may confer effective protection against nonspecific cellular nucleases, but its presence or absence does not affect the usual degradation pathway. Many experiments offer support for the latter role, while not necessarily being inconsistent with the former.

Several different kinds of studies have indicated that poly(A), or specifically the poly(A)–PABP complex, can confer protection on mRNAs. These include well-known studies in which the stabilities of polyadenylated and unadenylated mRNAs were compared after injection into *Xenopus* oocytes (reviewed in Bernstein and Ross, 1989; Jackson and Standart, 1990). The results of these studies were mixed, poly(A) increasing the stability of some transcripts but not others. Drummond *et al.* (1985) found that poly(A) had little immediate effect, but enhanced the long-term stability of all RNAs tested. In retrospect, it seems that the *Xenopus* oocyte does not provide a suitable environment for evaluation of mRNA stability. For example, all injected mRNAs, polyadenylated or not, are unusually stable in oocytes (Drummond *et al.*, 1985), and there are no known endogenous maternal mRNAs that are degraded following their deadenylation in oocytes. Introduction of mRNA molecules into other cell types by microinjection or electroporation is an alternative strategy. A reporter mRNA with a 200-nt tail was about two times more stable than one with a 60-nt tail when injected into *Xenopus* embryos, in which mRNAs are far less stable than those in oocytes (Harland and Misher, 1988). Likewise, a 50-nt poly(A) tract was found to moderately enhance the chemical stability of reporter mRNAs introduced into a variety of plant protoplasts and CHO (hamster) cells, although it had an even greater effect on translation (Gallie *et al.*, 1989; Gallie and Walbot, 1990). The 3′ terminus from the the tobacco mosaic virus (TMV) RNA, which forms a pseudoknot structure, was shown to provide equivalent stabilization. Neither the poly(A) nor the TMV terminus appears to act simply by providing a stable structural block to nonspecific 3′ exonucleases, since extraneous downstream sequences eliminate the stability-enhancing effect, and replacement with other highly structured RNA sequences provides no stabilization. A problem with all experiments in which naked RNAs are introduced into cells is not knowing the extent to which such molecules assemble to form authentic RNP complexes and polysomes.

In vitro studies have provided some support for a role for poly(A) in mRNA stability. Bernstein *et al.* (1989) demonstrated that the PABP is required to protect polyadenylated β-globin mRNA from a ribosome-associated nuclease *in vitro* (see Section III.D). While the nuclease(s) active in this assay may or may not be involved in the globin mRNA turnover pathway *in vivo*, these studies clearly illustrate the ability of the poly(A)–PABP complex to shield associated RNA from a nuclease that probably is physiologically relevant. There is also evidence that the absence of a long poly(A) tail can enhance the susceptibility to cleavage of nearby sequences in the 3′ UTR. In an examination of sites on rabbit β-globin mRNA that were particularly vulnerable to nuclease activities, it was found that oligo(A)-terminated globin mRNAs were more susceptible to specific 3′ cleavages by both endogenous and exogenous nucleases than were globin mRNAs with longer poly(A) tails (Albrecht *et al.*, 1984). In a cell-free mRNA decay system, the poly(A) tail of c-*myc* mRNA was shown to be mostly removed before the onset of degradation of the RNA body, whose decay is characterized by cleavages in the AU-rich regions of the 3′ UTR (Brewer and Ross, 1988, 1989). This sequence of events is consistent with poly(A) removal being a prerequisite for degradation.

However, evidence that the poly(A)–PABP complex actually contributes to the half-life of specific mRNAs *in vivo* has been difficult to obtain. Most *in vivo* studies have been merely correlative. Alterations in steady-state poly(A) length have been observed to correlate directly with changes in mRNA stability (Paek and Axel, 1987; Cochrane and Deeley, 1988). A correlation was made between a rapid rate of poly(A) loss and a short half-life for two adenovirus mRNAs when compared with host cell mRNA (Wilson *et al.*, 1978). A number of studies have demonstrated that measurable poly(A) loss precedes degradation for specific mRNAs, consistent with a sequential pathway (Mercer and Wake, 1985; Restifo and Guild, 1986; Green and Dove, 1988; Wilson and Treisman, 1988; Baker *et al.*, 1989; Swartwout and Kinniburgh, 1989: Laird-Offringa *et al.*, 1990). Inhibition of protein kinase C activity has been shown to cause both a 10-fold reduction in the half-life of tumor necrosis factor (TNF) mRNA and an accelerated rate of deadenylation of this mRNA in rat astrocytes and a mouse macrophage-like cell line, raising the possibility that poly(A) removal is the kinase-dependent step in TNF mRNA degradation (Lieberman *et al.*, 1992). Peppel and Baglioni (1991) have taken advantage of the finding that inhibition of protein synthesis in mouse cells constitutively expressing human β-interferon (IFN-β) results in the accumulation of stable IFN-β mRNAs of two distinct poly(A) length classes. After release of cells from cycloheximide inhibition, mRNAs carrying the shorter poly(A) tails disappeared more quickly than mRNAs carrying the longer poly(A) tails.

The main experimental limitation to addressing the relationship between poly(A) and stability *in vivo* has been the inability to experimentally alter the polyadenylation status of a particular mRNA. The generation of specific unadenylated mRNAs by mutation of the polyadenylation signal has not been possible due to the interdependence of 3′-end formation and polyadenylation. Cordycepin preferentially inhibits poly(A) synthesis and was used by Zeevi *et al.* (1982) to produce poly(A)-deficient molecules *in vivo*. They showed that adenovirus mRNAs synthesized in the presence of cordycepin were transported and formed polysomes fairly normally, but exhibited a reduced rate of accumulation in the cytoplasm, a finding most compatible with a decrease in the cytoplasmic stability of cordycepin-blocked mRNAs. However, the pleiotropic effects of this drug were a cause for caution in interpretation. Shyu *et al.* (1991) have overcome this problem by identifying the sequences of c-*fos* mRNA that direct its rapid poly(A) loss and transferring them to a β-globin mRNA. These studies revealed that sequences that confer rapid deadenylation also confer rapid degradation on the chimeric mRNAs. Two different regions of the c-*fos* mRNA were found to independently accelerate both poly(A)-shortening and mRNA degradation when inserted into a β-globin mRNA: an AU-rich region from the 3′ UTR and a 320-nt segment from the coding region. For all mRNA constructs, extensive poly(A) loss preceded the onset of mRNA degradation. Interestingly, while the coding region element functions to promote rapid deadenylation and mRNA degradation in serum-stimulated cells, it apparently does not function in actively growing cells. One experiment indicated that deadenylation and degradation can be temporally uncoupled. Three point mutations in the c-*fos* AU-rich element, each altering one AUUUA motif, had only a small effect on the ability of this element to facilitate rapid deadenylation of the globin mRNA ("BBB + ARE3"; Fig. 2B), but markedly inhibited its ability to destabilize the globin mRNA. This suggests that rapid deadenylation may be conferred by the AU-richness of the sequence, but that the AUUUA motifs may have another role in the subsequent degradation process (Shyu *et al.*, 1991). The ability to confer both mRNA instability and rapid, complete deadenylation on β-globin mRNA by specific c-*fos* sequences strongly supports a cause-and-effect relationship between deadenylation and degradation for the c-*fos* transcript.

Reports of similar studies on the c-*myc* mRNA destabilizing sequences should be available soon, and the comparison will be interesting. There is evidence that c-*myc* mRNA might exhibit both deadenylation-dependent and deadenylation-independent turnover; the kinetics of c-*myc* mRNA loss from polyadenylated and poly(A)-deficient fractions in HL-60 cells suggest that polyadenylated mRNA may be directly degraded in differentiating cells while it is deadenylated before degradation in growing cells (Swartwout and Kinniburgh, 1989). Both c-*fos* mRNA and c-*myc* mRNA are stabilized by inhibitors of translation, and Laird-Offinga *et al.* (1990) have provided evidence that poly(A)-shortening may be the actual translation-dependent step in the degradation of c-*myc* mRNA. Since the ability of protein synthesis inhibitors to stabilize c-*myc* mRNA is conferred by a coding region sequence (Wisdom and Lee, 1991), it will be interesting to see whether this element normally operates to stimulate rapid poly(A) removal.

Muhlrad and Parker (1992) have used a genetic screen to identify mutations that confer enhanced stability of the MFA2 mRNA in yeast. They found that all mutations in a particular region of the 3′ UTR which retard mRNA degradation also retard the rate of deadenylation. This finding provides additional strong evidence for linkage between these processes for some mRNAs. Because MFA2 3′ UTR sequences also function to stimulate rapid deadenylation *in vitro* using purified components (Lowell *et al.*, 1992; Section III.E), the mechanism by which these sequence elements act may be readily investigated in this system. It is not yet clear whether the process of terminal deadenylation described *in vitro* also occurs generally *in vivo*. The occurrence of a minimal poly(A) length for many mRNAs in yeast cells suggests either that terminal deadenylation does not always occur, or that it is followed by rapid degradation of the deadenylated mRNA (Muhlrad and Parker, 1992).

Another approach to altering the polyadenylation status of RNAs *in vivo* is now available with the isolation of yeast strains with temperature-sensitive mutations in poly(A) metabolism. Actin mRNA stability was examined in RNA14 and RNA15 mutants, which exhibit a rapid shortening of poly(A) at the restrictive temperature (see Section III.E). Transfer of cells to the restrictive temperature resulted in destabilization of actin mRNA accumulated before the temperature shift (Minvielle-Sebastia *et al.*, 1991). A yeast strain with a conditional mutation in poly(A) polymerase has recently been isolated. Preliminary studies indicate that unadenylated mRNAs produced by these cells at the restrictive temperature are indeed relatively unstable (Patel and Butler, 1992). Other mutants to be exploited include PABP mutants (Sachs and Davis, 1989) and a PAN mutant (Sachs and Deardorff, 1992). These mutants will permit studies on total RNA populations and will make it possible to address the interrelationship of poly(A) metabolism, mRNA stability, and translation.

B. Is Poly(A) Removal a Requisite Step in mRNA Degradation Pathways?

Several studies have demonstrated that there is not a direct relationship between the stability of an mRNA and the length of its poly(A) tail (Palatnik *et al.*, 1980; Krowczynska *et al.*, 1985; Santiago *et al.*, 1987; Shapiro *et al.*, 1988; Herrick *et al.*, 1989). In fact the opposite is true: stable mRNAs, in general, tend to be in the shorter steady-state poly(A) size classes. There is no evidence that the ubiquitous poly(A)-shortening process leads inevitably to molecules bearing poly(A) tails of "destabilizing" length. It seems likely that long poly(A) and short poly(A) tails are functionally equivalent with regard to stability, but it remains an open question whether more extensive or complete removal destabilizes an mRNA. mRNAs that have been identified as being subject to complete poly(A) removal (mammalian c-*fos* and c-*myc*, *Chlamydomonas* α- and β-tubulin) are also relatively unstable mRNA species. It is a reasonable hypothesis at this time to suggest that complete, or near complete, poly(A) removal renders an mRNA unstable, perhaps by loss of a minimal PABP binding site (Lowell *et al.*, 1992).

There are two different mechanisms by which complete poly(A) removal may occur: (1) a nuclease activity (either exo- or endo-) that attacks the poly(A) itself, or (2) an endonucleolytic cleavage in the 3′ UTR. The first case may be exemplified by the c-*fos* and c-*myc* mRNAs, and in such cases, digestion of poly(A) by a poly(A)-specific nuclease may be the rate-limiting step in mRNA degradation. In the second case, the rate-limiting event is due to a nonpoly(A)-specific nuclease activity. However, one outcome is the same, i.e., complete removal of poly(A). While 3′ UTR cleavages are not usually thought of as deadenylation events, it remains possible that poly(A) removal is the consequential destabilizing act.

Whatever the specific rate-limiting step, subsequent degradation of the mRNA body is very rapid, since intermediates in degradation are usually not detectable *in vivo*. Yet, it is not at all clear how initial rate-limiting cleavages render an mRNA so extremely unstable. It could be due to a variety of effects for different mRNAs: simple exposure of a nonproteinated 3′ end to an exonuclease, unmasking of other nuclease-sensitive sites, or loss of ability to initiate translation, which might secondarily expose other sensitive sites. Any of these effects could be mediated directly by loss of the poly(A)–PABP complex. If so, all stabilization conferred by the poly(A)–PABP complex could be considered "specific," since poly(A) removal would be an obligate step in destabilization.

Bergmann and Brawerman (1980) have described an endogenous nuclease activity that removes poly(A) in one step at or near the mRNA–poly(A) junction. There is now evidence that an early event in the degradation of at least some mRNAs, both stable and unstable, may be endonucleolytic cleavage in AU-rich regions of the 3′ UTR (Albrecht *et al.*, 1984; Brewer and Ross, 1988; Bakker *et al.*, 1988; Binder *et al.*, 1989; Kowalski and Denhardt, 1989; Stoeckle and Hanafusa, 1989; Bandyopadhyay *et al.*, 1990). In some cases, shortening of the poly(A) tail may expedite the cleavage (Brewer and Ross, 1988), but in other cases, such cleavages can occur without prior loss of poly(A) (Albrecht *et al.*, 1984; Binder *et al.*, 1989; Stoeckle and Hanafusa, 1989; Bandyopadhyay *et al.*, 1990; Stoeckle, 1992). Binder *et al.* (1989) have identified particular sequences, AAU or UAA, in probable single-stranded loop domains of the 3′ UTR as likely sites of cleavage during degradation of apolipoprotein II mRNA. They point out that these two sequences are present in all mRNAs that contain the AAUAAA hexamer, raising the possibility that the hexamer itself, if unprotected, might be a target site for cleavage. Figure 7 illustrates some possible relationships between poly(A) metabolism and mRNA degradation for different mRNAs.

mRNA stability	Rate & extent of poly(A) shortening	Probability of destabilizing cleavage: ↓ = low; ⬇ = high	rate limiting event in degradation
1 stable	slow or rapid, incomplete	↓ AAAAAAAAAAAAA ↓ AAA	internal cleavage
2 stable	slow, incomplete	↓ AAAAAAAAAAAAA ⬇ AAA	poly(A) shortening
3 stable	slow, complete	AAAAAAAAAAAAA [⬇]	poly(A) removal
4 unstable	slow or rapid, incomplete	⬇ AAAAAAAAAAAAA ⬇ AAA	internal cleavage
5 unstable	rapid, incomplete	↓ AAAAAAAAAAAAA ⬇ AAA	poly(A) shortening
6 unstable	rapid, complete	AAAAAAAAAAAAA [⬇]	poly(A) removal

FIGURE 7 Possible relationships between poly(A) metabolism and mRNA degradation for a variety of stable and unstable mRNAs. The following options are illustrated. (1,2,4,5) Incomplete poly(A) shortening does not directly destabilize an mRNA. (2,5) Incomplete poly(A) shortening enhances the probability of a destabilizing cleavage. The location of the destabilizing cleavage may or may not be in the 3′ UTR. Enhancement could be direct (e.g., via exposure of cleavage site) or indirect (e.g., via reduced efficiency of translation initiation). In either case the rate of poly(A) shortening could be the rate limiting event in degradation. (1,4) If the length of poly(A) has no effect on the probability of the destabilizing cleavage, then the rate limiting event in degradation is the cleavage itself. (3,6) Complete (or near complete) poly(A) loss either may directly destabilize an mRNA or may allow a destabilizing cleavage. For stable mRNAs, there is no evidence to date that any option other than 1 applies. For unstable mRNAs, there is evidence for each of the three options (4,5,6).

In order to demonstrate that poly(A) removal per se is destabilizing *in vivo*, it will be necessary to develop methods for producing specific unadenylated mRNAs without altering potentially critical 3′ UTR cleavage sites. Perhaps the use of ribozyme-containing transcripts designed to self-cleave their 3′ termini *in vivo* will be a workable approach (Eckner *et al.*, 1991).

VI. Concluding Comments

There is substantial evidence for mRNA-specific regulation of poly(A) length. It clearly occurs in the cytoplasm and may occur in the nucleus. The function of length regulation, if not the details of its mechanism of action, is manifest in only two situations. Evidence that a process of rapid poly(A) removal plays a role in the degradation pathway of c-*fos* mRNA, and probably c-*myc* mRNA and MFA2 mRNA, is quite convincing. In developing gametes and early embryos, programmed changes in poly(A) length represent a highly conserved mechanism for control of translation.

If mRNA-specific differences in poly(A) acquisition in the nucleus do occur, these differences may be reproducible in extracts, and the molecular basis for altered polyadenylation could then be explored using purified components. A detailed understanding of the actual process of nuclear polyadenylation is forthcoming. The complete reaction can be now resconstituted from highly purified mammalian components (Wahle, 1991b). cDNAs for the bovine and yeast poly(A) polymerases have been isolated and expressed *in vitro* to produce functional proteins (Raabe *et al.*, 1991; Wahle *et al.*, 1991; Lingner *et al.*, 1991b), and cloned copies of cDNAs for other components in the reaction will undoubtedly be available before long.

The functional significance of the ubiquitous cytoplasmic poly(A)-shortening process that gives rise to most poly(A) length heterogeneity remains unknown. The fact that yeast cells can survive without this process (Sachs and Davis, 1989) suggests that it may be a nonessential function. On the other hand, the gene encoding the poly(A) degrading enzyme PAN, which presumably catalyzes poly(A)-shortening in yeast, is essential for cell viability (Sachs and Deardorff, 1992). When considering the significance of cytoplasmic poly(A) addition and poly(A)-shortening reactions in somatic and vegetative cells, it is worth keeping in mind that, in oocytes, not only poly(A) tail length, but also the dynamic process of polyadenylation may be important in the function of some mRNAs. It could be proposed that the function of slow poly(A)-shortening is simply to counterbalance an essential poly(A) extension reaction. With the entrance of yeast genetics into this field, fundamental questions concerning the significance of poly(A) metabolism may finally be posed and answered.

There is evidence supporting the functional involvement of poly(A) in both mRNA turnover and translation, but how it functions in either capacity remains unknown. The process of translational recruitment that occurs in gametes and early embryos may be distinct from simple translation initiation in somatic and vegetative cells, so the application of findings obtained in developmental systems to control translation per se is not yet clear. Yet evidence is accumulating that poly(A), or the poly(A)–PABP complex, may be generally involved in the initiation of translation (Jackson and Standart, 1990; Munroe and Jacobson, 1990; Sachs, 1990; Sachs and Deardorff, 1992). It is possible that the two roles may be coupled, at least in some cases; inhibition of translation by poly(A) loss may secondarily effect destabilization of an mRNA. While there is no direct evidence that any mRNA is degraded by such a pathway *in vivo,* there is much evidence that the processes of translation and mRNA degradation are linked in a variety of ways for different mRNAs. As the research effort directed at understanding the functions of poly(A) continues to intensify, important new insights are assured in the near future.

References

Abraham, A. K., and Jacob, S. T. (1978). Hydrolysis of poly(A) to adenine nucleotides by purified poly(A) polymerase. *Proc. Natl. Acad. Sci. USA* **75,** 2085–2087.

Adam, S. A., Nakagawa, T., Swanson, M. S., Woodruff, T. K., and Dreyfuss, G. (1986). mRNA polyadenylate-binding protein: Gene isolation and sequencing and identification of a ribonucleoprotein consensus sequence. *Mol. Cell. Biol.* **6,** 2932–2943.

Adams, D. S., and Jeffrey, W. R. (1978). Poly(adenylic acid) degradation by two distinct process in the cytoplasmic RNA of *Physarum polycephalum. Biochemistry* **17,** 4519–4524.

Ahlquist, P., and Kaesberg, P. (1979). Determination of the length distribution of poly(A) at the 3′ terminus of the virion RNAs of EMC virus, poliovirus, rhinovirus, RAV-61 and CPMV and of mouse globin mRNA. *Nucleic Acids Res.* **7,** 1195–1204.

Ahringer, J., and Kimble, J. (1991). Control of the sperm-oocyte switch in *Caenorhabditis elegans* hermaphrodites by the *fem-3* 3′ untranslated region. *Nature (London)* **349,** 346–348.

Albrecht, G., Krowczynska, A., and Brawerman, G. (1984). Configuration of β-globin messenger RNA in rabbit reticulocytes. *J. Mol. Biol.* **178,** 881–896.

Åström, J., Åström, A., and Virtanen, A. (1991). In vitro deadenylation of mammalian mRNA by a HeLa cell 3′ exonuclease. *EMBO J.* **10,** 3067–3071.

Åström, J., Åström, A., and Virtanen, A. (1992). Properties of a HeLa cell 3′ exonuclease specific for degrading poly(A) tails of mammalian mRNA. *J. Biol. Chem.* **267,** 18154–18159.

Bachvarova, R., DeLeon, V., Johnson, A., Kaplan, G., and Paynton, B. V. (1985). Changes in total RNA, polyadenylated RNA, and actin mRNA during meiotic maturation of mouse oocytes. *Dev. Biol.* **108,** 325–331.

Baer, B. W., and Kornberg, R. D. (1980). Repeating structure of cytoplasmic poly(A)-ribonucleoprotein. *Proc. Natl. Acad. Sci. USA* **77,** 1890–1892.

Baer, B. W., and Kornberg, R. D. (1983). The protein responsible for the repeating structure of cytoplasmic poly(A)-ribonucleoprotein. *J. Cell Biol.* **96,** 717–721.

Baker, E. J., Diener, D. R., and Rosenbaum, J. L. (1989). Accelerated poly(A) loss on α-tubulin mRNAs during protein synthesis inhibition in *Chlamydomonas. J. Mol. Biol.* **207,** 771–781.

Bakker, O., Arnberg, A. C., Noteborn, M. H. M., Winter, A. J., and Ab, G. (1988). Turnover products of the apo very low density lipoprotein II messenger RNA from chicken liver. *Nucleic Acids Res.* **16,** 10109–10118.

Bandyopadhyay, R., and Brawerman, G. (1992). The 3′-terminal structure of total and "poly(A)-deficient" mouse beta-actin messenger RNA. *Biochimie* **74,** 1031–1034.

Bandyopadhyay, R., Coutts, M., Krowczynska, A., and Brawerman, G. (1990). Nuclease activity associated with mammalian mRNA in its native state: Possible basis for selectivity in mRNA decay. *Mol. Cell. Biol.* **10,** 2060–2069.

Bandziulis, R. J., Swanson, M. S., and Dreyfuss, G. (1989). RNA-binding proteins as developmental regulators. *Genes Dev.* **3,** 431–437.

Bardwell, V. J., Zarkower, D., Edmonds, M., and Wickens, M. (1990). The enzyme that adds poly(A) to mRNAs is a classical poly(A) polymerase. *Mol. Cell. Biol.* **10,** 846–849.

Berger, E. M., Vitek, M. P., and Morganelli, M. (1985). Transcript length heterogeneity at the small heat shock protein genes of *Drosophila*. *J. Mol. Biol.* **186,** 137–148.

Bergmann, I. E., and Brawerman, G. (1977). Control of breakdown of the polyadenylate sequence in mammalian polyribosomes: Role of poly(adenylic acid)–protein interactions. *Biochemistry* **16,** 259–264.

Bergmann, I. E., and Brawerman, G. (1980). Loss of the polyadenylate segment from mammalian messenger RNA. *J. Mol. Biol.* **139,** 439–454.

Bernstein, P., and Ross, J. (1989). Poly(A), poly(A) binding protein and the regulation of mRNA stability. *Trends Biochem. Sci.* **14,** 373–377.

Bernstein, P., Peltz, S. W., and Ross, J. (1989). The poly(A)–poly(A)-binding protein complex is a major determinant of mRNA stability *in vitro*. *Mol. Cell. Biol.* **9,** 659–670.

Bienroth, S., Wahle, E., Suter-Crazzolara, C., and Keller, W. (1991). Purification of the cleavage and polyadenylation factor involved in the 3′-processing of messenger RNA precursors. *J. Biol. Chem.* **266,** 19768–19776.

Binder, R., Hwang, S-P. L., Ratnasabapathy, R., and Williams, D. L. (1989). Degradation of apolipoprotein II mRNA occurs via endonucleolytic cleavage at 5′-AAU-3′/5′-UAA-3′ elements in a single-stranded loop domains of the 3′-noncoding region. *J. Biol. Chem.* **264,** 16910–16918.

Bloch, C., Perrin, F., and Lacroute, F. (1978). Yeast temperature-sensitive mutants specifically impaired in processing of poly(A)-containing RNAs. *Mol. Gen. Genet.* **165,** 123–127.

Braun, R. E., Peschon, J. J., Behringer, R. R., Brinster, R. L., and Palmiter, R. D. (1989). Protamine 3′-untranslated sequences regulate temporal translational control and subcellular localization of growth hormone in spermatids of transgenic mice. *Genes Dev.* **3,** 793–803.

Brawerman, G. (1973). Alterations in the size of the poly(A) segment in newly-synthesized messenger RNA of mouse sarcoma 180 ascites cells. *Mol. Biol. Rep.* **1,** 7–13.

Brawerman, G. (1981). The role of the poly(A) sequence in mammalian messenger RNA. *CRC Crit. Rev. Biochem.* **10,** 1–38.

Brawerman, G., and Diez, J. (1975). Metabolism of the polyadenylate sequence of nuclear RNA and messenger RNA in mammalian cells. *Cell* **5,** 271–280.

Brewer, G., and Ross, J. (1988). Poly(A) shortening and degradation of the 3′ A + U-rich sequences of human c-*myc* mRNA in a cell-free system. *Mol. Cell. Biol.* **8,** 1697–1708.

Brewer, G., and Ross, J. (1989). Regulation of c-*myc* mRNA stability in vitro by a labile destabilizer with an essential nucleic acid component. *Mol Cell. Biol.* **9,** 1991–2006.

Burd, C. G., Matunis, E. L., and Dreyfuss, G. (1991). The multiple RNA-binding domains of the mRNA poly(A)-binding protein have different RNA-binding activities. *Mol. Cell Biol.* **11,** 3419–3424.

Butler, J. S., and Platt, T. (1988). RNA processing generates the mature 3′ end of yeast CYC1 messenger RNA in vitro. *Science* **242,** 1270–1274.

Butler, J. S., Sadhale, P. P., and Platt, T. (1990). RNA processing in vitro produces mature 3′ ends of a variety of *Saccharomyces cerevisiae* mRNAs. *Mol. Cell. Biol.* **10,** 2599–2605.

Carrazana, E. J., Pasieka, K. B., and Majzoub, J. A. (1988). The vasopressin mRNA poly(A) tract is unusually long and increases during stimulation of vasopressin gene expression *in vivo*. *Mol. Cell. Biol.* **8,** 2267–2274.
Carter, D. A., and Murphy, D. (1989). Independent regulation of neuropeptide mRNA levels and poly(A) tail length. *J. Biol. Chem.* **264,** 6601–6603.
Carter, D. A., and Murphy, D. (1992). Nuclear mechanisms mediate rhythmic changes in vasopressin mRNA expression in the rat suprachiasmatic nucleus. *Mol. Brain Res.* **12,** 315–321.
Carter, K. C., Bryant, S., Gadson, P., and Papaconstantinou, J. (1989). Deadenylation of α1-acid glycoprotein mRNA in cultured hepatic cells during stimulation by dexamethasone. *J. Biol. Chem.* **7,** 4112–4119.
Christofori, G., and Keller, W. (1988). 3′ cleavage and polyadenylation of mRNA precursors in vitro requires a poly(A) polymerase, a cleavage factor, and a snRNP. *Cell* **54,** 875–889.
Christofori, G., and Keller, W. (1989). Poly(A) polymerase purified from HeLa cell nuclear extract is required for both cleavage and polyadenylation of pre-mRNA in vitro. *Mol. Cell. Biol.* **9,** 193–203.
Cochrane, A. W., and Deeley, R. G. (1988). Estrogen-dependent activation of the avian very low density apolipoprotein II and vitellogenin genes. *J. Mol. Biol.* **203,** 555–567.
Diez, J., and Brawerman, G. (1974). Elongation of the polyadenylate segment of messenger RNA in the cytoplasm of mammalian cells. *Proc. Natl. Acad. Sci. USA* **71,** 4091–4095.
Drawbridge, J., Grainger, J. L., and Winkler, M. M. (1990) Identification and characterization of the poly(A)-binding proteins from the sea urchin: A quantitative analysis. *Mol. Cell. Biol.* **10,** 3994–4006.
Drummond, D. R., Armstrong, J., and Colman, A. (1985). The effect of capping and polyadenylation on the stability, movement and translation of synthetic messenger RNAs in *Xenopus* oocytes. *Nucleic Acids Res.* **13,** 7375–7394.
Dworkin, M. B., and Dworkin-Rastl, E. (1985). Changes in RNA titers and polyadenylation during oogenesis and oocyte maturation in *Xenopus laevis*. *Dev. Biol.* **112,** 451–457.
Dworkin, M. B., Rudensay, L. M., and Infante, A. A. (1977). Cytoplasmic nonpolysomal ribonucleoprotein particles in sea urchin embryos and their relationship to protein synthesis. *Proc. Natl. Acad. Sci. USA* **6,** 2231–2235.
Eckner, R., Ellmeier, W., and Birnstiel, M. L. (1991). Mature mRNA 3′ end formation stimulates RNA export from the nucleus. *EMBO J.* **10,** 3513–3522.
Edmonds, M. (1990). Polyadenylate polymerases. *In* Methods in Enzymology (J. E. Dahlberg and J. N. Abelson, Eds.), Vol. 181, pp. 161–170. Academic Press, San Diego.
Fox, C. A., and Wickens, M. (1990). Poly(A) removal during oocyte maturation: A default reaction selectively prevented by speciic sequences in the 3′ UTR of certain maternal mRNAs. *Genes Dev.* **4,** 2287–2298.
Fox, C. A., Sheets, M., and Wickens, M. P. (1989). Poly(A) addition during maturation of frog oocytes: Distinct nuclear and cytoplasmic activities and regulation by the sequence UUUUUAU. *Genes Dev.* **3,** 2151–2162.
Fung, B. P., Brilliant, M. H., and Chikaraishi, D. M. (1991). Brain-specific polyA$^-$ transcripts are detected in polyA$^+$ RNA: Do complex polyA$^-$ brain RNAs really exist? *J. Neurosci.* **11,** 701–708.
Gallie, D. R., and Walbot, V. (1990). RNA pseudoknot domain of tobacco mosaic virus can functionally substitute for a poly(A) tail in plant and animal cells. *Genes Dev.* **4,** 1149–1157.
Gallie, D. R., Lucas W. J., and Walbot, V. (1989). Visualizing mRNA expression in plant protoplasts: Factors influencing efficient mRNA uptake and translation. *Plant Cell* **1,** 301–311.
Geoghegan, T. E., and McCoy, L. (1986). Biogenesis and cell cycle relationship of poly(A)-actin mRNA in mouse ascites cells. *Exp. Cell Res.* **162,** 175–182.

Gershon P., and Moss, B. (1992). Transition from rapid processive to slow nonprocessive polyadenylation by vaccinia virus poly(A) polymerase catalytic subunit is regulated by the net length of the poly(A) tail. *Genes Dev.* **6,** 1575–1586.

Gilmartin, G. M., and Nevins, J. R. (1989). An ordered pathway of assembly of components required for polyadenylation site recognition and processing. *Genes Dev.* **3,** 2180–2189.

Grange, T., Martins de Sa, C., Oddos, J., and Pictet, R. (1987). Human mRNA polyadenylate binding protein: Evolutionary conservation of a nucleic acid binding motif. *Nucleic Acids Res.* **15,** 4771–4787.

Green, L. L., and Dove, W. F. (1988). Correlation between mRNA stability and poly(A) length over the cell cycle of *Physarum polycephalum. J. Mol. Biol.* **72,** 91–98.

Greenberg, J. R. (1981). The polyribosomal mRNA–protein complex is a dynamic structure. *Proc. Natl. Acad. Sci. USA* **78,** 2923–2926.

Groner, B., and Phillips, S. L. (1975). Polyadenylate metabolism in the nuclei and cytoplasm of *Saccharomyces cerevisiae. J. Biol. Chem.* **250,** 5640–5646.

Harland, R., and Misher, L. (1988). Stability of RNA in developing embryos and identification of a destabilizing sequence in TFIIIA messenger RNA. *Development* **102,** 837–852.

Heidaran, M. A., and Kistler, W. S. (1987). Transcriptional and translational control of the message for transition protein 1, a major chromosomal protein of mammalian spermatids. *J. Biol. Chem.* **262,** 13309–13315.

Herrick, D., Parker, R., and Jacobson, A. (1990). Identification and comparison of stable and unstable mRNAs in *Saccharomyces cerevisiae. Mol. Cell. Biol.* **10,** 2269–2284.

Huarte, J., Belin, D., Vassalli, A., Strickland, S., and Vassalli, J-D. (1987). Meiotic maturation of mouse oocytes triggers the translation and polyadenylation of dormant tissue-type plasminogen activator mRNA. *Genes Dev.* **1,** 1201–1211.

Huarte, J., Stutz, A., O'Connell, M. L., Gubler, P., Belin, D., Darrow, A. L., Strickland, S., and Vassalli, J-D. (1992). Transient translational silencing by reversible mRNA deadenylation. *Cell* **69,** 1021–1030.

Humphrey, T., Sadhale, P., Platt, T., and Proudfoot, N. (1991). Homologous mRNA 3′ end formation in fission and budding yeast. *EMBO J.* **10,** 3503–3511.

Hunt, A. G., Chu, N. M., Odell, J. T., Nagy, F., and Chua, N-H. (1987). Plant cells do not properly recognize animal gene polyadenylation signals. *Plant Mol. Biol.* **8,** 23–35.

Hyman, L. E., and Wormington, W. M. (1988). Translational inactivation of ribosomal protein mRNAs during *Xenopus* oocyte maturation. *Genes Dev.* **2,** 598–605.

Iatrou, K., and Dixon, G. H. (1977). The distribution of poly(A)$^+$ and poly(A)$^-$ messenger RNA sequences in the developing trout testis. *Cell* **10,** 433–441.

Irniger, S., Egli, C. M., and Braus, G. H. (1991). Different cleasses of polyadenylation sites in the yeast *Saccharomyces cerevisiae. Mol. Cell. Biol.* **11,** 3060–3069.

Ivell, R., and Richter, D. (1984). The gene for the hypothalamic peptide hormone oxytocin is highly expressed in the bovine corpus luterum: Biosynthesis, structure and sequence analysis. *EMBO J.* **3,** 2351–2354.

Ivell, R., Schmale, H., Krisch, B., Nahke, P., and Richter, D. (1986). Expression of a mutant vasopressin gene: Differential polyadenylation and read-through of the mRNA 3′ end in a frame-shift mutant. *EMBO J.* **5,** 971–977.

Jackson, R. J., and Standart, N. (1990). Do the poly(A) tail and 3′ untranslated region control mRNA translation? *Cell* **62,** 15–24.

Jacob, S. T., and Rose, K. M. (1983). Poly(A) polymerase from eukaryotes. *In* Enzymes of Nucleic Acid Synthesis and Modification (S. T. Jacob, Ed.), Vol. 2, pp. 135–157. CRC Press, Boca Raton, Florida.

Jacob, S. T., Terns, M. P., Hengst-Zhang, J. A., and Vulapalli, R. S. (1990). Polyadenylation of mRNA and its control. *CRC Crit. Rev. Euk. Gene Exp.* **1,** 49–59.

Keller, W., Bienroth, S., Land, K. M., and Christofori, G. (1991). Cleavage and polyadenylation factor CPF specifically interacts with the pre-mRNA 3′ processing signal, AAUAAA. *EMBO J.* **10,** 4241–4249.

Kelly, J. M., and Cox, R. A. (1982). Periodicity in the length of 3'-poly(A) tails from native globin mRNA of rabbit. *Nucleic Acids Res.* **10,** 4173–4179.

Kleene, K. C. (1989). Poly(A) shortening accompanies the activation of translation of five mRNAs during spermiogenesis in the mouse. *Development* **106,** 367–373.

Kleene, K. C., Distel, R. J., and Hecht, N. B. (1984). Translational regulation and deadenylation of a protamine mRNA during spermiogenesis in the mouse. *Dev. Biol.* **105,** 71–79.

Kowalski, J., and Denhardt, D. (1989). Regulation of the mRNA for monocyte-derived neutrophil-activating peptide in differentiating HL60 promyelocytes. *Mol. Cell. Biol.* **9,** 1946–1957.

Krane, I. M., Spindel, E. R., and Chin, W. W. (1991). Thyroid hormone decreases the stability and the poly(A) tract length of rat thyrotropin β-subunit messenger RNA. *Mol. Endocrin.* **5,** 469–475.

Krowczynska, A., and Brawerman, G. (1986). Structural features in the 3'-terminal region of polyribosome-bound rabbit globin messenger RNAs. *J. Biol. Chem.* **261,** 397–402.

Krowczynska, A., Yenofsky, R., and Brawerman, G. (1985). Regulation of messenger RNA stability in mouse erythroleukemia cells. *J. Mol. Biol.* **181,** 231–239.

Lafrère, V., Vincent, A., and Amalric, F. (1990). Drosophila melanogaster poly(A)-binding protein: cDNA cloning reveals an unusually long 3'-untranslated region of the mRNA, also present in other eukaryotic species. *Gene* **96,** 219–225.

Laird-Offringa, I. A. (1992). What determines the instability of c-*myc* proto-oncogene mRNA? *BioEssays* **14,** 119–124.

Laird-Offringa, I. A., Elfferich, P., Knaken, H. J., de Ruiter, J., and van der Eb, A. J. (1989). Analysis of polyadenylation site usage of the c-*myc* oncogene. *Nucleic Acids Res.* **17,** 6499–6514.

Laird-Offringa, I. A., de Wit, C. L., Elfferich, P., and van der Eb, A. J. (1990). Poly(A) tail shortening is the translation-dependent step in c-*myc* mRNA degradation. *Mol. Cell. Biol.* **10,** 6132–6140.

Lieberman, A. P., Pitha, P. M., and Shin, M. L. (1992). Poly(A) removal is the kinase-regulated step in tumor necrosis factor mRNA decay. *J. Biol. Chem.* **267,** 2123–2126.

Lingner, J., Radtke, I., Wahle, E., and Keller, W. (1991a). Purification and characterization of poly(A) polymerase from *Saccharomyces cereviseae. J. Biol. Chem.* **266,** 8741–8746.

Lingner, J., Kellerman, J., and Keller, W. (1991b). Cloning and expression of the essential gene for poly(A) polymerase from *S. cerevisiae. Nature* (*London*) **354,** 496–498.

Littauer, U. Z., and Soreq, H. (1982). The regulatory function of polyA() and adjacent 3' sequences in translated RNA. *Prog. Nucleic Acid Res. Mol. Biol.* **27,** 53–83.

Lowell, J., Rudner, D., and Sachs, A. (1992). 3'-UTR-dependent deadenylation by the yeast poly(A) nuclease. *Genes Dev.* **6,** 2088–2099.

Manley, J. L. (1988). Polyadenylation of mRNA precursors. *Biochim. Biophys. Acta* **950,** 1–12.

Manrow, R. E., and Jacobson, A. (1986). Identification and characterization of developmentally regulated mRNP proteins of *Dictyostelium discoideum. Dev. Biol.* **116,** 213–227.

Mansbridge, J. N., Crossley, J. A., Lanyon, W. G., and Williamson, R. (1974). The poly(adenylic acid) sequence of mouse globin messenger RNA. *Eur. J. Biochem.* **44,** 261–269.

McGrew, L. L., and Richter, J. D. (1990). Translational control by cytoplasmic polyadenylation during *Xenopus* oocyte maturation: Characterization of *cis* and *trans* elements and regulation by cyclin/MPF. *EMBO J.* **9,** 3743–3751.

McGrew, L. L., Dworkin-Rastl, E., Dworkin, M., and Richter, J. D. (1989). Poly(A) elongation during *Xenopus* oocyte maturation is required for translational recruitment and is mediated by a short sequence element. *Genes Dev.* **3,** 803–815.

Medford, R. M., Wyndro, R. M., Nguyen, H. T., and Nadal-Ginard, B. (1980). Cytoplasmic processing of myosin heavy chain messenger RNA: Evidence provided by using a recombinant DNA plasmid. *Proc. Natl. Acad. Sci. USA* **77,** 5749–5753.

Mercer, J. F. B., and Wake, S. A. (1985). An analysis of the rate of metallothionein mRNA poly(A)-shortening using RNA blot hybridization. *Nucleic Acids Res.* **13,** 7229–7943.

Merkel, C. G., Wood, T. G., and Lingrel, J. B. (1976). Shortening of the poly(A) region of mouse globin messenger RNA. *J. Biol. Chem.* **251,** 5512–5515.

Minvielle-Sebastia, L., Winsor, B., Bonneaud, N., and Lacroute, F. (1991). Mutations in the yeast RNA14 and RNA15 genes result in an abnormal mRNA decay rate: Sequence analysis reveals an RNA-binding domain in the RNA15 protein. *Mol. Cell. Biol.* **11,** 3075–3087.

Moore, C. L., and Sharp, P. A. (1985). Accurate cleavage and polyadenylation of exogenous RNA substrate. *Cell* **41,** 845–855.

Morrison, M. R., Brodeur, R., Pardue, S., Baskin, F., Hall, C. L., and Rosenberg, R. (1979). Differences in the distribution of poly(A) size classes in individual messenger RNAs from neuroblastoma cells. *J. Biol. Chem.* **254,** 7675–7683.

Muhlrad, D., and Parker, R. (1992). Mutations affecting stability and deadenylation of the yeast MAF2 transcript. *Genes Dev.* **6,** 2100–2111.

Müller, W. E. G., Seibert, G., Steffen, R., and Zahn, R. K. (1976). Endoribonuclease IV, further investigation on the specificity. *Eur. J. Biochem.* **70,** 249–258.

Müller, W. E. G., Arendes, J., Zahn, R. K., and Schröder, H. C. (1978). Control of enzymatic hydrolysis of polyadenylate segment of messenger RNA: Role of polyadenylate-associated proteins. *Eur. J. Biochem.* **86,** 283–290.

Munroe, D., and Jacobson, A. (1990). Tales of poly(A): A review. *Gene* **91,** 151–158.

Murphy, D., and Carter, D. (1990). Vasopressin gene expression in the rodent hypothalamus: Transcriptional and posttranscriptional responses to physiological stimulation. *Mol. Endocrinol.* **4,** 1051–1059.

Murphy, D., Pardy, K., Seah, V., and Carter, D. (1992). Posttranscriptional regulation of rat growth hormone gene expression: Increased message stability and nuclear polyadenylation accompany thyroid hormone depletion. *Mol. Cell. Biol.* **12,** 2624–2632.

Muschel, R., Khoury, G., and Reid, L. M. (1986). Regulation of insulin mRNA abundance and adenylation: Dependence on hormones and matrix substrata. *Mol. Cell. Biol.* **6,** 337–341.

Nietfeld, W., Mentzel, H., and Pieler, T. (1990). The *Xenopus laevis* poly(A) binding protein is composed of multiple functionally independent RNA binding domains. *EMBO J.* **9,** 3699–3705.

Paek, I., and Axel, R. (1987). Glucocorticoids enhance stability of human growth hormone mRNA. *Mol. Cell. Biol.* **7,** 1496–1507.

Palatnik, C. M., Storti, R. V., and Jacobson, A. (1979). Fractionation and functional analysis of newly synthesized and decaying messenger RNAs from vegetative cells of *Dictyostelium discoideum*. *J. Mol Biol.* **128,** 371–395.

Palatnik, C. M., Storti, R. V., Capone, A. K., and Jacobson, A. (1980). Messenger RNA stability in *Dictyostelium discoideum*. *J. Mol. Biol.* **141,** 99–118.

Palatnik, C. M., Wilkins, C., and Jacobson, A. (1984). Translational control during early *Dictyostelium* development: Possible involvement of poly(A) sequences. *Cell* **36,** 1017–1025.

Paris, J., and Philippe, M. (1990). Poly(A) metabolism and polysomal recruitment of maternal mRNAs during early *Xenopus* development. *Dev. Biol.* **140,** 221–224.

Paris, J., and Richter, J. D. (1990). Maturation-specific polyadenylation and translational control: Diversity of cytoplasmic polyadenylation elements, influence of poly(A) tail size, and formation of stable polyadenylation complexes. *Mol. Cell. Biol.* **10,** 5634–5645.

Paris, J., Osborne, H. B., Couturier, A., Le Guellec, R., and Philippe, M. (1988). Changes in the polyadenylation of specific stable RNA during the early development of *Xenopus laevis*. *Gene* **72,** 169–176.

Paris, J., Swenson, K., Piwnica-Worms, H., and Richter, J. D. (1991). Maturation-specific polyadenylation: In vitro activation by $p34^{cdc2}$ and phosphorylation of a 58-kD CPE-binding protein. *Genes Dev.* **5,** 1697–1708.

Pastori, R. L., Moskaitis, J. E., Buzek, S. W., and Schoenberg, D. R. (1991). Coordinate estrogen-regulated instability of serum protein-coding messenger RNAs in *Xenopus laevis*. *Mol. Endocrinol.* **5,** 461–468.

Pastori, R. L., Moskaitis, J. E., Buzek, S. W., and Schoenberg, D. R. (1992). Differential regulation and polyadenylation of transferrin mRNA in *Xenopus* liver and oviduct. *J. Steroid Biochem. Mol. Biol.,* **42,** 649–657.

Patel, D., and Butler, J. S. (1992). Conditional defect in mRNA 3′-end processing caused by a mutation in the gene for poly(A) polymerase. *Mol. Cell. Biol.* **12,** 3297–3304.

Paynton, B. V., Rempel, R., and Bachvarova, R. (1988). Changes in state of adenylation and time course of degradation of maternal mRNAs during oocyte maturation and early embryonic development in the mouse. *Dev. Biol.* **129,** 304–314.

Peltz, S. W., Brewer, G., Kobs., G., and Ross, J. (1987). Substrate specificity of the exonuclease activity that degrades H4 histone mRNA. *J. Biol Chem.* **262,** 9382–9388.

Peppel, K., and Baglioni, C. (1991). Deadenylation and turnover of interferon-β mRNA. *J. Biol. Chem.* **266,** 6663–6666.

Proudfoot, N. (1991). Poly(A) signals. *Cell* **64,** 671–674.

Raabe, T., Bollum, F. J., and Manley, J. L. (1991). Primary structure and expression of bovine poly(A) polymerase. *Nature (London)* **353,** 229–234.

Restifo, L. L., and Guild, G. M. (1986). Poly(A) shortening of coregulated transcripts in *Drosophila*. *Dev. Biol.* **115,** 507–510.

Robinson, B. G., Frim, D. M., Schwartz, W. J., and Mazjoub, J. A. (1988). Vasopressin mRNA in the suprachiasmatic nuclei: Daily regulation of polyadenylate tail length. *Science* **241,** 342–344.

Rose, K. M., and Jacob, S. T. (1980). Phosphorylation of nuclear poly(adenylic acid) polymerase by protein kinase: Mechanism of enhanced poly(adenylic acid) synthesis. *Biochemistry* **19,** 1472–1477.

Rosenthal, E. T., and Ruderman, J. V. (1987). Widespread changes in the translation and adenylation of maternal messenger RNAs following fertilization of *Spisula* oocytes. *Dev. Biol.* **121,** 237–246.

Rosenthal, E. T., and Wilt, F. H. (1986). Patterns of maternal messenger RNA accumulation and adenylation during oogenesis in *Urechis caupo*. *Dev. Biol.* **117,** 55–63.

Ryner, L. C., Takagaki, Y., and Manley, J. L. (1989). Multiple forms of poly(A) polymerase purified from HeLa cells function in specific mRNA 3′-end formation. *Mol. Cell. Biol.* **9,** 4229–4238.

Sachs, A. B. (1990). The role of poly(A) in the translation and stability of mRNA. *Curr. Opinion Cell Biol.* **2,** 1092–1098.

Sachs, A. B., Bond, M. W., and Kornberg, R. D. (1986). A single gene from yeast for both nuclear and cytoplasmic polyadenylate-binding proteins: Domain structure and expression. *Cell* **45,** 827–835.

Sachs, A. B., and Davis, R. W. (1989). The poly(A) binding protein is required for poly(A) shortening and 60S ribosomal subunit-dependent translation initiation. *Cell* **58,** 857–867.

Sachs, A. B., Davis, R. W., and Kornberg, R. D. (1987). A single domain of yeast poly(A)-binding protein is necessary and sufficient for RNA binding and cell viability. *Mol. Cell. Biol.* **7,** 3268–3276.

Sachs, A., and Deardorff, J. (1992). Translation initiation requires the PAB-dependent poly(A) ribonuclease in yeast. *Cell* **70,** 961–973.

Sallés, F. J., Darrow, A. L., O'Connell, M. L., and Strickland, S. (1992). Isolation of novel murine maternal mRNAs regulated by cytoplasmic polyadenylation. *Genes Dev.,* **6,** 1202–1212.

Santiago, T. C., Bettany, A. J. E., Purvis, I. J., and Brown, A. J. P. (1987). Messenger RNA stability in *Saccharomyces cerevisiae*: The influence of translation and poly(A) tail length. *Nucleic Acids Res.* **15,** 2417–2429.

Sawicki, S. G., Jelinek, W., and Darnell, J. E. (1977). 3′-terminal addition to HeLa cell nuclear and cytoplasmic poly(A). *J. Mol. Biol.* **113,** 219–227.

Schäfer, M., Kuhn, R., Bosse, F., and Schäfer, U. (1990). A conserved element in the leader mediates post-meiotic translation as well as cytoplasmic polyadenylation of a *Drosophila* spermatocyte mRNA. *EMBO J.* **9,** 4519–4525.

Schoenberg, D. R., Moskaitis, J. E., Smith, L. H., and Pastori, R. L. (1989). Extranuclear estrogen-regulated destabilization of *Xenopus laevis* serum albumin mRNA. *Mol. Endocrinol.* **3,** 805–814.

Schröder, H. C., Dose, K., Zahn, R. K., and Müller, W. E. G. (1980). Purification and characterization of a poly(A)-specific exoribonuclease from calf thymus. *J. Biol. Chem.* **255,** 4535–4538.

Sehgal, P. B., Lyles, D. S., and Tamm, I. (1978). Superinduction of human fibroblast interferon production: Further evidence for increased stability of interferon mRNA. *Virology* **89,** 186–198.

Setyono, B., and Greenberg, J. R. (1981). Proteins associated with poly(A) and other regions of mRNA and hn RNA molecules as investigated by crosslinking. *Cell* **24,** 775–783.

Shapiro, R. A., Herrick, D., Manrow, R. E., Blinder, D., and Jacobson, A. (1988). Determinants of mRNA stability in *Dictyostelium discoideum* amoebae: Differences in poly(A) tail length, ribosome loading, and mRNA size cannot account for the heterogeneity of mRNA decay rates. *Mol. Cell. Biol.* **8,** 1957–1969.

Sheets, M. D., and Wickens, M. (1989). Two phases in the addition of a poly(A) tail. *Genes Dev.* **3,** 1401–1412.

Sheets, M. D., Ogg, S. C., and Wickens, M. P. (1990). Point mutations in AAUAAA and the poly(A) addition site: Effects on the accuracy and efficiency of cleavage and polyadenylation in vitro. *Nucleic Acids Res.* **18,** 5799–5805.

Sheiness, D., Puckett, L., and Darnell, J. E. (1975). Possible relationship of poly(A) shortening to mRNA turnover. *Proc. Natl. Acad. Sci. USA* **72,** 1077–1081.

Shyu, A-B., Belasco, J. G., and Greenberg, M. E. (1991) Two distinct destabilizing elements in the *c-fos* message trigger deadenylation as a first step in rapid mRNA decay. *Genes Dev.* **5,** 221–231.

Simon, R., and Richter, J. (1992). Translational control by poly(A) elongation during Xenopus development: Differential repression and enhancement by a novel cytoplasmic polyadenylation element. *Genes Dev.* (in press).

Sonenshein, G. E., Geoghegan, T. E., and Brawerman, G. (1976). A major species of mammalian messenger RNA lacking a polyadenylate segment. *Proc. Natl. Acad. Sci. USA* **73,** 3088–3092.

Strickland, S., Huarte, J., Belin, D., Vassalli, A., Rickles, R. J., and Vassalli, J-D. (1988). Antisense RNA directed against the 3′ noncoding region prevents dormant mRNA activation in mouse oocytes. *Science* **241,** 680–684.

Stoeckle, M. (1992). Removal of a 3′ non-coding sequence is an initial step in degradation of groα mRNA and is regulated by interleukin-1. *Nucl. Acids Res.* **20,** 1123–1127.

Stoeckle, M., and Hanafusa, H. (1989). Processing of 9E3 mRNA and regulation of its stability in normal and Rous sarcoma virus-transformed cells. *Mol. Cell. Biol.* **9,** 4738–4745.

Swartwout, S. G., and Kinniburgh, A. J. (1989) c-*myc* RNA degradation in growing and differentiating cells: Possible alternate pathways. *Mol. Cell. Biol.* **9,** 288–295.

Swartwout, S. G., Preisler, H., Guon, W., and Kinniburgh, A. J. (1987) Relatively stable population of c-*myc* RNA that lacks long poly(A). *Mol. Cell. Biol.* **7,** 2052–2058.

Swiderski, R. E., and Richter, J. D. (1988). Photocrosslinking of proteins to maternal mRNA in *Xenopus* oocytes. *Dev. Biol.* **128,** 349–358.

Takagaki, Y., Ryner, L. C., and Manley, J. L. (1989). Four factors are required for 3′-end cleavage of pre-mRNAs. *Genes Dev.* **3,** 1711–1724.

Varnum, S. M., and Wormington, W. M. (1990). Deadenylation of maternal mRNAs during *Xenopus* oocyte maturation does not require specific *cis*-sequences: A default mechanism for translational control. *Genes Dev.* **4,** 2278–2286.

Varnum, S. M., Hurney, C. A., and Wormington, W. M. (1992) Maturation-specific deadenylation in *Xenopus* oocytes requires nuclear and cytoplasmic factors. *Dev. Biol.*, **153,** 283–290.

Vassalli, J-D., Huarte, J., Belin, D., Gubler, P., Vassalli, A., O'Connell, M. L., Parton, L. A., Rickles, R. J., and Strickland, S. (1989). Regulated polyadenylation controls mRNA translation during meiotic maturation of mouse oocytes. *Genes Dev.* **3,** 2163–2171.

Wahle, E. (1991a). Purification and characterization of a mammalian polyadenylate polymerase involved in the 3′ end processing of messenger RNA precursors. *J. Biol. Chem.* **266,** 3131–3139.

Wahle, E. (1991b). A novel poly(A)-binding protein acts as a specificity factor in the second phase of messenger RNA polyadenylation. *Cell* **66,** 759–768.

Wahle, E. (1992). The end of the message: 3′-end processing leading to polyadenylated messenger RNA. *BioEssays* **14,** 113–118.

Wahle, E., Martin G., Schiltz, E., and Keller, W. (1991). Isolation and expression of cDNA clones encoding mammalian poly(A) polymerase. *EMBO J* **10,** 4251–4257.

Wang, M., Cutler, M., Karimpour, I., and Kleene, K. (1992). Nucleotide sequence of a mouse testis poly(A) binding protein cDNA. *Nucl. Acids Res.* **20,** 3519.

Wharton, R. P., and Struhl, G. (1991). RNA regulatory elements mediate control of Drosophila body pattern by the posterior morphogen *nanos*. *Cell* **67,** 955–967.

Whittemore, L-A., and Maniatis, T. (1990). Postinduction turnoff of beta-interferon gene expression. *Mol. Cell. Biol.* **10,** 1329–1337.

Wickens, M. (1990a). How the messenger got its tail: Addition of poly(A) in the nucleus. *Trends Biochem. Sci.* **15,** 277–281.

Wickens, M. (1990b). In the beginning is the end: Regulation of poly(A) addition and removal during early development. *Trends Biochem. Sci.* **15,** 320–324.

Wigley, P. L., Sheets, M. D., Zarkower, D. A., Whitner, M. E., and Wickens, M. (1990). Polyadenylation of mRNA: Minimal substrates and a requirement for the 2′ hydroxyl of the U in AAUAAA. *Mol. Cell Biol.* **10,** 1705–1713.

Wilson, M. C., Sawicki, S. G., White, P. A., and Darnell, J. E. (1978). A correlation between the rate of poly(A) shortening and half-life of messenger RNA in adenovirus transformed cells. *J. Mol. Biol.* **126,** 23–36.

Wilson, T., and Treisman, R. (1988). Removal of poly(A) and consequent degradation of c-*fos* mRNA facilitated by 3′ AU-rich sequences. *Nature (London)* **336,** 396–399.

Wisdom, R., and Lee, W. (1991) The protein-coding region of c-*myc* mRNA contains a sequence that specifies rapid mRNA turnover and induction by protein synthesis inhibitors. *Genes Dev.* **5,** 232–243.

Wu, J. C., and Miller, W. L. (1991). Progresterone shortens poly(A) tails of the mRNAs for alpha and beta subunits of ovine leuteinizing hormone. *Biol. Reprod.* **45,** 215–220.

Yaffe, D., Nudel, U., Mayer, Y., and Neuman, S. (1985). Highly conserved sequences in the 3′ untranslated region of mRNAs coding for homologous proteins in distantly related species. *Nucleic Acids Res.* **13,** 3723–3737.

Zeevi, M., Nevins, J. R., and Darnell, J. E. (1982). Newly formed mRNA lacking polyadenylic acid enters the cytoplasm and the polyribosomes but has a shorter half-life in the absence of polyadenylic acid. *Mol. Cell. Biol.* **2,** 517–525.

Zelus, B. D., Giebelhaus, D. H., Eib, D. W., Kenner, K. A., and Moon, R. T. (1989). Expression of the poly(A)-binding protein during development of *Xenopus laevis*. *Mol. Cell. Biol.* **9,** 2756–2760.

Zingg, H. H., and Lefebvre, D. L. (1989). Oxytocin mRNA: Increase of polyadenylate tail size during preganancy and lactation. *Mol. Cell. Endocrinol.* **65,** 59–62.

Zingg, H. H., Lefebvre, D. L., and Almazan, G. (1988). Regulation of poly(A) tail size of vasopressin mRNA. J. Biol. Chem. **263,** 11041–11043.

16

mRNA Decay in Cell-Free Systems

JEFF ROSS

I. Introduction

An important recent advance in the analysis of mRNA stability is the use of *in vitro* systems for studying mRNA decay. This chapter discusses the rationale for using such systems and emphasizes how the scarcity of genetic methodology in many higher organisms almost necessitates devising *in vitro* models of cytoplasmic mRNA turnover. Several cell-free systems are described, highlighting the advantages and limitations of each and describing how they have been used to answer specific questions. Finally, some future directions and important questions amenable to *in vitro* analyses are discussed.

II. Rationale

A. Genetic Limitations, mRNA Stability, and *in Vitro* Systems from Higher Organisms

Three sorts of observations have brought general agreement that mRNA turnover influences gene expression in most or all cells (reviewed in Shapiro *et al.*, 1987; Brawerman, 1989; Cleveland and Yen, 1989; Peltz *et al.*, 1991). First, the steady-state levels of many mRNAs correlate more closely with their cytoplasmic half-lives than with the rates at which their genes are transcribed. Second, the half-lives of some mRNAs change in response to the nutritional status of the cell, its position in the replication cycle, or its stage of differentiation. Thus, cells frequently regulate gene expression by changing the half-lives of specific mRNAs. Third, aberrant regulation of mRNA stability can apparently influence the initiation and promotion of carcinogenesis and the phenotype of certain neoplastic cells. It is thus essential to identify the factors that influence mRNA half-life and to understand how changing the half-lives of certain mRNAs affects cell growth, differentiation, and neoplastic transformation. A number of questions must be addressed. What enzymes degrade mRNAs, and where do they first begin to attack mRNA molecules? What degradation intermediates are generated? What sequences and/or structures determine mRNA half-life? How does the rate of translation or the cytoplasmic localization of an mRNA affect its turnover? What trans-acting factors regulate mRNA turnover, and how do they function?

Analyses with intact cells have provided abundant information on which mRNAs are stable or unstable and how mRNA sequences influence mRNA half-life. However, in-depth analysis of mRNA turnover in cells is limited for several reasons. First, transcription inhibitors are often used to measure mRNA half-life but rarely, if ever, affect transcription specifically. Many also slow translation and other metabolic processes. As a result, the inhibitors themselves can influence the results. The measured half-life of an mRNA can differ significantly, depending on whether transcription is repressed by adding a transcription inhibitor or by withdrawing a gene-specific inducing agent, such as a hormone or growth factor (Schwartz, 1973; Steinberg *et al.*, 1975; Khochbin *et al.*, 1988; Müllner and Kühn, 1988; Goldberg *et al.*, 1991). Both manipulations block transcription, but removing an inducer influences only one or a few genes and inhibits macromolecular metabolism minimally. Transcriptional inhibitors are far less specific. Therefore, half-life measurements with transcriptional inhibitors are sometimes unreliable.

A second limitation of studies with intact cells involves the scarcity of genetic approaches available in many higher organisms. Consider the deceptively simple question of how poly(A) is removed from a particular mRNA. Using various methods, investigators have shown that poly(A) is sometimes removed from a population of mRNA molecules prior to their further degradation (see Chapters 9 and 15). However, because of the difficulty of radiolabeling an mRNA to a sufficient specific activity in nucleated cells, a definitive and quantitative analysis of the poly(A) removal pathway has not been possible. To contrast this situation, consider the following example involving prokaryotic mRNA metabolism. Regnier and Hajnsdorf (1991) investigated the processing/degradation pathway of a bicistronic mRNA from the *rpsO–pnp* operon of *Escherichia coli,* which encodes ribosomal protein S15 and polynucleotide phosphorylase. Only full-length monocistronic mRNAs were observed in wild-type strains, but the bicistronic mRNA was detected in strains deficient in RNase E, RNase III, or both. Using combinations of strains lacking one or both enzymes, Regnier and Hajnsdorf mapped cleavage sites in the bicistronic mRNA and determined which RNase generated which cleavage. The genetic approach thus permitted the investigators to monitor RNA processing events in cells, to detect mRNA decay intermediates, and to identify which enzymes generated them.

As is the case in wild-type *E. coli* cells, it is difficult or impossible to detect discrete mRNA decay intermediates in exponentially growing cells from higher organisms. Unfortunately, the prospects for identifying all of the intermediates in a decay pathway are bleak, because RNase-deficient cell lines are not available. As a result, little is known about which enzymes degrade mRNAs, and even less is understood about how regulatory factors influence cytoplasmic mRNA stability. These deficiencies highlight the necessity of exploiting *in vitro* mRNA decay systems.

B. Advantages of Analyzing mRNA Stability in Cell-Free Extracts

1. Half-lives can be measured without transcription inhibitors, thereby avoiding toxic and nonspecific side effects. Some *in vitro* mRNA decay systems seem to provide good half-life correlations: mRNAs that are long-lived in cells are also relatively stable *in vitro,* and unstable mRNAs are relatively unstable in both.

2. mRNA decay intermediates that are difficult to detect in intact cells because of their lability can sometimes be detected *in vitro,* where they are more stable.

3. Messenger ribonucleases and trans-acting factors that regulate mRNA stability can be assayed and purified.

4. mRNA sequences and structures can be modified in various ways *in vitro,* and the effects of specific mutations on mRNA half-life can be investigated. In contrast, in intact cells a chimeric gene encoding the mRNA of interest must also contain sequences required for transcription, processing, polyadenylation, transport, and, perhaps, translation. These requirements limit the sorts of modified substrates it is possible to test.

C. Limitations of Cell-Free mRNA Decay Systems

1. The constraints imposed on interpreting experiments with any cell-free system (transcription, translation, etc.) apply also to *in vitro* mRNA decay. For example, an mRNA might be degraded exonucleolytically 3′ to 5′ *in vitro* and in cells, but there is no assurance that the same enzyme is involved in both cases. The enzyme must be purified and analyzed further in order to link *in vitro* data with intracellular events.

2. Cell fractionation destroys cytoarchitecture, and extracts from higher organisms are often translationally inefficient or inert. Therefore, any links between the half-life of an mRNA and its cytoplasmic localization or its translation might be disrupted. Methods for overcoming this limitation are discussed later.

III. Useful Approaches for Analyzing mRNA Stability *in Vitro*

A. General Considerations

All *in vitro* mRNA decay systems currently in use contain some cytoplasmic components plus some reagents usually required for cell-free translation such as ATP, GTP, a monovalent cation, and magnesium. However, no unique set of *in vitro* reaction conditions can be considered optimal for every mRNA. In fact, the following observations suggest that various *in vitro* systems will be required in order to analyze the turnover of different mRNAs. (1) Some mRNAs are degraded efficiently only while they are associated with ribosomes (Ernest, 1982; Capasso *et al.*, 1987; Graves *et al.*, 1987; Blume *et al.*, 1989; Gay *et al.*, 1989; Wisdom and Lee, 1991); others are degraded only in specific cytoplasmic locations or compartments (Mason *et al.*, 1988; Jäck *et al.*, 1989; Zambetti *et al.*, 1990). For example, if histone mRNA is modified by inserting a nonsense codon close enough to its translation initiation site so that few or no ribosomes associate with the mRNA, its decay is not regulated properly. Unmodified histone mRNA is destabilized at the end of the S-phase and is barely detectable during the G1-phase, while the modified mRNA is easily detectable in G1-phase cells (Capasso *et al.*, 1987; Graves *et al.*, 1987). The half-lives of immunoglobulin and histone mRNAs vary by 5-fold, depending on whether the mRNAs are associated with free or membrane-bound polysomes (Mason *et al.*, 1988; Jäck *et al.*, 1989; Zambetti *et al.*, 1990). To analyze the regulation of these mRNAs *in vitro*, the reaction mixtures should probably include the relevant subcytoplasmic fractions (membranes, membrane-bound polysomes, etc.) and should be translationally active. Only in this way will the *in vitro* conditions reflect the intracellular environment relevant to the particular mRNA being investigated. (2) The half-lives of some mRNAs fluctuate depending on the cell type, growth conditions, etc. (Brewer and Ross, 1989; Harris *et al.*, 1991). For example, estrogen treatment of hepatocytes changes the half-life of vitellogenin mRNA by 10-fold or more (Brock and Shapiro, 1983; Cochrane and Deeley, 1988; Gordon *et al.*, 1988). Therefore, in order to understand how estrogen influences mRNA half-life, it might be necessary to prepare extracts from cells exposed to the hormone for appropriate times.

Several tests can be used to gauge the validity of an *in vitro* mRNA decay system. The system must reflect relative intracellular half-lives. If two mRNAs are degraded at rates that differ by 20-fold in cells but by less than 2-fold *in vitro*, then the cell-free system does not accurately reflect intracellular mRNA decay processes. Half-lives measured *in vitro* might be longer than those observed in cells, but relative decay rates should be preserved. A cell-free system should generate mRNA decay intermediates that are identical to those detected in cells. It should also reflect the intracellular regulation of mRNA half-life. If an mRNA is stabilized in hormone-treated cells, extracts from hormone-treated cells should degrade the mRNA more slowly than extracts from untreated cells. Several *in vitro* mRNA decay systems have already been shown to satisfy these criteria. Therefore, these systems should be useful for identifying and purifying messenger RNases and trans-acting regulatory factors (see Section IV.C).

B. The Substrate

1. *Endogenous Substrate (mRNP)*

Polysomes have been the most frequently used source of endogenous substrate for *in vitro* mRNA decay reactions. By definition, polysomes contain the substrate messenger ribonucleoprotein (mRNP). Depending on the cell lysis conditions, they might also contain mRNases.

Advantages. First, assays using polysomes are relatively straightforward. The polysomes are simply incubated with the required buffer, salts, and other components; RNA is extracted; and mRNA half-life is assessed by molecular hybridization. Second, the substrate (mRNP) is synthesized by the cell and must be competent for translation, since it is polysome-associated. Therefore, within the constraints imposed on any subcellular fraction prepared in the laboratory, endogenous mRNP is as close as possible to being an authentic substrate. Third, trans-acting regulatory factors can affect the stability of polysome-associated mRNP *in vitro* (see later). This observation is particularly important, since these same factors often have no effect on the *in vitro* half-life of exogenous, protein-free mRNA substrates. Perhaps the factors regulate mRN*P*, not mRN*A*.

Disadvantages. First, if the mRNA is scarce, large amounts of polysomes might be required for each assay. Isolation of sufficient polysomes containing undegraded mRNA can be difficult, especially from primary animal tissues, which seem to contain significantly higher nonspecific RNase activity than cell lines. [I define nonspecific RNase in a cell extract as an enzyme(s) capable of degrading all RNAs (rRNA, tRNA, mRNA, etc.) at similar rates. The term is not meant to imply lack of specific function, only our ignorance of what the function might be]. Second, as mentioned above, synthesis of interesting mutant mRNA substrates might be difficult or impossible in intact cells.

2. Exogenous RNA Substrate

The half-lives of several sorts of exogenous substrates, including protein-free mRNA extracted from the cell, mRNP isolated from the cell, or *in vitro* synthesized RNA, can be analyzed in cell-free mRNA decay reactions.

Advantages. First, the detection of the substrate mRNA can be relatively easy, if the substrate is synthesized *in vitro* and is radiolabeled. Second, *in vitro* synthesized exogenous transcripts can be more easily modified than endogenous mRNAs.

Disadvantages. The relative half-lives of protein-free mRNA substrates might not reflect those measured in intact cells. For example, endogenous, polysome-associated c-*myc* mRNP is at least 40-fold less stable than γ-globin mRNP *in vitro*, reflecting their relative intracellular half-lives (Ross and Kobs, 1986; Brewer and Ross, 1988). In contrast, *in vitro* synthesized (exogenous) c-*myc* mRNA is at most two- to threefold less stable than *in vitro* synthesized γ-globin mRNA incubated under the same cell-free conditions (J. Ross, unpublished observations). There are several possible explanations for the failure to observe differential decay with the exogenous, protein-free substrates. They might not form the same RNP structures as their natural, endogenous counterparts. They are unlikely to associate efficiently with ribosomes in extracts from most nucleated mammalian cells, and they might not be precise replicas of endogenous mRNAs. For example, the exogenous mRNA might be undermethylated. These differences between mRNP and mRNA might also influence the capacity of the RNA substrates to respond to regulatory factors *in vitro*. For example, polysome-associated c-*myc* mRNP is destabilized *in vitro* by a cytosolic factor, but *in vitro* synthesized c-*myc* mRNA is not destabilized under identical conditions (Brewer and Ross, 1989; G. Brewer and J. Ross, unpublished observations). These observations suggest that some mRNAs must be complexed with RNA-binding proteins and/or be associated with ribosomes in order to be regulated and degraded properly *in vitro*.

C. Extracts and Reaction Conditions

Approximately six different extracts have been described for investigating mRNA decay *in vitro* (Table 1). Each seems to have certain advantages and disadvantages. For example, mRNA is translated poorly or not at all in crude extracts from nucleated cells but is translated in reticulocyte extracts. Insufficient information is available to determine which extracts reflect most accurately the known relative intracellular half-lives of different mRNAs. It

TABLE 1 ***In Vitro*** **mRNA Decay Extracts**

Extract	References	Characteristics
Crude, postnuclear supernatant	Ross and Kobs, 1986 Pei and Calame, 1988 Wager and Assoian, 1990	Use low to moderate salt concentrations Some methods include nonionic detergents in lysis buffers
Crude, postnuclear supernatant (lysolecithin extract)	Brown *et al.*, 1983 Baker and Lai, 1990 Krikorian and Read, 1990	Used with adherent tissue culture cells Some translational initiation
Polysomes and postpolysomal supernatant	Ross and Kobs, 1986 Bandyopadhyay *et al.*, 1990	RNase activity in both fractions mRNase activity in polysomes
Ribosomal salt wash	Ross *et al.*, 1987 Sunitha and Slobin, 1987 Liang and Jost, 1991	Ribosome-free high-salt eluate
mRNP	Sunitha and Slobin, 1987 Bandyopadhyay *et al.*, 1990	Prepared by differential centrifugation Starvation of cells increases mRNP recovery
Soluble chloroplast extracts	Stern and Gruissem, 1987 Adams *et al.*, 1989 Adams and Stern, 1990	Chloroplast system capable of transcribing exogenous DNA, processing exogenous pre-mRNA, and degrading exogenous mRNA
Reticulocyte extract	Wreschner and Rechavi, 1988 Hepler *et al.*, 1990	Significant translation

will probably be necessary to use different extracts and combinations of extracts, so that each cell-free system will reflect a given set of cell growth conditions relevant to the mRNA being studied (see Section V).

1. *Crude Postnuclear Supernatant*

Extracts are prepared by lysing cells in buffers with or without detergents, followed by low-speed centrifugation to remove nuclei and membranous material. Postnuclear supernatants are harvested and incubated at 30 to 37°C, usually in the presence of an ATP-generating system, a monovalent cation, magnesium, and, in some cases, amino acids and an RNase inhibitor. Although some amino acid incorporation occurs in these extracts, translation rates are very low. It is unclear whether extracts from RNase-rich tissues can be adapted to these systems. Several investigators have observed differential mRNA decay in cytoplasmic extracts from *Chlamydomonas rein-*

hardtii (E. Baker, personal communication) and oat seedlings (D. Byrne and J. Colbert, personal communication).

A variation of this method involves permeabilizing adherent tissue culture cells by exposing them briefly to a buffer containing lysolecithin. Cells are then scraped from the dish, nuclei are removed by low-speed centrifugation, and the supernatant is used for mRNA decay assays. Translation rates are significantly faster in lysolecithin extracts than in other sorts of crude extracts. Moreover, exogenous mRNA can be translated for a limited period.

Postnuclear supernatant extracts from tissue culture cells are easy to prepare. On the other hand, since they include most of the cell cytoplasm, they are not particularly useful for identifying the trans-acting factors that regulate mRNA stability. For this reason, some investigators have preferred fractionated systems (see below).

2. *Polysomes and Postpolysomal Supernatant*

Polysome preparations may contain some of the nucleases responsible for degrading mRNAs. Therefore, the decay of endogenous, polysome-associated mRNP or exogenous, protein-free mRNA can be analyzed, and the effects of soluble, trans-acting factors can be assayed by mixing polysomes with crude or fractionated postpolysomal supernatant (Table 2). Polysome-containing reactions are frequently performed under conditions that are compatible with translation, although little translation may occur.

3. *High Salt Extract of Ribosomes*

Some or most of the RNases associated with polysomes can be solubilized and separated from the polysomes by extraction with high salt (0.3–1.0 *M*), followed by ultracentrifugation to pellet the salt-washed polysomes. The enzymes present in the resulting supernatant or ribosomal salt wash (RSW) are highly active on protein-free mRNA substrates, and the RSW provides a potential starting material for mRNase purification. It is unclear at this time whether RSW will be useful for analyzing differential decay of exogenous, protein-free mRNA substrates (see Section III.B2b).

4. *mRNP*

Several investigators have analyzed mRNA decay using mRNP as both a substrate and a source of mRNase. These extracts are free of or deficient in ribosomes, suggesting that the mRNPs themselves contain mRNA-degrading nucleases.

5. *Combined Transcription-Decay System*

Transcription, processing, and mRNA stability have been analyzed concurrently in crude extracts from plant chloroplasts. Exogenous RNA

TABLE 2 Putative mRNA Stability Regulatory Factors Identified in Cell-Free Systems

Factor	Properties	Reference
p66	Detected in avian hepatocytes Binds to 5′-UTR of vitellogenin II mRNA Might prolong half-life of vitellogenin II mRNA Induced by estrogen to associate with ribosomes	Liang and Jost, 1991
Labile cytosolic destabilizing factors	Detected in postpolysomal and postmitochondrial supernatants of cultured cells Prolong half-lives of c-*myc*, c-*myb*, and urokinase-type plasminogen activator mRNAs Inactivated when translation is blocked One might be a ribonucleoprotein	Brewer and Ross, 1989 Altus and Nagamine, 1991
Auf (AU-rich element–poly(U)-binding/degradation factor)	Detected in postpolysomal supernatant of cultured cells Binds to AU-rich regions of mRNAs Binds in 3′-UTR of c-*myc* mRNA and destabilizes the mRNA Composed of two proteins, p37 and p40	Brewer, 1991
Duplex RNA	Accumulates in many virus-infected cells Activates cascade which induces RNase L Might lead to localized degradation of nascent viral RNA	Baglioni *et al.*, 1984
Virion host shutoff protein of HSV	Carried into cells with the virion Induces indiscriminate destabilization of most host cell and viral mRNAs A 57-kDa protein	Krikorian and Reed, 1990 Sorenson *et al.*, 1991
TPA-inducible stabilizer of TGF-β1 mRNA	Detected in cytosol of TPA-treated cells Prolongs half-life of TGF-β1 mRNA	Wager and Assoian, 1990
Inhibitor(s) of "nonspecific" RNases	Detected in most or all eukaryotic cells Required to maintain integrity of mRNAs in cell-free extracts	Stolle and Benz, 1988 Bandyopadhyay *et al.*, 1990

substrates can also be processed and then degraded differentially in these extracts.

6. Rabbit Reticulocyte Lysates

Exogenous mRNAs are incubated with crude or micrococcal nuclease-treated lysates under protein synthesizing conditions. Although differential mRNA decay in reticulocyte extracts has been reported, it is unclear whether a significant percentage of the input mRNA was actually recruited into polysomes and translated. A significant advantage of reticulocyte extracts is that the link between mRNA stability and translation can be investigated directly. For example, following incubation of the extract with micrococcal nuclease to destroy endogenous mRNA, exogenous, radiolabeled mRNAs can be translated, and the polysomes can then be isolated and incubated with subcellular fractions from nucleated cells. Mixed, translation-competent extracts of this sort might also be useful for detecting trans-acting regulatory factors and for mapping mRNA stability determinants.

IV. mRNA Decay Pathways, mRNases, and Regulatory Factors Identified in Cell-Free mRNA Decay Systems

A. Enzymes and mRNA Decay Intermediates

Discrete ribonucleases with a proven role in mRNA degradation have not been identified in the cells of higher organisms, nor has the complete decay pathway of a specific mRNA been determined. mRNAs do get degraded in cell-free systems, and some steps in the *in vitro* decay pathway are identical to those in the intracellular pathway (see below). Therefore, it is likely that authentic mRNases carry out these reactions *in vitro*. However, none of the putative mRNases have as yet been purified and characterized. The catalytic properties and specificity of the mRNases have been inferred only by analyzing the decay products generated during *in vitro* incubations. RNase L, which is activated in response to interferon, might effect mRNA destabilization in virus-infected cells, but it is unclear to what extent RNase L distinguishes viral from cellular mRNAs (Baglioni *et al.*, 1984; Cohrs *et al.*, 1988). The following information must be obtained in order to describe completely the decay of a specific mRNA: (1) identification of *all* of the decay products generated during mRNA degradation; (2) purification of any RNases that could account for the appearance of these products; (3) identification of the relevant enzyme using genetic tests and/or immunologic or other reagents that block the RNase activity specifically. If a certain 3′ to 5′ exoribonuclease were responsible for the initial steps in histone mRNA degradation (see subsequent text and Fig. 1), histone mRNA should not be degraded in reactions containing an antibody raised against the enzyme.

1. *Evidence for Exonuclease Activity*

Histone mRNA appears to be degraded in a 3′ to 5′ direction *in vitro* and in intact cells, and two properties of the earliest detectable decay intermediates suggest that a 3′ to 5′ exonuclease degrades the mRNA at

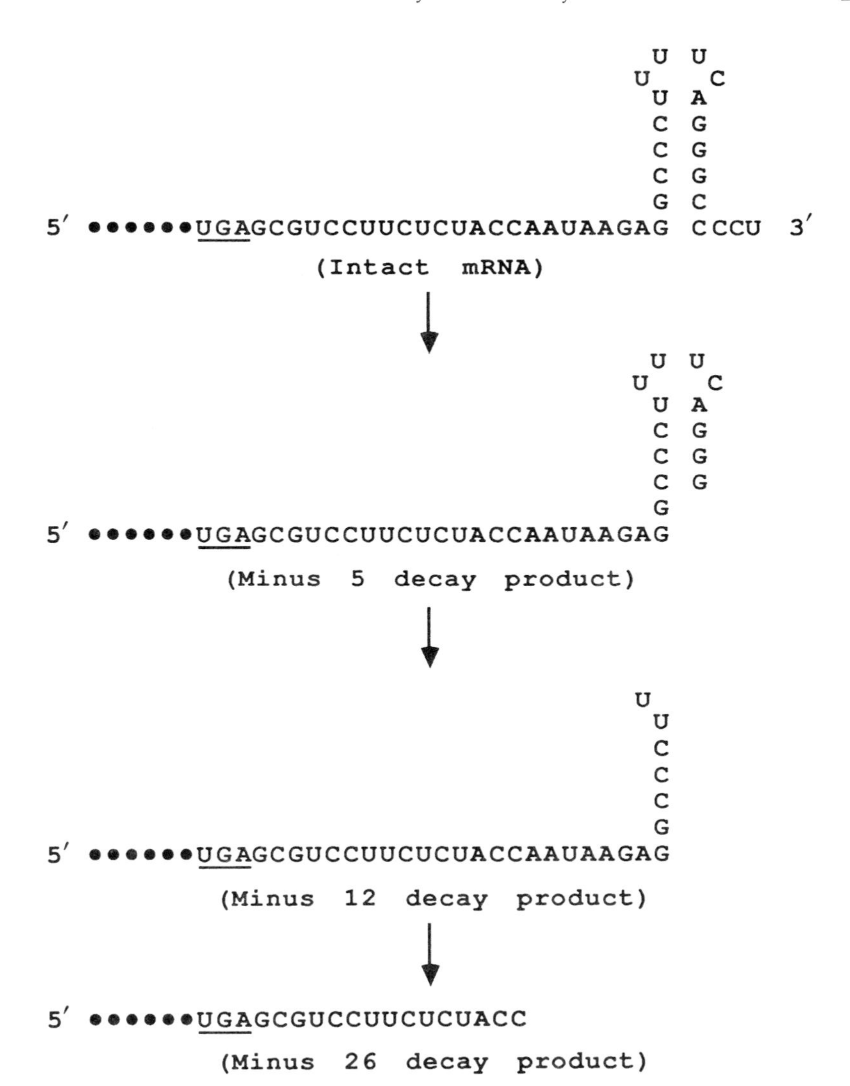

FIGURE 1 Diagram of a proposed pathway for histone mRNA decay in cells and in cell-free extracts. The termination codon (underlined) and 3′-UTR are shown. The minus 5 and minus 12 decay products have been observed both *in vitro* and in intact cells (Ross and Kobs, 1986; Ross *et al.*, 1986). The minus 26 product has been observed consistently *in vitro* and has been observed in some experiments with intact cells (J. Ross, unpublished observations).

a nonuniform rate. First, only small segments of the mRNA (mono- or oligonucleotides) are removed at each step (Ross and Kobs, 1986). Second, discrete decay products are detected, suggesting that the putative exonuclease pauses at specific sites (Ross *et al.*, 1986) (Fig. 1). If the exonuclease removed mononucleotides at a constant rate, it would have generated a uniform ladder of decay products, not discrete ones. Since the RNase responsible for these early degradation events has not been purified, it is possible that the cleavages result from an endonuclease that removes oligonucleotides (see subsequent text and Fig. 1) and that proceeds in a 3′ to 5′ direction.

The first detectable histone mRNA decay product lacks approximately 5 nucleotides (nt) from the 3′ end. A second group of decay products, lacking 5 to 7 additional nt, appears at later times in the decay reaction, suggesting a sequential pathway. These *in vitro* observations are particularly significant, because identical decay products are detected in cells that are rapidly degrading histone mRNA (Ross *et al.*, 1986). Therefore, the *in vitro* system accurately reproduces at least the initial steps in the intracellular pathway of histone mRNA degradation. With time, the *in vitro* decay products continue to shorten, implying further 3′ to 5′ degradation. The 3′ terminus of the other major decay product is located approximately 15 nt downstream of the translation termination site (J. Ross, unpublished observations). Perhaps a ribosome that is paused at or near the termination codon retards the exonuclease. It is unclear how many enzymes are required for complete destruction of each histone mRNA molecule.

Poly(A) shortening appears to be the first step in the degradation of short-lived, polysome-associated c-*myc* mRNA *in vitro* (Brewer and Ross, 1988). The shortening reaction seems to occur gradually over time, implying exonucleolytic decay. However, because the poly(A) tracts of the mRNA analyzed in these experiments were heterogeneous, it was not possible to monitor the poly(A) removal pathway precisely. Random endonucleolytic cleavages might have generated a similar pattern of decay products.

More direct evidence for exonucleolytic poly(A) shortening was obtained with yeast cell extracts (Sachs and Deardorff, 1992; Lowell *et al.*, 1992). A partially purified 135-kDa nuclease degrades poly(A) specifically, but only in reactions containing poly(A) binding protein (PABP). The enzyme, which is called PAN for PABP-dependent poly(A) nuclease, degrades a 100-nt-long poly(A) substrate to 15 to nt in the presence of PABP. When the poly(A) is located downstream of the 3′-untranslated region (UTR) of unstable yeast α-factor mRNA, all of the poly(A) tract is destroyed. Therefore, PAN is dependent on an RNA-binding protein (PABP), and the extent of PAN digestion is modulated by sequences in the 3′ UTR. These data are significant with respect to poly(A) metabolism and to the regulation of mRNA stability (see Section IV.B).

Indirect evidence for exonucleolytic mRNA processing or decay has been obtained using chloroplast extracts. Exogenous mRNA substrates with a 3′ stem–loop are 20-fold more stable than those lacking a stem–loop, implying that the duplex region blocks a 3′ to 5′ exonuclease (Stern and Gruissem, 1987; Stern *et al.*, 1989) (see Chapter 14).

In summary, many of the mRNAs that have been studied thus far are degraded by 3′ to 5′ exonucleases. On the other hand, the possibility that mRNAs can be degraded in a 5′ to 3′ direction has not been excluded. Uncapped mRNAs are less stable in cell-free translation and mRNA decay systems than their capped counterparts (Furuichi *et al.*, 1977; Shimotono *et al.*, 1977; McCrae and Woodland, 1981; Drummond *et al.*, 1985; Peltz *et al.*, 1987). Therefore, the 5′ region can influence mRNA decay rates. Moreover, a magnesium-dependent 5′ exonuclease has recently been purified from mouse ascites cell cytoplasmic extracts (Coutts *et al.*, 1993a and b). This RNase is as active on a capped RNA substrate as on one with a 5′-triphosphate. It will be most interesting to learn whether this enzyme attacks unstable mRNAs more readily than stable ones and/or whether its function is to degrade endonuclease cleavage products. Degradation in a 5′ to 3′ direction would probably be more economical than 3′ to 5′ decay, since complete translation products are produced if a 5′ to 3′ exonuclease follows the last translating ribosome in its journey toward the termination codon. In contrast, as soon as a 3′ to 5′ exonuclease reaches the coding region of the mRNA, synthesis of full-length protein is precluded.

2. *Evidence for Endonuclease Activity*

Decay products that appear to arise from endonucleolytic cleavages have been observed in several cell-free mRNA decay systems. Relatively stable decay products seem to result from discrete internal cleavages in eEF-Tu and β-globin mRNPs (Sunitha and Slobin, 1987; Bandyopadhyay *et al.*, 1990). In cell-free extracts containing polysomes, the 3′ termini of discrete c-*myc* mRNA decay intermediates are located within the 3′ UTR (Brewer and Ross, 1988). However, only the 5′ portions of the c-*myc* decay products were detected, so the possibility of pausing during 3′ to 5′ exonucleolytic decay was not excluded. Direct evidence for endonucleolytic cleavage within the c-*myc* coding region is presented in Section IV.B.

mRNAs are cleaved endonucleolytically in a combined transcription/processing/stability system prepared from chloroplasts (Adams and Stern, 1990). Many chloroplast genes are transcribed into mRNA precursors that are then processed by a 3′ to 5′ exonuclease to mRNAs with 3′-terminal stem–loop structures. The resulting 3′-terminal duplex region of the mRNA probably blocks further processing by the exonuclease. Once synthesized, the half-life of the mRNA itself seems to be determined by the susceptibility of the loop region to endonucleolytic attack. Purine-rich loops are cleaved more readily than pyrimidine-rich loops. A divalent cation-independent endoribonuclease has been identified and partially purified from chloroplasts (Hsu-Ching and Stern, 1991). However, this enzyme cleaves primarily at the translation termination site, not within the loop, so its role in mRNA turnover is unclear.

An endonuclease might be responsible for destabilizing albumin and other serum protein-coding mRNAs in male *Xenopus* hepatocytes exposed to estrogen (Pastori *et al.*, 1991). Polysomes were isolated from the livers of untreated and estrogen-treated animals, and some of the ribosome-associated proteins were solubilized by high salt extraction. The ribosomal salt wash from treated animals contains a magnesium-independent endoribonuclease activity that degrades albumin mRNA faster than ferritin mRNA *in vitro*, consistent with observations in the whole animal. Estrogen does not change the intracellular level or activity of the RNase. Rather, it seems to induce the enzyme to translocate from an unknown site in the cytoplasm to either polysomes or rough endoplasmic reticulum. In other words, little of the enzyme cosediments with polysomes in untreated cells, while most of it is polysome-associated following estrogen treatment. These interesting observations confirm the notion that regulation of mRNA stability is sometimes linked with the cytoplasmic location of mRNA, mRNases, or both (see Section V.D).

Indirect evidence suggests that a possible endonucleolytic cleavage site exists in the 5′ UTR of chicken vitellogenin mRNA (Liang and Jost, 1991). A 66-kDa liver cell protein binds to an RNA fragment from this region. When the fragment is incubated with an extract from rooster liver cytoplasm, it is stabilized at least fourfold in the presence of the protein, implying that p66 protects the fragment from endonucleolytic attack.

B. *RNA-Binding Proteins and mRNA Stability* in Vitro

A number of RNA-binding proteins have been described (reviewed in Bandziulis *et al.*, 1989). Some have been implicated in regulating mRNA function and/or metabolism, but few have been purified and assayed for their effect on mRNA turnover *in vitro*. These proteins and their effects on mRNA stability are described in this section. Section C describes some trans-acting factors that regulate mRNA half-lives. The distinction between these two groups of regulatory factors has been made on the basis of the known properties of the factors but is somewhat arbitrary. The reader should note that many or all of the so-called trans-acting factors described in section C might turn out to be RNA-binding proteins.

PABP is a well-characterized RNA-binding protein that is known to influence translation and poly(A) size in cells and mRNA stability *in vitro* (reviewed in Bernstein and Ross, 1989; Sachs, 1990). Some *in vitro* experiments indicate a role for the poly(A)–PABP complex in determining mRNA stability. First, cell extracts were depleted of most or all PABP, either by mixing the extracts with a large excess of exogenous poly(A) or by passing the extracts over a poly(A)–Sepharose column. Then, the extracts were incubated with *in vitro* synthesized, radiolabeled mRNAs. Polyadenylated mRNAs, which are normally associated with PABP and are relatively stable in complete extracts, are degraded 3- to 10-fold faster in PABP-depleted extracts (Bernstein *et al.*, 1989). They are degraded in two steps: deadenylation, followed by degradation of the mRNA body. When exogenous, purified yeast PABP is added to PABP-depleted extracts, the mRNAs are stabilized and are not rapidly deadenylated. PABP has no effect on the stability of nonpolyadenylated mRNAs.

These results imply that poly(A), in conjunction with PABP, can protect some mammalian cell mRNAs from RNases. If so, then PABP must bind less avidly to unstable than to stable mRNAs. If PABP binds to poly(A) and protects it from nuclease attack, if poly(A) shortening must precede further degradation steps [as seems to be the case for at least some mRNAs (Mercer and Wake, 1985; Swartwout and Kinniburgh, 1989; Minvielle-Sebastia *et al.*, 1991)], and if PABP were bound as tightly to unstable mRNAs as to stable ones, then all mRNAs should be similarly protected and have similar half-lives, which is not the case. Perhaps sequences within the body of each mRNA can influence the affinity of PABP for the poly(A) tract (Bernstein *et al.*, 1989). Sequences within unstable mRNAs might reduce the binding affinity, increase the tendency of PABP to dissociate from the poly(A), and thereby shorten the mRNA half-life.

It seems likely that the function of poly(A) and PABP will become clearer when *in vitro* technology is combined with the analysis of mutant strains of yeast. *Saccharomyces cerevisiae* cells lacking PABP are not viable (Sachs and Davis, 1989) but can be rescued by suppressor gene products that influence the synthesis or function of the large ribosomal subunit (Sachs and Davis, 1990). Analyses of the PABP-deficient mutants and of the extragenic suppressor strains have revealed at least two major functions for poly(A) and its associated protein in yeast: translation initiation and poly(A) shortening. *In vitro* poly(A) shortening catalyzed by PAN plus PABP is influenced by sequences in the mRNA 3′ UTR (see Section IV.A.1). Therefore, in order to remove poly(A) completely from an mRNA, a combination of the PABP–poly(A) complex plus a deadenylation element in the 3′ UTR might be required (Sachs and Deardorff, 1992; Lowell *et al.*, 1992). The ability to assay PAN activity *in vitro* on a variety of polyadenylated templates should reveal how sequences within the mRNA body influence poly(A) shortening rates.

It is important to note that the steady-state size of poly(A) in PABP-deficient yeast cells is actually longer than that in wild-type cells (Sachs and Davis, 1989). Moreover, PABP is required for the degradation of poly(A) by PAN. These results indicate that a PABP–poly(A) complex is the substrate for poly(A) removal in yeast and seems to be inconsistent with the *in vitro* experiments using mammalian cell extracts, which indicated that PABP protects the poly(A) from destruction (Bernstein *et al.*, 1989). [The reader should consult the reviews by Bernstein and Ross (1989) and Sachs (1990) for more detailed discussions of the data and their interpretation.] Additional studies of mRNA decay *in vitro* and of other yeast proteins that regulate RNA transport, poly(A) metabolism, and mRNA stability should provide additional insights into poly(A) function (Piper and Aamand, 1989; Minvielle-Sebastia *et al.*, 1991; Leeds *et al.*, 1991).

A 66-kDa protein solubilized from chicken liver polysomes binds to the 5′ UTR of chicken vitellogenin II mRNA (Liang and Jost, 1991) (see Chapter 8). When a radioactive RNA fragment from this region is incubated in a crude rooster liver cytoplasmic extract, the RNA is destroyed within 30 min. However, when partially purified p66 is added to the reactions, some of the RNA fragment remains intact after 90 min, suggesting that p66 binds to and protects this region of the mRNA. Estrogen influences both the quantity per hepatocyte of p66, increasing it approximately 10-fold, and its intracellular distribution. Following 3 days of estrogen exposure, the protein is located on polysomes and within the nucleus; in untreated cells it is primarily cytosolic and presumably not associated with polysomes (H. Liang and J.-P. Jost, unpublished observations). It would be interesting to know whether the initial mRNA–protein interaction occurs within the nucleus and, if so, whether the interaction affects mRNA transport and translation, as well as turnover.

A polysome-associated protein of approximately 75 kDa binds *in vitro* to the terminal 182 nt of the c-*myc* mRNA coding region (Bernstein *et al.*, 1992). When an exogenous RNA fragment corresponding to this region is added to an *in vitro* mRNA decay system containing polysomes, endogenous c-*myc* mRNP is specifically and rapidly destabilized. The first degradation step appears to be endonucleolytic cleavage within the 182-nt coding region determinant. One model to account for these observations is that the 75-kDa protein is normally bound to polysome-associated c-*myc* mRNP and functions as a stabilizing factor, protecting it from endonuclease attack. Adding exogenous competitor RNA titrates away the protein, thereby exposing the previously protected region to the endonuclease. The function, if any, of this protein in regulating c-*myc* mRNA half-lives during the growth and differentiation of various types of cells needs to be investigated.

Taken together, these experiments reveal an important link between mRNA-binding proteins and mRNA stability. Some of the trans-acting regulators that influence the half-lives of specific mRNAs are also RNA-binding proteins (Section IV.C). It will be important to determine whether other RNA-binding proteins, including the AU-binding factor and the iron-responsive element binding protein, affect mRNA stability *in vitro* (Müllner and Kühn, 1988; Casey *et al.*, 1989; Ratnasabapathy *et al.*, 1990; Bohjanen *et al.*, 1991; Malter and Hong, 1991).

C. trans-Acting Regulators

As discussed earlier, the scarcity of genetic methods for analyzing mRNA turnover in some higher organisms has, with few exceptions, hampered the identification of factors regulating mRNA stability. Fortunately, some trans-acting regulators can be detected using cell-free mRNA decay systems (Table 2).

*1. Constitutive, Cytosolic Destabilizers of c-*myc *and Urokinase-Type Plasminogen Activator mRNAs*

When S130 (postribosomal supernatant) from exponentially growing erythroleukemia cells is mixed with polysomes, polysome-bound c-*myc* mRNA is degraded approximately fourfold faster than in reactions lacking the S130 (Brewer and Ross, 1989). S130 derived from cells pretreated with cycloheximide has no effect on c-*myc* mRNA stability. These data suggest that cells contain a cytosolic mRNA destabilizing factor that is inactivated or destroyed when translation slows or stops in response to a nonspecific translational inhibitor. The destabilizer is not a generic RNase, because it does not affect most other mRNAs. It is inactivated by micrococcal nuclease, suggesting that it contains a nucleic acid component, but is not inactivated by proteinase K. Perhaps it is a proteinase-resistant ribonucleoprotein.

A presumably different c-*myc* mRNA destabilizing factor has been purified from erythroleukemia cells by fractionating the S130 in a sucrose gradient (Brewer, 1991). The factor was initially identified by its ability to bind to an RNA fragment containing the AU-rich element (ARE) from the 3′ UTR of human c-*myc* mRNA. When the factor is mixed with polysomes in cell-free mRNA decay reactions, c-*myc* mRNA is destabilized, whereas globin mRNA stability is unaffected (Fig. 2). The factor forms stable complexes with the AREs from c-*myc* and colony stimulating factor mRNAs and with poly(U). Therefore, it was named Auf, or ARE-poly(U)-binding/degradation factor. Purified Auf consists of two proteins, 37 and 40 kDa. Treatment of purified Auf with proteinase K destroys the p40 component and the capacity of Auf to destabilize c-*myc* mRNA. Surprisingly, p37 is not degraded by proteinase K and is still able to bind to an ARE. Perhaps only p37 actually binds to the ARE, but both p37 and p40 are required to induce mRNA destabilization. It will be important to determine whether Auf destabilizes all or only some ARE-containing mRNAs and to assess the relationship, if any, between it and the cycloheximide-sensitive cytosolic destabilizer described above.

The *in vitro* half-life of urokinase-type plasminogen activator (uPA) mRNA is also shortened in response to a labile factor whose activity disappears from extracts prepared from cycloheximide-treated cells (Altus and Nagamine, 1991). A postmitochondrial supernatant fraction was prepared

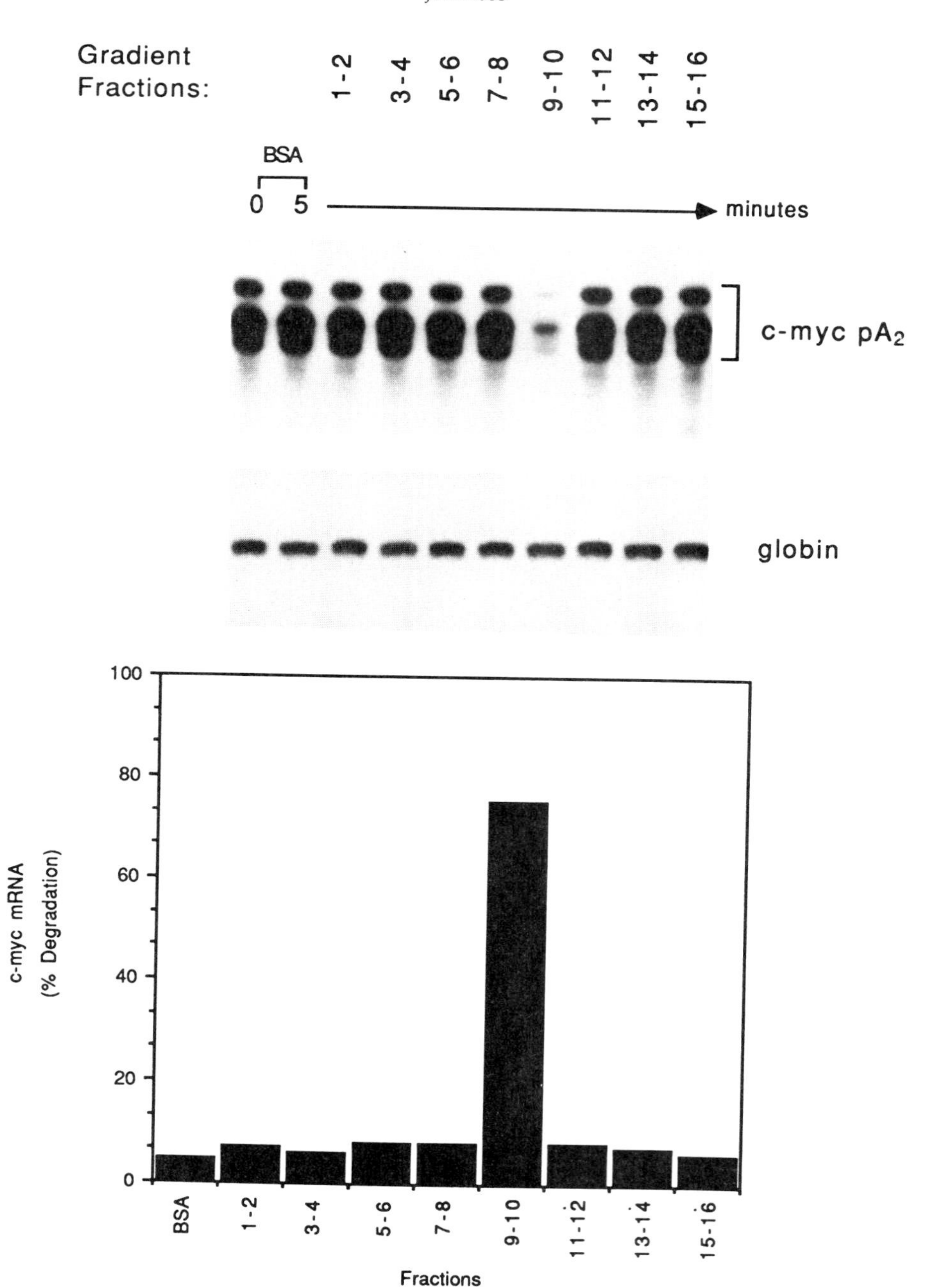
Gradient Fractions:
1-2
3-4
5-6
7-8
9-10
11-12
13-14
15-16
BSA
0 5
minutes
c-myc pA2
globin
c-myc mRNA (% Degradation)
100
80
60
40
20
0
BSA
1-2
3-4
5-6
7-8
9-10
11-12
13-14
15-16
Fractions

from LLC-PK_1 cells, which are derived from pig kidney. When the supernatant is incubated *in vitro* for varying times, the half-life of endogenous uPA mRNA is 0.5 hr (Fig. 3). This result is consistent with the observed short half-life of uPA mRNA in growing cells. When the supernatant is prepared from cells exposed for 60 to 90 min to cycloheximide, the *in vitro* half-life of uPA mRNA is greater than 20 hr, which again is consistent with observations in intact cells. Therefore, uPA mRNA stability is regulated by a labile factor whose activity depends on continuous protein synthesis. The factor does not appear to be a soluble protein.

Some indirect evidence suggests that a trans-acting factor influences c-*myc* mRNA stability by interacting with sequences in the 5′ UTR. A truncated, *in vitro* synthesized c-*myc* mRNA substrate, lacking most of the 5′ UTR, is degraded two- to threefold less rapidly than full-length exogenous c-*myc* mRNA in a cell-free system containing postnuclear supernatant from mouse plasmacytoma cells (Pei and Calame, 1988). Addition of competitor RNA containing the 5′ UTR prolongs the half-life of both substrates but affects the full-length one to a greater extent. In other words, the half-life ratios of truncated to full-length substrates are 2 : 1 to 3 : 1 without competitor and 1 : 1 with competitor. These data were interpreted to mean that a trans-acting factor destabilizes the full-length mRNA by binding within its 5′ UTR. However, deleting the 5′ UTR does not stabilize c-*myc* mRNA in intact cells (Bonnieu *et al.*, 1988), a finding that is inconsistent with the *in vitro* data.

2. *Interferon, RNase L, and the Stability of Viral mRNAs*

Viral infection induces interferon, which, in turn, stimulates the synthesis of 2′–5′-oligoadenylate [2′–5′(A)] synthetase. The synthetase is activated by double-stranded RNA and generates 2′–5′(A), which binds to and thereby activates a cellular endoribonuclease, RNase L (reviewed in Silverman *et al.*, 1988; Salehzada *et al.*, 1991). RNase L activation might occur at localized sites where viral RNA is being replicated, accounting for its presumed specificity for viral mRNA (Baglioni *et al.*, 1984). When postnuclear supernatant extracts from interferon-treated HeLa cells are incubated in the presence of 2′–5′(A), rRNA degradation is enhanced. In parallel

FIGURE 2 Effect of purified Auf on the decay of c-*myc* mRNA *in vitro*. The S130 (postribosomal supernatant) of exponentially growing K562 erythroleukemia cells was fractionated by three consecutive sucrose gradients, to obtain highly purified Auf. Pooled pairs of fractions from the third gradient were concentrated and added to mRNA decay reactions containing K562 cell polysomes. Following incubation for 0 or 5 min, RNA was extracted, and the amounts of c-*myc* and γ-globin mRNAs were determined by RNase protection assays. The data show that purified Auf is located in fractions 9 + 10. Auf specifically induces the destabilization of c-*myc* mRNA, because the stable control mRNA (γ-globin) is not degraded in reactions containing fractions 9 + 10. [Reprinted by permission of the American Society for Microbiology; from Brewer, (1991).]

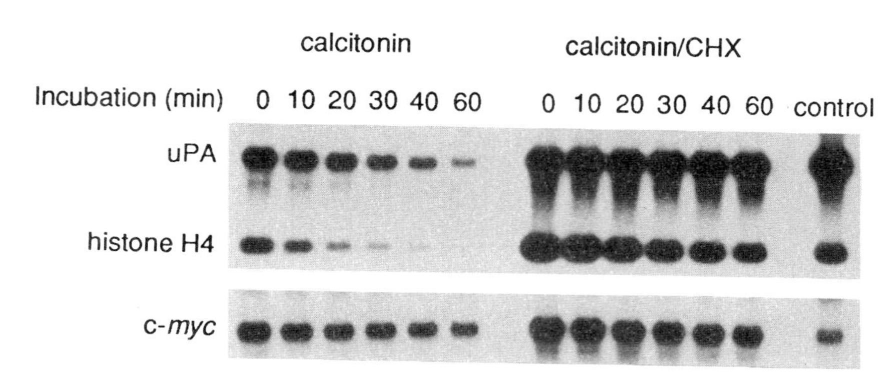

FIGURE 3 Regulation of urokinase-type plasminogen activator (uPA) mRNA stability *in vitro* by a labile regulatory factor. LLC-PK_1 tissue culture cells were treated with calcitonin, to induce uPA mRNA, in the presence or absence of cycloheximide. Postmitochondrial supernatants and polysomes were prepared and were incubated in separate cell-free mRNA decay reactions. RNA was extracted, electrophoresed in a denaturing gel, blotted to a nylon membrane, and hybridized to probes for uPA, histone, and c-*myc* mRNAs. The half-life of uPA mRNA differs by a maximum of twofold in reactions containing polysomes from control or cycloheximide-treated cells (Altus and Nagamine, 1991; data not shown here). In contrast, the uPA mRNA half-life is significantly shorter in reactions containing postmitochondrial supernatant from calcitonin-treated cells (left side). The mRNA is quite stable with postmitochondrial supernatant from cycloheximide-treated cells (right side), indicating that translational inhibition influences the activity of an mRNA stability regulatory factor with some specificity for uPA mRNA. [Reprinted by permission of the American Society for Biochemistry and Molecular Biology; from Altus and Nagamine (1991).]

reactions lacking exogenous 2′–5′(A), rRNA remains intact, suggesting that 2′–5′(A) activates an RNase activity, which is presumably RNase L. To determine whether viral mRNA is specifically targeted for degradation, the yield of reovirus RNA transcribed in extracts from control and interferon-treated cells was compared. The yield is lower in extracts from interferon-treated cells, probably because of enhanced degradation. The reduction in viral RNA requires 2′–5′(A), since inhibitors of 2′–5′(A) synthesis increase viral RNA levels. These results were interpreted to mean that RNase L activation plays a direct role in destabilizing nascent viral RNA. A model suggesting localized activation of RNase L was based on the observation that viral replication intermediates contain a long single-stranded region (template) plus a smaller duplex region (template–nascent RNA hybrid). The double-stranded region might activate 2′–5′(A) synthetase, leading to 2′–5′(A) production and RNase L activation at sites of viral RNA replication. As a result, the nascent, partially duplex RNA would be selectively destroyed.

In vitro systems are being used to assess the differential stability of specific viral mRNAs. Late mRNA of human papillomavirus type 16 (HPV 16) is apparently inherently unstable in infected cells, because of a sequence located in its 3′ UTR (Kennedy *et al.*, 1990). When protein-free late HPV mRNA is incubated with HeLa cell polysomes, it is destroyed within 30 min (Kennedy *et al.*, 1991). In contrast, late mRNA lacking the instability element is stable *in vitro* for at least 60 min. It would be most interesting to know whether the half-life of this mRNA is longer in differentiating keratinocytes than in undifferentiated skin epithelial cells, which are nonpermissive for HPV replication.

3. Host Cell mRNA Destabilization Induced by the Virion Host Shutoff (vhs) Gene of Herpes Simplex Virus

Almost every host mRNA is degraded soon after infection of cells with herpes simplex virus (HSV) types I or II (reviewed in Read and Frenkel, 1983; Fenwick and Everett, 1990; Smibert and Smiley, 1990). mRNA destabilization requires a virion component, the virion host shutoff (*vhs*) protein, since mutants with defects in the gene encoding this protein fail to induce mRNA destabilization at early times postinfection.

Independent experiments with different systems indicate that the HSV destabilizer is active *in vitro*. In one case, postnuclear supernatant was prepared by exposing cells briefly to lysolecithin, homogenizing the cells, and removing nuclei by low-speed centrifugation (Krikorian and Read, 1990). The half-lives of endogenous cell and viral mRNAs were compared in extracts from uninfected cells and from cells infected with wild-type HSV or HSV encoding a defective *vhs* protein. The half-life of glyceraldehyde-3-phosphate dehydrogenase (GAPDH) mRNA is greater than 4 hr in lysates from uninfected or mutant virus-infected cells but is at most 2 hr in lysates from cells infected with wild-type virus. Viral mRNAs and exogenous, protein-free mRNAs are also destabilized in extracts from virus-infected cells. Some translation occurs in these extracts, although it is unclear whether translation is required for destabilization. Destabilization occurs only in reactions containing relatively high magnesium concentrations (20 m*M*), and both heat and proteinase K inactivate the destabilizing activity.

In related experiments, polysomes and postpolysomal supernatant (S130) fractions were prepared from mock-infected and HSV-infected murine erythroleukemia cells (Sorenson *et al.*, 1991). Each fraction was then incubated with polysomes from uninfected human erythroleukemia cells, which served as a target for the destabilizer. Human γ-globin mRNA is stable for 10 hr or longer in reactions supplemented with S130 from uninfected cells, from mock-infected cells, or from cells infected with *vhs* mutant viruses. In contrast, it is almost completely degraded after 1 hr in reactions containing S130 from virus-infected cells (Fig. 4). The *vhs* protein, either partially purified from virions or synthesized by cell-free translation, also

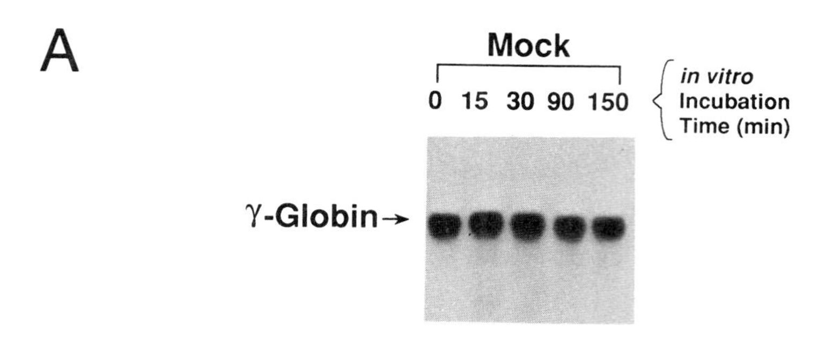

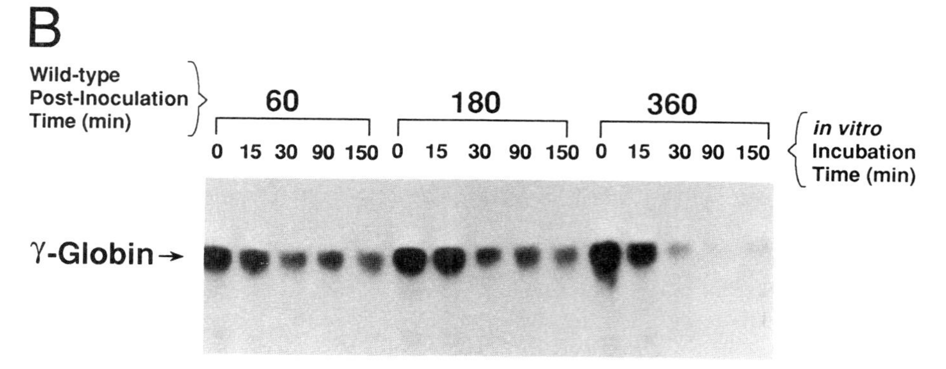

FIGURE 4 Analysis of the HSV-1-encoded mRNA destabilizing factor in the postpolysomal supernatant (S130) of infected cells. Polysomes from uninfected human cells were mixed with S130 from mock-infected (A) or HSV-infected (B) mouse cells and incubated in cell-free mRNA decay reactions for the indicated times. RNA was extracted and hybridized with an end-labeled probe for human γ-globin mRNA. Reactions were then treated with S1 nuclease, and nuclease-resistant fragments were electrophoresed. The times above the brackets in B indicate the times postinfection when the cells were harvested. The data indicate that the S130 of HSV-infected cells contains an activity capable of destabilizing an mRNA that, in uninfected cells, is very stable. [Reprinted by permission of Oxford University Press; from Sorenson *et al.*, (1991).]

induces mRNA destabilization in the presence of uninfected cell polysomes and S130 (C. Sorenson and J. Ross, unpublished observations). Since the protein is inactive without the S130, it presumably functions in conjunction with one or more cytosolic components.

The mechanism of this mRNA destabilization process is unknown. The *vhs* protein might activate a latent RNase, similar to the effect of interferon on RNase L. It might inactivate RNase inhibitors, permitting latent or nonspecific RNases to attack mRNAs indiscriminately. Alternatively, it might modify the structures of either the mRNAs or their associated proteins, thereby making them more susceptible to RNase attack. The *vhs* protein is not an RNase, because solubilized virions containing active *vhs* protein lack detectable RNase activity on deproteinized RNA substrates (J. Ross, unpublished observations). Regardless of its mechanism of action, it will be important to determine how *vhs* protein selectively targets mRNA for destruction while sparing rRNA and tRNA. Perhaps the *vhs* protein or a factor through which it acts recognizes distinctive features of mRNAs, such as their caps or poly(A) tracts.

4. mRNA Stabilization Factor Induced by Phorbol Esters

The pleiotropic effects of 12-tetradecanoylphorbol-13-acetate (TPA) on cell growth, differentiation, and carcinogenesis in mammalian tissue culture cell lines and in animal tissues, especially skin, are well documented (reviewed in Pitot, 1986). Primarily through their interactions with protein kinase C, TPA and related phorbol esters influence gene expression by changing gene transcription rates or by altering cytoplasmic mRNA half-lives. For example, TPA stabilizes cytokine mRNAs by as much as 40-fold (Thorens *et al.*, 1987).

TPA also stabilizes transforming growth factor-β1 (TGF-β1) mRNA by approximately 5-fold in the U-937 promonocyte cell line (Wager and Assoian, 1990). To determine the stabilization mechanism, postnuclear extracts were prepared from untreated and TPA-treated cells using a nonionic detergent and low-speed centrifugation. Total cell, deproteinized RNA was added to the extracts, which were then incubated for various times. TGF-β1 mRNA is degraded approximately 10-fold faster than GAPDH mRNA in extracts from untreated cells. More importantly, TGF-β1 mRNA is degraded approximately 10-fold less rapidly in extracts from TPA-treated cells than in extracts from untreated cells. If anything, GADPH and rRNA are degraded more rapidly in extracts from TPA-treated cells. Therefore, TPA induces a cytoplasmic stabilizer with some specificity for TGF-β1 mRNA. This result is consistent with the long-term increase in TGF-β1 expression associated with TPA-induced differentiation of hematopoietic cells and cell lines. It will be important to fractionate the extract in order to characterize the stabilizing factor(s).

5. Ribonuclease Inhibitors, Nonspecific RNases, and mRNA Stability in Vitro

Most, if not all, mammalian cells contain ribonuclease inhibitors, whose functions are unknown. The best characterized of these are the 50-kDa proteins that have been purified from human placenta and rodent liver and that inhibit the RNase A family of ribonucleases (Blackburn and Jailkhani, 1979; Lee *et al.*, 1989). The functions of those inhibitors in cells are unknown, but the following experiments suggest that they block nonspecific RNases from indiscriminately degrading cellular mRNAs *in vitro*. Deproteinized

RNAs incubated with polysomes or crude cytoplasmic extracts of tissue culture cells are degraded rapidly and nonspecifically (Stolle and Benz, 1988; Bandyopadhyay *et al.*, 1990). However, when a whole cell extract or a high-speed cytosolic supernatant fraction is added to the reactions, the RNAs are stabilized, implying the existence of a potent RNase inhibitor in the cytosol. Globin mRNA is approximately 40-fold more stable than c-*myc* mRNA in intact cells, but both mRNAs are rapidly degraded in cell-free reactions containing polysomes but lacking any RNase inhibitors (J. Ross and G. Kobs, unpublished observations; Ross and Kobs, 1986). In contrast, in the presence of human placental RNase inhibitor, globin mRNA is very stable *in vitro*, whereas c-*myc* mRNA is degraded rapidly. These results raise two interesting questions. What properties of mRNases render them insensitive to at least some RNase inhibitors? What are the functions of the "nonspecific" RNases and of the RNase inhibitors that block their activity?

V. Future Directions

It seems that *in vitro* mRNA decay systems will be useful for determining which enzymes degrade mRNAs, how trans-acting factors regulate mRNA stability, and how growth factors, hormones, viral infection, heat shock, carcinogens, and various environmental stimuli regulate the regulators. Since the *in vitro* turnover of some mRNAs is trans-influenced by mixing fractionated extracts from different cells (Section IV.C), at least some regulatory factors maintain their activity in cell extracts and should be amenable to purification.

A. Identification and Characterization of Messenger Ribonucleases

Various cytoplasmic RNases have been and will continue to be purified, but it is important to determine which are authentic mRNases and which are nonspecific nucleases capable of degrading mRNAs *in vitro* but have no role in intracellular mRNA decay. Investigators should be able to exploit cell-free mRNA decay systems to distinguish authentic mRNases from nonspecific nucleases. For example, the initial steps in histone mRNA degradation are apparently catalyzed by a 3′–5′ exonuclease that is bound to polysomes or to mRNP isolated from cells homogenized in buffers containing low to moderate salt. This enzyme is released from polysomes in active form by high salt, is magnesium-dependent, and is insensitive to placental RNase inhibitor (Ross *et al.*, 1987). In contrast, other nucleases in the ribosomal salt wash are inhibited by the RNase inhibitor. The role of any highly purified 3′–5′ exonuclease in histone mRNA turnover can be assessed by determining whether it requires magnesium and is resistant to placental RNase inhibitor and by asking whether antibodies that block its enzymatic activity also inhibit histone mRNA decay *in vitro*. Specificity can be confirmed by showing that the decay of other mRNAs that are normally attacked by endonucleases is unaffected by the anti-exonuclease antibody.

After cloning mRNase genes, it might be interesting to ask whether overexpression of specific mRNases influences cell growth and/or differentiation. The role of mRNA turnover in cell differentiation is particularly intriguing, since the half-lives of some mRNAs change by 10-fold or more soon after undifferentiated cells are exposed to differentiation factors (reviewed in Shapiro *et al.*, 1987). Will overproduction of an mRNase accelerate or depress the differentiation process?

B. Elucidation of mRNA Decay Pathways

mRNA decay intermediates are so short-lived in intact cells that it is difficult to delineate mRNA decay pathways. Since degradation processes generally occur more slowly *in vitro* than in cells, it should be possible to exploit *in vitro* systems to identify mRNA decay intermediates. Such experiments will have two important consequences: (1) They might focus attention on a specific region of an mRNA as a likely site of mRNAse attack. For example, *in vitro* assays first identified 3′-terminal decay intermediates of histone mRNA (Ross and Kobs, 1986). Having a small 3′ region on which to focus, probes were designed specifically to detect the same intermediates in intact cells (Ross *et al.*, 1986). (2) By following mRNA decay pathways *in vitro*, it should be possible to confirm which enzymes carry out which steps. If an mRNA were degraded sequentially, first by exonuclease attack on the poly(A) tract and then by endonuclease attack within the mRNA body, it should be feasible to analyze each step separately using RNase-specific antibodies.

C. Analysis of the Relationship between mRNA Stability and Translation

The half-lives of many eukaryotic mRNAs increase when protein synthesis is inhibited, but the stabilization mechanism is poorly understood. Is stabilization cis-induced by blocking mRNA translocation or by altering mRNA or polysome structure? Is it trans-induced by activating RNase inhibitors or by inactivating RNases or mRNA destabilizing factors, as postulated for the translation-dependent cytosolic factor that facilitates the degradation of c-*myc* mRNA?

The answer will probably vary, depending on the mRNA. For example, tubulin mRNA half-life is autoregulated by tubulin monomers, but only when the mRNA is associated with polysomes and is being actively translated (Gay *et al.*, 1989) (see Chapter 10). As the nascent tubulin peptide emerges from the ribosome, its first four amino acids might interact somehow with tubulin monomers. This interaction, in turn, might activate a ribosome-associated mRNase, leading to tubulin mRNA breakdown. *In vitro,* histone mRNAs, like tubulin mRNA, are autoregulated (destabilized) by the proteins they encode. However, histone autoregulation seems to occur independent of the translational elongation of histone mRNA (Peltz and Ross, 1987). According to one model, histone proteins accumulate transiently in the cytoplasm at the end of the S-phase, and, after reaching a sufficient level, induce histone mRNA degradation. Exposure of growing cells to cycloheximide does result in the stabilization of ribosome-associated histone mRNA. However, the model suggests that stabilization occurs not because histone mRNA translocation along the polysome is blocked. Rather, it occurs because histone proteins (the required trans-acting factors) do not accumulate to a sufficient level to trigger autoregulation in cycloheximide-treated cells. c-*myc* mRNA is also stabilized in exponentially growing cells exposed to cycloheximide. In this case, *in vitro* data suggest that stabilization results from the inactivation of a constitutive cytosolic destabilizing factor, not from a direct (cis-acting) effect of cycloheximide on c-*myc* mRNA translation (Brewer and Ross, 1989).

It thus seems that translational inhibitors can stabilize mRNAs in various ways, and *in vitro* mRNA decay systems will probably be useful in sorting out the details. For example, it is frequently noted that c-*myc* mRNA levels increase in cycloheximide-treated cells because the mRNA is "stabilized", implying the induction of an mRNA stabilization factor. However, the *in vitro* data cited above suggest that c-*myc* mRNA levels rise in cycloheximide-treated cells because a destabilizing factor is inactivated. The distinction between these two mechanisms is not trivial.

It is also important to distinguish the notion of ribosome association from translational elongation. The stability of a particular mRNA might be regulated only when elongation or translocation of the mRNA occurs, as suggested for tubulin autoregulation. Conversely, histone mRNA regulation might require only that the mRNA be ribosome-associated, not that it be elongated. It has been difficult to interpret unambiguously the results of translation arrest experiments in intact cells, because the primary effect (cis-, trans-, or both) of a translational inhibitor on mRNA half-life usually cannot be determined with absolute certainty. At sufficient levels, cycloheximide slows or stops mRNA translocation, but it also blocks the synthesis of trans-acting factors that might regulate mRNA stability. Therefore, stabilization of an mRNA in cycloheximide-treated cells could occur because the mRNA is not being translated (a cis-effect) or because a separate regulatory factor is not synthesized (a trans-effect).

In vitro mRNA decay systems offer a way to resolve some of these dilemmas, because the effects of ribosome association, elongation, and trans-acting factors on mRNA stability can be investigated separately. Therefore, it will be important to assess which *in vitro* reaction conditions permit both efficient translation and mRNA turnover to occur. The most promising systems, the lysolecithin and reticulocyte extracts, might also be useful for investigating why mRNAs with nonsense codons are less stable than their wild-type counterparts. Experiments in transgenic mice indicate that the destabilization of globin mRNAs with nonsense mutations occurs in the cytoplasm and that premature translational termination might activate a ribosome-bound endonuclease (Lim *et al.*, 1989; L. Maquat, personal communication). Perhaps *in vitro* systems can be used to decipher the connection between premature termination and mRNA instability.

D. Correlation between mRNA Stability and Cytoplasmic Location

Several experiments have documented how the cytoplasmic location of an mRNA can influence its stability. Immunoglobulin mRNA is approximately fivefold more stable in plasma cells than in B cells, in part because a significantly larger percentage of the mRNA is associated with membrane-bound polysomes in the plasma cell (Mason *et al.*, 1988). If the leader peptide sequence is altered, so that most of the mRNA remains on "free" polysomes in the plasma cell, the mRNA is not stabilized. Since the quantity of endoplasmic reticulum increases significantly as B cells differentiate to plasma cells, stabilization during B cell differentiation might correlate with the percentage of immunoglobulin mRNA associated with the endoplasmic reticulum. Histone mRNA is normally associated with "free" polysomes and is destabilized when DNA synthesis ceases. However, when the mRNA is targeted to membrane-bound polysomes, it remains stable when DNA synthesis is interrupted (Zambetti *et al.*, 1987).

In vitro systems might be useful for analyzing these localization effects, especially if it proves feasible to link mRNA stability and translation. Perhaps the existing systems can be modified to reflect more accurately the intracellular conditions to which an mRNA is normally exposed. For example, by adding pancreatic microsomes to reticulocyte extracts, the *in vitro* half-life of an mRNA that is translated and associated with endoplasmic reticulum can be investigated (Baker and Lai, 1990).

E. Identification and Characterization of Trans-Acting mRNA Stability Regulators

It should be possible to use *in vitro* systems to identify and characterize trans-acting factors that influence cell growth, differentiation, and neoplastic transformation by modulating the half-lives of specific mRNAs. Several such factors have already been detected. Others, some of which are RNA-binding proteins, might regulate both translation and mRNA turnover. For example, it will be most interesting to know whether the AU-binding factor stabilizes mRNAs, as suggested by the correlation between its level and the stabilities of certain mRNAs (Malter and Hong, 1991). Based on their studies with PAN in yeast extracts, Sachs and Lowell have suggested that RNA-binding proteins might influence mRNA stability by attracting nonspecific RNases to or repelling them from specific sites within the mRNA (Lowell *et al.*, 1992) . In other words, protein binding would generate "landmarks" for the RNases. This idea is especially attractive, because it designates functions for both the binding proteins and the nonspecific RNases of the cell. *In vitro* systems can be exploited to characterize how these factors might interact with the mRNA decay machinery.

Cell-free systems can also be used to characterize gene products that are first identified genetically as regulators of mRNA function and metabolism. As discussed in Section IV.B, genetic experiments in yeast and cell-free experiments with a mammalian mRNA decay system have provided interesting, albeit conflicting, data on how PABP influences poly(A) shortening, translation, and mRNA turnover. Additional interesting yeast mutants are becoming available for resolving these questions and for answering others. For example, cells which overexpress a double-strand specific RNase fail to conjugate and sporulate (Xu *et al.*, 1990; Iino *et al.*, 1991), while cells deficient in a protein encoded by the *UPF1* gene do not rapidly degrade mRNAs with premature translation termination codons (Leeds *et al.*, 1991) (see Chapter 13). Do these or related proteins influence mRNA stability in cell-free extracts? If so, it might be feasible to correlate the mutant phenotype with a specific biochemical defect and, thereby, to combine observations in genetically tractable organisms with analyses of mRNA turnover *in vitro*.

References

Adams, C. C., and Stern, D. B. (1990). Control of mRNA stability in chloroplasts by 3′ inverted repeats: Effects of stem and loop mutations on degradation of *psbA* mRNA *in vitro*. *Nucleic Acids Res.* **18,** 6003–6010.

Altus, M. S., and Nagamine, Y. (1991). Protein synthesis inhibition stabilizes urokinase-type plasminogen activtor mRNA: Studies *in vivo* and in cell-free decay reactions. *J. Biol. Chem.* **266,** 21190–21196.

Baglioni, C., de Benedetti, A., and Williams, G. J. (1984). Cleavage of nascent reovirus mRNA by localized activation of the 2′–5′-oligoadenylate-dependent endoribonuclease. *J. Virol.* **52,** 865–871.

Baker, S. C., and Lai, M. M. C. (1990). An *in vitro* system for the leader-primed transcription of coronavirus mRNAs. *EMBO J.* **9,** 4173–4179.

Bandyopadhyay, R., Coutts, M., Krowczynska, A., and Brawerman, G. (1990). Nuclease activity associated with mammalian mRNA in its native state: possible basis for selectivity in mRNA decay. *Mol. Cell. Biol.* **10,** 2060–2069.

Bandziulis, R. J., Swanson, M. S., and Dreyfuss, G. (1989). RNA-binding proteins as developmental regulators. *Genes Dev.* **3,** 431–437.

Bernstein, P., and Ross, J. (1989). Poly(A), poly(A)-binding protein and the regulation of mRNA stability. *Trends Biochem. Sci.* **14,** 373–377.

Bernstein, P. L., Herrick, D. J., Prokipcak, R. D., and Ross, J. (1992). Control of c-*myc* mRNA half-life *in vitro* by a protein capable of binding to a coding region stability determinant. *Genes Dev.*, **6,** 642–654.

Bernstein, P., Peltz, S. W., and Ross, J. (1989). The poly(A)–poly(A)-binding protein complex is a major determinant of mRNA stability *in vitro*. *Mol. Cell. Biol.* **9,** 659–670.

Blackburn, P., and Jailkhani, B. L. (1979). Ribonuclease inhibitor from human placenta: Interaction with derivatives of ribonuclease A. *J. Biol. Chem.* **254,** 12488–12493.

Blume, J. E., and Shapiro, D. J. (1989). Ribosome loading, but not protein synthesis, is required for estrogen stabilization of *Xenopus laevis* vitellogenin mRNA. *Nucleic Acids Res.* **17,** 9003–9014.

Bohjanen, P. R., Petryniak, B., June, C. H., Thompson, C. B., and Lindsten, T. (1991). An inducible cytoplasmic factor (AU-B) binds selectively to AUUUA multimers in the 3′ untranslated region of lymphokine mRNA. *Mol. Cell. Biol.* **11,** 3288–3295.

Bonnieu, A., Piechaczyk, M., Marty, L., Cuny, M., Blanchard, J.-M., Fort, P., and Jeanteur, P. (1988). Sequence determinants of c-*myc* mRNA turn-over: Influence of 3′ and 5′ non-coding regions. *Oncogene Res.* **3,** 155–166.

Brawerman, G. (1989). mRNA decay: Finding the right targets. *Cell* **57,** 9–10.

Brewer, G. (1991). An A + U-rich element RNA-binding factor regulates c-*myc* mRNA stability *in vitro*. *Mol. Cell. Biol.* **11,** 2460–2466.

Brewer, G., and Ross, J. (1988). Poly(A) shortening and degradation of the 3′ AU-rich sequences of human c-*myc* mRNA in a cell-free system. *Mol. Cell. Biol.* **8,** 1697–1708.

Brewer, G., and Ross, J. (1989). Regulation of c-*myc* mRNA stability *in vitro* by a labile destabilizer with an essential nucleic acid component. *Mol. Cell. Biol.* **9,** 1996–2006.

Brock, M. L., and Shapiro, D. J. (1983). Estrogen stabilizes vitellogenin mRNA against cytoplasmic degradation. *Cell* **34,** 207–214.

Brown, G. D., Peluso, R. W., Moyer, S. A., and Moyer, R. W. (1983). A simple method for the preparation of extracts from animal cells which catalyze efficient *in vitro* protein synthesis. *J. Biol. Chem.* **258,** 14309–14314.

Capasso, O., Bleecker, G. C., and Heintz, N. (1987). Sequences controlling histone H4 mRNA abundance. *EMBO J.* **6,** 1825–1831.

Casey, J. L., Koeller, D. M., Ramin, V. C., Klausner, R. D., and Harford, J. B. (1989). Iron regulation of transferrin receptor mRNA levels requires iron-responsive elements and a rapid turnover determinant in the 3′ untranslated region of the mRNA. *EMBO J.* **8,** 3693–3699.

Cleveland, D. W., and Yen, T. J. (1989). Multiple determinants of eukaryotic mRNA stability. *New Biol.* **1,** 121–126.

Cochrane, A. W., and Deeley, R. G. (1988). Estrogen-dependent activation of the avian very low density apolipoprotein II and vitellogenin genes: Transient alterations in mRNA polyadenylation and stability during induction. *J. Mol. Biol.* **203,** 555–567.

Cohrs, R. J., Goswami, B. B., and Sharma, O. K. (1988). Occurrence of 2-5A and RNA degradation in the chick oviduct during rapid estrogen withdrawal. *Biochemistry* **27,** 3246–3252.

Coutts, M., and Brawerman, G. (1993a). A 5′ exonuclease from cytoplasmic extracts of mouse sarcoma 180 ascites cells. *Biochim. Biophys.* Acta (in press).

Coutts, M., Krowczynska, A., and Brawerman, G. (1993b). Protection of mRNA against nucleases in cytoplasmic extracts of mouse sarcoma ascites cells. *Biochim. Biophys. Acta* (in press).

Drummond, D. R., Armstrong, J., and Colman, A. (1985). The effect of capping and polyadenylation on the stability, movement and translation of synthetic messenger RNAs in *Xenopus* oocytes. *Nucleic Acids Res.* **13,** 7375–7394.

Ernest, M. J. (1982). Regulation of tyrosine aminotransferase messenger ribonucleic acid in rat liver: Effect of cycloheximide on messenger ribonucleic acid turnover. *Biochemistry* **21,** 6761–6767.

Fenwick, M. L., and Everett, R. D. (1990). Inactivation of the shutoff gene (UL41) of herpes simplex types 1 and 2. *J. Gen. Virol.* **71,** 2961–2967.

Furuichi, Y., LaFiandra, A., and Shatkin, A. J. (1977). 5′-terminal structure and mRNA stability. *Nature (London)* **266,** 235–239.

Gay, D. A., Sisodia, S. S., and Cleveland, D. W. (1989). Autoregulatory control of β-tubulin mRNA stability is linked to translation elongation. *Proc. Natl. Acad. Sci. USA* **86,** 5763–5767.

Goldberg, M. A., Gaut, C. C., and Bunn, H. F. (1991). Erythropoietin mRNA levels are governed by both the rate of gene transcription and posttranscriptional events. *Blood* **77,** 271–277.

Gordon, D. A., Shelness, G. S., Nicosia, M., and Williams, D. L. (1988). Estrogen-induced destabilization of yolk precursor protein mRNAs in avian liver. *J. Biol. Chem.* **263,** 2625–2631.

Graves, R. A., Pandey, N. B., Chodchoy, N., and Marzluff, W. F. (1987). Translation is required for regulation of histone mRNA degradation. *Cell* **48,** 615–626.

Harris, M. E., Bohni, R., Schneiderman, M. H., Ramamurthy, L., Schumperli, D., and Marzluff, W. F. (1991). Regulation of histone mRNA in the unperturbed cell cycle: Evidence suggesting control at two posttranscriptional steps. *Mol. Cell. Biol.* **11,** 2416–2424.

Hepler, J. E., Van Wyk, J. J., and Lund, P. K. (1990). Different half-lives of insulin-like growth factor I mRNAs that differ in length of 3′-untranslated sequence. *Endocrinology* **127,** 1550–1552.

Hsu-Ching, C., and Stern, D. B. (1991). Specific ribonuclease activities in spinach chloroplasts promote mRNA maturation and degradation. *J. Biol. Chem.,* **266,** 24205–24211.

Iino, Y., Sugimoto, A., and Yamamoto, M. (1991). *S. pombe* pac1 +, whose overexpression inhibits sexual development, encodes a ribonuclease III-like RNase. *EMBO J.* **10,** 221–226.

Jäck, H.-M., Berg, J., and Wabl, M. (1989). Translation affects immunoglobulin mRNA stability. *Eur. J. Biochem.* **19,** 843–847.

Kennedy, I. M., Haddow, J. K., and Clements, J. B. (1990). Analysis of human papillomavirus type 16 late mRNA 3′ processing signals *in vitro* and *in vivo*. *J. Virol.* **64,** 1825–1829.

Kennedy, I. M., Haddow, J. K., and Clements, J. B. (1991). A negative regulatory element in the human papillomavirus 16 genome acts at the level of late mRNA stability. *J. Virol.* **65,** 93–97.

Khochbin, S., Principaud, E., Chabanas, A., and Lawrence, J.-J. (1988). Early events in murine erythroleukemia cells induced to differentiate: Accumulation and gene expression of the transformation-associated cellular protein p53. *J. Mol. Biol.* **200,** 55–64.

Krikorian, C. R., and Read, G. S. (1990). *In vitro* mRNA degradation system to study the virion host shutoff function of herpes simplex virus. *J. Virol.* **65,** 112–122.

Lee, F. S., Fox, E. A., Zhou, H.-M., Strydom, D. J., and Vallee, B. L. (1989). Primary structure of human placental ribonuclease inhibitor. *Biochemistry* **27,** 8545–8553.

Leeds, P., Peltz, S. W., Jacobson, A., and Culbertson, M. R. (1991). The product of the yeast *UPF1* gene is required for rapid turnover of mRNAs containing a premature translational termination codon. *Genes Dev.,* **5,** 2303–2314.

Liang, H., and Jost, J.-P. (1991). An estrogen-dependent polysomal protein binds to the 5′ untranslated region of the chicken vitellogenin mRNA. *Nucleic Acids Res.* **19,** 2289–2294.

Lim, S., Mullins, J. J., Chen, C.-M., Gross, K. W., and Maquat, L. E. (1989). Novel metabolism of several β^{o}-thalassemic β-globin mRNAs in the erythroid tissues of transgenic mice. *EMBO J.* **8,** 2613–2619.

Lowell, J. E., Rudner, D. Z., and Sachs, A. B. (1992). 3′-UTR-dependent deadenylation by the yeast poly(A) nuclease. *Genes Dev.* **6,** 2088–2099.

Malter, J. S., and Hong, Y. (1991). A redox switch and phosphorylation are involved in the post-translational up-regulation of the adenosine–uridine binding factor by phorbol ester and ionophore. *J. Biol. Chem.* **266,** 3167–3171.

Mason, J. O., Williams, G. T., and Neuberger, M. S. (1988). The half-life of immunoglobulin mRNA increases during B-cell differentiation: A possible role for targeting to membrane-bound polysomes. *Genes Dev.* **2,** 1003–1011.

McCrae, M. A., and Woodland, H. R. (1981). Stability of non-polyadenylated viral mRNAs injected into frog oocytes. *Eur. J. Biochem.* **116,** 467–470.

Mercer, J. F. B., and Wake, S. A. (1985). An analysis of the rate of metallothionein poly(A)-shortening using RNA blot hybridization. *Nucleic Acids Res.* **13,** 7929–7943.

Minvielle-Sebastia, L., Winsor, B., Bonneaud, N., and Lacroute, F. (1991). Mutations if the yeast *rna14* and *rna15* genes result in an abnormal mRNA decay rate: Sequence analysis reveals an RNA-binding domain in the rna15 protein. *Mol. Cell. Biol.* **11,** 3075–3087.

Müllner, E. W., and Kühn, L. C. (1988). A stem–loop in the 3′ untranslated region mediates iron-dependent regulation of transferrin receptor mRNA stability in the cytoplasm. *Cell* **53,** 815–825.

Pastori, R. L., Moskaitis, J. E., and Schoenberg, D. R. (1991). Estrogen-induced ribonuclease activity in *Xenopus* liver. *Biochemistry,* **30,** 10490–10498.

Pei, R., and Calame, K. (1988). Differential stability of c-*myc* mRNAs in a cell-free system. *Mol. Cell. Biol.* **8,** 2860–2868.

Peltz, S. W., Brewer, G., Bernstein, P., Hart, P. A., and Ross, J. (1991). Regulation of mRNA turnover in eukaryotic cells. *In* Critical Reviews in Eukaryotic Gene Expression (G. S. Stein, J. L. Stein, and J. B. Lian, Eds.), pp. 99–126. CRC Press, Ann Arbor, Michigan.

Peltz, S. W., Brewer, G., Kobs, G., and Ross, J. (1987). Substrate specificity of the exonuclease activity that degrades H4 histone mRNA. *J. Biol. Chem.* **262,** 9382–9388.

Peltz, S. W., and Ross, J. (1987). Autogenous regulation of histone mRNA decay by histone proteins in a cell-free system. *Mol. Cell Biol.* **7,** 4345–4356.

Piper, P. W., and Aamand, J. L. (1989). Yeast mutation thought to arrest mRNA transport markedly increases the length of the 3′ poly(A) on polyadenylated RNA. *J. Mol. Biol.* **8,** 697–700.

Pitot, H. C. (1986). Fundamentals of Oncology. Dekker, New York.

Ratnasabapathy, R., Hwang, S.-P. L., and Williams, D. L. (1990). The 3′-untranslated region of apolipoprotein II mRNA contains two independent domains that bind distinct cytosolic factors. *J. Biol. Chem.* **265,** 14050–14055.

Read, G. S., and Frenkel, N. (1983). Herpes simplex virus mutants defective in the virion-associated shutoff of host polypeptide synthesis and exhibiting abnormal synthesis of alpha (immediate early) viral polypeptides. *J. Virol.* **46,** 498–512.

Regnier, P., and Hajnsdorf, E. (1991). Decay of mRNA encoding ribosomal protein S15 of *Escherichia coli* is initiated by an RNase E-dependent endonucleolytic cleavage that removes the 3′ stabilizing stem and loop structure. *J. Mol. Biol.* **217,** 283–292.

Ross, J., and Kobs, G. (1986). H4 histone messenger RNA decay in cell-free extracts initiates at or near the 3′ terminus and proceeds 3′ to 5′. *J. Mol. Biol.* **188,** 579–593.

Ross, J., Kobs, G., Brewer, G., and Peltz, S. W. (1987). Properties of the exonuclease activity that degrades H4 histone mRNA. *J. Biol. Chem.* **262,** 9374–9381.

Ross, J., Peltz, S. W., Kobs, G., and Brewer, G. (1986). Histone mRNA degradation *in vivo:* The first detectable step occurs at or near the 3′ terminus. *Mol. Cell. Biol.* **6,** 4362–4371.

Ross, J., and Pizarro, A. (1983). Human beta and delta globin messenger RNAs turn over at different rates. *J. Mol. Biol.* **167,** 607–617.

Sachs, A. (1990). The role of poly(A) in the translation and stability of mRNA. *Curr. Opinion Cell Biol.* **2,** 1092–1098.

Sachs, A. B., and Davis, R.W. (1989). The poly(A) binding protein is required for poly(A) shortening and 60S ribosomal subunit-dependent translation initiation. *Cell* **58,** 857–867.

Sachs, A. B., and Davis, R. W. (1990). Translation initiation and ribosomal biogenesis: Involvement of a putative rRNA helicase and rpL46. *Science* **247,** 1077–1079.

Sachs, A. B., and Deardorff, J. A. (1992). Translation initiation requires the PAB-dependent poly(A) ribonuclease in yeast. *Cell* **70,** 961–973.

Salehzada, T., Silhol, M., Lebleu, B., and Bisbal, C. (1991). Polyclonal antibodies against RNase L: Subcellular localization of this enzyme in mouse cells. *J. Biol. Chem.* **266,** 5808–5813.

Schwartz, R. J. (1973). Control of glutamine synthetase synthesis in the embryonic chick neural retina: A caution on the use of actinomycin D. *J. Biol. Chem.* **248,** 6426–6435.
Shapiro, D. J., Blume, J. E., and Nielsen, D. A. (1987). Regulation of messenger RNA stability in eukaryotic cells. *BioEssays* **6,** 221–226.
Shimotohno, K., Kodama, Y., Hashimoto, J., and Miura, K. (1977). Importance of 5′-terminal blocking structure to stabilize mRNA in eukaryotic protein synthesis. *Proc. Natl. Acad. Sci., USA* **74,** 2734–2738.
Silverman, R. H., Jung, D. D., Nolan-Sorden, Dieffenbach, C. W., Kedar, V. P., and SenGupta, D. N. (1988). Purification and analysis of murine 2-5A-dependent RNase. *J. Biol. Chem.* **263,** 7336–7341.
Smibert, C. A., and Smiley, J. R. (1990). Differential regulation of endogenous and transduced β-globin genes during infection of erythroid cells with a herpes simplex virus type 1 recombinant. *J. Virol.* **64,** 3882–3894.
Sorenson, C. M., Hart, P. A., and Ross, J. (1991). Analysis of herpes simplex virus-induced mRNA destabilizing activity using an *in vitro* mRNA decay system. *Nucleic Acids Res.* **19,** 4459–4465.
Steinberg, R. A., Levinson, B. B., and Tomkins, G. M. (1975). "Superinduction" of tyrosine aminotransferase by actinomycin D: A reevaluation. *Cell* **5,** 29–35.
Stern, D. B., and Gruissem, W. (1987). Control of plastid gene expression: 3′ inverted repeats act as mRNA processing and stabilizing elements, but do not terminate transcription. *Cell* **51,** 1145–1157.
Stern, D. B., Jones, H., and Gruissem, W. (1989). Function of plastid mRNA 3′ inverted repeats: RNA stabilization and gene-specific protein binding. *J. Biol. Chem.* **264,** 18742–18750.
Swartwout, S. G., and Kinniburgh, A. J. (1989). c-*myc* RNA degradation in growing and differentiating cells: Possible alternate pathways. *Mol. Cell. Biol.* **9,** 288–295.
Thorens, B., Mermod, J., and Vassalli, P. (1987). Phagocytosis and inflammatory stimuli induce GM-CSF mRNA in macrophages through posttranscriptional regulation. *Cell* **48,** 671–679.
Wager, R. E., and Assoian, R. K. (1990). A phorbol ester-regulated ribonuclease system controlling transforming growth factor β1 gene expression in hematopoietic cells. *Mol. Cell. Biol.* **10,** 5893–5990.
Wisdom, R., and Lee, W. (1991). The protein-coding region of c-*myc* mRNA contains a sequence that specific rapid mRNA turnover and induction by protein synthesis inhibitors. *Genes Dev.* **5,** 232–243.
Wreschner, D. H., and Rechavi, G. (1988). Differential mRNA stability to reticulocyte ribonucleases correlates with 3′ non-coding $(U)-_nA$ sequences. *Eur. J. Biochem.* **172,** 333–340.
Xu, H.-P., Rodgers, L., and Wigler, M. (1990). A gene from *S. pombe* with homology to *E. coli* RNase III blocks conjugation and sporulation when overexpressed in wild type cells. *Nucleic Acids Res.* **18,** 5304.
Zambetti, G., Stein, J., and Stein, G. (1990). Role of messenger RNA subcellular localization in the posttranscriptional regulation of human histone gene expression. *J. Cell. Physiol.* **144,** 175–182.

17

Eukaryotic Nucleases and mRNA Turnover

AUDREY STEVENS

I. Introduction

mRNA turnover is a process that plays a major role, along with mRNA transcription and processing, in determining the mRNA and protein levels of cells. Although many proteins involved in transcription and processing (e.g., adenylation) have been described, specific enzymes involved in mRNA turnover remain largely a mystery. The purpose of this chapter is to describe the characterization of selected RNases that may be involved in mRNA turnover. They were chosen largely on the basis of specificity, and most of those described have been extensively purified. Only short descriptions of the rationale for their playing a role in turnover are presented, since other chapters of this book describe features of turnover pathways that implicate certain types of enzymes.

Figure 1 shows three basic pathways of turnover of the mRNA chain and cap structure and the type(s) of RNases that may be involved. Pathways 1 and 2 involve hydrolysis of the mRNA chains by exoribonucleases, yielding 5′-mononucleotides. Both types of exoribonucleases, those having a 5′→3′ mode and those having a 3′→5′ mode of hydrolysis, are described in Section III. In pathway 1, it is possible that the mRNA cap structure would have to be removed before exoribonuclease (5′→3′) hydrolysis could occur; therefore, mRNA decapping enzymes are described first in Section II. Pathway 3 involves initial endonucleolytic cleavage(s) with mRNA fragment production, followed by further degradation (probably exoribonucleolytic) of the resulting fragments. Although many endoribonucleases have been described, it is difficult to relate them to the mRNA turnover process;

1. 5′→3′ Exoribonuclease Hydrolysis of mRNA Chain

Step 1: Hydrolysis of cap structure

$$m^7G(5')pp{\downarrow}pNpN(pN)_n \xrightarrow[\text{decapping enzyme}]{\text{mRNA}} pp(m^7G) + pNpN(pN)_n$$

Step 2: Exoribonuclease hydrolysis of chain

$$pN\overset{1}{\downarrow}pN\overset{2}{\downarrow}(pN)_n \xrightarrow[\text{exoribonuclease}]{5'\text{-}3'} n(pN)$$

2. 3′→5′ Exoribonuclease Hydrolysis of mRNA Chain

Step 1: Exoribonuclease hydrolysis of chain

$$m^7G(5')pppN(pN)_n\overset{2}{\downarrow}pN\overset{1}{\downarrow}pN \xrightarrow[\text{exoribonuclease}]{3'\text{-}5'} n(pN) + m^7G(5')pppN$$

Step 2: Hydrolysis of cap structure

$$m^7G(5')pppN \longrightarrow p(m^7G) + ppN \text{ or } pp(m^7G) + pN$$

3. Initial Endonucleolytic Cleavage(s) of mRNA Chain

Step 1: Endonucleolytic cleavage of chain (at one or possibly many sites)

$$m^7G(5')pppN(pN)_npN{\downarrow}pN(pN)_n \xrightarrow{\text{endonuclease}} m^7G(5')pppN(pN)_npN + pN(pN)_n$$

Step 2: 5′→3′ Exoribonuclease hydrolysis of fragments (as in pathway 1)

or 3′→5′ exoribonuclease hydrolysis of fragments (as in pathway 2)

FIGURE 1 Possible pathways of mRNA chain turnover.

therefore, only two reported endonuclease activities are described in Section IV. mRNA chains contain poly(A) tails, which have for some time been postulated to play a role in mRNA stability (see references below). Deadenylation is not included in the pathways shown in Fig. 1, but it may be a first step in the turnover of some mRNAs. Recently discovered enzymes that specifically remove poly(A) termini are described in Section V, and they may play a key role in mRNA turnover.

More than one pathway of mRNA turnover may be found in cells, possibly involving the same mRNA species and responding to specific stimuli (Shyu *et al.*, 1989; Swartwout and Kinniburgh, 1989). As suggested by some of the studies described below, a specific mRNA sequence may control an initial inactivating event, such as deadenylation, decapping, or endonucleolytic cleavage. The modification may render the mRNA chain subject to further hydrolysis by several types of RNases.

II. Hydrolysis of mRNA Cap Structures

A 5′-terminal cap structure, $m^7G(5')ppp(5')N$ (m^7GpppN), is a feature of almost all eukaryotic mRNAs (see review articles by Shatkin, 1976; Banerjee, 1980). The cap structure is important in the process of translation (Shatkin, 1985), and, as described below (Section III.A), its specific or indiscriminate removal may labilize mRNAs to the action of 5′→3′ exoribonucleases. Hydrolysis of the cap structure is essential for complete turnover of the mRNA chain, and enzymes that may catalyze the hydrolysis are described here. These enzymes and their use in cap structure analysis have recently been reviewed by Furuichi and Shatkin (1989). Figure 2 shows the cap structure and the sites of hydrolysis by the enzymes described here.

A. HeLa Cell m^7G-Specific Pyrophosphatase

Nuss *et al.* (1975) reported the detection in HeLa cell extracts of a 7-methylguanosine-specific enzyme activity that cleaves m^7GpppN^m or m^7GpppN to m^7GMP and NDP (N^mDP). GpppG and the ring-opened derivative of m^7GpppG^m were not hydrolyzed. m^7GpppG^m at the 5′ end of intact reovirus mRNA was not cleaved, but oligonucleotides of chain length up to 10 nucleotides containing m^7GpppG^m at the 5′ end were substrates.

In a subsequent report, Nuss and Furuichi (1977) characterized the HeLa cell enzyme further. Results of DEAE–cellulose chromatography and nondenaturing polyacrylamide gel electrophoresis (PAGE) strongly suggested that HeLa cell extracts contain only one activity capable of hydrolyzing m^7GpppN to yield m^7GMP and NDP (where N is an unmethylated or 2′-0-methylated ribonucleoside, or an oligonucleotide of up to 8 to 10 nucleotides in length). Centrifugal analysis of the extracts showed that about 95% of the activity was in the cytoplasmic fraction, and it appeared not to be associated with lysosomes. About 25% of the activity was ribosome-associated. The authors stated that m^7GpppN- cleaving activity was also detectable in cell-free extracts prepared from wheat embryos, *Artemia salina*, mouse lymphocytes, and rabbit reticulocytes.

Analysis of the physical properties of a DEAE–cellulose enzyme fraction from HeLa cells by centrifugation and molecular-sieve chromatography showed that the enzyme has a sedimentation coefficient of approximately

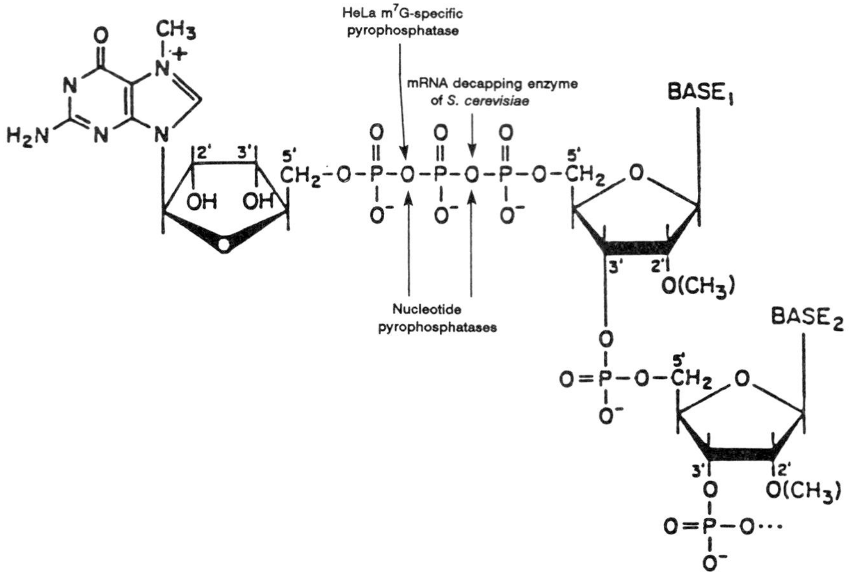

FIGURE 2 mRNA cap structure, showing sites of pyrophosphatase cleavage. Capped mRNAs all contain m^{7}G, but the ribose residues of the initial nucleotides of the RNA chain are methylated to different extents on the 2′-OH. The following cap nomenclature is used: m^{7}GpppN (cap 0); m^{7}GpppN^m (cap 1); m^{7}GpppN^{m}pN^m (cap 2); m^{7}GpppN^{m}pN^{m}pN^m (cap 3); m^{7}GpppN^{m}pN^{m}pN^{m}pN^m (cap 4).

4.9 S and a Stokes radius of 3.9 nm, leading to a calculated value of 81,000 as the approximate molecular weight.

No divalent cation requirement was found. The enzyme shows activity over a broad pH range from 4.5 to 10, with an optimum at 7.5. It exhibits a low K_m for m^{7}GpppN (1.7 μM).

Further studies of substrate specificity suggested that the enzyme recognizes the N^7 methyl group or at least the partial positive charge at the N^7 position of the guanosine moiety of m^{7}GpppN. GpppG and the ring-opened derivative of m^{7}GpppG^m are not hydrolyzed by the partially purified enzyme. Prospective substrates methylated at the N^7 position are very effective inhibitors of m^{7}GpppG hydrolysis. m^{7}GDP and m^{7}GTP inhibit the cleavage by 50% at a concentration equal to that of m^{7}GpppN. However, m^{2,2,7}GpppN, the 5′-terminal structure of snRNA in most eukaryotic cells (Busch *et al.*, 1982), is not hydrolyzed. GMP, GDP, NAD, PPi, GTP, GpppG, and GppppG inhibit only at very high concentrations (5–200$\times 10^3$ times higher than that of m^{7}GpppN).

To investigate the importance of the 2-amino group for substrate recognition, a comparison was made of the inhibitory properties of guanosine, inosine, 7-methylguanosine, and 7-methylinosine. Neither guanosine nor inosine is inhibitory at concentrations up to 10 mM, 7-methylguanosine showed an I_{50} of approximately 25 μM, and 7-methylinosine is not inhibitory. The results showed that the 2-amino residue may also play a role in substrate recognition by the enzyme.

The HeLa cell m^7G-specific pyrophosphatase may be involved in the hydrolysis of cap structures and capped oligonucleotides resulting from mRNA turnover by pathways 2 or 3 in Fig. 1. It is possible that it may cleave m^7GMP from intact mRNA bound in polysomes *in vivo*. Certainly, further studies to investigate this possibility would be interesting. Also, whether a similar enzyme is found in most eukaryotic cells has not been clearly shown, and characterization of more enzymes that hydrolyze cap structures may show interesting additional features.

B. mRNA Decapping Enzyme of *Saccharomyces cerevisiae*

Stevens (1980) described the detection of a pyrophosphatase that hydrolyzes mRNA cap structures in a partially purified high-salt wash fraction of ribosomes of *S. cerevisiae*. [^{3}H]Methyl-5′-capped mRNA of yeast was used as a substrate for the enzyme fraction, which catalyzed hydrolysis to [^{3}H]m^7GDP. This product was not hydrolyzed further, nor were $m^7GpppA(G)$ and UDP glucose, suggesting that the enzyme is a unique pyrophosphatase. Upon hydroxylapatite chromatography of high-salt wash fractions of yeast ribosomes, the decapping enzyme elutes closely with 5′→3′ exoribonuclease activity [described as a second exoribonuclease by Larimer and Stevens (1990) and discussed briefly in Section III.A below]. It can be separated from the exoribonuclease activity by heparin–agarose chromatography, and its overall purification is as much as 10,000-fold (Stevens, 1988).

Centrifugal and molecular-sieve chromatographic analyses showed that the enzyme has a sedimentation coefficient of 5.0 S and a Stokes radius of 3.8 nm, allowing an estimate of 79,000 as the apparent molecular weight. The optimum pH for activity of the decapping enzyme is 7.5 to 8.5. A divalent cation is required for activity; EDTA inhibits the reaction completely. Mg^{2+} stimulates maximally at 1 mM and Mn^{2+}, 85% as well, at 1 mM. The K_m for yeast mRNA is 12.5 μM.

Use of various capped RNA substrates with the yeast enzyme led to the following observations: (1) Yeast mRNA treated with high concentrations of RNase A, nuclease P1, or micrococcal nuclease is inactive as a substrate. (2) Gel electrophoretic analysis of reaction mixtures shows that long RNA chains (a synthetic capped RNA of 540 nucleotides was used) are substrates and that the RNA is not degraded before or after decapping. The products of the reaction are m^7GDP and 5′-pRNA. (3) Treatment of a synthetic capped

RNA (540 nucleotides) with increasing concentrations of RNase A showed a decline in activity with shortening of the RNA chain. The use of synthetic capped RNAs of different sizes (50 to 540 nucleotides) as substrates showed a significant decrease in decapping rate with a decrease in chain length. (4) GpppG–RNA is hydrolyzed at the same rate as m^{7}GpppG–RNA; thus, the enzyme is not m^{7}G-specific. (5) The pyrophosphate bonds of a synthetic RNA containing a 5′-triphosphate end group were not hydrolyzed.

The yeast mRNA decapping enzyme was detected in partially purified extracts of yeast during the purification of a 5′→3′ exoribonuclease, as described in Section III.A below. It is possible that a similar decapping activity is also present in higher eukaryotic cells, but that the total RNase activity of crude extracts has precluded its detection. That the decapping enzyme purifies closely with a 5′→3′ exoribonuclease could be of considerable interest, since two such enzymes may be involved in mRNA turnover as shown in Fig. 1, pathway 1. It will be of interest to compare the decapping rates of yeast mRNAs with different turnover rates, analyzing the effect of sequence elements, particularly destabilizing elements (Herrick *et al.*, 1990; Parker and Jacobson, 1990), on the decapping rate.

C. Nucleotide Pyrophosphatases

Several nonspecific pyrophosphatases have been reported to cleave the pyrophosphate linkages of the mRNA cap structure. Snake venom nucleotide pyrophosphatase is often used for structural analysis (see Furuichi and Shatkin, 1989), but it is contaminated with traces of phosphodiesterase activity. Nucleotide pyrophosphatases purified from tobacco tissue culture cells (Shinshi *et al.*, 1976a, 1976b) and potato (Kole *et al.*, 1976; Bartkiewicz *et al.*, 1984) are nuclease-free and hydrolyze the mRNA cap structure without cleaving the RNA chain. The products of the reaction with the nucleotide pyrophosphatases are m^{7}GMP, Pi, and 5′-pRNA.

The potato and tobacco nucleotide pyrophosphatases have been purified to homogeneity. Both have a molecular weight of 75,000 as determined by SDS–PAGE. They also hydrolyze pyrophosphate groups at the 3′- and 5′-OH of nucleosides and phosphodiester linkages in synthetic aryl esters of 5′- and 3′-NMP and phosphate. Both are most active at pH 6.0 and do not require a divalent cation. A careful kinetic analysis of the substrate specificity of the potato enzyme shows that it exhibits highest affinities for NAD, A(5′)pp(5′)A, A(5′)ppp(5′)A, and m^{7}GpppG^m. Other substrates such as ATP, dATP, and dT(5′)pp(5′)dT have about a 10-fold higher K_m. It is possible that if eukaryotic cells lack a specific decapping enzyme active on mRNA or on short 5′-terminal capped mRNA structures, the nonspecific nucleotide pyrophosphatases may be involved in the hydrolysis of cap structures.

III. Exoribonucleases

As shown in pathways 1 and 2 of Fig. 1, exoribonucleases may be involved in pathways of mRNA turnover by being the primary catalyst, by hydrolyzing the entire mRNA chain, or by acting as a secondary catalyst, hydrolyzing mRNA fragments produced by initial endonucleolytic cleavage. It is likely that the exoribonucleases described below may be very important in the turnover process.

A. 5′→3′ Exoribonucleases

The first evidence for the occurrence of enzymes hydrolyzing RNA by a 5′→3′ mode with the production of 5′-mononucleotides was provided by Furuichi *et al.* (1977) and Shimotohno *et al.* (1977). Furuichi *et al.* (1977) found that GpppG- and m^7GpppG^m-capped reovirus mRNAs are considerably more stable than mRNAs with unblocked termini when injected into *Xenopus laevis* oocytes or incubated in cell-free protein-synthesizing extracts of wheat germ and mouse L cells. Since GpppG-capped mRNA was as stable as m^7GpppG^m-capped mRNA, the authors concluded that the greater stability did not depend on translation, but seemed to result from protection against 5′-exonucleolytic degradation. With UMP-labeled RNA, 5′-UMP was identified as the major product with the wheat germ extracts. Shimotohno *et al.* (1977) showed that several mRNAs lacking the 5′-cap structure are less stable than their capped counterparts when incubated with wheat germ extracts and that a high-speed supernatant fraction has the same discriminatory hydrolytic activity. Again, mononucleotides were the primary products.

Shimotohno and Miura (1977) briefly reported a 20-fold purification of an enzyme fraction from wheat germ extracts that hydrolyzed 5′-mono, di-, or triphosphorylated RNAs, but not 5′-capped RNA or 5′-OH RNA, to 5′-mononucleotides. A 5′-terminal label was released more rapidly than a 3′-terminal label, suggesting a 5′→3′ mode of hydrolysis. No further description of the wheat germ enzyme has appeared.

The first 5′→3′ exoribonuclease to be highly purified was from *S. cerevisiae* (Stevens, 1978). In recent studies, described below, it has been designated XRN1 (eXoRiboNuclease 1). Several 5′→3′ exoribonucleases have subsequently been purified from nuclear fractions. Lasater and Eichler (1984) isolated and described a 5′→3′ exoribonuclease from Ehrlich ascites tumor cell nucleoli, and Stevens and Maupin (1987a) described such an enzyme purified from human placenta nuclei. More recently, Murthy *et al.* (1991) described a HeLa cell nuclear enzyme. The enzymes isolated from nuclear fractions are not described here. The presence of cytoplasmic 5′→3′ exoribonucleases is suggested by the results showing greater lability of decapped mRNA upon injection into *Xenopus* oocytes [as discussed earlier

(Furuichi *et al.*, 1977)] and by the finding (Peltz *et al.*, 1987) of the instability of decapped mRNAs with a cytoplasmic extract of Freund erythroleukemia cells. Existing data suggest that XRN1 is a cytoplasmic activity, and it is now described in more detail.

Stevens (1978) reported that XRN1 purified 200-fold from a high-salt wash fraction of ribosomes of *S. cerevisiae* hydrolyzed capped RNAs and 5′-OH RNAs considerably more slowly than RNAs with 5′-phosphate termini. 5′-Mononucleotides were the products. A subsequent report (Stevens, 1979) showed that rRNA is hydrolyzed by XRN1 in a 5′→3′ mode as assayed with 5′- and 3′-terminally labeled molecules. XRN1 was then highly purified (1000-fold), and its mode of hydrolysis was studied in detail using poly(A), rRNA, and oligo(A) as substrates (Stevens, 1980). The enzyme has a broad pH optimum around 8.0, requires a divalent cation, and is stimulated by monovalent cations. The degree of stimulation by particular cations is dependent on the substrate used. With either poly(A) or rRNA as substrate, the enzyme has a processive mode of hydrolysis. The oligonucleotides $(pA)_{3-5}$ are hydrolyzed by the enzyme, and the hydrolysis is dependent on a 5′-phosphate end group. Phosphorylation of the 3′ end has little effect on the rate of hydrolysis. Further evidence for a 5′→3′ mode of hydrolysis comes from a paper chromatographic analysis of the products of the hydrolysis of $[^3H](pA)_5$ labeled at the 5′ terminus with ^{32}P. The results of the experiment are shown in Fig. 3. No ^{32}P label appears in the $[^3H](pA)_2$ that accumulates as an intermediate, showing that the hydrolysis proceeds with 5′→3′ polarity.

Further studies (Stevens and Maupin, 1987b) using either denaturing PAGE or physical characterization showed that XRN1 has a molecular weight of about 160,000. It had little or no activity with single- or double-stranded T5 DNA as a substrate, but synthetic polydeoxyribonucleotides were strong competitive inhibitors of the hydrolysis of synthetic ribopolymers. The enzyme hydrolyzed the poly(A) moiety of poly(A)–poly(U) and poly(A)–poly(dT) at about 50% of the rate of poly(A).

As a step toward determining the metabolic role of XRN1, a yeast gene designated *XRN1* was first cloned and then disrupted to determine the effect on yeast cell growth (Larimer and Stevens, 1990). The studies showed that the gene is not essential, but that its absence markedly affects the yeast cell doubling time. Sequencing of the *XRN1* gene showed an open reading frame of 4584 bp, capable of encoding 1528 amino acids (Larimer *et al.*, 1992; Larimer, GenBank Accession No. M62423). By comparison with GenBank sequences, it was found that the *XRN1* gene is identical to the yeast DST2 gene (Dykstra *et al.*, 1991), encoding a protein with DNA strand exchange activity (Dykstra *et al.*, 1990). The report on the *DST2* gene appeared with a paper describing the cloning and characterization of the *SEP1* gene (Tishkoff *et al.*, 1991), also encoding a DNA strand exchange protein (Kolodner *et al.*, 1987), and the two genes are identical. Sequence data also

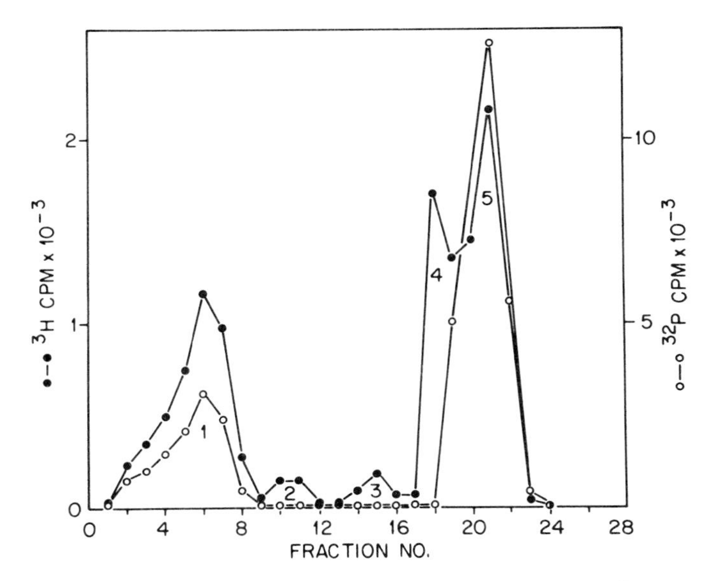

FIGURE 3 Paper chromatography of the products of the hydrolysis of $[^3H](pA)_5$, labeled at the 5′ terminus with ^{32}P, by XRN1 of *S. cerevisiae*. The experiment, including the paper chromatographic analysis of the hydrolysis products, was as described in Fig. 6 of Stevens (1980). Peaks 1 to 5, $(pA)_5$, $(pA)_4$, $(pA)_3$, $(pA)_2$, and 5′-AMP, respectively.

showed that the *KEM1* gene, described by Kim *et al.* (1990) and affecting yeast nuclear fusion, is the same. A recent report (Kipling *et al.*, 1991) indicates that the *RAR5* gene affecting plasmid stability is also identical. A commentary on the gene and its cloning in five laboratories on the basis of different functional characteristics has recently appeared (Kearsey and Kipling, 1991).

In studies of their protein, Johnson and Kolodner (1991) have found that it also has low-level 5′→3′ DNase activity. RNase activity was not measured at that time, but a comparison of the RNase and DNase activities (A. Johnson and R. Kolodner, personal communication) now shows that single-stranded RNA is the best substrate (single-stranded RNase/single-stranded DNase activity on homopolymers = 10–15). It is difficult to discuss the significance of the different enzymatic activities found to be associated with the same protein at this time since the studies are still incomplete.

Studies of the role of the enzyme in RNA metabolism and mRNA turnover are in progress. The portion of the internal transcribed spacer-1 found on 20S pre-rRNA accumulates in yeast lacking XRN1 (Stevens *et al.*, 1991), strongly suggesting that it is a cytoplasmic RNase involved in the hydrolysis of the pre-rRNA spacer fragment. Results show that the levels and rates of synthesis of protein bands resolved by one-dimensional PAGE are in substantial disarray (Larimer *et al.*, 1992). These findings as well as measurements of several specific mRNA levels by Northern blot analysis suggest a disparity in mRNA levels. Results show longer (x2–4) half-lives of specific short-lived mRNAs (*CYC1, RP51A, MFα1*). Further studies with *XRN1* gene disruptants (C. L. Hsu and A. Stevens, unpublished data) show that cells lacking XRN1 have a high level (40–75%) of deadenylated mRNA species (based on Northern blot analysis of RNA fractions able or unable to bind to oligo(dT)–cellulose) compared with wild-type cells (5–20%). The results suggest that mRNA turnover is affected in the *XRN1* gene disruptants, but it is not possible to say whether the turnover is affected as a primary or secondary consequence of loss of the enzyme. The confusing array of phenotypes found to be associated with the *XRN1* gene-deficient yeast could result from XRN1 being involved in mRNA turnover.

A second 5′→3′ exoribonuclease of yeast was detected during the purification of the mRNA decapping enzyme described above (Stevens, 1988). It was also detected in the *XRN1*-gene-deficient yeast (Larimer and Stevens, 1990), but it has not been as highly characterized as XRN1. The protein is encoded by an essential gene that bears homology to *XRN1* (Kenna *et al.*, in press).

A search of cytoplasmic extracts of mouse sarcoma 180 ascites cells for nuclease activity has led to the finding and partial purification of an enzyme that hydrolyzes RNA to 5′-mononucleotides with faster release of label from the 5′ terminus than from the 3′ terminus (Coutts *et al.*, in press; Coutts and Brawerman, in press). It is the predominant nuclease in the extracts, being resistant to an RNase inhibitor found in the extracts. Experimental findings include: (1) The enzyme releases only labeled 5′-CMP from a synthetic RNA labeled with [α-^{32}P]CTP. (2) 3′-[^{32}P]CMP end label of 5S RNA is released as the mononucleotide while 5′-label is released as the 5′-terminal mononucleotide (5′-UMP) and the 5′-terminal dinucleotide (pUpG). (Two minor products remain unidentified.) The results suggest that the second (and possibly, third) phosphodiester bond from the 5′ terminus is cleaved as well as, or better than, the first. (3) 5′-pUpG label of 5S RNA is released faster than 3′-CMP label. (4) 5′-capped RNA and RNA with a 5′-triphosphate end group are hydrolyzed at close to the same rate as RNA with a 5′-phosphomonoester end group, cleavage occurring in both cases predominantly at the second phosphodiester bond from the 5′ end. Coutts and Brawerman feel that the enzyme may be a unique 5′→3′ exoribonuclease that could play a role in mRNA turnover by directly hydrolyzing capped mRNA molecules from the 5′ terminus and that the resistance of capped mRNA to hydrolysis that they have observed in crude extracts may be due to certain mRNA binding proteins not specific for m^{7}GpppG.

Further investigations of the incidence and type of 5′→3′ exoribonucleases present in cytoplasmic fractions are very important in that they may determine the exact role that the enzymes may play in mRNA turnover.

The purification from yeast extracts of an mRNA decapping enzyme, yielding mRNA chains with 5′-phosphate termini, and of 5′→3′ exoribonucleases requiring such a 5′ structure led to the suggestion (Stevens, 1988) that two such enzymes could together play a key role in mRNA turnover. Higher eukaryotes may contain a 5′→3′ exoribonuclease as just described for ascites cells, with capped mRNA being hydrolyzed when protective factors (such as binding proteins) are absent.

B. 3′→5′ Exoribonucleases

Exonucleases that degrade RNA by a 3′→5′ mode, yielding 5′-mononucleotides, have been highly purified from nuclei (Lazarus and Sporn, 1967) and nucleoli (Eichler and Eales, 1985). Such enzymes have not been highly purified from the cytoplasm of eukaryotic cells; however, studies by Ross and Kobs (1986) of H4 histone mRNA decay using an *in vitro* polysome system from the human erythroleukemia cell line K562 have led to the identification of an apparent exonuclease acting at the 3′ terminus of histone mRNA.

The studies of Ross *et al.* (1986) showed that, both in cells and in the cell-free decay system, the first detectable step in histone mRNA decay is at the 3′ terminus (see Chapter 16). With the polysome system (Ross and Kobs, 1986), the decay of endogenous histone mRNA, analyzed by S1 nuclease mapping using suitable probes, occurred by shortening of the chains at the 3′ terminus. The largest of the detectable decay products was only 6–15 nucleotides smaller than the intact mRNA. Five or six additional discrete decay products that had lost 30–90 nucleotides from the 3′ terminus appeared after longer incubation times. The smallest products observed were 90–100 nucleotides shorter, and no additional products were observed. Decay of *in vitro* synthesized H4 histone mRNA was analyzed, and the authors concluded that degradation also begins at or near the 3′ terminus. The studies suggested that the decay of the histone mRNA results from the action of a 3′→5′ exoribonuclease, although an endonuclease that cleaves small segments in the 3′→5′ direction was not excluded. If an exoribonuclease is involved, the gradual shortening of the chains with generation of discrete smaller mRNA species would mean that pause sites are present.

The properties and substrate specificity of the 3′→5′ degrading activity were described in subsequent reports (Ross *et al.*, 1987; Peltz *et al.*, 1987). An RNase inhibitor, such as that from placenta, was needed in these assays to block nonspecific RNase activity and to permit different mRNAs to be degraded at different rates that correlated with *in vivo* turnover rates. For degradation of endogenous histone mRNA, divalent cation (Mg^{2+}) is required, and the enzyme is active at monovalent cation concentrations ranging from 0.5 to 200 m*M*. The enzyme is bound to ribosomes isolated from low-salt cell extracts, but it is eluted from the ribosomes by exposing them to 0.3 *M* KCl. The high-salt wash extract rapidly degrades exogenous histone mRNA.

Analysis of the substrate specificity of the polysome-associated enzyme was carried out using five synthetic RNAs lacking poly(A) at the 3′ terminus (including histone mRNA) and one (histone mRNA) containing poly(A) (Peltz *et al.*, 1987). The results of the studies can be summarized as follows: (1) The five mRNAs lacking poly(A) are all degraded rapidly. (2) Polyadenylated histone mRNA is degraded at 10% of the rate of unmodified histone mRNA. (3) Single-stranded DNA is degraded at about 5% of the rate of RNA lacking poly(A). Double-stranded DNA and RNA are very stable. (4) RNA with a 3′-phosphate group is degraded at about 5% of the rate of RNA containing a 3′-hydroxyl group. The latter finding is in accord with the enzymatic activity being a 3′→5′ exoribonuclease.

Using the polysome high-salt wash fraction depleted of poly(A) binding protein (PAB) either by adding excess poly(A) competitor or by passing the extracts over a poly(A)–Sepharose column, Bernstein *et al.* (1989) investigated the effect of poly(A) and the poly(A) binding protein (PAB) on mRNA turnover. They showed that polyadenylated mRNAs for β-globin, chloramphenicol acetyltransferase, and simian virus 40 virion protein are degraded 3 to 10 times faster in reactions lacking PAB than in those containing it. The poly(A) tract was the first mRNA segment to be degraded in the PAB-depleted reactions, but the remainder of the RNA was also then degraded. The decay rates of nonpolyadenylated RNAs were unaffected by PAB. The results suggest that PAB binding to poly(A) blocks the 3′→5′ exonuclease, which, in the absence of PAB, degrades the entire RNA molecule including the poly(A) tract.

Further purification of the purported exoribonuclease from the high-salt wash fraction of the K562 polysome fraction should be possible, allowing very interesting studies of its mode of degradation. Searches for 3′→5′ exoribonucleases need to be done with other cytoplasmic extracts, so that more such enzymes can be purified and characterized.

IV. Endonucleases

Although cells contain many endonuclease activities, none has been definitely correlated with mRNA turnover. As shown in Fig. 1, pathway 3, endonuclease cleavage may be the first step that occurs in turnover, followed by exonucleolytic action on the free termini of the resulting fragments. The fragments may also be degraded by an endoribonuclease that lacks specificity and yields mononucleotides as final products. That fragments of mRNA are not readily detected *in vivo* suggests that, if endonucleolytic cleavage is the first step in mRNA turnover, subsequent degradation is rapid; however, fragments possibly resulting from endonucleolytic cleavage have been found with several mRNAs. Albrecht *et al.* (1984) found that β-globin mRNA fragments are detectable in rabbit reticulocytes. More recently, in the cases of apolipoprotein II mRNA (Binder *et al.*, 1989), a maternal homeo-box mRNA (Brown and Harland, 1990), 9E3 mRNA (Stoeckle and Hanafusa, 1989), and IGF-II mRNA (Meinsma *et al.*, 1991), fragments both 3′ and 5′ to a localized cleavage site were identified. The cleavage sites in these mRNAs were in or adjacent to the 3′ untranslated region (3′ UTR). The studies suggest that endonucleolytic hydrolysis plays a role in some mRNA turnover pathways.

Destabilizing sequences described in a recent review (Cleveland and Yen, 1989) also suggest possible endonuclease action. The most intensely studied instability element is found in mRNAs for protooncogenes, lymphokines, and cytokines. These mRNAs have very short half-lives and contain long (30–80 nucleotide) AU-rich domains in the 3′ UTR (Shaw and Kamen, 1986). The sequences may be subject to endonuclease cleavage or involved in the control of endonucleolytic cleavage. Studies of deadenylation (see later) suggest that they could also act by binding proteins that trigger enzymes acting at the termini of mRNA chains.

A. mRNP-Bound RNase

Bandyopadhyay *et al.* (1990) described RNase activity associated with polysome and messenger ribonucleoprotein particles (mRNP) from mouse sarcoma 180 (S-180) ascites cells and rabbit reticulocytes. The activity remained associated with polysomes after a high-salt wash and with mRNPs released from polysomes by treatment with EDTA. That no effect on the rate of endogenous mRNA degradation was found with serial dilution of the fractions suggested that the RNase(s) is closely associated with the mRNA. The activity was inhibited by the cytosolic fraction, which appeared to contain an inhibitor similar to the placental RNase inhibitor. The latter protein also inhibited the mRNP-associated RNase activity, suggesting that the activity is endonucleolytic. Cleavage of exogenous RNA (5S rRNA and β-globin mRNA) also occurred. With purified reticulocyte polysomes, endogenous β-globin mRNA was cleaved at sites containing an AU dinucleotide sequence in the 3′-terminal region. Similar sites were previously reported as possible *in vivo* cleavage sites (Albrecht *et al.*, 1984). Cleavages at the same sites occurred when isolated reticulocyte mRNA was incubated with S-180 polysomes. There were additional cleavages in the 5′-terminal region under the latter conditions. Examination of cleavage sites in endogenous mRNAs following incubation with S-180 high-salt-washed polysomes showed that a P40 mRNA was cleaved primarily in the 3′ UTR, while a P21 mRNA was cleaved in the 5′ noncoding region only. Actin mRNA was cleaved in an internal region. The results suggest an mRNA-bound endonuclease activity that cleaves mRNAs at specific sites. It is possible that exoribonuclease digestion of one or more fragments resulting from limited endonuclease action could have generated some of the fragments detected in these studies.

With the *in vitro* polysome system described above under Section III.B, Brewer and Ross (1988) found that following deadenylation, c-*myc* mRNA hydrolysis generated decay products with 3′ termini located within the AU-rich region of the 3′ UTR. In contrast, the 5′-terminal region was stable. A destabilizing RNA binding factor that may regulate the stability of c-*myc* mRNA was described by Brewer and Ross (1989) and Brewer (1991).

Sunitha and Slobin (1987) found that RNase activity is associated with mRNP particles of Friend erythroleukemia cells. A divalent cation-dependent RNase could be washed off mRNPs with 0.5 *M* NaCl, and the mRNAs associated with the salt-washed fraction were stable.

More *in vitro* studies should be done using synthetic substrates containing a purported cleavage site previously identified *in vivo* (see earlier for a description of such sites). Enzymes that cleave specifically at a discrete site might be found.

B. RNase L

One mediator of interferon action is RNase L, a $p_n(A2'p)_nA[2',5'$-oligo(A)]-dependent endoribonuclease. Since my purpose here is to describe the purification and characterization of the enzyme, many of the studies related to its action will not be described; however, there is further information in review articles that have appeared in the last 10 years (Lengyel, 1982; Baglioni, 1984; Lengyel, 1987). RNase L is involved in viral mRNA degradation in RNA-virus-infected cells and may, under specific conditions and/or with specific mRNAs, be involved in the turnover of cellular mRNA. RNase L is present in all mammalian cells examined (Nilsen *et al.*, 1981), but needs activation by 2′,5′-oligo(A) (see review articles just cited). The 2′,5′-oligo(A) synthetases that catalyze the synthesis of the oligo(A) are activated by double-stranded RNA, and there are several types of these synthetases in cells (see Chebath *et al.*, 1987, and references therein), making an overall analysis of the function of RNase L difficult. A large variety of stimuli result in an increase in the level of 2′, 5′-oligo(A) synthetase(s) in cells, including heat shock treatment, glucocorticoids, growth arrest, and differentiation (Lengyel, 1987; Johnston and Torrence, 1984; Krause *et al.*, 1985). It is possible that an increase in the level of one of the oligo(A) synthetases and/or of double-stranded RNA structures in cellular mRNA may render certain mRNAs susceptible to degradation by RNase L (Dani *et al.*, 1985).

RNase L activity was first described by Brown *et al.* (1976), and many studies deal with its purification and properties. Extensive purification from mouse Ehrlich ascites tumor cells was reported (Floyd-Smith and Lengyel, 1986), and the enzyme was recently purified 5×10^5-fold to homogeneity by Silverman *et al.* (1988) from mouse liver extracts. The purification involved using an affinity resin consisting of $(A2'p)_3A$ covalently attached to cellulose and also electroelution from a polyacrylamide gel. The enzyme is found in the postribosomal supernatant. It has a molecular weight of 80,000, and the data of Silverman *et al.* (1988) suggest that it may be the only 2′,5′-oligo(A)-dependent RNase in mouse liver, in contrast to the multiple species of 2′,5′-oligo(A) synthetases.

RNase L is latent unless activated by 2′,5′-oligo(A). The oligo(A) must be a trimer or tetramer to activate RNase L (Williams *et al.*, 1979; Martin *et al.*, 1979), and other structural features affecting activity have been carefully studied (Krause *et al.*, 1986). Substitution of a 3′–5′ phosphodiester bond for one of the 2′–5′ bonds of $2',5'\text{-oligo(A)}_3$ greatly reduces the activation of RNase L (Lesiah *et al.*, 1983). The activation of RNase L by 2′,5′-oligo(A) is reversible (Slattery *et al.*, 1979). The 2′,5′-oligo(A) binding site of the mouse liver enzyme has a K_a of $2.5 \times 10^{10}\ M^{-1}$ (Silverman *et al.*, 1988).

The sequence specificity of RNase L was determined by Floyd-Smith *et al.* (1981) and Wreschner *et al.* (1981) using both synthetic polyribonucleotides and natural RNAs. Floyd-Smith *et al.* reported that RNase L purified from Ehrlich ascites tumor cells cleaved RNA most frequently after UA, UG, and UU. Wreschner *et al.* (1981) showed that RNase L in extracts of rabbit reticulocytes, mouse ascites tumor cells, and human lymphoblastoid cells cleaved on the 3′ side of UN sequences to yield UpNp-terminated products. The predominant cleavages were at UA and UU sequences.

A second murine 2′,5′-oligo(A)-dependent endoribonuclease having a size of 40 kDa has recently been described (Bisbal *et al.*, 1989). Whether the second enzyme is a breakdown product of RNase L remains unclear.

Agonists and antagonists (2′,5′-oligo(A) analogs) are being used to activate or inhibit cellular RNase L activity in order to evaluate its biological role (Johnston and Torrence, 1984; Watling *et al.*, 1985; Bisbal *et al.*, 1987). The many facets of the RNase L story make this a difficult task.

V. mRNA Deadenylating Enzymes

Many studies suggest that there is a correlation between mRNA stability and the stability of the 3′ poly(A) tails of certain mRNAs [for reviews, see Brawerman (1981) and Ross (1988)]. Sequence elements of specific mRNAs allow complete deadenylation and then rapid degradation of the remainder of the mRNA chains (Shaw and Kamen, 1986; Cleveland and Yen, 1989; Wilson and Treisman, 1988; Shyu *et al.*, 1991; Laird-Offringa *et al.*, 1990). The role of PAB in controlling mRNA stability has also received much attention (see Bernstein and Ross, 1989). PAB structure has been highly conserved in nature, suggesting a similar function in all cells (Sachs, 1990). Work by Sachs, Davis, and Kornberg (Sachs *et al.*, 1987; Sachs and Davis, 1989; Sachs and Davis, 1990) shows that PAB is essential in yeast and involved in both translation initiation and poly(A) tail shortening.

Two enzymes that specifically degrade poly(A) chains in an apparent exonucleolytic manner have recently been found. Åström *et al.* (1991) described a 3′ exonuclease of HeLa cells that deadenylates mRNA and leaves the mRNA body intact after poly(A) removal. The enzyme was found in a DEAE–Sephacel fraction of a nuclear extract of HeLa cells, but the authors speculate that the enzyme could be associated with readily sedimentable cytoplasmic components and thus contaminate the nuclear fraction. Only 3′ poly(A) tails of synthetic RNA substrates are removed [poly(U), poly(G), or poly(C) tails were not, or very poorly, removed]. That the enzyme liberates 5′-AMP and requires a 3′-hydroxyl terminus on single-stranded poly(A) chains was recently reported (Åström *et al.*, 1992).

Using the finding that PAB is required for poly(A) tail degradation in yeast, Sachs and Deardorff (1992) and Lowell *et al.* (1992) have identified and purified a poly(A) nuclease (PAN) that requires PAB for activity. A yeast strain deficient in PAB, but viable due to an alteration of its 60 S ribosomal subunit (Sachs and Davis, 1990), was used to identify and purify PAN from cell-free extracts. They first found that addition of PAB to a high speed supernatant (S100) fraction results in a shortening of endogenous mRNA poly(A) tails to lengths below 50 nucleotides. Little or no activity was found in the absence of PAB. The S100 extracts were also found to hydrolyze poly(A) to acid-soluble nucleotides only in the presence of PAB.

PAN has been purified 100,000-fold from yeast S100 extracts by DEAE–fast flow, phosphocellulose, poly(U)-Sepharose, and CM-Sepharose chromatography, and the gene has been cloned. The gene is essential and encodes a 161-kDa protein with distinct domains containing repeated sequence elements. PAN hydrolyzes poly(A) chains to 5′-AMP, and the enzyme is not nucleotide specific if there is a PAB-binding site less than about 30 nt from the 3′-terminus being hydrolyzed. A distributive hydrolysis is found in most cases. With synthetic poly(A)-containing substrates and, possibly, with mRNA lacking a PAB-binding sequence, the hydrolysis of the final 10–25 nt of the poly(A) chains (terminal deadenylation) is slow and does not require PAB.

Lowell *et al.* (1992) have also examined the effect of adding PAB on the deadenylation of several specific mRNAs with PAN and find that the yeast a-factor mRNA (*MFA2*) is degraded significantly faster. This mRNA contains a sequence element in its 3′ UTR that is similar to the AU-rich destabilizing sequences of higher eukaryotes (see Chapter 9) (Cleveland and Yen, 1989). The investigations of Lowell *et al.* (1992) suggest that such sequence elements, by binding PAB or, possibly, other proteins that have been detected in higher eukaryotes (Gillis and Malter, 1991; Brewer, 1991; Bohjanen *et al.*, 1991; Vakalopoulou *et al.*, 1991), may be involved in the deadenylation reaction catalyzed by PAN. A synthetic substrate containing the 3′ UTR is completely deadenylated by PAN, while with an inverted sequence, the terminal deadenylation reaction is inhibited. Details of mRNA sequence effects on PAN can be found in Lowell *et al.* (1992).

The finding of a specific poly(A) hydrolyzing enzyme stimulated by poly(A) binding protein and its possible response to mRNA sequence elements opens up an exciting new area of investigation. Further studies of PAN-like activities in other cells will be very interesting.

VI. Concluding Remarks

Table I shows the ribonucleases described here and a brief description of their specificity and possible role in mRNA turnover. It is difficult to evaluate the possible role that each type of ribonuclease may play in mRNA turnover. The studies of yeast cells lacking XRN1 suggest that pathway 1 in Fig. 1 may play a role in mRNA turnover in yeast. The mRNA decapping enzyme of yeast may expose the mRNA chains to 5′→3′ exoribonucleolytic hydrolysis. In higher eukaryotic cells, exoribonucleases that hydrolyze the cap structure as well as the mRNA chain may be predominant enzymes. It seems quite likely, too, that an mRNA decapping enzyme may be found in cells other than yeast. Genetic studies of the yeast enzyme and the use of cap analogs in other systems may contribute to an evaluation of the role of pathway 1 in turnover. Analysis of the effect of destabilizing sequences and poly(A) tails on the rate of decapping by the yeast enzyme will be interesting.

The results on deadenylation and PAN suggest that poly(A) tail removal may be the first key step in some turnover pathways. Removal of the poly(A) tail by an enzyme such as PAN may then expose the mRNA chains to further hydrolysis by a 3′→5′ exoribonuclease such as the polysome-associated enzyme described above and depicted in pathway 2 of Fig. 1. It will be important to purify and characterize cytoplasmic exoribonucleases from different types of cells. Purification of the enzymes also will allow gene cloning and mutagenesis to aid in the evaluation of function.

That fragments, both 5′ and 3′ to a cleavage site, have been detected with several mRNAs *in vivo* suggests that endonucleolytic cleavage of mRNA can occur; whether it is an initial inactivating step and, thus, a key step in turnover as depicted in pathway 3 of Fig. 1 is not known. It is also possible that certain mRNAs are first inactivated by decapping or deadenylating enzymes (such as the yeast decapping enzyme and PAN) and that the subsequent turnover steps involve cleavage by relatively nonspecific endoribonucleases. Although it can be difficult to study specific endonucleases in cell-free systems because of nonspecific RNase contamina-

TABLE 1 Summary of RNases Described Here

Enzyme	Reaction and specificity	Possible role in mRNA turnover
HeLa m^7G-specific pyrophosphatase	Hydrolyzes the cap structure itself and the cap structure of oligonucleotides of up to 10 nucleotides in length. The product is m^7GMP.	Hydrolysis of the cap structure or short oligonucleotides containing it left as degradative products of turnover
mRNA decapping enzyme of *S. cerevisiae*	Hydrolyzes the cap structure of mRNA molecules, yielding m^7GDP. The cap structure itself or short molecules are not hydrolyzed.	Removal of the cap structure to allow 5′→3′ exoribonuclease hydrolysis of the mRNA chain.
Nucleotide pyrophosphatases (tobacco, potato)	Hydrolyzes pyrophosphate linkages of the cap structure itself and the mRNA cap structure.	Hydrolysis of the cap structure left as a degradative product or removal of the mRNA cap structure.
5′→3′ exoribonuclease 1 of *S. cerevisiae*	Hydrolyzes RNA by a 5′→3′ mode, yielding 5′-NMP. 5′-phosphomonoester end groups are required for rapid hydrolysis.	Hydrolysis of decapped mRNA molecules from the 5′ terminus.
5′→3′ ascites cell exoribonuclease	Hydrolyzes RNA to 5′-NMP by a 5′→3′ mode, with the second 5′ phosphodiester bond being cleaved first in some cases. Capped RNA is hydrolyzed also.	Hydrolysis of mRNA from the 5′ terminus
3′→5′ exoribonuclease of human erythroleukemia cells	Hydrolyzes RNA from the 3′ terminus with likely production of 5′-NMP.	Hydrolysis of mRNA from the 3′ terminus.
mRNP-bound RNase(s) [mouse sarcoma 180 (S-180) ascites cells and rabbit reticulocytes]	Endonucleotytic fragmentation of mRNA occurs and the enzyme is closely associated with the RNA.	Endonuclease cleavage of mRNA may be a first degradative step, followed by exoribonuclease digestion of fragments.
RNase L (all mammalian cells)	Cleaves RNA on the 3′ side of UN sequences to yield UpNp-terminated oligonucleotide products. Latent unless activated by (2′–5′)oligo(A).	Under special conditions, may be involved in the endonucleolytic hydrolysis of mRNA.
Poly(A) exonuclease of HeLa cells	Cleaves poly(A) chains at MRNA 3′ terminus to 5′-AMP. The mRNA body is poorly cleaved.	Removes poly(A) tails, possibly facilitating further turnover.
Poly(A) nuclease of *S. cerevisiae* (PAN)	Cleaves poly(A) chains to 5′-AMP and requires PAB for activity. May be controlled partially by sequence elements.	As just above.

tion, destabilizing sequences of certain mRNAs may be useful for detecting and purifying specific endonuclease activities that act only on mRNA chains containing these sequences.

Acknowledgments

The work in my laboratory that is described in this review was supported by the Office of Health and Environmental Research, U.S. Department of Energy, under Contract DE-AC05-84OR21400 with Martin Marietta Energy Systems, Inc., and by grants from the National Institutes of Health. I thank S. K. Niyogi, W. Cohn, and K. K. Niyogi for helpful comments on the manuscript and the authors who sent reprints and/or preprints of their work.

References

Albrecht, G., Krowczynska, A., and Brawerman, G. (1984). Configuration of β-globin messenger RNA in rabbit reticulocytes. *J. Mol. Biol.* **178,** 881–896.

Åström, J., Åström, A., and Virtanen, A. (1991). *In vitro* deadenylation of mammalian mRNA by a HeLa cell 3′ exonuclease. *EMBO J.* **10,** 3067–3071.

Åström, J., Åström, A., and Virtanen, A. (1992). Properties of a HeLa cell 3′ exonuclease specific for degrading poly(A) tails of mammalian mRNA. *J. Biol. Chem.* **267,** 18154–18159.

Baglioni, C., and Nilsen, T. W. (1984). Mechanisms of antiviral action of interferon. *In* "Interferon" (I. Gresser, Ed.), Vol. 5, pp. 23–42. Academic Press, New York.

Bandyopadhyay, R., Coutts, M., Krowczynska, A., and Brawerman, G. (1990). Nuclease activity associated with mammalian mRNA in its native state: Possible basis for selectivity in mRNA decay. *Mol. Cell. Biol.* **10,** 2060–2069.

Banerjee, A. K. (1980). 5′-terminal cap structure in eucaryotic messenger ribonucleic acids. *Microbiol. Rev.* **44,** 175–205.

Bartkiewicz, M., Sierakowska, H., and Shugar, D. (1984). Nucleotide pyrophosphatase from potato tubers. *Eur. J. Biochem.* **143,** 419–426.

Bernstein, P., and Ross, J. (1989). Poly(A), poly(A) binding protein and the regulation of mRNA stability. *Trends Biochem. Sci.* **14,** 373–377.

Bernstein, P., Peltz, S. W., and Ross, J. (1989). The poly(A)–poly(A)-binding protein complex is a major determinant of mRNA stability *in vitro*. *Mol. Cell. Biol.* **9,** 659–670.

Binder, R., Hwang, S. P. L., Ratnasabapathy, R., and Williams, D. L. (1989). Degradation of apolipoprotein II mRNA occurs via endonucleolytic cleavage at 5′-AAU-3′/5′-UAA-3′ elements in single-stranded loop domains of the 3′-noncoding region. *J. Biol. Chem.* **264,** 16910–16918.

Bisbal, C., Silhol, M., Lemaitre, M., Bayard, B., Salehzada, T., and Lebleu, B. (1987). 5′-modified agonist and antagonist $(2'\text{-}5')(A)_n$ analogues: Synthesis and biological activity. *Biochemistry* **26,** 5172–5178.

Bisbal, C., Salehzada, T., Lebleu, B., and Bayard, B. (1989). Characterization of two murine $(2'\text{–}5')(A)_n$-dependent endonucleases of different molecular mass. *Eur. J. Biochem.* **179,** 595–602.

Bohjanen, P. R., Petryniak, B., June, C. H., Thompson, C. B., and Lindsten, T. (1991). An inducible cytoplasmic factor (AU-B) binds selectively to AUUUA multimers in the 3′ untranslated region of lymphokine mRNA. *Mol. Cell. Biol.* **11,** 3288–3295.

Brawerman, G. (1981). The role of the poly(A) sequence in mammalian messenger RNA. *CRC Crit. Rev. Biochem.* **10,** 1–38.

Brewer, G. (1991). An A + U-rich element RNA-binding factor regulates c-*myc* mRNA stability *in vitro*. *Mol. Cell. Biol.* **11,** 2460–2466.

Brewer, G., and Ross, J. (1988). Poly(A) shortening and degradation of the 3′ A+ U-rich sequences of human c-*myc* RNA in a cell-free system. *Mol. Cell. Biol.* **8,** 1697–1708.

Brewer, G., and Ross, J. (1989). Regulation of c-*myc* mRNA stability *in vitro* by a labile destabilizer with an essential nucleic acid component. *Mol. Cell. Biol.* **9,** 1996–2006.

Brown, B. D., and Harland, R. M. (1990). Endonucleolytic cleavage of a maternal homeo box mRNA in *Xenopus* oocytes. *Genes Dev.* **4,** 1925–1935.

Brown, G. E., Lebleu, B., Kawakita, M., Shaila, S., Sen, G. C., and Lengyel, P. (1976). Increased endonuclease activity in an extract from mouse Ehrlich ascites tumor cells which had been treated with a partially-purified interferon preparation: Dependence on double-stranded RNA. *Biochem. Biophys. Res. Commun.* **69,** 114–122.

Busch, H., Reddy, R., Rothblum, L., and Choi, Y. C. (1982). SnRNAs, SnRNPs, and RNA processing. *Annu. Rev. Biochem.* **51,** 617–654.

Chebath, J., Benech, P., Hovanessian, A., Galabru, J., and Revel, M. (1987). Four different forms of interferon-induced 2′,5′-oligo(A) synthetase identified by immunoblotting in human cells. *J. Biol. Chem.* **262,** 3852–3857.

Cleveland, D. W., and Yen, T. J. (1989). Multiple determinants of eukaryotic mRNA stability. *The New Biol.* **1,** 121–126.

Coutts, M., Krowczynska, A., and Brawerman, G. (1993). Protection of mRNA against nucleases in cytoplasmic extracts of mouse sarcoma ascites cells. *Biochim. Biophys. Acta,* in press.

Coutts, M., and Brawerman, G. (1993). A 5′ exoribonuclease from cytoplasmic extracts of mouse sarcoma 180 ascites cells. *Biochim. Biophys. Acta,* in press.

Dani, C., Mechti, N., Piechaczyk, M., Lebleu, B., Jeanteur, P., and Blanchard, J. M. (1985). Increased rate of degradation of c-*myc* mRNA in interferon-treated Daudi cells. *Proc. Natl. Acad. Sci. USA* **82,** 4896–4899.

Dykstra, C. C., Hamatake, R. K., and Sugino, A. (1990). DNA strand transfer protein β from yeast mitotic cells differs from strand transfer protein α from meiotic cells. *J. Biol. Chem.* **265,** 10968–10973.

Dykstra, C. C., Kitada, K., Clark, A. B., Hamatake, R. K., and Sugino, A. (1991). Cloning and characterization of *DST2*, the gene for DNA strand transfer protein β from *Saccharomyces cerevisiae*. *Mol. Cell. Biol.* **11,** 2583–2592.

Eichler, D. C., and Eales, S. J. (1985). Purification and properties of a novel nucleolar exoribonuclease capable of degrading both single-stranded and double-stranded RNA. *Biochemistry* **24,** 686–691.

Floyd-Smith, G., and Lengyel, P. (1986). RNase L, a (2′–5′)-oligoadenylate-dependent endoribonuclease: Assays and purification of the enzyme. Cross-linking to a (2′–5′)-oligoadenylate derivative. *In* Methods in Enzymology (S. Pestka, Ed.), Vol. 119, pp. 489–499. Academic Press, San Diego.

Floyd-Smith, G., Slattery, E., and Lengyel, P. (1981). Interferon action: RNA cleavage pattern of a (2′–5′)oligoadenylate-dependent endonuclease. *Science* **212,** 1030–1032.

Furuichi, Y., LaFiandra, A., and Shatkin, A. J. (1977). 5′-terminal structure and mRNA stability. *Nature* (*London*) **266,** 235–239.

Furuichi, Y., and Shatkin, A. J. (1989). Characterization of cap structures. *In* Methods in Enzymology (J. E. Dahlberg and J. N. Abelson, Eds.), Vol. 180, pp. 164–176. Academic Press, San Diego.

Gillis, P., and Malter, J. S. (1991). The adenosine–uridine binding factor recognizes the AU-rich elements of cytokine, lymphokine, and oncogene mRNAs. *J. Biol. Chem.* **266,** 3172–3177.

Herrick, D., Parker, R., and Jacobson, A. (1990). Identification and comparison of stable and unstable mRNAs in *Saccharomyces cerevisiae*. *Mol. Cell. Biol.* **10,** 2269–2284.

Johnson, A. W., and Kolodner, R. D. (1991). Strand exchange protein 1 from *Saccharomyces cerevisiae:* A novel multifunctional protein that contains DNA strand exchange and exonuclease activities. *J. Biol. Chem.* **266,** 14046–14054.

Johnston, M. I., and Torrence, P. F. (1984). The role of interferon induced proteins, double-stranded RNA and 2′,5′-oligoadenylate in the interferon-mediated inhibition of viral translation. *In* Interferon (R. M. Friedman, Ed.), Vol. 3, pp. 189–298. Elsevier, Amsterdam.

Kearsey, S., and Kipling, D. (1991). Recombination and RNA processing: A common strand. *Trends Cell Biol.* **1,** 110–112.

Kenna, M., Stevens, A., McCammon, M., and Douglas, M. G. (1993). An essential yeast gene with homology to the exonuclease-encoding *XRN1/KEM1* gene also encodes a protein with exoribonuclease activity. *Mol. Cell. Biol.,* in press.

Kim, J., Ljungdahl, P. O., and Fink, G. R. (1990). *kem* mutations affect nuclear fusion in *Saccharomyces cerevisiae. Genetics* **126,** 799–812.

Kipling, D., Tambini, C., and Kearsey, S. E. (1991). *rar* mutations which increase artificial chromosome stability in *Saccharomyces cerevisiae* identify transcription and recombination proteins. *Nucleic Acids Res.* **19,** 1385–1391.

Kole, R., Sierakowska, H., and Shugar, D. (1976). Novel activity of potato nucleotide pyrophosphatase. *Biochim. Biophys. Acta* **438,** 540–550.

Kolodner, R., Evans, D. H., and Morrison, P. T. (1987). Purification and characterization of an activity from *Saccharomyces cerevisiae* that catalyzes homologous pairing and strand exchange. *Proc. Natl. Acad. Sci. USA* **84,** 5560–5564.

Krause, D., Panet, A., Arad, G., Dieffenbach, C. W., and Silverman, R. H. (1985). Independent regulation of ppp(A2′p)$_n$A-dependent RNase in NIH 3T3, clone 1 cells by growth arrest and interferon treatment. *J. Biol. Chem.* **260,** 9501–9507.

Krause, D., Lesiak, K., Imai, J., Sawai, H., Torrence, P. F., and Silverman, R. H. (1986). Activation of 2-5A-dependent RNase by analogs of 2-5A(5′–0-triphosphoryladenylyl(2′–5′)adenylyl(2′→5′)adenosine) using 2′,5′-tetraadenylate (core)–cellulose. *J. Biol. Chem.* **261,** 6836–6839.

Laird-Offringa, I. A., DeWit, C. L., Elfferich, P., and Van Der Eb, A. J. (1990). Poly(A) tail shortening is the translation-dependent step in c-*myc* mRNA degradation. *Mol. Cell. Biol.* **10,** 6132–6140.

Larimer, F. W., and Stevens, A. (1990). Disruption of the gene *XRN1,* coding for a 5′→3′ exoribonuclease, restricts yeast cell growth. *Gene* **95,** 85–90.

Larimer, F. W., Hsu, C. L., Maupin, M. K., and Stevens, A. (1992). Characterization of the *XRN1* gene encoding a 5′→3′ exoribonuclease: Sequence data and analysis of disparate protein and mRNA levels of gene-disrupted cells. *Gene,* **120,** 51–57.

Lasater, L. S., and Eichler, D. C. (1984). Isolation and properties of a single-strand 5′→3′ exoribonuclease from Erhlich ascites tumor cell nucleoli. *Biochemistry* **23,** 4367–4373.

Lazarus, H. M., and Sporn, M. B. (1967). Purification and properties of a nuclear exoribonuclease from Ehrlich ascites tumor cells. *Proc. Natl. Acad. Sci. USA* **57,** 1386–1393.

Lengyel, P. (1982). Biochemistry of interferons and their actions. *Annu. Rev. Biochem.* **51,** 251–282.

Lengyel, P. (1987). Double-stranded RNA and interferon action. *J. Interferon Res.* **7,** 511–519.

Lesiak, K., Imai, J., Floyd-Smith, G., and Torrence, P. F. (1983). Biological activities of phosphodiester linkage isomers of 2-5A. *J. Biol. Chem.* **258,** 13082–13088.

Lowell, J. E., Rudner, D. Z., and Sachs, A. B. (1992). 3′-UTR-dependent deadenylation by the yeast poly(A) nuclease. *Genes Dev.* **6,** 2088–2099.

Martin, E. M., Birdsall, N. J. M., Brown, R. E., and Kerr, I. M. (1979). Enzymic synthesis, characterisation and nuclear-magnetic-resonance spectra of pppA2′p5′A2′p5′A and related oligonucleotides: Comparison with chemically synthesized material. *Eur. J. Biochem.* **95,** 295–307.

Meinsma, D., Holthuizen, P. E., Van den Brande, J. L., and Sussenbach, J. S. (1991). Specific endonucleolytic cleavage of IGF-II mRNAs. *Biochem. Biophys. Res. Commun.* **179,** 1509–1516.

Murthy, K. G. K., Park, P., and Manley, J. L. (1991). A nuclear micrococcal-sensitive, ATP-dependent exoribonuclease degrades uncapped but not capped RNA substrates. *Nucleic Acids Res.* **19,** 2685–2692.

Nilsen, T. W., Wood, D. L., and Baglioni, C. (1981). 2′,5′-oligo(A)-activated endoribonuclease: Tissue distribution and characterization with a binding assay. *J. Biol. Chem.* **256,** 10751–10754.

Nuss, D. L., Furuichi, Y., Koch, G., and Shatkin, A. J. (1975). Detection in HeLa cell extracts of a 7-methyl guanosine specific enzyme activity that cleaves m^7GpppN^m. *Cell* **6,** 21–27.

Nuss, D. L., and Furuichi, Y. (1977). Characterization of the $m^7G(5')pppN$-pyrophosphatase activity from HeLa cells. *J. Biol. Chem.* **252,** 2815–2821.

Parker, R., and Jacobson, A. (1990). Translation and a 42-nucleotide segment within the coding region of the mRNA encoded by the *MATα1* gene are involved in promoting rapid mRNA decay in yeast. *Proc. Natl. Acad. Sci. USA* **87,** 2780–2784.

Peltz, S. W., Brewer, G., Kobs, G., and Ross, J. (1987). Substrate specificity of the exonuclease activity that degrades H4 histone mRNA. *J. Biol. Chem.* **262,** 9382–9388.

Ross, J. (1988). Messenger RNA turnover in eukaryotic cells. *Mol. Biol. Med.* **5,** 1–14.

Ross, J., and Kobs, G. (1986). H4 histone messenger RNA decay in cell-free extracts initiates at or near the 3′ terminus and proceeds 3′→5′. *J. Mol. Biol.* **188,** 579–593.

Ross, J., Kobs, G., Brewer, G., and Peltz, S. W. (1987). Properties of the exonuclease activity that degrades H4 histone mRNA. *J. Biol. Chem.* **262,** 9374–9381.

Ross, J., Peltz, S. W., Kobs, G., and Brewer, G. (1986). Histone mRNA degradation *in vivo:* The first detectable step occurs at or near the 3′ terminus. *Mol. Cell. Biol.* **6,** 4362–4371.

Sachs, A. B. (1990). The role of poly(A) in the translation and stability of mRNA. *Curr. Opinion Cell Biol.* **2,** 1092–1098.

Sachs, A. B., and Davis, R. W. (1989). The poly(A) binding protein is required for poly(A) shortening and 60S ribosomal subunit-dependent translation initiation. *Cell* **58,** 857–867.

Sachs, A. B., and Davis, R. W. (1990). Translation initiation and ribosomal biogenesis: Involvement of a putative rRNA helicase and RPL46. *Science* **247,** 1077–1079.

Sachs, A. B., Davis, R. W., and Kornberg, R. D. (1987). A single domain of yeast poly(A)-binding protein is necessary and sufficient for RNA binding and cell viability. *Mol. Cell. Biol.* **7,** 3268–3276.

Sachs, A. B., and Deardorff, J. A. (1992). Translation initiation requires the PAB-dependent poly(A) ribonuclease in yeast. *Cell* **70,** 961–973.

Shatkin, A. J. (1976). Capping of eukaryotic mRNAs. *Cell* **9,** 645–653.

Shatkin, A. J. (1985). mRNA cap binding proteins: Essential factors for initiating translation. *Cell* **40,** 223–224.

Shaw, G., and Kamen, R. (1986). A conserved AU sequence from the 3′ untranslated region of GM-CSF mRNA mediates selective mRNA degradation. *Cell* **46,** 659–667.

Shimotohno, K., and Miura, K. (1977). A novel 5′-exonuclease which degrades uncapped mRNA. *In* Proceedings of the 1977 Molecular Biology Meeting of Japan, pp. 83–85.

Shimotohno, K., Kodama, Y., Hashimoto, J., and Miura, K. (1977). Importance of 5′-terminal blocking structure to stabilize mRNA in eukaryotic protein synthesis. *Proc. Natl. Acad. Sci. USA* **74,** 2734–2738.

Shinshi, H., Miwa, M., Kato, K., Noguchi, M., Matsushima, T., and Sugimura, T. (1976a). A novel phosphodiesterase from cultured tobacco cells. *Biochemistry* **15,** 2185–2190.

Shinshi, H., Miwa, M., and Sugimura, T. (1976b). Enzyme cleaving the 5′-terminal methylated blocked structure of messenger RNA. *FEBS Lett.* **65,** 254–257.

Shyu, A. B., Belasco, J. G., and Greenberg, M. E. (1991). Two distinct destabilizing elements in the c-*fos* message trigger deadenylation as a first step in rapid mRNA decay. *Genes Dev.* **5,** 221–231.

Shyu, A. B., Greenberg, M. E., and Belasco, J. G. (1989). The c-*fos* transcript is targeted for rapid decay by two distinct mRNA degradation pathways. *Genes Dev.* **3,** 60–72.
Silverman, R. H., Jung, D. D., Nolan-Sorden, N. L., Dieffenbach, C. W., Kedar, V. P., and SenGupta, D. N. (1988). Purification and analysis of murine 2-5A-dependent RNase. *J. Biol. Chem.* **263,** 7336–7341.
Slattery, E., Ghosh, N., Samanta, H., and Lengyel, P. (1979). Interferon, double-stranded RNA, and RNA degradation: Activation of an endonuclease by (2′–5′)A_n. *Proc. Natl. Acad. Sci. USA* **76,** 4778–4782.
Stevens, A. (1978). An exoribonuclease from *Saccharomyces cerevisiae*: Effect of modifications of 5′ end groups on the hydrolysis of substrates to 5′ mononucleotides. *Biochem. Biophys. Res. Commun.* **81,** 656–661.
Stevens, A. (1979). Evidence for a 5′→3′ direction of hydrolysis by a 5′ mononucleotide-producing exoribonuclease from *Saccharomyces cerevisiae*. *Biochem. Biophys. Res. Commun.* **86,** 1126–1132.
Stevens, A. (1980). Purification and characterization of a *Saccharomyces cerevisiae* exoribonuclease which yields 5′-mononucleotides by a 5′→3′ mode of hydrolysis. *J. Biol. Chem.* **255,** 3080–3085.
Stevens, A. (1980). An mRNA decapping enzyme from ribosomes of *Saccharomyces cerevisiae*. *Biochem. Biophys. Res. Commun.* **96,** 1150–1155.
Stevens, A. (1988). mRNA-decapping enzyme from *Saccharomyces cerevisiae:* Purification and unique specificity for long RNA chains. *Mol. Cell. Biol.* **8,** 2005–2010.
Stevens, A., and Maupin, M. K. (1987a). A 5′→3′ exoribonuclease of human placental nuclei: Purification and substrate specificity. *Nucleic Acids Res.* **15,** 695–708.
Stevens, A., and Maupin, M. K. (1987b). A 5′→3′ exoribonuclease of *Saccharomyces cerevisiae:* Size and novel substrate specificity. *Arch. Biochem. Biophys.* **252,** 339–347.
Stevens, A., Hsu, C. L., Isham, K. R., and Larimer, F. W. (1991). Fragments of the internal transcribed spacer 1 of pre-rRNA accumulate in *Saccharomyces cerevisiae* lacking 5′→3′ exoribonuclease 1. *J. Bacteriol.* **173,** 7024–7028.
Stoeckle, M. Y., and Hanafusa, H. (1989). Processing of 9E3 mRNA and regulation of its stability in normal and Rous sarcoma virus-transformed cells. *Mol. Cell. Biol.* **9,** 4738–4745.
Sunitha, I., and Slobin, L. I. (1987). An *in vitro* system derived from Friend erythroleukemia cells to study messenger RNA stability. *Biochem. Biophys. Res. Commun.* **144,** 560–568.
Swartwout, S. G., and Kinniburgh, A. J. (1989). c-*myc* RNA degradation in growing and differentiating cells: Possible alternate pathways. *Mol. Cell. Biol.* **9,** 288–295.
Tishkoff, D. X., Johnson, A. W., and Kolodner, R. D. (1991). Molecular and genetic analysis of the gene encoding the *Saccharomyces cerevisiae* strand exchange protein Sep1. *Mol. Cell. Biol.* **11,** 2593–2608.
Vakalopoulou, E., Schaack, J., and Shenk, T. (1991). A 32-kilodalton protein binds to AU-rich domains in the 3′ untranslated regions of rapidly degraded mRNAs. *Mol. Cell. Biol.* **11,** 3355–3364.
Watling, D., Serafinowska, H. T., Reese, C. B., and Kerr, I. M. (1985) Analogue inhibitor of 2-5A action: Effect on the interferon-mediated inhibition of encephalomyocarditis virus replication. *EMBO J.* **4,** 431–436.
Williams, B. R. G., Golgher, R. R., Brown, R. E., Gilbert, C. S., and Kerr, I. M. (1979). Natural occurrence of 2-5A in interferon-treated EMC virus-infected L cells. *Nature (London)* **282,** 582–586.
Wilson, T., and Treisman, R. (1988). Removal of poly(A) and consequent degradation of c-*fos* mRNA facilitated by 3′ AU-rich sequences. *Nature (London)* **336,** 396–399.
Wreschner, D. H., McCauley, J. W., Skehel, J. J., and Kerr, I. M. (1981). Interferon action—Sequence specificity of the ppp(A2′p)$_n$A-dependent ribonuclease. *Nature (London)* **289,** 414–417.

PART III

METHODS OF ANALYSIS

18

Experimental Approaches to the Study of mRNA Decay

JOEL G. BELASCO AND GEORGE BRAWERMAN

I. Kinetics of mRNA Decay

The degradation of mRNA *in vivo* usually obeys first-order kinetics; that is, the decay rate of a message (v) is proportional to both the rate constant for decay (k_{decay}) and its cellular concentration ([mRNA]).

$$\text{mRNA} \xrightarrow{k_{decay}} \text{degradation products}$$

$$v = -d[\text{mRNA}]/dt = k_{decay}\,[\text{mRNA}]$$

Because it is generally more convenient to think about mRNA decay in terms of an mRNA lifetime rather than a decay rate constant, the stability of a message *in vivo* is usually reported as a half-life, the time required for half of the existing mRNA molecules to be degraded. The half-life of an mRNA ($t_{1/2}$) is inversely proportional to its decay rate constant.

$$t_{1/2} = \ln 2\ /k_{decay}$$

When mRNA turnover is measured in growing cells, there is a contribution to the apparent turnover rate constant (k_{obs}) from dilution due to cell growth:

$$k_{obs} = \ln 2\ /\ t_{obs} = \ln 2\ (1/t_{1/2} + 1/t_D), \qquad (1)$$

where t_{obs} is the effective mRNA half-life and t_D is the cell doubling time. This contribution from the rate of cell growth is negligible when the mRNA half-life is much shorter than the cell doubling time.

The conformity of mRNA decay to first-order kinetics implies that the cellular machinery for mRNA degradation cannot distinguish among molecules of a given type that differ in age. This generally appears to be the case, but exceptions are known. For example, the time-dependent removal of the poly(A) tail of some mRNAs (such as the mammalian c-*fos* and c-*myc* mRNAs; see Chapter 9) appears to be a prerequisite for their degradation. Consequently, after a short burst of c-*fos* or c-*myc* transcription, the newly synthesized mRNA molecules decay with biphasic kinetics, comprising a lag phase during which deadenylation occurs and a first-order phase during which the body of the mRNA is degraded (see Chapter 15). In such cases of complex decay kinetics, an mRNA lifetime cannot meaningfully be described merely in terms of a half-life.

II. Measurement of Decay Rates

A. mRNA Half-Life

The metabolic stability of mRNA can be defined by its half-life in the cell. Depending upon the method used for mRNA detection, the half-life can represent anything from the rate of initial cleavage of the RNA chain to the rate at which it is converted to small fragments. This distinction can be of practical significance. Although the initial cleavage is the rate-determining step for overall degradation of the many messages that decay in an all-or-none fashion, decay-resistant fragments of some mRNAs can persist *in vivo* as relatively long-lived decay intermediates (Chen *et al.*, 1988; Stoeckle and Hanafusa, 1989).

Half-lives most frequently are measured by blocking mRNA synthesis (see later), isolating cytoplasmic RNA at time intervals, and monitoring the rate of loss of a particular message by analyzing equal amounts of these samples with a message-specific probe. The decay rate constant is then obtained from the slope of a semilogarithmic plot of mRNA concentration as a function of time (Fig. 1). The plot is defined by

$$\ln([\mathrm{mRNA}]_t/[\mathrm{mRNA}]_0) = -k_{\mathrm{obs}}t, \tag{2}$$

where $[\mathrm{mRNA}]_t$ is the mRNA concentration at time t, $[\mathrm{mRNA}]_0$ is the initial mRNA concentration, and k_{obs} is as defined in Eq. 1. Ideally, measurement of an mRNA half-life involves the analysis of several RNA samples collected over a period of two to three half-lives, followed by linear regression (least-squares) analysis to identify the line that best fits the data. A similar analysis can be performed on pulse-labeled mRNA in cells subjected to a chase in the absence of labeled RNA precursor (see later).

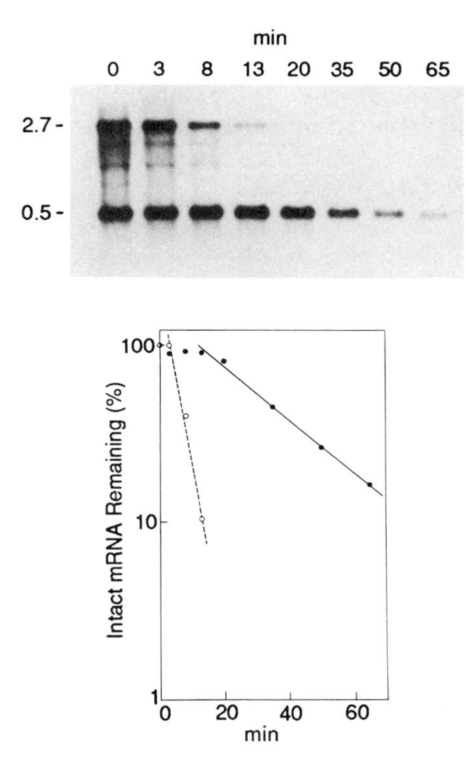

FIGURE 1 Northern blot analysis of bacterial mRNA degradation. RNA samples were isolated from *Rhodobacter capsulatus* after transcription inhibition with rifampicin. Bands correspond to two different *puf* operon mRNAs. Beneath the autoradiogram is a semilogarithmic graph of the decay of each of these mRNAs. Because the 0.5-kb *puf* mRNA (●) is a processing product of the 2.7-kb transcript (○), it exhibits an apparent lag in decay until its precursor is largely degraded. [Reproduced with permission from Chen *et al.* (1988).]

B. Rate of Loss of mRNA after Interruption of Transcription

Both global and message-specific methods are available for blocking mRNA synthesis in order to allow measurements of the subsequent rate of loss of the mRNA of interest.

1. *Inhibition of RNA Polymerase in Prokaryotes*

The inhibitor commonly used for this purpose is rifampicin, which binds specifically to the β subunit of bacterial RNA polymerase and prevents further transcription initiation. Inhibition with this drug is fast and easy. Entry of rifampicin into cells and its binding to RNA polymerase generally occur within seconds, although it takes somewhat longer for the last round of ongoing transcription to be completed. Even though rifampicin treatment arrests cell growth and eventually leads to total depletion of the cellular mRNA pool, it is generally assumed that mRNA metabolism is not affected during the relatively short time of exposure to the drug.

2. *Inhibition of Transcription in Yeast*

The most commonly used methods of nonspecific transcription inhibition in this organism involve the thermal inactivation of a temperature-sensitive RNA polymerase II (*rpbl-1*) or the use of inhibitory drugs (e.g., thiolutin or 1,10-phenanthroline) (Santiago *et al.*, 1986; Nonet *et al.*, 1987; Herrick *et al.*, 1990) (see Chapter 13). The principal disadvantage of these methods is that prolonged inhibition of global mRNA synthesis may have a severe impact on cellular physiology. In addition, use of the temperature-sensitive RNA polymerase requires that cells be heat-shocked (i.e., subjected to a temperature upshift from 24 to 36°C), while thiolutin and 1,10-phenanthroline have pleiotropic effects on certain genes and metabolic pathways. Fortunately, relative mRNA stabilities measured using thiolutin or the *rpbl-1* allele are similar (Herrick *et al.*, 1990). Thus it is likely that in most instances these two methods provide accurate mRNA half-life values.

3. *Inhibition of RNA Polymerase II in Higher Eukaryotes*

Global inhibition of mRNA synthesis is usually accomplished by treatment either with actinomycin D, which interferes with transcription by intercalating into the DNA, or with 5,6-dichloro-1β-D-ribofuranosylbenzimidazole (DRB), which interacts directly with the RNA polymerase II transcription apparatus (Zandomeni *et al.*, 198; Maderious and Chen-Kiang, 1984). These drugs are best used with caution, as they have a severe impact on cellular physiology. Furthermore, treatment with these drugs has been shown in a number of cases to lead to inhibition of cellular pathways for mRNA degradation (Mullner and Kuhn, 1988; Shyu *et al.*, 1989).

4. *Use of Promoters Subject to Transient Induction or Repression*

The advantage of this approach is that it does not disrupt normal cell growth. One *Escherichia coli* promoter that has been used for this purpose is the *trp* promoter, which can be swiftly repressed by adding tryptophan to the medium (Morse *et al.*, 1969). Similarly, in yeast, transcription from the *GAL1* promoter can be inhibited by adding glucose to the medium (see Chapter 13). In mammalian cells, the *c-fos* promoter has been valuable for this purpose (Shyu *et al.*, 1989). Transcription from the *c-fos* promoter is transiently induced in a variety of cell types by stimulation with growth factors, membrane-depolarizing agents, and neurotransmitters (Rivera and Greenberg, 1990). Within 30–40 min after induction, transcription from this promoter abruptly ceases, thereby allowing the degradation of a nearly synchronous population of newly synthesized mRNAs to be monitored.

Repressible promoters have been used to study the decay of heterologous mRNAs by constructing appropriate gene fusions (Shyu *et al.*, 1989) (see also Chapter 13). In the case of the *c-fos* promoter, transient inducibility by growth factor stimulation can be achieved simply by inserting a small DNA element (the serum-response element) upstream of heterologous genes (Rivera and Greenberg, 1990).

5. *Measurements of mRNA Levels*

These are determined most commonly either by gel electrophoresis followed by blotting or by nuclease protection analysis. A half-life measured by the Northern blot procedure reflects the rate of initial cleavage of the full-length RNA molecule. However, the resolution of this method may not be adequate to detect initial cleavage near one of the mRNA termini (e.g., within a short 5′ untranslated region or within a poly(A) tail).

Measurements by S1- or RNase-protection analysis allow the detection of degradative events within a specific region of the mRNA, using a probe that is complementary to that region (von Gabain *et al.*, 1983; Ross *et al.*, 1986). However, by focusing on a limited segment of the mRNA molecule, important features of the overall degradation pathway may be missed.

mRNA decay can also be monitored by dot- or slot-blotting. These methods yield the rate of conversion of an RNA chain to fragments that are too short to bind to nitrocellulose or nylon membranes or to form stable RNA–DNA or RNA–RNA duplexes with the probe. Although this procedure may be convenient in cases where the initial cleavage in mRNA is followed by rapid digestion of the resulting fragments, it is less informative than Northern blotting or nuclease protection analysis. It is also more susceptible to error due to nonspecific hybridization of the probe.

C. Rate of Loss of Newly Synthesized (Pulse-Labeled) mRNA

Another method for measuring the rate of mRNA decay is pulse–chase analysis, in which the RNA, rather than the probe, is radiolabeled. The cellular RNA is labeled for a relatively brief period, usually with tritiated uridine, and the loss of labeled RNA molecules is followed after depletion of the radioactive precursor pool for RNA synthesis. Since there is no need to interrupt transcription, the cell remains under steady-state conditions, and physiological perturbations that may affect mRNA stability are avoided. Although this might seem to be the method of choice for measuring mRNA decay rates, it presents various technical difficulties. It is also limited to the analysis of relatively stable mRNAs because of the length of time required to deplete the labeled precursor pool (see later).

1. *Chase Procedure*

Since the decay of newly synthesized mRNA can be measured only after cessation of synthesis of radioactive RNA molecules, it is essential to ensure that the specific radioactivity of the precursors for RNA synthesis has become negligible. In bacteria, this is generally achieved by adding a large excess of unlabeled precursor to the medium (Belasco *et al.*, 1985). Because of the relatively slow equilibration of the precursor pool, this chase may be combined with transcription inhibition by rifampicin (Belasco *et al.*, 1985), although this might seem to defeat the major benefit of the pulse–chase procedure.

In the case of mammalian and insect cells, the medium containing the labeled precursor is usually replaced by nonradioactive medium. The depletion of the labeled precursor pool in these cells takes hours (Krowczynska *et al.*, 1985). High levels of nonradioactive uridine and cytidine are commonly included in the chase medium to accelerate this process, but it has been observed that the presence of high levels of nucleosides can affect normal RNA metabolism in mammalian cells (Lowenhaupt and Lingrel, 1978). Glucosamine, which combines with UTP to form UDP-*N*-acetylhexosamine, has been used to shrink the intracellular UTP pool for RNA synthesis, and the combination of glucosamine and cold uridine and cytidine can result in very rapid depletion of labeled precursors, thus permitting the analysis of short-lived RNAs (Levis and Penman, 1977). The timing of pool depletion can be monitored by following the course of uptake of label into total cytoplasmic RNA, which is predominantly rRNA (Krowczynska *et al.*, 1985), or into tRNA (Levis and Penman, 1977).

2. *Measuring Radioactivity in Individual mRNA Species*

To quantitate mRNA radioactivity, total cytoplasmic RNA is isolated and hybridized to complementary DNA immobilized on small pieces of nitrocellulose or other membranes. Before scintillation counting, the hybrids must be washed extensively to eliminate the vast amount of radioactivity from contaminating RNA species (Harpold *et al.*, 1979; Brock and Shapiro, 1983, Krowczynska *et al.*, 1985; Paek and Axel, 1987). Ribonuclease treatment can be included to help eliminate any adsorbed RNA not present in true RNA–DNA hybrids. A convenient way to express the radioactivity data is as a percentage of the total RNA (which is primarily stable rRNA) present in the samples used for hybridization.

The procedure does not distinguish between intact mRNA molecules and fragments long enough to form stable hybrids with the immobilized DNA. Size fractionation of the RNA prior to hybridization can minimize this problem, provided that the RNA fractions can be recovered quantitatively (Belasco *et al.*, 1985).

3. Measuring Radioactivity in Total mRNA

Information on the overall rate of decay of total cellular mRNA can be obtained by selectively labeling the mRNA, and then following the loss of radioactivity in total RNA. In bacteria, preferential labeling can be achieved by using a very short labeling period, since the stable RNAs are synthesized much more slowly than the mRNA in these cells (Donovan and Kushner, 1986). The decay rate of bulk mRNA is measured simply by monitoring the loss of radioactivity from acid-insoluble cellular RNA during the chase period.

In mammalian cells, it is possible to block the synthesis of rRNA selectively by exposing the cells to low levels of actinomycin D (Penman *et al.*, 1968). Under these conditions, the radioactivity present in polyribosomes is due mostly to mRNA, and loss of radioactivity from the polyribosomal RNA fraction can be monitored as in the case of bacterial RNA. However, overall mRNA decay in eukaryotic cells can be measured more precisely by following the loss of radioactivity from the cytoplasmic RNA fraction that binds to oligo(dT)–cellulose, which represents the poly(A)-containing mRNA (Singer and Penman, 1973).

D. Rate of Loss of Functional mRNA

The metabolic stability of an mRNA species can also be estimated by following the loss of its capacity to serve as a template for protein synthesis. Two methods have been used in bacteria to measure the "functional half-life" of mRNAs. One involves inhibiting further mRNA synthesis and monitoring the ensuing loss of protein synthetic capacity by pulse labeling newly synthesized proteins with a radioactive amino acid at time intervals thereafter (Fig. 2). The functional half-life of a specific message can then be determined by isolating from each sample the protein that the mRNA encodes (e.g., by immunoprecipitation) and plotting the changes in its specific radioactivity (cpm/mg) as a function of time (Fig. 2). Because the data from such an experiment should obey Eq. 2, the slope of the semilogarithmic plot provides the rate constant for decay of functional mRNA.

The other method for determining functional mRNA half-life is applicable to mRNAs derived from transcriptionally inducible genes and is especially useful for estimating the decay rates of messages that encode readily assayable protein products. It involves first inducing a brief burst of mRNA synthesis. Further transcription is then blocked abruptly, and the subsequent accumulation of the induced protein is monitored by immunoassay or by activity assay (Blundell *et al.*, 1972). The rate of synthesis of the protein will decline gradually as the mRNA that encodes it is degraded (Fig. 3). If the protein is stable and not diluted by cell growth, its cellular concentration will approach a maximum value at a rate that is governed by the rate

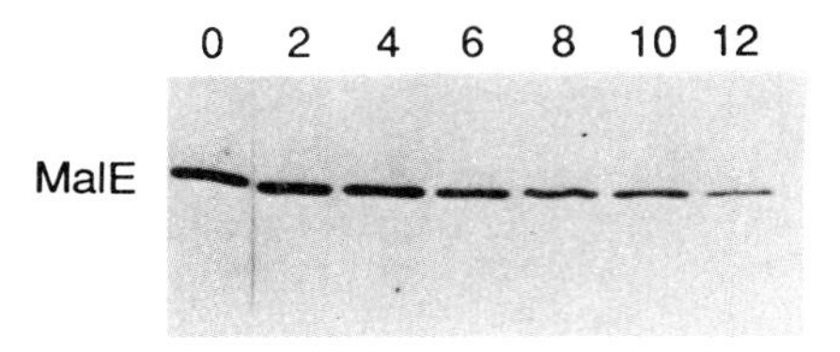

FIGURE 2 Functional assay of bacterial mRNA decay after transcription inhibition. Samples of an *Escherichia coli* culture taken at the indicated times (min) after treatment with rifampicin were pulse-labeled with a radioactive amino acid. The protein product of *malE* mRNA was then immunoprecipitated, subjected to electrophoresis, and detected by autoradiography. [Reproduced with permission from Newbury *et al.* (1987).]

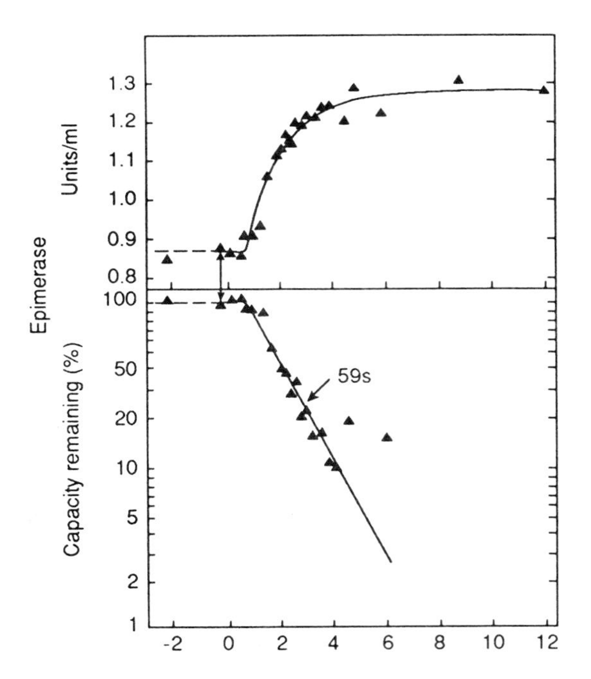

FIGURE 3 Assay for functional inactivation of bacterial mRNA after a brief burst of transcription. An *Escherichia coli* culture was exposed to fucose for 90 sec to induce the *gal* operon and then treated with rifampicin to inhibit transcription. Samples were taken at the indicated times (min) and assayed for UDP glucose-4-epimerase, the *galE* gene product. (top) Accumulation of enzyme activity after the burst of transcription. (bottom) Data replotted according to Eq. 3 to obtain the decay rate of *galE* mRNA. [Reproduced with permission from Achord and Kennell (1974).]

constant (k) for functional inactivation of the mRNA. This rate constant can be obtained from

$$\ln (1 - [P]_t/[P]_\infty) = -kt, \tag{3}$$

where $[P]_t$ is the protein concentration at time t and $[P]_\infty$ is the limiting protein concentration achieved long after transcription ceases.

In principle, the functional assay can be very effective for detecting small mRNA truncations that destroy the capacity of the mRNA to direct translation. Such small alterations in mRNA length might be missed by the Northern blot or RNase protection assays. However, a cell's capacity to synthesize a given protein is not necessarily a measure of the concentration of the corresponding mRNA (see Chapter 6). For mRNAs under translational control, the extent of translation *in vivo* may depend on accessory factors (either repressors or activators) that may be present in rate-determining amounts. This potential problem could be avoided by isolating the RNA and translating it in a cell-free system (Cereghini *et al.*, 1979). The interpretation of a functional half-life can also be complicated if a single polypeptide is encoded by multiples mRNAs that decay at different rates and are translated with different efficiencies.

E. Rate of Accumulation of mRNA

1. *Kinetics of Approach to Steady-State Labeling*

The metabolic stability of mRNA can also be determined by measuring the rate at which it accumulates to a steady-state level. This has been done in mammalian cells by exposing them to continuous labeling with a radioactive RNA precursor (e.g., uridine) and following the time course of accumulation of radioactivity in the mRNA under study (Fig. 4). The half-time for reaching maximum labeling depends on the mRNA half-life and on the cell doubling time. The rate constant (k_{obs}) is derived from the following equation (see Greenberg, 1972):

$$\ln (1 - [\mathrm{mRNA}]_t/[\mathrm{mRNA}]_\infty) = -k_{obs}t, \tag{4}$$

where $[\mathrm{mRNA}]_t$ represents mRNA specific radioactivity at time t, and $[\mathrm{mRNA}]_\infty$ is the limiting specific radioactivity that is approached asymptotically [see Eq. 1 for the relation of k_{obs} to $t_{1/2}$].

This method is valid only if the specific radioactivity of the nucleotide precursor pool remains constant during the course of the experiment. This is usually ensured by keeping the supply of the radioactive precursor constant and can be verified directly by measuring the radioactivity in the cellular nucleotide pool.

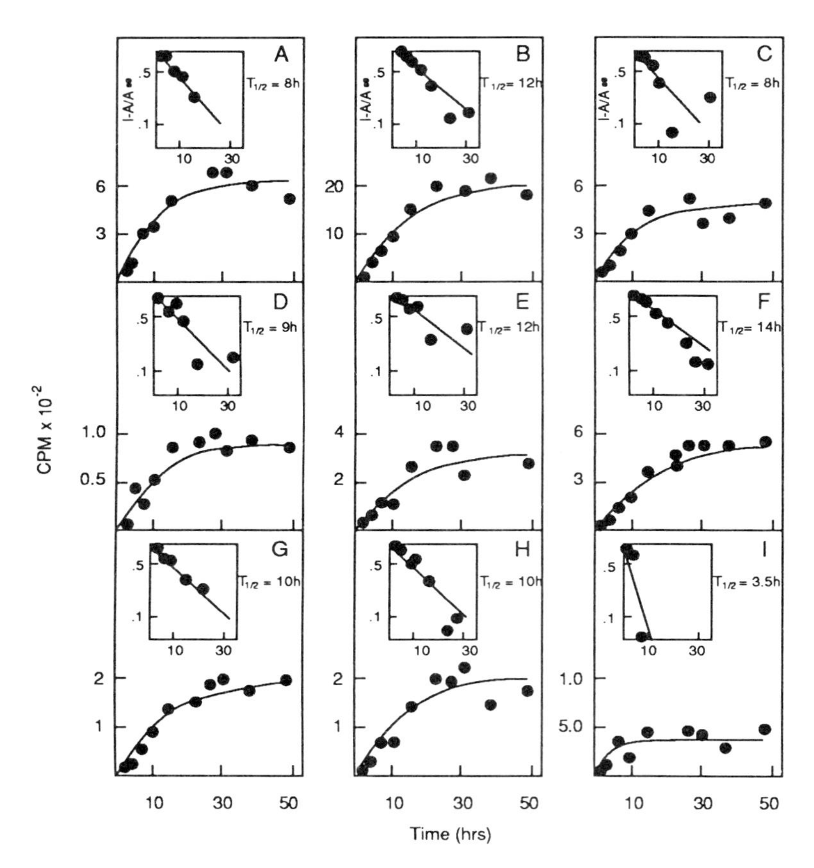

FIGURE 4 Decay rates of mammalian mRNAs as determined by the kinetics of approach to steady-state labeling. Cultured Chinese hamster ovary cells were incubated in medium containing tritiated uridine. The accumulation of radioactivity in individual mRNA species was monitored by hybridizing RNA isolated at various times with each of nine different filter-immobilized cDNAs. Insets contain data replotted according to Eq. 4 to obtain mRNA decay rates. [Reproduced with permission from Harpold *et al.* (1981).]

2. *Kinetics of Induction of mRNA Accumulation*

A similar approach can be used to determine the metabolic stability of an mRNA species during an induction process that leads to accumulation of the mRNA to a new steady-state level. This method is applicable to mRNAs derived from transcriptionally inducible genes and should be useful for measuring the half-lives of messages whose synthesis cannot be inhibited readily. The half-life of the induced message is calculated by plotting the data according to Eq. 4, with $[\text{mRNA}]_t$ and $[\text{mRNA}]_\infty$ representing the mRNA concentration at time t and the limiting mRNA concentration, respectively. For induction processes that may involve changes in both transcription and mRNA decay, this kind of analysis could be used to determine the rates of these two processes after induction. These parameters can be derived from

$$d[\text{mRNA}]/dt = T - k_{\text{obs}} [\text{mRNA}], \tag{5}$$

where T represents the rate of mRNA synthesis, and k_{obs} is as defined in Eq. 1 (Guyette *et al.*, 1979). Thus in a plot of $d[\text{mRNA}]/dt$ versus [mRNA], the slope of the straight line will yield the mRNA decay rate, and extrapolation of the line to [mRNA] = 0 will yield the rate of synthesis.

III. Estimation of Changes in mRNA Stability by Comparison of Transcription Rates and Relative mRNA Levels

Since steady-state mRNA levels are a function of both rates of synthesis and rates of decay, relative decay rates could, in theory, be deduced by measuring mRNA levels if the corresponding rates of synthesis were known. Faced with difficulty in measuring decay rates directly, many investigators have used this approach for the purpose of determining whether an induction process involves stabilization or destabilization of a given mRNA species. This approach has also been used extensively to identify structural elements in mRNA that control its metabolic stability (Fig. 5). Relative gene transcription rates, as determined by nuclear run-on assays, are used as a measure of changes in rates of mRNA synthesis. Thus a 10-fold increase in the concentration of a given mRNA species coupled with a 2-fold increase in the rate of transcription might be taken as an indication of a 5-fold increase in mRNA stability. However, while measurements of relative mRNA levels can be quite precise, nuclear run-on transcription rate measurements are subject to considerable error, and small discrepancies between changes in measured mRNA levels and changes in transcription rates should not be taken as evidence for altered rates of mRNA decay.

Furthermore, this method depends on the assumption that the rate of transcription provides a reliable measure of the rate of formation of mature cytoplasmic mRNA. However, as mRNA production can be strongly affected by post-transcriptional events in the nucleus, reliance on this approach can lead to incorrect conclusions. This potential pitfall is illustrated by the observation that the concentration of the mammalian triosephosphate isomerase mRNA is reduced as a result of insertion of a premature termination codon at certain locations within the gene. Although the initial conclusion was that premature termination of translation leads to destabilization of this mRNA (Daar and Maquat, 1988), subsequent studies have indicated that these mutations instead cause a defect in mRNA splicing or

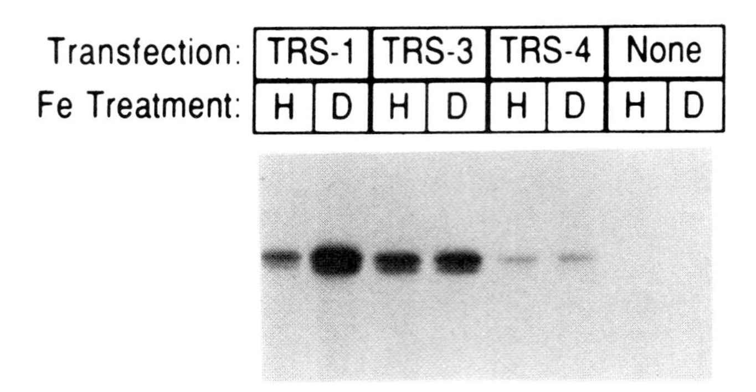

FIGURE 5 Estimation of a change in mRNA stability from a change in steady-state mRNA concentration. Mouse fibroblasts were transfected with one of three plasmids encoding variants of transferrin receptor mRNA (TRS-1, TRS-2, TRS-3) or with a plasmid containing no transferrin receptor gene sequences (none). The transfected cells were treated with hemin (H), an iron source, or with desferrioxamine (D), an iron chelator, and total cytoplasmic RNA was isolated and analyzed by Northern blotting. The difference in the concentration of TRS-1 mRNA under the two growth conditions suggests an increase in stability for TRS-1 mRNA under low-iron conditions whereas no such change is evident for TRS-3 and TRS-4 mRNA. Furthermore, the relatively high level of TRS-3 mRNA under both conditions was taken to indicate that this mRNA is stable even when iron is abundant, and the low level of TRS-4 mRNA under both conditions was taken to indicate that this mRNA remains labile even when iron is scarce. These conclusions have since been confirmed by direct half-life measurements after transcription inhibition (see Chapter 11). [Reproduced with permission from Casey *et al.* (1989).]

transport (Cheng *et al.*, 1990). Direct half-life measurements on the mRNA molecules that have reached the cytoplasm show the decay rate to be unchanged. Similarly, studies of the effect of nonsense mutations on dihydrofolate reductase mRNA have shown that, although the mutant mRNA molecules accumulate to a reduced level in the cytoplasm, their rate of decay in this cell compartment is unaffected (Urlaub *et al.*, 1989). As in the case of the triosephosphate isomerase mRNA, these findings led to the conclusion that the mutations cause a defect in splicing.

IV. Measurements of Poly(A) Sizes

A. Fate of the Poly(A) Tail

The poly(A) sequence of eukaryotic mRNAs becomes gradually shorter after the mRNA molecules enter the cytoplasm. Although the steady-state size distributions for individual species are rather heterogeneous, some species undergo a shortening or lengthening of their poly(A) tail in connection with certain physiological processes (see Chapters 8, 9, and 15). Precise measurements of poly(A) sizes under such conditions may eventually lead to understanding the functional significance of poly(A) size diversity. Determining whether the poly(A) tail normally can be removed in its entirety and whether some mRNA species can survive in the "poly(A)-less" state also requires precise methods for analyzing the 3′ terminus of mRNA chains.

B. Analysis of Poly(A) Size Distribution

Shifts in electrophoretic mobility of a given mRNA species, as measured by Northern blot analysis, are usually taken as an indication of a change in the size distribution of its poly(A) tail. In order to be certain that the apparent change in size is due to alteration of this sequence, RNA samples are deadenylated *in vitro* by annealing the poly(A) in the preparations with oligo(dT) and then cleaving the RNA strand of the DNA–RNA hybrids with RNase H. The resulting poly(A)-depleted mRNA chains should no longer differ in their relative electrophoretic mobility.

Since the poly(A) tails of most mRNAs represent a relatively small portion of the mRNAs' total length, size measurements on intact mRNA molecules are usually not sensitive enough to detect small changes due to shifts in poly(A) size. The sensitivity of the procedure can be greatly enhanced by estimating the length of a defined 3′-terminal fragment of the mRNA. For this purpose, site-specific fragmentation of the mRNA can be accomplished by oligonucleotide-directed RNase H cleavage (Fig. 6). The mRNA is annealed with an oligodeoxynucleotide complementary to a site 0.3–0.4 kb upstream of the poly(A) tail and is then cleaved with RNase H at the internal site that has formed a DNA–RNA hybrid (Shyu *et al.*, 1991). The length of the resulting 3′ RNA fragment, which still carries the poly(A) tail, is compared with that of the same fragment after deadenylation with oligo(dT) and RNase H. This analysis yields relatively precise estimates of poly(A) tail lengths in addition to information on the extent of the decrease or increase in length (Fig. 6).

Another procedure for estimating poly(A) length involves first annealing mRNA with a uniformly radiolabeled RNA that is complementary to the 3′-terminal segment of the transcribed portion of the message (Wilson and Treisman, 1988). Subsequent digestion with RNase T1, which is specific for unpaired guanosine residues, removes all of the mRNA and probe sequences flanking the duplex region, except for the poly(A). The electrophoretic mobility of the resulting tailed duplex on a nondenaturing gel reflects the length of both the RNA–RNA duplex and the attached poly(A) tail. The contribution of the poly(A) to mobility can be determined by comparison with an RNA sample deadenylated *in vitro* by prior treatment with oligo(dT) and RNase H. However, because electrophoresis must be

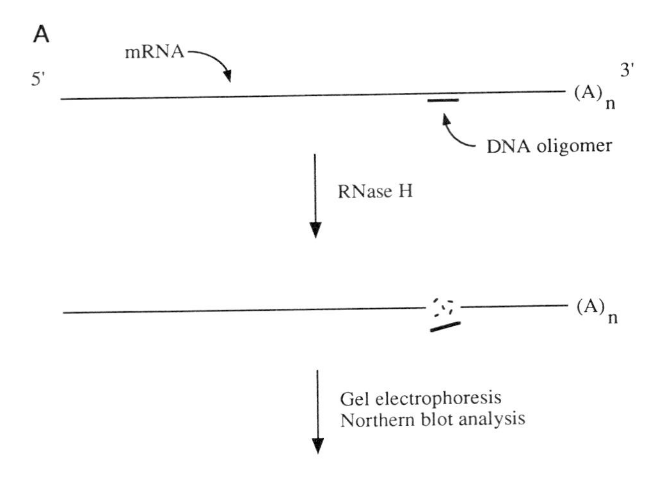

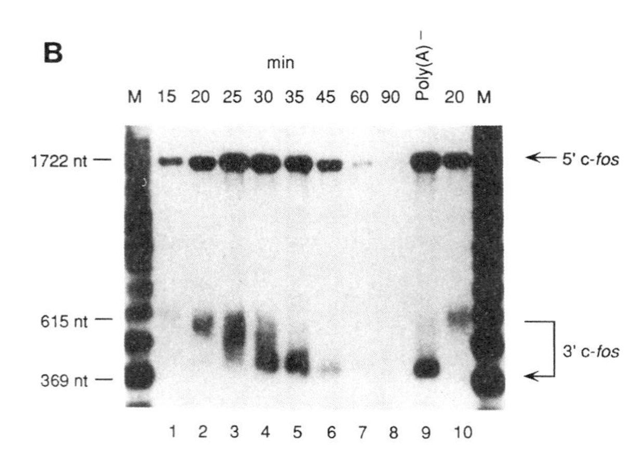

FIGURE 6 Analysis of poly(A) length after RNase H cleavage of mRNA. (A) Diagram illustrating the experimental procedure. RNase H is used to cleave an RNA–DNA hybrid formed by annealing cytoplasmic RNA to an oligodeoxynucleotide that is complementary to a specific mRNA segment. The resulting mRNA fragments are analyzed by Northern blotting. (B) Deadenylation and decay of *c-fos* mRNA in mouse fibroblasts. Cytoplasmic RNA isolated at time intervals after exposure to serum (to effect transient stimulation of *c-fos* gene transcription) was analyzed by the procedure described in A. The upper band represents the 5′ fragment of *c-fos* mRNA, and the lower band represents the 3′ fragment, with the poly(A) tail decreasing in length. Lane 9 contains mRNA that was deadenylated *in vitro* prior to gel electrophoresis of the RNA fragments, to obtain the size of the 3′-terminal fragment without any poly(A) tail. Lanes M contain molecular size markers. [Reproduced with permission from Shyu *et al.* (1991).]

performed under nondenaturing conditions, this method is not as effective in yielding precise poly(A) lengths.

C. Identification of mRNA Chains without Poly(A)

Failure of mRNA chains to bind to oligo(dT)–cellulose or to poly(U)–Sepharose is commonly interpreted as an indication that they lack a poly(A) tail. However, this procedure cannot distinguish whether the tail is completely missing or just too short to form stable hybrids. mRNA chains with poly(A) tails comprising as many as 20–30 adenylate residues fail to bind to oligo(dT)–cellulose (Krowczynska and Brawerman, 1986). Although poly(U)–Sepharose can bind shorter poly(A) segments, the mRNA chains in the resulting unadsorbed fraction can still carry tails of 15–20 bases (Kaufmann *et al.*, 1977). Since poly(A) sequences of such lengths may still be capable of forming a complex with poly(A)-binding protein (Sachs *et al.*, 1987), they could represent functionally competent tails.

It is possible to make precise measurements of very short tails by using a cDNA probe that is complementary to a relatively short 3′-terminal region of an mRNA species and that carries a poly(dT) sequence complementary to the beginning of the poly(A) sequence. Hybridization followed by S1 nuclease digestion yields protected DNA fragments whose poly(dT) length corresponds to the sizes of the poly(A) on the mRNA (Krowczynska and Brawerman, 1986). The use of this S1-protection assay has shown that essentially all the globin mRNA chains in reticulocytes carry a poly(A) tail (Krowczynska and Brawerman, 1986). The substantial portions of the mammalian β-actin mRNA molecules that were believed to be poly(A)$^-$ (Hunter and Garrels, 1977; Kaufman *et al.*, 1977; Geoghegan *et al.*, 1978) have also been found to contain a poly(A) tail of considerable size (Bandyopadhyay and Brawerman, 1992).

V. Identification of Structural Elements That Affect mRNA Stability

A. Destabilizing or Stabilizing Sequences in the mRNA

cis-acting RNA sequences that determine mRNA half-life fall into two categories. One is the class of mRNA destabilizing elements. These may be either ribonuclease target sites or facilitators of ribonuclease attack elsewhere in the message. In principle, mutational inactivation of such an element should prolong the lifetime of an otherwise short-lived message. This criterion sometimes is difficult to satisfy because of the possible occurrence of more than one instability element in the same mRNA (Kabnick and Housman, 1988; Shyu *et al.*, 1989). In such cases, a destabilizing element can be recognized instead by its ability to reduce the half-life of the heterologous long-lived messages into which it is introduced.

Conversely, elements that protect mRNA from degradation can be classified as mRNA stabilizers. These may be ubiquitous structures at mRNA termini (e.g., 3′ stem–loops in prokaryotes, 5′ caps in eukaryotes) that prevent rapid destruction of the RNA molecules by nonspecific nucleases. There are also special elements, such as the *E. coli ompA* 5′ UTR (Emory and Belasco, 1990), that are responsible for the unusual longevity of the limited set of long-lived mRNAs in which they occur. The key property that operationally defines an mRNA stabilizer is its ability to increase the longevity of the heterologous mRNAs into which it is introduced. Alternatively, mutational inactivation of an mRNA stabilizer can reduce the lifetime of an otherwise long-lived message that contains such an element.

In principle, inserting a putative stability or instability determinant into a heterologous transcript might alter the stability of the recipient mRNA simply by interrupting the normal sequence at the site of insertion. Therefore, an important control in such an analysis is to test the effect of introducing an inert RNA segment at the same location.

It is also important to bear in mind that some mRNA structural elements cannot be eliminated simply by deleting the corresponding gene segment. For example, removal of a prokaryotic transcription terminator generally will not prevent a stem–loop from forming at the mRNA 3′ end. It will merely cause transcription to terminate at an often unpredictable distal site, which itself is likely to encode a stem–loop. Similarly, poly(A)-lacking mRNA molecules cannot be generated in eukaryotic cells by removal of the normal polyadenylation signal.

B. Sequences That Affect mRNA Structure or Translation

Sometimes, a determinant of mRNA stability or instability may consist of a structure formed by base pairing between different sequence elements (Mullner and Kuhn, 1988; Casey *et al.*, 1989; Emory *et al.*, 1992). In such cases, RNA mutations created to define stabilizing or destabilizing sequences can have unforeseen structural consequences that will complicate the interpretation of the resulting half-life data. Consequently, the interpretation of such data may require RNA structural analysis.

Mutations that disrupt translation can radically alter the mechanism and kinetics of mRNA decay. Therefore, unless the purpose of an experiment is to examine the effects of translation on mRNA decay, it usually is best to engineer mutations so as not to disturb translation. Thus, mRNA deletions and fusions within the coding region that alter the reading frame should be avoided, as such alterations could either prevent translation of mRNA segments that normally are translated or cause translation of normally untranslated RNA segments. Also, mutations within the 5′ and 3′ UTRs may inadvertently affect the translation initiation process. In this case, interpretation of the decay data in terms of defined nuclease target sites may be misleading, and it may be important to examine the effect of the alterations on the capacity of the mRNA to be translated. Conversely, in studies designed to examine the relation between mRNA stability and translatability, it may not be sufficient to eliminate translation initiation signals such as the AUG codon or the Shine–Dalgarno sequence, since these alterations may lead to initiation at cryptic sites.

References

Achord, D., and Kennell, D. (1974). Metabolism of messenger RNA from the *gal* operon of *Escherichia coli*. *J. Mol. Biol.* **90,** 581–599.

Bandyopadhyay, R., and Brawerman, P. (1992). Secondary structure at the beginning of the poly(A) sequence of mouse β-actin messenger RNA. *Biochimie* **74,** 1031–1034.

Belasco, J. G., Beatty, J. T., Adams, C. W. von Gabain, A., and Cohen, S. N. (1985). Differential

Blundell, M., Craig, E., and Kennell, D. (1972). Decay rates of different mRNA in *E. coli* and models of decay. *Nature New Biol.* **238,** 46–49.

Brock, M. L., and Shapiro, D. J. (1983). Estrogen stabilizes vitellogenin mRNA against cytoplasmic degradation. *Cell* **34,** 207–214.

Casey, J. L., Koeller, D. M., Ramin, V. C., Klausner, R. D., and Harford, J. B. (1989). Iron regulation of transferrin receptor mRNA levels requires iron-responsive elements and a rapid turnover determinant in the 3′ untranslated region of the mRNA. *EMBO J.* **8,** 3693–3699.

Chen, C. A, Beatty, J. T., Cohen, S. N., and Belasco, J. G. (1988). An intercistronic stem–loop structure functions as an mRNA decay terminator necessary but insufficient for *puf* mRNA stability. *Cell* **52,** 609–619.

Cheng, J., Fogel-Petrovic, M., and Maquat, L. E. (1990). Translation to near the distal end of the penultimate exon is required for normal levels of spliced triosephosphate isomerase mRNA. *Mol. Cell. Biol.* **10,** 5215–5225.

Daar, I. O., and Maquat, L. E. (1988). Premature translation termination mediates triosephosphate isomerase mRNA degradation. *Mol. Cell. Biol.* **8,** 802–813.

Donovan, W. P., and Kushner, S. R. (1986). Polynucleotide phosphorylase and ribonuclease II are required for cell viability and mRNA turnover in *Escherichia coli* K-12. *Proc. Natl. Acad. Sci. USA* **83,** 120–124

Emory, S. A., and Belasco, J. G. (1990). The *ompA* 5′ untranslated RNA segment functions in *Escherichia coli* as a growth-rate-regulated mRNA stabilizer whose activity is unrelated to translational efficiency. *J. Bacteriol.* **172,** 4472–4481.

Emory, S. A., Bouvet, P., and Belasco, J. G. (1992). A 5′-terminal stem–loop structure can stabilize mRNA in *Escherichia coli*. *Genes and Dev.* **6,** 135–148.

Geoghegan, T. E., Sonenshein, G. E., and Brawerman, G. (1978). Characteristics and polyadenylate content of the actin messenger RNA of mouse sarcoma-180 ascites cells. *Biochemistry* **17,** 4200–4207.

Greenberg, J. R. (1972). High stability of messenger RNA in growing cultured cells. *Nature (London)* **240,** 102–104.
Guyette, W. A., Matusik, R. J., and Rosen, J. M. (1979). Prolactin-mediated transcriptional and post-transcriptional control of casein gene expression. *Cell* **17,** 1013–1023.
Harpold, M. M., Wilson, M. C., and Darnell, J. E., Jr. (1981). Chinese hamster polyadenylated messenger ribonucleic acid: Relationship to non-polyadenylated sequences and relative conservation during messenger ribonucleic acid processing. *Mol. Cell. Biol.* **1,** 188–198.
Herrick, D., Parker, R., and Jacobson, A. (1990). Identification and comparison of stable and unstable mRNAs in *Saccharomyces cerevisiae. Mol. Cell. Biol.* **10,** 2269–2284.
Hunter, T., and Garrels, J. I. (1977). Characterization of the mRNAs for α, β and γ-actin. *Cell* **12,** 767–781.
Kabnick, K. S., and Housman, D. E. (1988). Determinants that contribute to cytoplasmic stability of human *c-fos* and β-globin mRNAs are located at several sites in each mRNA. *Mol. Cell. Biol.* **8,** 3244–3250.
Kaufmann, Y., Milcarek, C., Berissi, H., and Penman, S. (1977). HeLa cell poly(A)$^-$ mRNA codes for a subset of poly(A)$^+$ mRNA-directed proteins with an actin as the major product. *Proc. Natl. Acad. Sci. USA* **74,** 4801–4805.
Krowczynska, A., and Brawerman, G. (1986). Structural features in the 3′-terminal region of polyribosome-bound rabbit globin messenger RNA. *J. Biol. Chem.* **261,** 397–402.
Krowczynska, A., Yenofsky, R., and Brawerman, G. (1985). Regulation of messenger RNA stability in mouse erythroleukemia cells. *J. Mol. Biol.* **181,** 231–239.
Levis, R., and Penman, S. (1977). The metabolism of poly(A)$^+$ and poly(A)$^-$ hnRNA in cultured Drosophila cells studied with a rapid uridine pulse–chase. *Cell* **11,** 105–113.
Maderious, A., and Chen-Kiang, S. (1984). Pausing and premature termination of human RNA polymerase II during transcription of adenosines in vivo and in vitro. *Proc. Natl. Acad. Sci. USA* **81,** 5931–5935.
Morse, D. E., Moselle, R. D., and Yanofsky, C. (1969). Dynamics of synthesis, translation, and degradation of *trp* operon messenger RNA in *E. coli. Cold Spring Harbor Symp. Quant. Biol.* **34,** 725–739.
Mullner, E. W., and Kuhn, L. C. (1988). A stem–loop in the 3′ untranslated region mediates iron-dependent regulation of transferrin receptor mRNA stability in the cytoplasm. *Cell* **53,** 815–825.
Newbury, S. F., Smith, N. H., Robinson, E. C., Hiles, I. D., and Higgins, C. F. (1987). Stabilization of translationally active mRNA by prokaryotic REP sequences. *Cell* **48,** 297–310.
Paek, I., and Axel, R. (1987). Glucocorticoids enhance stability of human growth hormone mRNA. *Mol. Cell. Biol.* **7,** 1496–1507.
Penman, S., Vesco, C., and Penman, M. (1968). Localization and kinetics of formation of nuclear heterodisperse RNA, cytoplasmic heterodisperse RNA and polyribosome-associated messenger RNA in HeLa cells. *J. Mol. Biol.* **34,** 49–69
Rivera, V. M., and Greenberg, M. E. (1990). Growth factor-induced gene expression: The ups and downs of *c-fos* regulation. *New Biol.* **2,** 751–758.
Ross, J., Peltz, S. W., Kobs, G., and Brewer, G. (1986). Histone mRNA degradation in vivo: The first detectable step occurs at or near the 3′ terminus. *Mol. Cell. Biol.* **6,** 4362–4371.
Sachs, A. B., Davis, R. W., and Kornberg, R. D. (1987). A single domain of yeast poly(A)-binding protein is necessary and sufficient for RNA binding and cell viability. *Mol. Cell. Biol.* **7,** 3268–3276.
Shyu, A. B., Greenberg, M. E., and Belasco, J. G. (1989). The *c-fos* transcript is targeted for rapid decay by two distinct mRNA degradation pathways. *Genes Dev.* **3,** 60–72.
Stoeckle, M. Y., and Hanafusa, H. (1989). Processing of 9E3 mRNA and regulation of its stability in normal and Rous sarcoma virus-transformed cells. *Mol. Cell. Biol.* **9,** 4738–4745.

Singer, R. H., and Penman, S. (1973). Messenger RNA in HeLa cells: Kinetics of formation and decay. *J. Mol. Biol.* **78,** 321–334.
Urlaub, G., Mitchell, P. J., Ciudad, C. J., and Chasin, L. A. (1989). Nonsense mutations in the dihydrofolate reductase gene affect RNA processing. *Mol. Cell. Biol.* **9,** 2868–2880.
Wilson, T., and Treisman, R. (1988). Removal of poly(A) and consequent degradation of *c-fos* mRNA facilitated by 3′ AU-rich sequences. *Nature* **336,** 396–399.
Zandomeni, R., Mittleman, B., Bunick, D., Ackerman, S., and Weinman, R. (1982). Mechanism of action of dichloro-α-D-ribofuranosylbenzimidazole: Effect on in vitro transcription. *Proc. Natl. Acad. Sci. USA* **79,** 3167–3170.

Index